Crop Genetics: Techniques for Improvement

Crop Genetics: Techniques for Improvement

Edited by Alexis Sanders

SYRAWOOD
PUBLISHING HOUSE

New York

Published by Syrawood Publishing House,
750 Third Avenue, 9th Floor,
New York, NY 10017, USA
www.syrawoodpublishinghouse.com

Crop Genetics: Techniques for Improvement
Edited by Alexis Sanders

International Standard Book Number: 978-1-68286-580-4 (Hardback)

Cataloging-in-Publication Data

Crop genetics : techniques for improvement / edited by Alexis Sanders.
       p. cm.
Includes bibliographical references and index.
ISBN 978-1-68286-580-4
1. Crops--Genetics. 2. Plant genetics. 3. Crop improvement. I. Sanders, Alexis.
SB106.G46 C76 2018
631.523 3--dc23

TABLE OF CONTENTS

Permissions

List of Contributors

Index

PREFACE

Crop genetics is a practice which improves the productivity of crops. Mostly a trait is introduced in the crop which does not exist in it naturally. It aims at increasing yield, tolerance towards abiotic stresses and to alter the composition of the food. The different types of genetic modifications that exist are transgenic, subgenic and cisgenic. The book presents researches and studies performed by experts across the globe. For all those who are interested in crop genetics, this book can prove to be an essential guide.

This book is a comprehensive compilation of works of different researchers from varied parts of the world. It includes valuable experiences of the researchers with the sole objective of providing the readers (learners) with a proper knowledge of the concerned field. This book will be beneficial in evoking inspiration and enhancing the knowledge of the interested readers.

In the end, I would like to extend my heartiest thanks to the authors who worked with great determination on their chapters. I also appreciate the publisher's support in the course of the book. I would also like to deeply acknowledge my family who stood by me as a source of inspiration during the project.

Editor

The Shade Avoidance Syndrome in Arabidopsis: The Antagonistic Role of Phytochrome A and B Differentiates Vegetation Proximity and Canopy Shade

Jaime F. Martínez-García[1,2]*, Marçal Gallemí[2], María José Molina-Contreras[2], Briardo Llorente[2], Maycon R. R. Bevilaqua[2,3¤], Peter H. Quail[4,5]

1 Institució Catalana de Recerca i Estudis Avançats, Barcelona, Spain, **2** Centre for Research in Agricultural Genomics (CRAG), Consortium CSIC-IRTA-UAB-UB, Barcelona, Spain, **3** CAPES foundation, Ministry of Education of Brazil, Brasilia - DF, Brazil, **4** Department of Plant and Microbial Biology, University of California, Berkeley, California, United States of America, **5** US Department of Agriculture/Agriculture Research Service, Plant Gene Expression Center, Albany, California, United States of America

Abstract

Light limitation caused by dense vegetation is one of the greatest threats to plant survival in natural environments. Plants detect such neighboring vegetation as a reduction in the red to far-red ratio (R:FR) of the incoming light. The low R:FR signal, perceived by phytochromes, initiates a set of responses collectively known as the shade avoidance syndrome, intended to reduce the degree of current or future shade from neighbors by overtopping such competitors or inducing flowering to ensure seed production. At the seedling stage these responses include increased hypocotyl elongation. We have systematically analyzed the Arabidopsis seedling response and the contribution of phyA and phyB to perception of decreased R:FR, at three different levels of photosynthetically active radiation. Our results show that the shade avoidance syndrome, induced by phyB deactivation, is gradually antagonized by phyA, operating through the so-called FR-High Irradiance Response, in response to high FR levels in a range that simulates plant canopy shade. The data indicate that the R:FR signal distinguishes between the presence of proximal, but non-shading, neighbors and direct foliar shade, via a intrafamily photosensory attenuation mechanism that acts to suppress excessive reversion toward skotomorphogenic development under prolonged direct vegetation shade.

Editor: Enamul Huq, University of Texas at Austin, United States of America

Funding: MG and MJMC received FPI fellowships from the Spanish Ministry of Economy and Competitivity (MINECO). BL received a Marie Curie IIF fellowship (CarotenActors) from the European Union FP7. JFMG received a 1-year fellowship from the Spanish Ministry of Education for a sabbatical stay in PHQ laboratory. The authors' research is supported by grants from the Generalitat de Catalunya and Spanish MINECO - Fondo Europeo de Desarrollo Regional (FEDER) to JFMG (2009-SGR697, Xarba and BIO2011-23489) and by National Institutes of Health (2R01 GM-047475), Department of Energy (DEFG03-87ER13742), and United States Department of Agriculture Agricultural Research Service Current Research Information System (5335-21000-032-00D) to PHQ. The funders had no role in study design, data collection and analysis, decision to publish, or preparation of the manuscript.

Competing Interests: The authors have declared that no competing interests exist.

* Email: jaume.martinez@cragenomica.es

¤ Current address: UEM Universidade Estadual de Maringá, Departamento de Biotecnologia, Biologia Celular e Genética, Maringá, Paraná, Brazil

Introduction

The shade avoidance syndrome (SAS) refers to a set of plant responses aimed at adapting plant growth and development to high plant density environments, like those found in forests, prairies or orchard communities. Two related but different situations can occur in these environments: plant proximity (without direct vegetative shading) and direct plant canopy shade [1–3]. Because vegetation preferentially reflects far-red (FR) light compared to other wavelengths, plant proximity generates a reduction in the red (R, about 600–700 nm) to far-red (FR, between 700–800 nm) ratio (R:FR) in the light impinging on neighbors. By contrast, under a plant canopy, light from the visible region (called photosynthetically active radiation or PAR, between 400–700 nm) is strongly absorbed by the chlorophyll and carotenoid photosynthetic pigments whereas FR, which is poorly absorbed by the leaves, is transmitted through (or reflected from)

vegetation. As a consequence, under direct plant canopy shade both the amount of PAR (light quantity) and R:FR (light quality) are greatly reduced, in the latter case mostly by the selective depletion of R light caused by the filtering of sunlight through the leaves [1,2,4–6].

This low R:FR signal is perceived by the phytochrome (phy) photoreceptors. Phys detect the R and FR part of the spectrum and have a major role in controlling several adaptive responses such as seed germination, stem elongation, leaf expansion and flowering time. In Arabidopsis (*Arabidopsis thaliana*), a small gene family of five members encodes the phys (*PHYA-PHYE*) [7]. Although phyB is the major phy controlling the SAS, genetic and physiological analyses have shown that other phys act redundantly with phyB in the control of some aspects of SAS-driven development, such as flowering time (phyD, phyE), petiole elongation (phyD, phyE) and internode elongation between rosette leaves (phyE) [2,4]. The photolabile phyA has the unique capacity to function as a FR-light sensor through a mechanism termed the FR-High Irradiance Response (HIR) [2,8,9]. In contrast to the other phys, phyA has an antagonistic negative role in the SAS hypocotyl response, although varying degrees of regulation have been reported: *phyA* mutant seedlings growing under low R:FR light showed from moderately [10] to extremely long hypocotyls [11]. An antagonistic activity between phyA and phyB has also been shown in seedlings exposed to varying ratios of monochromatic R and FR [12,13]. However, it has been argued that the adaptive significance of this phyA antagonism is limited and may instead be an inevitable consequence of the intrinsic properties of phyA selected for their role in seedling deetiolation [2].

Phys exist in two photoconvertible forms, an inactive R-absorbing Pr form and an active FR-absorbing Pfr form. In light-grown plants, the steady-state ratio of Pr and Pfr conformers depends on the R:FR ratio. Under high R:FR the photoequilibrium is displaced towards the active Pfr form and the SAS is suppressed. Under low R:FR the photoequilibrium is displaced towards the inactive form and SAS is induced. This induction is regulated at least partly by the interaction of active phys with various PHYTOCHROME INTERACTING FACTORs (PIFs) [14–16], which results in rapid changes in the expression of dozens of *PHYTOCHROME RAPIDLY REGULATED* (*PAR*) genes, postulated to be instrumental in implementing the SAS responses [17–21]. Because many of these *PAR* genes encode transcriptional regulators, it is assumed that shade responses are a consequence of the phy regulation of a complex transcriptional network, as postulated for seedling de-etiolation [22,23], that seems to be organized in functional modules [24]. Genetic analyses demonstrated positive and negative roles in SAS regulation for several *PAR* genes encoding transcriptional regulators, including members of the homeodomain leucine zipper class II (*ATHB2, ATHB4, HAT1, HAT2* and *HAT3*), basic-helix-loop-helix (*BEE1, BEE2, BIM1, BIM2, HFR1, PAR1, PAR2* and *PIL1*) and B-BOX CONTAINING (BBX) families of proteins [17–19,24–30]. Most of these studies were done analyzing hypocotyl elongation. Therefore, low R:FR perception rapidly changes the balance of positive and negative factors, resulting in the appropriate SAS responses, i.e., eventually causing hypocotyls to elongate. Evidence for the involvement of several of these factors in controlling auxin levels and sensitivity in mediating this elongation response has been reported [16,31,32].

The light treatments used to induce the SAS vary among laboratories, resulting in differences in the extent of the responses (usually hypocotyl length) reported for the same genotype. For instance, a review of several papers in the field reported that Arabidopsis Col-0 hypocotyls elongate in response to low R:FR (under laboratory conditions usually provided by white light supplemented with FR light, W+FR) from a minimum of about 2.5 mm to a maximum of ca. 9 mm [15,20,26,28,33–35]. In addition to media composition, variations in the timing and nature of the W+FR treatment might also explain some of the observed differences. For instance, the reported effect of the negative SAS regulator HFR1 on the shade-induced hypocotyl elongation ranged from mild [25] to very strong [27]. We have noted previously that the very strong phenotype of *hfr1* mutant seedlings was observed under shade conditions that reduced both R:FR and PAR (400–700 nm), whereas the mild phenotype occurred under shade conditions where only the R:FR ratio was reduced without significantly affecting the PAR [25]. Indeed, although the SAS is generally considered to be mainly induced by light of reduced R:FR, other light parameters, such as low-intensity light of the whole PAR spectrum and low blue light (which is part of the PAR spectrum), are also known to contribute to these responses [6,36–

38]. Together, these observations highlight the fact that different shade conditions (such as variable PAR and/or R:FR) employed by different labs might account for some of the observed variability in the SAS response.

In this paper we have investigated the effect of both the level of PAR and supplemental FR (which results in different R:FR ratios without altering PAR) in the incoming light on hypocotyl elongation. To address the contribution of the two major phys in this response, we have systematically analyzed wild-type, *phyA* and *phyB* mutant seedlings. We observe that, independently of the PAR level employed, the R:FR ratio strongly and differentially affects elongation of wild-type, *phyA* and *phyB* hypocotyls. Our results indicate that quantitative variation in the R:FR ratio provides a dual signal with a likely different meaning in nature: when the R:FR is moderately lowered, it mimics plant proximity without direct shading, whereas when it is very low, it mimics direct plant canopy shade. In addition, the effects of these two environmental conditions can be distinguished genetically, with phyA and phyB having different roles in transducing the signals, as shown previously for seedling de-etiolation.

Materials and Methods

Plant material and growth conditions

Arabidopsis (*Arabidopsis thaliana*) plants for seed production were grown in the greenhouse as described [39]. The *phyA-501* (SALK_014575) [40] and *phyB-9* [41] mutant lines are in Col-0 ecotype. The *phyA* and *phyB* mutant lines in L*er* have been described previously [42]. Homozygous *phyA-501* plants were genotyped as indicated in Figure S1 by using specific oligos: MSO31 (5′-TAG-AGC-ACC-GCA-CAG-CTG-CC-3′), MSO32 (5′-GAA-GCT-ATC-TCC-TGC-AGG-TGG–3′) and LBb1 (5′-GCG-TGG-ACC-GCT-TGC-TGC-AAC-T-3′).

All the experiments were performed with seeds surface-sterilized and sown on Petri dishes with solid growth medium without sucrose (GM–; 0.215% (w/v) MS salts plus vitamins, 0.025% (w/v) MES pH 5.8) [17]. After stratification (3–6 days), plates were incubated in growth chambers at 22°C under continuous W that was provided by 2–4 cool-white horizontal fluorescent tubes (Figures 1–3), unless otherwise stated. These tubes delivered different amounts of photosynthetically active radiation (PAR), and a R:FR of about 2.5. In Figures 4 and 5, plates were incubated in growth chambers at 22°C under continuous W provided by 4 cool-white vertical fluorescent tubes (PAR of 20–25 μmol·m^{-2}·s^{-1}, R:FR of about 2.5). Simulated shade (W+FR) was generated by enriching W with supplementary FR provided by LED lamps (www.quantumdev.com; or www.philips.com/horti). Unless otherwise stated, fluence rates and PAR were measured with a LI-1800 spectroradiometer (Li-Cor Inc., www.licor.com); to calculate the R:FR, windows of 30 nm around the R (640–670 nm) and FR (720–750 nm) peaks were employed. For Figure 4, 5, and S4, fluence rates were measured with a Spectrosense2 meter associated with a 4-channel sensor (Skye Instruments Ltd., www.skyeinstruments.com), which measures PAR (400–700 nm) and 10 nm windows in the R (664–674 nm) and FR (725–735 nm) regions.

Hypocotyl length measurements

The National Institutes of Health ImageJ software (Bethesda, MD, USA; http://rsb.info.nih.gov/) was used on digital images to measure the length of hypocotyl seedlings (after laying out seedlings flat on agar plates) as indicated elsewhere [26]. At least 25 seedlings were used for each treatment. Experiments were repeated 3–5 times and a representative one is shown. Statistical

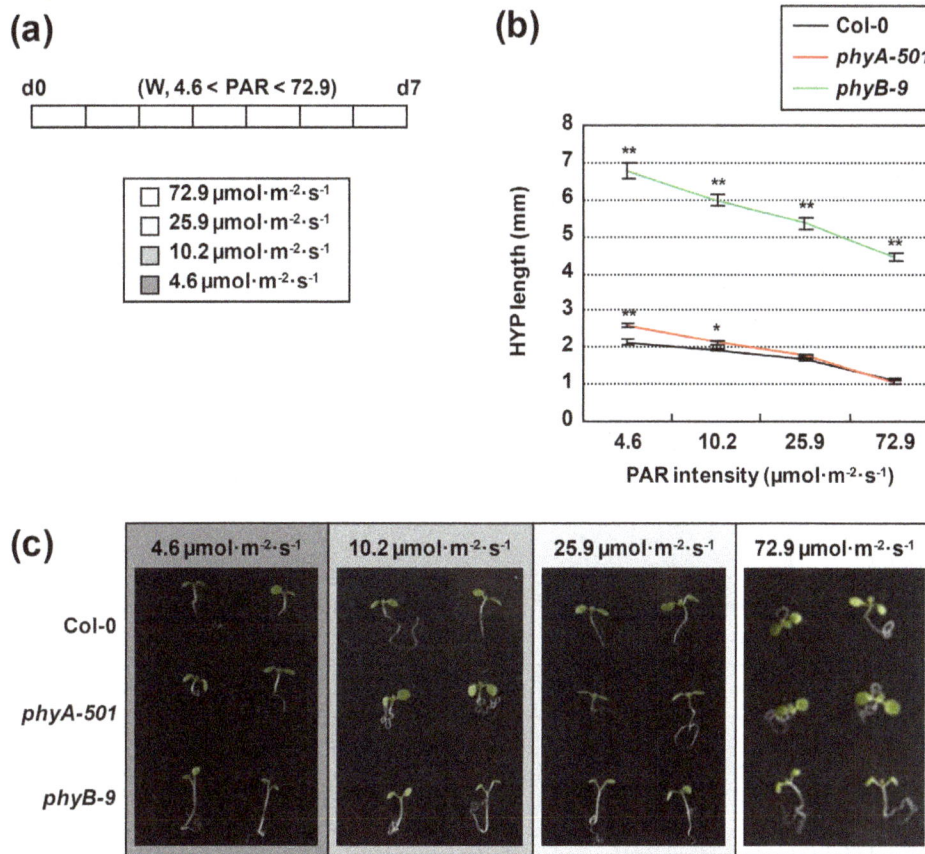

Figure 1. Effect of increasing intensities of white light on hypocotyl elongation of wild-type, *phyA-501* and *phyB-9* seedlings. (a) Seedlings of the indicated genotypes were grown from the day of germination until day 7 under W of increasing intensities (photosynthetic active radiation, PAR, between 4.6 and 72.9 $\mu mol \cdot m^{-2} \cdot s^{-1}$; R:FR>2.0). **(b)** Hypocotyl length of seedlings grown as indicated in **a**. Values are means $\pm$ SE of at least 25 hypocotyls for each light treatment. Asterisks indicate significant differences (*$P<0.05$, **$P<0.01$) relative to the control grown under the same light intensity. **(c)** Representative seedlings, grown as indicated in **a**, are shown for the three genotypes analyzed.

analyses of the data (t-test) were performed using GraphPad Prism version 4.00 for Windows (www.graphpad.com/) or Excel.

Protein extraction and Western blot analyses

Extracts were prepared following the direct extract protocol indicated elsewhere [43] with the modifications described below. Extracts shown in Figure S1 were prepared from Arabidopsis Col-0 and *phyA-501* seedlings germinated and grown in the dark for 5 days. Fifteen seedlings from each genotype were harvested and placed in 1.5 mL Eppendorf tubes containing 150 μL of (1$\times$) Laemmli buffer. Extracts shown in other Figures were prepared from Arabidopsis Col-0 seedlings germinated and grown in the dark for 5 days and then exposed to 4 and 8 h of W or W+FR. On the day of harvest, 20 seedlings from each treatment were harvested and placed in 1.5 mL Eppendorf tubes containing 300 μL of Laemmli buffer supplemented with protease inhibitors (10 μg/mL Aprotinin, 1 μg/mL E-64, 10 μg/mL Leupeptin, 1 μg/mL Pepstatin A, 10 μM PMSF). Plant material was ground using disposable grinders in the Eppendorf tube at room temperature until the mixture was homogeneous (usually less than 15 s). Once all the samples were prepared, tubes were placed in boiling water for 3 minutes. Tubes were centrifuged in a microfuge at maximum speed (13000 g, 10 min) immediately before loading. Fifteen μL of each extract, equivalent to about 1.5 (Figure S1) or 1 seedling (Figures 5, S4), were loaded per lane in an

SDS - 8% PAGE. Immunoblot analyses of phyA and TUB were performed as indicated [43] with some minor changes. Mouse monoclonal antibody (mAb) 0.73D, that recognizes phyA from both monocots and dicots [43], were used at 1:5000 dilutions. Membranes were stripped and rehybridized with a commercial mouse mAb against α-tubulin (www.sigmaaldrich.com) at a 1:10000 dilution. Anti-mouse horseradish peroxidase-conjugated antibody (www.promega.com) was used as a secondary antibody. ECL or ECL-plus chemiluminescence kits (www3.gehealthcare.com) were used for detection.

Accession numbers

Sequence data from this paper can be found in the Arabidopsis Genome Initiative or GenBank/EMBL databases under the following accession numbers: *PHYA* (At1g09570) and *PHYB* (At2g18790).

Results

The intensity of continuous white light affects seedling morphology

For this study, we have systematically analyzed the hypocotyl response of wild-type (Col-0), *phyA* and *phyB* mutant seedlings to different light conditions. As a phyB-deficient line, we have employed *phyB-9*, a well characterized line [41]. As a phyA-deficient line we

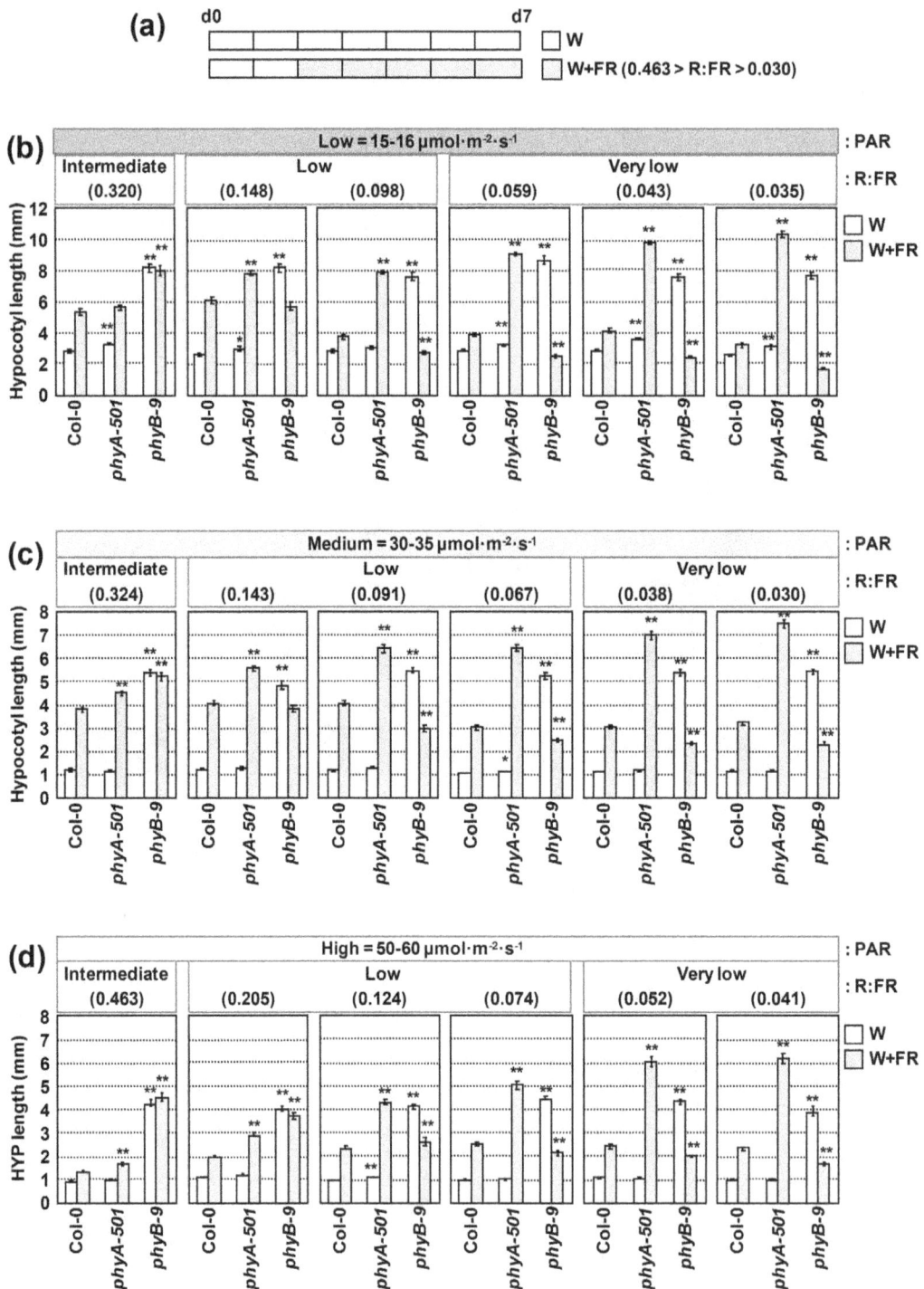

Figure 2. Effect of different R:FR on hypocotyl elongation of wild-type, *phyA-501* and *phyB-9* seedlings under low, medium or high PAR. (a) Seedlings were germinated and grown for 2 days under W light and then either kept in W or transferred to W supplemented with increasing amounts of FR for 5 more days. Hypocotyl length of seedlings grown as indicated in **a** under (**b**) low, (**c**) medium and high (**d**) PAR. The amount of PAR is given at the top of each section. The type of R:FR applied (nomenclature provided in Table S1) in the given W+FR treatments is indicated at the top, of the graphs; the R:FR value of each experiment is indicated at the top of each graph. In **b**, **c** and **d**, values are means ± SE of at least 25 hypocotyls for each light treatment. Asterisks indicate significant differences (*P<0.05, **P<0.01) relative to the control (Col-0) grown under the same light conditions.

have employed a SALK line (SALK_014575) that contains a T-DNA insertion in the middle of the first intron of the *PHYA* gene, about 1260 bp downstream of the ATG start codon (Figure S1a) [40]. In etiolated mutant seedlings, no levels of phyA were detected, whereas

tubulin levels were similar to those of wild type (Figure S1b). This line was blind to continuous monochromatic FR, whereas it was as responsive to monochromatic R light as the wild type (Figure S1c).

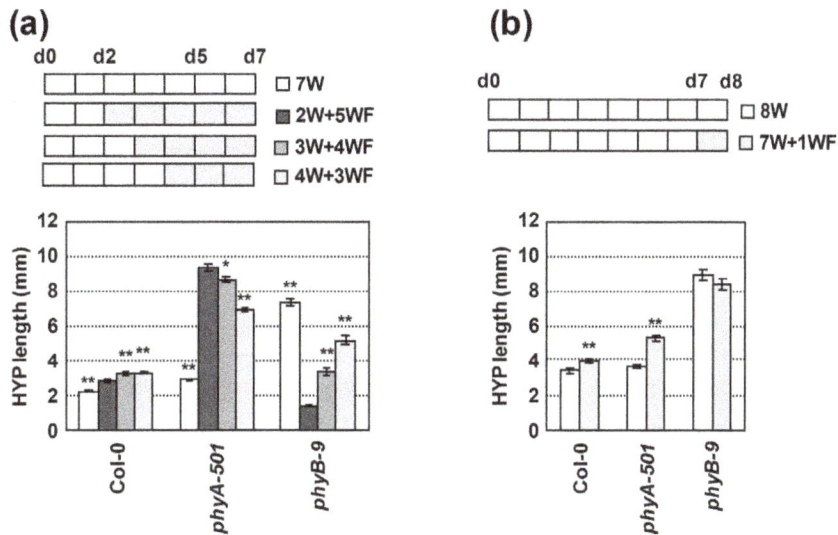

Figure 3. Effect of time of W+FR treatment on hypocotyl elongation of wild-type, *phyA-501* and *phyB-9* seedlings under low PAR. (a) Hypocotyl length of seedlings germinated and grown for 2, 3, 4 or 7 days under W light and then transferred to W+FR for 5, 4, 3 or 0 more days, respectively. **(b)** Hypocotyl length of seedlings germinated and grown for 7 days under W light and then either kept in W or transferred to W+FR for 1 more day. PAR was of 15–16 $\mu mol \cdot m^{-2} \cdot s^{-1}$ and R:FR of 0.059. Values are means $\pm$ SE of at least 25 hypocotyls for each light treatment. Asterisks indicate significant differences (*P<0.05, **P<0.01) relative to the same genotype (Col-0) grown for 2 days under W and 5 days under W+FR.

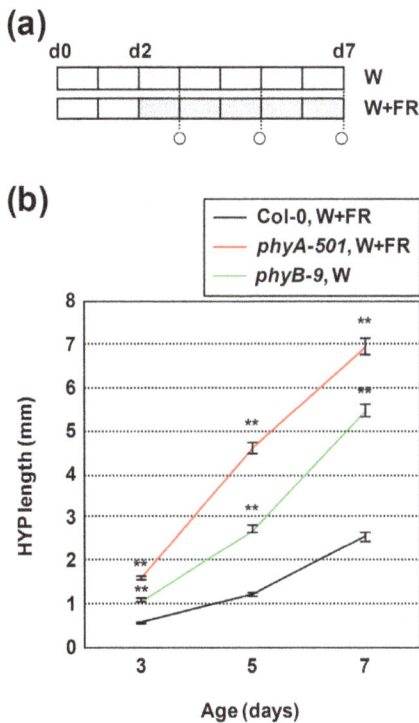

Figure 4. Effect of *phyA* and *phyB* mutations on the temporal evolution of the hypocotyl length. (a) Seeds were germinated and grown for 2 days under W (PAR was of 20–25 $\mu mol \cdot m^{-2} \cdot s^{-1}$) and then either kept under W (*phyB* seedlings) or transferred to W+FR (R:FR = 0.038) for 5 more days (Col-0 and *phyA* seedlings). Circles indicate the days on which hypocotyls were measured. **(b)** Hypocotyl length of seedlings grown as indicated in **a**. Values are means $\pm$ SE of at least 25 hypocotyls for each light treatment. Asterisks indicate significant differences (*P<0.05, **P<0.01) relative to the wild type seedlings grown under the corresponding light conditions.

Together, these results confirm that this T-DNA insertion line is a null mutant for *PHYA*. We named this new allele as *phyA-501*.

It is already known that light intensity affects seedling development, particularly hypocotyl length [20]. To get different light intensities, we employed neutral filters that reduced the intensity of white (W) light provided by 4 fluorescent tubes. As a result, PAR ranged from a minimum of 4.6 $\mu mol \cdot m^{-2} \cdot s^{-1}$ to a maximum of 72.9 $\mu mol \cdot m^{-2} \cdot s^{-1}$ (with no filters) (Figure 1a). As expected, hypocotyl elongation of wild-type, *phyA* and *phyB* seedlings was decreased when PAR amount was increased. Hypocotyls of *phyB* seedlings were always longer than those of wild-type and *phyA* seedlings under all different light intensities, as previously reported [41]. By contrast, *phyA* hypocotyls were generally longer than Col-0 at lower light intensities, whereas at high light intensities the differences in length were reduced or abolished (Figure 1b). Additional differences were evident in other morphological traits of the seedlings of the three genotypes analyzed: low light intensities reduced cotyledon expansion and delayed primary leaf development (Figure 1c). No higher PAR conditions were applied because under the highest light intensity employed here (i.e., about 73 $\mu mol \cdot m^{-2} \cdot s^{-1}$ from the beginning of germination) some seedlings showed signs of stress, such as a purple color and small size (data not shown).

Light of different R:FR differentially affects the hypocotyl elongation of wild-type, phyA and phyB seedlings

Next, we addressed the influence of light with varying reductions in R:FR on hypocotyl length. To manipulate the R:FR, W light of a fixed PAR was enriched with increasing fluence rates of FR light. We started our experiments with a relatively low level of W light, 15–16 $\mu mol \cdot m^{-2} \cdot s^{-1}$ in the PAR region. For simplicity, from now on we will refer to this intensity of W light as "low PAR". When supplementing with FR, the applied R:FR ranged from 0.320 to 0.035 depending on the amount of FR provided (Figure 2a). Whereas W light provided a high R:FR (> 1.5), W+FR provided light with moderate (intermediate R:FR of 0.5-0.3), substantial (low R:FR of 0.29-0.06) and very large (very

(a)

(b)

(c)

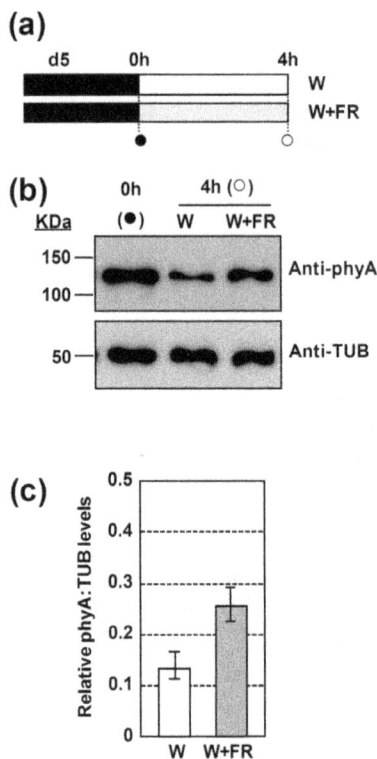

Figure 5. Phytochrome A is stabilized by white light of very low R:FR. (**a**) phyA levels were assayed in extracts from Col-0 seedlings grown in darkness for 5 days and then either transferred to W or W+FR for 4 hours. Circles indicate the harvest time of the plant material. (**b**) Representative steady-state levels of phyA (upper panel) and tubulin (TUB, lower panel) in extracts from seedlings grown as indicated in **a**,. Bands were detected by immunoblot using the phyA-specific mAb 073D or a TUB-specific mAb. TUB was used as a loading control. (**c**) Relative levels of phyA normalized to TUB in seedlings differentially grown for 4 h under W or W+FR, as indicated in **a**; n = 3 independent biological replicas; the P-value between the W and W+FR treated samples was 0.053.

low R:FR of 0.05-0.03) decrease in R:FR (see Table S1 for nomenclature used here). At the highest R:FR applied (0.320), W+FR light strongly induced hypocotyl elongation of Col-0 (about 5 mm) compared to the W-grown seedlings (about 2.5 mm). Further reductions in the R:FR were first slightly more effective (about 6 mm, R:FR of 0.148), and then rapidly less effective, in promoting hypocotyl elongation (about 4 mm for R:FR between 0.098-0.043), until shade-induced hypocotyl elongation was almost abolished (about 3.3 mm for the lowest R:FR tested of 0.035) (Figures 2b, S2a). These results indicate that quantitative variation in the R:FR provides a dual signal: when the R:FR is moderately or substantially lowered (R:FR of 0.320-0.148), it strongly induces hypocotyl elongation of Col-0, whereas when it is strongly reduced (R:FR<0.043; we define this range of R:FR as "very low R:FR") it is less effective in promoting the elongation of Col-0 hypocotyl elongation. At the highest R:FR tested (0.320), W+FR light induced the hypocotyl length of *phyA* to an extent similar to that observed for Col-0. But in striking contrast to the behavior of the Col-0 seedlings, the progressive reduction of the R:FR resulted in a gradual and strong promotion of hypocotyl length in the *phyA* seedlings (from about 7.8 mm at R:FR = 0.148 to more than 10 mm at the lowest R:FR tested of 0.035) (Figures 2b, S2a). These results agree with the reported negative role of phyA in

shade-induced hypocotyl elongation [10], that was most apparent at very low R:FR. W+FR light had a contrasting effect on hypocotyl elongation of *phyB* seedlings: at the highest R:FR (0.320) it did not increase the already W-grown long hypocotyls (about 8 mm). At lower R:FR (0.148 and 0.035), W+FR inhibited (rather than promoted) hypocotyl elongation compared to W-grown seedlings. As a result, at very low R:FR (i.e., the lowest R:FR used of 0.035), W+FR-grown *phyB* hypocotyls were even shorter than those of Col-0 growing under W (Figures 2b, S2a). A mild inhibition of the *phyB* hypocotyl elongation by W+FR has also been observed previously, an effect attributed to the phyA-imposed inhibition of hypocotyl elongation in W+FR, quite apparent in the absence of phyB [17,44,45].

We next used an intensity of W of 30–35 $\mu mol \cdot m^{-2} \cdot s^{-1}$ in the PAR region (from now on, we will refer to this light as of "medium PAR"). The R:FR applied ranged from 0.324 (intermediate R:FR) to 0.030 (very low R:FR) (Figure 2c). W+FR light strongly induced hypocotyl elongation of Col-0 compared to the W-grown seedlings (about 4 mm) for R:FR between 0.324-0.091. Further reductions in the R:FR (i.e., low and very low R:FR) were less effective in promoting hypocotyl elongation (about 3 mm for R:FR between 0.067-0.030) (Figures 2c, S2b). W+FR light also strongly induced hypocotyl length of *phyA* seedlings; the progressive reduction of the R:FR resulted in a very strong promotion of hypocotyl length (from about 4.5 mm at R:FR = 0.324 to more than 7 mm at the lowest R:FR tested of 0.030). For *phyB* seedlings, W+FR light did not affect hypocotyl length at the highest R:FR used (0.324) compared to W-grown seedlings, but it was progressively more effective in inhibiting hypocotyl elongation at lower R:FR (from 0.143 to 0.038) (Figures 2c, S2b).

Finally, we used an intensity of W of 50–60 $\mu mol \cdot m^{-2} \cdot s^{-1}$ in the PAR region (we will refer to this light as of "high PAR"). The applied R:FR ranged from 0.463 to 0.041 (Figures 2d). As for the low and medium PAR experiments, W+FR light induced hypocotyl elongation of Col-0 compared to the W-grown seedlings depending on the R:FR applied: the progressive reduction of the R:FR resulted in longer hypocotyls (from about 1.5 mm at a R:FR = 0.463 to almost 2.5 mm at a R:FR of 0.074). Further reductions in the R:FR were equally effective in promoting hypocotyl elongation (for R:FR<0.074) (Figures 2d, S2c). W+FR light affected hypocotyl length of *phyA* and *phyB* seedlings essentially as described for low and medium PAR cases: the reduction of the R:FR resulted in a progressive promotion of *phyA* hypocotyl length (from almost 2 mm at R:FR = 0.463 to more than 6 mm at the lowest R:FR tested of 0.041) and a progressive inhibition of *phyB* hypocotyls (it did not affect the already long hypocotyls at R:FR of 0.463 and 0.205, but it inhibited hypocotyl elongation at low and very low R:FR of 0.124 to 0.041) (Figures 2d, S2c).

We next analyzed the hypocotyl response to very low R:FR light (very low R:FR of 0.043, low PAR) of *phyA* and *phyB* mutant seedlings in the L*er* background. As shown in Figure S3, the hypocotyl elongation response of these L*er* genotypes was similar to the one observed in Col-0, confirming that similar effects were observed in other genetic backgrounds. Together, these results led us to conclude that (1) the reported negative role of phyA in the shade-induced hypocotyl elongation [10,13] becomes more apparent at very low R:FR (0.05-0.03) and (2) it is qualitatively independent of the range of PAR intensity tested.

The timing of treatment with very low R:FR strongly affects the hypocotyl elongation response

We also addressed the influence of the timing and duration of the W+FR treatment on the hypocotyl elongation response of Col-0, *phyA* and *phyB* seedlings. Seedlings were grown at low PAR and submitted to a low R:FR of 0.059, conditions that allow distinction between the hypocotyl phenotypes of all three genotypes (Figure 2b). When W+FR was applied from day 2 after germination (2W+5WF), hypocotyl length of the analyzed genotypes was affected as observed before: Col-0 hypocotyls were poorly responsive, and *phyA* and *phyB* hypocotyl elongation was strongly promoted and inhibited respectively. When W+FR was applied from days 3 or 4 from germination (3W+4WF and 4W+3WF), Col-0 and *phyB* hypocotyls elongated significantly more than those grown under 2W+5WF, the promotion of elongation being more obvious for *phyB* seedlings. By contrast, *phyA* hypocotyls elongated significantly less than those grown under 2W+5WF (Figure 3a). When W+FR was applied from day 7 after germination (7W+1WF), hypocotyl length of Col-0 and *phyA* seedlings was modestly promoted although *phyA* hypocotyls still were more responsive than those of Col-0 seedlings; by contrast *phyB* seedlings were unaffected (Figure 3b). These results indicated that shade-induced phyA repression of hypocotyl elongation was operative during the entire time of exposure to W+FR (from days 2 to 8). However, based on the effect of the timing of the W+FR application on the *phyA* and *phyB* final hypocotyl elongation, it seems that the repression was stronger at the early stages of seedling development, i.e., from days 2 to 4.

We also investigated whether the elongation in the *phyA* and *phyB* mutants occurred at the same time along the course of seedling development. Wild-type and *phyA* seedlings were grown at a PAR of 20–25 $\mu mol \cdot m^{-2} \cdot s^{-1}$ (medium PAR) and exposed to W+FR of R:FR of 0.083 (low R:FR) from day 2 to day 7 after germination, and *phyB* seedlings were grown from germination under W of the same PAR. Under these conditions we expected a noticeable hypocotyl elongation of all three genotypes. As shown in Figure 4a, hypocotyl length was recorded in 3-, 5- and 7-day-old seedlings. *phyA* hypocotyls were already longer than those of *phyB* and Col-0 on day 3 (in our growth conditions, 2-day-old hypocotyls of all three genotypes are still emerging from the seeds and, therefore, their length is close to 0 mm). Although elongation in both *phyA* and *phyB* seedlings was sustained along the whole period, *phyA* hypocotyls elongated more from days 3 to 5 (d3–d5, 3.01 mm; d5–d7, 2.35 mm), whereas *phyB* hypocotyls elongated substantially more from days 5 to 7 (d3–d5, 1.65 mm; d5–d7, 2.75 mm). In addition, Col-0 hypocotyls elongated more from days 5 to 7 (d3–d5, 0.63 mm; d5–d7, 1.36 mm), suggesting that under the W+FR conditions applied its elongation was strongly inhibited from days 2 to 5, the same time window in which *phyA* hypocotyls elongated more. These results are consistent with our previous conclusion that phyA-mediated repression was stronger at the early stages of seedling development.

Treatment with very low R:FR stabilizes phyA levels

phyA is abundant in etiolated tissues but is light-labile and so is rapidly depleted in light-treated tissues. The long hypocotyl phenotype of *phyA* seedlings grown under W+FR of very low R:FR suggested that phyA may be more abundant under very low R:FR in Col-0 than under high R:FR light. To test this possibility, phyA levels were analyzed by western blot in 5-day-old etiolated Col-0 seedlings exposed to W (PAR of 20–25 $\mu mol \cdot m^{-2} \cdot s^{-1}$, R:FR = 2.5) and W+FR (same PAR, very low R:FR of 0.038) for 4 h (Figure 5a). As shown in Figure 5b, phyA levels declined after exposure to either W or W+FR light. However, in extracts from W+FR-exposed seedlings, phyA levels were higher (Figure 5c). Longer exposure times of W+FR (8 h) showed a similar tendency (Figure S4). Altogether, our data indicate that the balance between phyA synthesis and degradation in seedlings is affected by the

R:FR of the incoming light, whereby light of very low R:FR has a milder destabilizing effect compared to W light of high R:FR.

Discussion

As mentioned above, nearby vegetation selectively reflects FR, thus lowering the R:FR, a signal that induces plant responses in anticipation of neighboring vegetation becoming a competitive threat [46]. If, despite these responses, neighboring vegetation directly shades the plant, light quantity becomes limiting, i.e., there is a reduction in the amount of radiation active in photosynthesis (i.e., PAR, between 400–700 nm), resulting in additional or more dramatic SAS responses [27]. Measurements by different authors of different natural light environments agree with this view [1–4,6,47,48]. In the laboratory, conditions that mimic plant proximity before actual canopy shading occurs (i.e., with lowered R:FR only, without changing PAR) have been termed *simulated shade*, and those that mimic natural situations when canopy closure occurs (which reduce both R:FR and PAR), have been termed *canopy shade* [25,28]. We have shown here that perception of these types of light-environment are genetically distinguishable by analyzing the hypocotyl elongation of *phyA* and *phyB* mutant seedlings in response to light of different R:FR, even without altering the PAR intensity: the inhibitory effect of phyA is readily observed under very low R:FR by (1) the conspicuous hypocotyl elongation of the *phyA* seedlings compared to the wild-type, and (2) the strong inhibition of the long-hypocotyl phenotype of the *phyB* seedlings. However, our data reveal a dichotomy in responsiveness across the range of R:FR tested. As mentioned, these phenotypes are observed under the lowest R:FR ratios (e.g. about 0.05-0.03) in our W+FR experiments here (Figure 2). However, these responses are essentially absent under our intermediate R:FR treatments (about 0.5-0.3), whereby *phyA* hypocotyls behave almost as wild-type ones and the long *phyB* hypocotyls are unaffected by the simulated shade signal (Figures 2, S2, 6a).

The antagonistic role of phyA in the SAS regulation has been noted previously when examining seedling deetiolation using natural shade conditions (provided by densely-grown plants of common wheat), and it was shown to be important for seedling establishment and survival under these specific conditions: *phyA* deficient mutants displayed extreme elongation growth, poor cotyledon development (phenotypes similar to those observed here under very low R:FR, i.e. R:FR = 0.05-0.03) (Figures 2, S3) and a lower survival rate than wild-type seedlings [11]. Our data show that the inhibitory role of phyA occurs independently of the range of PAR levels tested here (Figure 2), suggesting that the level of PAR in the incident light, although broadly recognized to modulate plant development (Figure 1), does not alone contain the essential differential information between these two extreme types of R:FR conditions (i.e. simulated and canopy shade). Hence, the R:FR signal alone seems sufficient to differentiate between plant proximity (mild reductions in the R:FR, without PAR decrease, defined here as "intermediate R:FR") and dense, direct canopy shade (strong reductions in the R:FR signal, defined here as "very low R:FR"). In this regard, *phyA* hypocotyls progressively elongated from intermediate to very low R:FR, and at the lowest levels tested they were much longer than those of *phyB* seedlings under W (high R:FR) (Figures 2, S2, S3), providing additional evidence for the effectiveness of the information contained in the R:FR signal. In addition, it suggests that the strong promotion of hypocotyl elongation observed in the *phyA* mutant background is due to the inactivation of the Pfr form of phyB and other photostable phys induced by the very low R:FR

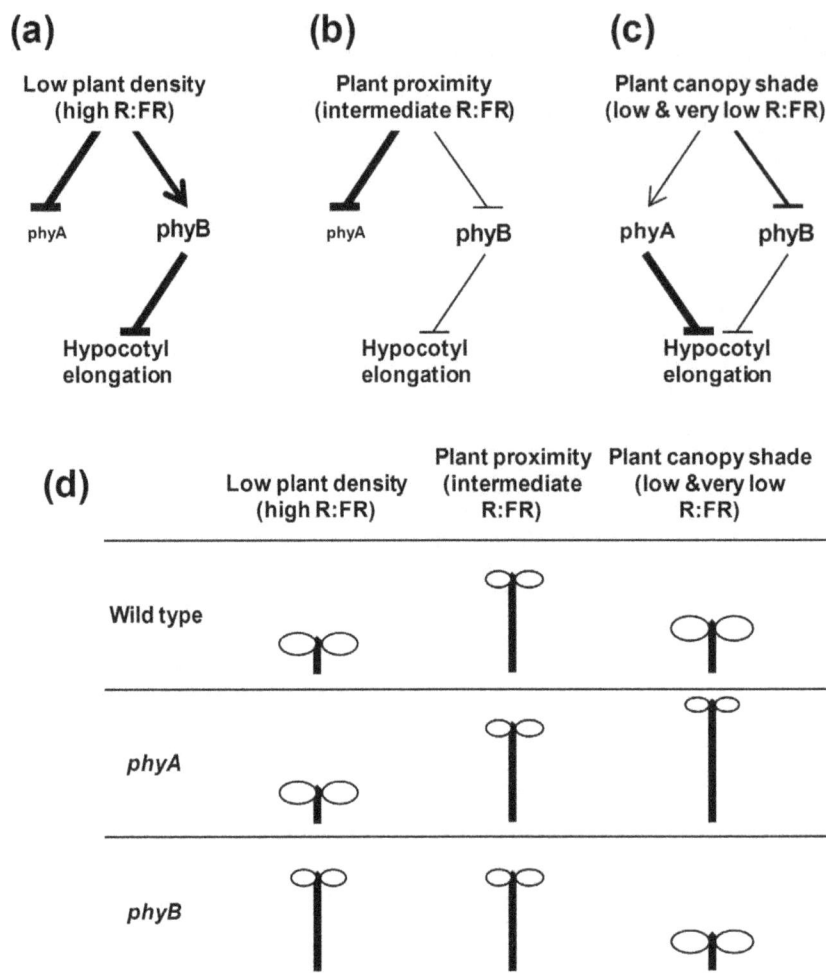

Figure 6. Model depicting the antagonistic effect of phyA and phyB on the shade-induced hypocotyl elongation. (a) Under low plant density, the high R:FR induces phyA degradation and stimulates the phyB active Pfr form, which strongly inhibits hypocotyl elongation. (b) In close proximity of vegetation, phyA degradation still occurs, but the low R:FR displaces the photoequilibrium of phyB towards the inactive Pr form, causing hypocotyls to elongate. (c) Under a plant canopy, the low or very low R:FR still displaces the photoequilibrium of phyB towards the inactive Pr form that stimulates hypocotyls to elongate. However, under these conditions phyA is stabilized, particularly at the beginning of the seedling emergence; as a consequence, phyA signaling is enhanced, thereby counteracting the inhibitory effect of the absence of active phyB, so that hypocotyls elongate only moderately. (d) Summary of phenotypes shown by the wild-type, *phyA* and *phyB* seedlings growing under the light conditions indicated in **a, b** and **c**.

light treatment. It might be argued that the PAR intensities employed in these experiments are low compared to what might be observed in nature (e.g., 750 $\mu mol \cdot m^{-2} \cdot s^{-1}$ in a clear midday in Buenos Aires, Argentina) [11]. Nonetheless, under natural dense canopies, PAR intensity might be reduced to less than 0.5% at the solar zenith, reaching even lower intensities than the ones employed in this work (1 $\mu mol \cdot m^{-2} \cdot s^{-1}$). Under these low PAR intensities, the R:FR results in values similar to those described here as low (0.10±0.02; Table S1) [11,49,50]. Therefore, lower R:FR values might be reached at other times of the day or latitudes. It is interesting to note that in chlorophyll rich organs, such as leaves, light absorption from the ultraviolet to the visible region by chlorophyll a leads to the emission (by fluorescence) of FR light [51]. Therefore, leaf chlorophylls can actively contribute to create low and very low R:FR signals in natural deep-shaded environments.

During the first week of seedling emergence and development, the inhibitory role of phyA is very apparent and easily observed at the beginning of this period (from days 2 to 5) when the potential to elongate is very high. Once this potential is diminished, the role of phyA also becomes less relevant (Figures 3, 4) [13]. Seedlings grown under photoperiodic conditions also respond to transient (2 h) low R:FR, which has revealed that there is also a circadian component in phyA action: whereas simulated shade given at subjective dusk increases hypocotyl length, when given at subjective dawn leads to a small inhibition of hypocotyl elongation (compared to untreated seedlings), an antagonistic effect of low R:FR shown to be dependent on phyA [18]. PhyA also (1) inhibits hypocotyl elongation under short-day conditions [10], under continuous R [52,53], or under continuous W of high R:FR but low PAR (Figure 1) and (2) promotes cotyledon expansion [52,54], likely also at the early stages of seedling development. However, the role of phyA in light-grown plants is not restricted to these early stages of the seedling development. In adult plants grown under short-day conditions, phyA has also a role in suppressing internode growth and leaf elongation [54].

The various roles of phyA in light-grown seedlings and plants are consistent with the evidence that, although phyA levels have been strongly decreased in these light-exposed tissues, they are not reduced to zero [3,55]. During seedling de-etiolation, phyA activation by FR results from the fact that continuous monochromatic FR light establishes and maintains a small fraction of the phyA population in the Pfr form, operating via the FR-HIR, over an extended period [3,23,56]. Indeed, although phyA is photolabile, FR-grown seedlings retain phyA at higher levels than R-grown seedlings as determined by western analyses blot analyses [55,57]. Also, under light/dark cycles, phyA accumulates during the night and is rapidly degraded during the day [55], which might explain the long hypocotyl phenotype of phyA seedlings grown under short-days conditions [10] and the absence of growth inhibition in phyA seedlings when a transient low R:FR treatment is given at subjective dawn [18]. We have observed that phyA is degraded under W in a R:FR-dependent manner: under very low R:FR conditions phyA is more stable than under high R:FR (Figures 5, S4). The rapid increase in the expression of PHYA in response to low R:FR light very likely contributes to the observed maintenance of high phyA protein levels [44]. Therefore, it seems likely that in fully de-etiolated seedlings the activation of phyA by W+FR of very low R:FR maintains a small fraction of phyA cycling in the active Pfr form that likely results in the observed suppression of the hypocotyl elongation [13,54,58].

It has been discussed in the literature that certain aspects of seedling SAS and de-etiolation affect the same traits but in opposite directions, such as accelerated hypocotyl cell elongation, retarded cotyledon expansion and reduced photosynthetic pigment accumulation. Indeed, gene-expression analyses provided initial molecular evidence for this view [17] and the continuous modulation by the phy-PIF signaling system has been implicated [21]. The analyses of seedling de-etiolation using Arabidopsis phyA and phyB mutants have established that both phytochromes have roles in seedling de-etiolation: phyA is the only photoreceptor responsible for the response of seedlings to continuous monochromatic FR, whereas phyB is mainly responsible for the responses of seedlings to continuous monochromatic R (Figure S5). The complementary actions of phyA and phyB in this process has been considered to provide optimum regulation of seedling growth after emergence from the soil [23]. We show here that during the SAS response of seedlings, phyB is deactivated by shade of intermediate, low and very low R:FR, whereas phyA is only strongly activated by shade of low, and very low R:FR, partly because of its higher levels. As a result, the phyA-Pfr produced and sustained in a cycling state strongly inhibits hypocotyl elongation via the FR-HIR activity of this phy [13,56] (Figure 6d). The differential effects of the phyA and phyB mutants on this process genetically defines the operation of two different pathways in SAS regulation [13], an additional similarity between the SAS and de-etiolation responses (Figures 6, S5). In the natural environment, continuous monitoring of the R:FR will determine the participation of phyA in the response to shade; when the R:FR is very low (such as in deep shade), phyA activation will prevent seedlings from exhibiting excessive elongation mediated principally by deactivation of phyB [58]. Whereas the overlapping actions of phyA and phyB will substantially promote de-etiolation in sparse vegetation [12], the antagonistic action of phyA and phyB will ensure the optimum elongation under deep shade, conditions in which R:FR can be strongly reduced partly due to the active emission of FR by the chlorophyll from the leaves [51].

Collectively, our data provide evidence that phyA functions in natural light environments to attenuate the SAS in response to direct canopy shading, but not to simple neighbor-proximity. This deduction refines the existing concept that phyA can "antagonize" the SAS via the FR-HIR [2,10,13], and supports the notion that

plants have evolved a sophisticated intrafamily photosensory attenuation mechanism that can discriminate between the threat and imposition of competition for PAR by neighboring vegetation. This dual-track mechanism provides young seedlings with the capacity for both rapid elongation upon sensing of impending competition (intermediate R:FR) (the "neighbor-detection response mode"), or attenuation of potentially deleterious excessive elongation upon direct interception of canopy shade (low or very low R:FR) (the "direct-shade response mode").

Supporting Information

Figure S1 The phyA-501 line is deficient in phyA. (a) Scheme of the PHYA genomic structure and the site of insertion of the T-DNA in the SALK_014575 line; arrows indicate the approximate location of primers used for genotyping (LBb1, MSO31 and MSO32). (b) Steady-state levels of phyA measured by protein blot. Immunoblot detection of phyA (upper panel) and tubulin (TUB, lower panel) levels in extracts from Col-0 and phyA-501 seedlings grown in darkness for 5 days. Bands were detected as indicated in Figure 5b. (c) Hypocotyl length in 4-day-old Col-0 and phyA-501 seedlings grown in darkness, continuous FR ($3.7 \ \mu mol \cdot m^{-2} \cdot s^{-1}$) and R ($12.8 \ \mu mol \cdot m^{-2} \cdot s^{-1}$), as shown in the upper part of the panel. Mean and SE values represent at least 25 seedlings from each treatment. Asterisks indicate significant differences (*$P<0.05$; **$P<0.01$) relative to control seedlings (Col-0) grown under the same conditions.

Figure S2 Effect of different R:FR on hypocotyl elongation of wild-type, phyA-501 and phyB-9 seedlings under low (a), medium (b) or high (c) PAR represented as the difference (left panels) or the ratio (right panels) between values under W+FR and W. Data were recalculated from the experiments generated for Figure 2. Values are the mean and SD from the 3 independent experiments.

Figure S3 Effect of very low R:FR ratios on hypocotyl elongation of wild-type (Ler), phyA and phyB seedlings under high light intensity. Seedlings were germinated and grown (PAR was of $15–16 \ \mu mol \cdot m^{-2} \cdot s^{-1}$) as indicated in Figure 2a. Under W+FR, R:FR was 0.043. Mean and SE values represent at least 25 seedlings from each light treatment. Asterisks indicate significant differences (*$P<0.05$; **$P<0.01$) relative to control seedlings (Ler) grown under the same light conditions.

Figure S4 Levels of phyA are stabilized by very low R:FR treatments. (a) phyA levels were detected in extracts from Col-0 seedlings grown in darkness for 5 days and then either transferred to W or W+FR for 4 and 8 hours. Symbols indicate the harvest time of the plant material. (b) Immunoblot detection of steady-state levels of phyA (upper panel) and TUB (lower panel) in extracts from seedlings grown as indicated in (a). Bands were detected as indicated in Figure 5b.

Figure S5 Schematic summary of the phenotypes of wild-type, phyA and phyB seedlings grown in the dark or monochromatic R o FR light (indicated at the top).

Table S1 Terminology of the various R:FR regimes applied in this work and its proposed equivalence under natural conditions.

Acknowledgments

We are grateful to the greenhouse services of CRAG and PGEC for plant care; NASC, for providing seed stocks; and Pablo Leivar, Anne Pfeiffer and Manuel Rodríguez-Concepción for comments on the manuscript.

References

1. Casal (2012) Shade Avoidance. The Arabidopsis Book.
2. Franklin KA (2008) Shade avoidance. New Phytol 179: 930–944.
3. Smith H (1995) Physiological and ecological function within the phytochrome family. Annual Review of Plant Physiology and Plant Molecular Biology 46: 289–315.
4. Martinez-Garcia JF, Galstyan A, Salla-Martret M, Cifuentes-Esquivel N, Gallemí M, et al. (2010) Regulatory components of shade avoidance syndrome. Advances in Botanical Research 53: 65–116.
5. Ruberti I, Sessa G, Ciolfi A, Possenti M, Carabelli M, et al. (2012) Plant adaptation to dynamically changing environment: The shade avoidance response. Biotechnology Advances 30: 1047–1058.
6. Casal JJ (2013) Photoreceptor signaling networks in plant responses to shade. Annu Rev Plant Biol 64: 403–427.
7. Bae G, Choi G (2008) Decoding of light signals by plant phytochromes and their interacting proteins. Annual Review of Plant Biology 59: 281–311.
8. Franklin KA, Quail PH (2010) Phytochrome functions in Arabidopsis development. J Exp Bot 61: 11–24.
9. Hennig L, Poppe C, Sweere U, Martin A, Schafer E (2001) Negative interference of endogenous phytochrome B with phytochrome A function in Arabidopsis. Plant Physiol 125: 1036–1044.
10. Johnson E, Bradley M, Harberd NP, Whitelam GC (1994) Photoresponses of Light-Grown phyA Mutants of Arabidopsis (Phytochrome A Is Required for the Perception of Daylength Extensions). Plant Physiology 105: 141–149.
11. Yanovsky MJ, Casal JJ, Whitelam GC (1995) Phytochrome A, phytochrome B and HY4 are invovled in hypocotyl growth responses to natural radiation in Arabidopsis: Weak de-etiolation of the phyA mutant under dense canopies. Plant, Cell and Environment.
12. Smith H, Xu Y, Quail PH (1997) Antagonistic but complementary actions of phytochromes A and B allow seedling de-etiolation. Plant Physiol 114: 637–641.
13. McCormac AC, Whitelam GC, Boylan MT, Quail PH, Smith H (1992) Contrasting Responses of Etiolated and Light-Adapted Seedlings to Red: Far-Red Ratio: A Comparison of Wild Type, Mutant and Transgenic Plants has Revealed Differential Functions of Members of the Phytochrome family. Journal of Plant Physiology 140: 707–714.
14. Martinez-Garcia JF, Huq E, Quail PH (2000) Direct targeting of light signals to a promoter element-bound transcription factor. Science 288: 859–863.
15. Lorrain S, Allen T, Duek PD, Whitelam GC, Fankhauser C (2008) Phytochrome-mediated inhibition of shade avoidance involves degradation of growth-promoting bHLH transcription factors. The Plant Journal 53: 312–323.
16. Li L, Ljung K, Breton G, Schmitz RJ, Pruneda-Paz J, et al. (2012) Linking photoreceptor excitation to changes in plant architecture. Genes and Development 26: 785–790.
17. Roig-Villanova I, Bou J, Sorin C, Devlin PF, Martinez-Garcia JF (2006) Identification of primary target genes of phytochrome signaling. Early transcriptional control during shade avoidance responses in Arabidopsis. Plant Physiology 141: 85–96.
18. Salter MG, Franklin KA, Whitelam GC (2003) Gating of the rapid shade-avoidance response by the circadian clock in plants. Nature 426: 680–683.
19. Steindler C, Matteucci A, Sessa G, Weimar T, Ohgishi M, et al. (1999) Shade avoidance responses are mediated by the ATHB-2 HD-zip protein, a negative regulator of gene expression. Development 126: 4235–4245.
20. Hornitschek P, Kohnen MV, Lorrain S, Rougemont J, Ljung K, et al. (2012) Phytochrome interacting factors 4 and 5 control seedling growth in changing light conditions by directly controlling auxin signaling. The Plant Journal 71: 699–711.
21. Leivar P, Tepperman JM, Cohn MM, Monte E, Al-Sady B, et al. (2012) Dynamic Antagonism between Phytochromes and PIF Family Basic Helix-Loop-Helix Factors Induces Selective Reciprocal Responses to Light and Shade in a Rapidly Responsive Transcriptional Network in Arabidopsis. The Plant Cell 24: 1398–1419.
22. Jiao Y, Lau OS, Deng XW (2007) Light-regulated transcriptional networks in higher plants. Nat Rev Genet 8: 217–230.
23. Quail PH (2002) Phytochrome Photosensory Signalling Networks. Nature Reviews Molecular Cell Biology 3: 85–93.
24. Cifuentes-Esquivel N, Bou-Torrent J, Galstyan A, Gallemi M, Sessa G, et al. (2013) The bHLH proteins BEE and BIM positively modulate the shade avoidance syndrome in Arabidopsis seedlings. The Plant Journal 75: 989–1002.
25. Roig-Villanova I, Bou-Torrent J, Galstyan A, Carretero-Paulet L, Portoles S, et al. (2007) Interaction of shade avoidance and auxin responses: a role for two novel atypical bHLH proteins. The EMBO Journal 26: 4756–4767.
26. Sorin C, Salla-Martret M, Bou-Torrent J, Roig-Villanova I, Martinez-Garcia JF (2009) ATHB4, a regulator of shade avoidance, modulates hormone response in Arabidopsis seedlings. The Plant Journal 59: 266–277.
27. Sessa G, Carabelli M, Sassi M, Ciolfi A, Possenti M, et al. (2005) A dynamic balance between gene activation and repression regulates the shade avoidance response in Arabidopsis. Genes and Development 19: 2811–2815.
28. Crocco CD, Holm M, Yanovsky MJ, Botto JF (2010) AtBBX21 and COP1 genetically interact in the regulation of shade avoidance. The Plant Journal 64: 551–562.
29. Gangappa SN, Crocco CD, Johansson H, Datta S, Hettiarachchi C, et al. (2013) The Arabidopsis B-BOX Protein BBX25 Interacts with HY5, Negatively Regulating BBX22 Expression to Suppress Seedling Photomorphogenesis. The Plant Cell 25: 1243–1257.
30. Galstyan A, Cifuentes-Esquivel N, Bou-Torrent J, Martinez-Garcia JF (2011) The shade avoidance syndrome in Arabidopsis: a fundamental role for atypical basic helix-loop-helix proteins as transcriptional cofactors. The Plant Journal 66: 258–267.
31. Hersch M, Lorrain S, de Wit M, Trevisan M, Ljung K, et al. (2014) Light intensity modulates the regulatory network of the shade avoidance response in Arabidopsis. Proc Natl Acad Sci U S A 111: 6515–6520.
32. Bou-Torrent J, Galstyan A, Gallemi M, Cifuentes-Esquivel N, Molina-Contreras MJ, et al. (2014) Plant proximity perception dynamically modulates hormone levels and sensitivity in Arabidopsis. J Exp Bot.
33. Keuskamp DH, Pollmann S, Voesenek LA, Peeters AJ, Pierik R (2010) Auxin transport through PIN-FORMED 3 (PIN3) controls shade avoidance and fitness during competition. Proceedings of the National Academy of Sciences, USA 107: 22740–22744.
34. Tao Y, Ferrer JL, Ljung K, Pojer F, Hong F, et al. (2008) Rapid synthesis of auxin via a new tryptophan-dependent pathway is required for shade avoidance in plants. Cell 133: 164–176.
35. Galstyan A, Bou-Torrent J, Roig-Villanova I, Martinez-Garcia JF (2012) A dual mechanism controls nuclear localization in the atypical basic-helix-loop-helix protein PAR1 of Arabidopsis thaliana. Molecular Plant 5: 669–677.
36. Keuskamp DH, Keller MM, Ballare CL, Pierik R (2012) Blue light regulated shade avoidance. Plant Signal Behav 7: 514–517.
37. Keller MM, Jaillais Y, Pedmale UV, Moreno JE, Chory J, et al. (2011) Cryptochrome 1 and phytochrome B control shade-avoidance responses in Arabidopsis via partially independent hormonal cascades. The Plant Journal 67: 195–207.
38. Tsukaya H, Kozuka T, Kim GT (2002) Genetic control of petiole length in Arabidopsis thaliana. Plant Cell Physiol 43: 1221–1228.
39. Martinez-Garcia JF, Virgos-Soler A, Prat S (2002) Control of photoperiod-regulated tuberization in potato by the Arabidopsis flowering-time gene CONSTANS. Proceedings of the National Academy of Sciences, USA 99: 15211–15216.
40. Ruckle ME, DeMarco SM, Larkin RM (2007) Plastid signals remodel light signaling networks and are essential for efficient chloroplast biogenesis in Arabidopsis. The Plant Cell 19: 3944–3960.
41. Reed JW, Nagpal P, Poole DS, Furuya M, Chory J (1993) Mutations in the gene for the red/far-red light receptor phytochrome B alter cell elongation and physiological responses throughout Arabidopsis development. The Plant Cell 5: 147–157.
42. Al-Sady B, Ni W, Kircher S, Schäfer E, Quail PH (2006) Photoactivated Phytochrome Induces Rapid PIF3 Phosphorylation Prior to Proteasome-Mediated Degradation. Molecular Cell 23: 439–446.
43. Martinez-Garcia JF, Monte E, Quail PH (1999) A simple, rapid and quantitative method for preparing Arabidopsis protein extracts for immunoblot analysis. The Plant Journal 20: 251–257.
44. Devlin PF, Yanovsky MJ, Kay SA (2003) A genomic analysis of the shade avoidance response in Arabidopsis. Plant Physiology 133: 1617–1629.
45. Leivar P, Monte E, Cohn MM, Quail PH (2012) Phytochrome Signaling in Green Arabidopsis Seedlings: Impact Assessment of a Mutually Negative phyB-PIF Feedback Loop. Molecular Plant 5: 734–749.
46. Ballaré CL, Sánchez RA, Scopel AL, Casal JJ, Ghersa CM (1987) Early detection of neighbour plants by phytochrome perception of spectral changes in reflected sunlight. Plant, Cell and Environment 10: 551–557.
47. Vandenbussche F, Verbelen J-P, Van Der Straeten D (2005) Of light and length: Regulation of hypocotyl growth in Arabidopsis. Bioessays 27: 275–284.
48. Smith H (1982) Light quality, photoperception, and plant strategy. Annual Review of Plant Physiology 33: 481–518.
49. Lei TT, Nilsen ET, Semones SW (2006) Light environment under Rhododendron maximum thickets and estimated carbon gain of regenerated forest tree seedlings. Plant Ecology 184: 143–156.
50. Lee DW (1987) The spectral Distribution of Radiation in Two Neotropical Rainforsts. Biotropica 19: 161–166.
51. Thornber JP (1975) Chlorophyll-proteins: light-harvesting and reaction center components of plants. Annual Review of Plant Physiology 26: 127–158.

Author Contributions

Conceived and designed the experiments: JFMG PHQ MG. Performed the experiments: JFMG MG MJMC BL MB. Analyzed the data: JFMG MG MJMC BL MB PHQ. Contributed reagents/materials/analysis tools: JFMG PHQ. Wrote the paper: JFMG PHQ.

52. Reed JW, Nagatani A, Elich TD, Fagan M, Chory J (1994) Phytochrome A and Phytochrome B Have Overlapping but Distinct Functions in Arabidopsis Development. Plant Physiol 104: 1139–1149.

53. Casal JJ, Mazzella MA (1998) Conditional synergism between cryptochrome 1 and phytochrome B is shown by the analysis of phyA, phyB, and hy4 simple, double, and triple mutants in Arabidopsis. Plant Physiol 118: 19–25.

54. Franklin KA, Praekelt U, Stoddart WM, Billingham OE, Halliday KJ, et al. (2003) Phytochromes B, D, and E Act Redundantly to Control Multiple Physiological Responses in Arabidopsis. Plant Physiology 131: 1340–1346.

55. Sharrock RA, Clack T (2002) Patterns of Expression and Normalized Levels of the Five Arabidopsis Phytochromes. Plant Physiology 130: 442–456.

56. Rausenberger J, Tscheuschler A, Nordmeier W, Wüst F, Timmer J, et al. (2011) Photoconversion and Nuclear Trafficking Cycles Determine Phytochrome A's Response Profile to Far-Red Light. Cell 146: 813–825.

57. Hudson M, Ringli C, Boylan MT, Quail PH (1999) The FAR1 locus encodes a novel nuclear protein specific to phytochrome A signaling. Genes Dev 13: 2017–2027.

58. Smith H, Whitelam G (1997) The shade avoidance syndrome: multiple responses mediated by multiple phytochromes. Plant, Cell and Environment 20: 840–844.

2

Informal "Seed" Systems and the Management of Gene Flow in Traditional Agroecosystems: The Case of Cassava in Cauca, Colombia

George A. Dyer[1]*, Carolina González[2,3], Diana Carolina Lopera[4]

1 The James Hutton Institute, Aberdeen, United Kingdom, **2** LACBiosafety Project, International Center for Tropical Agriculture (CIAT), Cali, Colombia, **3** International Food Policy Research Institute (IFPRI), Washington, D.C., United States of America, **4** LACBiosafety Project, International Center for Tropical Agriculture (CIAT), Cali, Colombia

Abstract

Our ability to manage gene flow within traditional agroecosystems and their repercussions requires understanding the biology of crops, including farming practices' role in crop ecology. That these practices' effects on crop population genetics have not been quantified bespeaks lack of an appropriate analytical framework. We use a model that construes seed-management practices as part of a crop's demography to describe the dynamics of cassava (*Manihot esculenta* Crantz) in Cauca, Colombia. We quantify several management practices for cassava—the first estimates of their kind for a vegetatively-propagated crop—describe their demographic repercussions, and compare them to those of maize, a sexually-reproduced grain crop. We discuss the implications for gene flow, the conservation of cassava diversity, and the biosafety of vegetatively-propagated crops in centers of diversity. Cassava populations are surprisingly open and dynamic: farmers exchange germplasm across localities, particularly improved varieties, and distribute it among neighbors at extremely high rates vis-à-vis maize. This implies that a large portion of cassava populations consists of non-local germplasm, often grown in mixed stands with local varieties. Gene flow from this germplasm into local seed banks and gene pools via pollen has been documented, but its extent remains uncertain. In sum, cassava's biology and vegetative propagation might facilitate pre-release confinement of genetically-modified varieties, as expected, but simultaneously contribute to their diffusion across traditional agroecosystems if released. Genetically-modified cassava is unlikely to displace landraces or compromise their diversity; but rapid diffusion of improved germplasm and subsequent incorporation into cassava landraces, seed banks or wild populations could obstruct the tracking and eradication of deleterious transgenes. Attempts to regulate traditional farming practices to reduce the risks could compromise cassava populations' adaptive potential and ultimately prove ineffectual.

Editor: Randall P. Niedz, United States Department of Agriculture, United States of America

Funding: This work was supported by the LACBiosafety Project (http://www.lacbiosafety.org/). The funders had no role in study design, data collection and analysis, decision to publish, or preparation of the manuscript.

Competing Interests: The authors have declared that no competing interests exist.

* E-mail: georgie.dyer@gmail.com

Introduction

The applicability of ecological concepts and methods to environmental management is perhaps nowhere as clear as in agroecosystems, particularly in centers of crop origin and diversity, where farmers' management of seed is an integral part of the ecology of crops and their wild relatives [1,2]. Crop landraces have been described as managed populations—open and dynamic systems that evolve in response to gene flow and selection [3,4]. Exchange of planting materials among farmers is considered a major selective pressure and partly responsible for these populations' diversity [5,6]. Scientists also recognize that seed exchange made domestication a more complex process than once thought. Yet, the complexity of farmers' past and present role in crop evolution is not fully appreciated. On one hand, farmer management does not reduce to seed exchange; cassava farmers, for instance, exercise frequency-dependent selection when conserving rare seed [5] and selection for heterozygosity when protecting volunteer seedlings [7]. But neither does management reduce to a selection pressure. In fact, seed exchange often is

regarded as a random force, more akin to genetic drift than to selection [8,9]. More generally, seed management and other farming practices constitute a set of forces that are intrinsic to a crop's demography and thus have quantifiable effects on gene flow and frequencies [9,10]. That these effects have not been quantified bespeaks lack of an appropriate analytical framework.

Seed exchange can have unintended consequences, including the propagation of crop diseases. Arguably, it was also the main mechanism for the spread of biological innovations during the onset of farming—a role that it continues to play in developing areas to date [10]. Agricultural innovations more generally have influenced every aspect of farming throughout history, from input use to land-use patterns. Innovations embodied in the seed, including genes of agronomic value and more recently transgenes, also entail *sui generis* risks [11]. Although studied extensively in industrialized countries, the unintended implications of biotechnology cannot yet be fully ascertained in centers of crop diversity, where transgenes could introgress into landraces and their wild relatives [12]. This applies to vegetatively-propagated crops, e.g., cassava and potato, which have been largely absent from discussions on biosafety [1,13]. Managing gene flow and other repercussions of farmer practices requires unraveling these practices' intimate association with crop demography.

In this paper we use a demographic model that construes seed-management practices as events in the life history of crops to describe the dynamics of cassava (*Manihot esculenta* Crantz). We estimate various demographic parameters for cassava in Cauca, Colombia—the first estimates of their kind for a vegetatively-propagated crop—and compare cassava's population dynamics with those of maize, a sexually-reproduced grain crop. We discuss the implications for gene flow, the conservation of cassava diversity, and the biosafety of vegetatively-propagated crops in centers of diversity.

Analytical framework

Cassava is a perennial shrub whose starchy, tuberous roots are a major source of carbohydrates in tropical countries. Numerous landraces of this crop are maintained in farming communities throughout the Amazon basin's rim, including Colombia [5,14]. In the wild, the species' diversity is highest in south-western Brazil, cassava's center of origin. Clonal propagation has not isolated cassava reproductively, so its genes (and transgenes) can introgress into wild *Manihot* wherever their distributions overlap [1]. Wild *M. esculenta* is absent in Colombia, but four other potentially intercrossing *Manihot* species are present [14]. A possible containment strategy thus could require restricting commercial release of genetically-modified cassava wherever *Manihot* diversity is present. An alternative strategy could be based on a detailed understanding of gene flow in this species [1].

Crop scientists recognize that farmer practices have implications for gene flow; but rather than studying these practices directly, they have opted to make inferences on them based on genetic data [8,15–16]. This is a sensible approach, perhaps the only one possible, when the focus is on historic populations, but not when seed management can be observed directly. Genetic data can be used confidently to test seed management's effects on gene flow and frequencies, but this requires sensible hypotheses that both recognize and understand the numerous practices involved.

It is generally taken for granted that seed management in traditional farming systems is well understood and amply documented. Our knowledge derives, in fact, from a handful of case studies that are not representative of a crop, a region or a farming system. Common generalizations have little empirical support. It is widely believed, for instance, that traditional farmers generally maintain a portfolio of crop diversity, when in fact it is most common to grow a single variety [10,17]. A similar misunderstanding, in the context of seed exchange, is that fields are sown using seed from a single, familiar but otherwise random source [9]. Analyses of seed management also are largely descriptive and seldom based on a quantitative analytical framework [16]. Thus, landrace management most often is characterized simply as more or less dynamic, seed exchange as more or less frequent or widespread, and landrace populations themselves as more or less open [4,8].

Seed and pollen exchange are both essential for the dispersal and persistence of alleles in cross-pollinated crops; but in contrast to pollen, seed is long lived and can be exchanged across long distances. "Seed" exchange (i.e., exchange of planting materials, including stakes and tubers) has an even greater weight on the gene flow of vegetatively-propagated crops, since "true seed" (i.e., fertilized ovules) and hence pollen often play no role in either formal or informal "seed" systems. Yet, seed movement rarely figures in models of gene flow in crops. Current models of transgene dispersal, for instance, focus exclusively on pollen [18].

These models are well suited to industrialized agriculture, where improved seed is replaced every cropping cycle, but not to traditional agriculture where farmers maintain landraces on farm [19,20]. Farmers exchange seed within their own communities but also introduce seed from other localities. Sometimes they replace this seed for their own but may also mix both. All of these practices can be construed as events in the life history of a crop and articulated into a demographic model to shed light on management's role in its population dynamics (see Methods).

Results

Seed replacement

A log-linear model tested the effect of seed type and the locality's elevation on seed replacement rates $(1 - p)$ [21]. G-tests for goodness of fit revealed significant differences in replacement across seed types ($P<0.01$) and elevation ($P<0.01$) (Table 1A). Landrace seed is replaced at significantly lower rates $(1 - p = 0.25)$ than improved varieties (0.35). Freeman-Tukey deviates (not shown) revealed nevertheless that these differences are present only at low and intermediate elevations. Altitudinal differences in seed replacement are not significant for improved varieties; but differences are evident for landraces, whose seed is replaced at the highest rate at high elevations $(1 - p = 0.33)$ and the lowest at intermediate elevations (0.15). The latter is the lowest rate of replacement of all type-by-altitude combinations.

Seed diffusion

None of the newly-acquired seed in the sample was purchased from a commercial source, and only 2% was obtained from a non-governmental organization or directly from the Center for Tropical Agriculture (CIAT). The rest was acquired from other farmers. A log linear model tested the effect of seed type and origin on diffusion rates, i.e., the probability that a seed lot is distributed to other farmers. The model revealed no differences across either seed types ($P=0.41$) or origin ($P=0.43$; Table 1B). Inspection of rates across categories suggested a possible interaction of the effects of seed-type and origin, with landraces diffused more than improved varieties when seed is local but less when it is introduced. However, the interaction is not significant; a test of complete independence could not be rejected ($P=0.75$), although possibly due to a reduced sample size (see below).

A separate log-linear model tested the effect of elevation and seed type on diffusion rates. In contrast to seed replacement, no systematic differences in diffusion rates were found across elevations ($P=0.19$; Table 2A); but controlling for elevation revealed the effect of seed type on seed diffusion. Diffusion is higher for landraces than for improved varieties: $q = 0.92$ and 0.84, respectively ($P<0.01$). In this case too, differences between seed types are present only at low and intermediate elevations. Despite the absence of systematic differences in diffusion across elevations, Freeman-Tukey deviates showed that improved seed at high elevations is diffused at higher rates than elsewhere, while landrace seed is diffused at lower rates.

A third model showed that diffusion depends significantly on whether seed has been saved across cycles (i.e., farmers' own seed) or acquired during the last cycle (i.e., new seed) (Table 2B). Farmers' own seed is diffused at higher rates than new seed: 0.95 vs. 0.76, respectively ($P<0.001$). Again, no systematic differences in diffusion across elevations were found ($P=0.25$), but Freeman-Tukey deviates showed significant altitudinal differences for seed saved across cycles. Own seed is replaced at the highest rate at high elevations and the lowest rate at low elevations.

Table 1. Various seed-management rates for cassava in Cauca, Colombia.[1]

	Type of seed		
	Landraces	**Improved**	**Total**
A. Replacement by elevation (N = 655)			
High	0.33	0.34	0.33
Intermediate	0.15	0.33	0.19
Low	0.21	0.38	0.28
Total	0.25	0.35	0.28
G elevation effect	17.4* (4 df)		
G type effect	12.0 *(3 df)		
B. Diffusion by origin[2] (N = 165)			
Local	0.78	0.65	0.75
Introduced	0.72	0.88	0.79
Total	0.77	0.72	0.76
G origin effect	1.7 (2 df)		
G type effect	1.8 (2 df)		
G complete independence	1.9 (4 df)		
C. Introduction by elevation (N = 170)			
High	0.12	0.27	0.15
Intermediate	0.26	0.79	0.46
Low	0.14	0.13	0.13
Total	0.15	0.35	0.21
G elevation effect	19.4* (4 df)		
G type effect	11.9* (3 df)		
D. Mixing by origin (N = 165)			
Local	0.44	0.32	0.41
Introduced	0.61	0.39	0.50
Total	0.47	0.35	0.43
G origin effect	1.0 (2 df)		
G type effect	1.6 (2 df)		
G complete independence	2.1 (4 df)		

Significance at the 0.05 level is indicated by *.
1. Expressed as a ratio (varying between 0 and 1), replacement rates imply that seed is not saved across cycles; diffusion rates entail the exchange of saved seed; introduction rates mean that seed is brought into a locality.
2. Seed is "local" if acquired from neighbors and "introduced" if acquired in another locality.

A final log-linear model found marginally insignificant differences in the diffusion of seed of local origin and introduced ($P = 0.08$; Table 2C). In this case too, no significant differences across elevations were found ($P = 0.15$). However, a test of complete independence is rejected ($P<0.001$). Freeman-Tukey deviates showed that introduced seed is diffused at higher rates at intermediate elevations than elsewhere. At intermediate elevations, introduced seed also is diffused at higher rates than seed acquired locally. No significant differences are present between other origin-by-elevation combinations.

Seed introduction and mixing

Almost one fourth of all new seed in the sample, i.e., seed acquired in the last cycle, is introduced. A log-linear model was used to analyze differences in the rate of introduction across elevations and seed types (Table 1C). Significant seed-type (P

<0.01) and altitudinal ($P <0.05$) effects are present. The rate of introduction of landraces is less than half that of improved varieties: $r = 0.15$ and 0.35, respectively. However, this is not true at every elevation; no differences between types are present at low elevations. Introduction of both landraces and improved varieties is highest at intermediate elevations. Improved varieties at intermediate elevations have the highest introduction rates of all type-by-altitude combinations. They are introduced at rates three times those of landraces at the same elevation and six times those of improved varieties at low elevations.

Farmers maintain an average 1.62 varieties of cassava. Landraces and improved varieties represent 71 and 23% of seed lots recorded in the sample, respectively. An additional 6% were identified by farmers as a mix of a landrace and an improved variety; but 43% were reported as having been grown mixed with other varieties at some point. The percentage is higher for landraces and introduced seed than for improved varieties and local seed, but differences are not significant (Table 1D). A log-linear model finds no significant effect of seed type ($P = 0.45$) or origin ($P = 0.62$) on mixing. A test of complete independence could not be rejected ($P = 0.73$).

Population growth rates

Growth rates of several cassava populations were estimated based on the parameters described above [10,19]. A growth rate equal to 1 would imply that the population's size is constant across cycles; a rate above/below 1 would imply that the population grows/declines in numbers. The estimated rate of growth of improved varieties across Cauca is $\lambda = 3.82$, while landraces grow at a slightly lower rate: $\lambda = 3.72$. Growth rates are highest at mid elevations for both cassava landraces ($\lambda = 4.06$) and improved varieties ($\lambda = 4.02$). But while landrace populations grow the least at high elevations ($\lambda = 3.61$), improved varieties experience their lowest growth at low elevations ($\lambda = 3.35$). Finally, the growth rates of improved varieties at high elevations and landraces at low elevations are 3.76 and 3.90, respectively.

Discussion

Biotechnology is expected to have a greater impact than classical crop breeding on vegetatively-propagated crops [22–24]. Genetically-modified (GM) cassava could spread widely across developing areas where farmers still rely on landraces. Surprisingly, discussions on biosafety have largely neglected the implications of vegetative propagation for current strategies to contain transgenes. In 1996, shortly before the commercial release of GM maize in the United States, experts took for granted that this germplasm would spread quickly, carried by farm workers across international borders from areas of industrialized agriculture into maize's center of diversity in Mexico [25]. Farming practices would then facilitate the diffusion and introgression of transgenes into maize landraces. Indeed, transgenes were detected in Mexican maize landraces in 2001 [26], and they had spread widely by 2002 [19].

Cassava's biology and mode of propagation is believed to reduce the risk of unintentional transgene spread and establishment vis-à-vis grain crops [27]. Flowering times, genetic compatibility factors, low fecundity and dormancy all seem to limit gene flow in cassava. It has been suggested that while appropriate isolation distances would reduce the risk of outcrossing, cassava's clonal propagation and herbicides (to remove volunteers) could prevent novel traits from being passed on if outcrossing were to occur [27]. Confined handling and transport of cassava stakes also seems less challenging than that of grains.

Table 2. Seed diffusion rates for cassava in Cauca, Colombia.

Elevation	A. Diffusion by type (N = 633)			B. Diffusion by source[1] (N = 691)			C. Diffusion by origin (N = 189)		
	landrace	improved	total	own	new	total	local	introduced	total
High	0.90	0.90	0.90	0.97	0.75	0.90	0.76	0.64	0.74
Intermediate	0.94	0.81	0.90	0.94	0.81	0.91	0.70	0.95	0.82
Low	0.95	0.83	0.92	0.93	0.72	0.87	0.66	0.80	0.68
Total	0.92	0.84	0.90	0.95	0.76	0.89	0.71	0.83	0.74
G elevation effect	6.1 (4 df)			5.4 (4 df)			6.7 (4 df)		
G origin effect							6.8 (3 df)		
G complete indep							26.4* (4 df)		
G source effect				50.6* (3 df)					
G type effect	14.8* (3 df)								

Significance at the 0.05 level is indicated by *.
1. Seed acquired during the current cycle is "new;" seed saved by the farmer from a previous cycle is his/her "own".

Cassava's multiplication potential via stakes is only 1/2,250 that of maize via seed [14]; moreover, stakes lose viability quickly, have no dormancy, are less likely to be lost, and are easily prevented from establishing and surviving in the environment [14,27]. The consequences of containment failures, it is argued, are thus of less concern with cassava than with other crops.

Although little is known about GM germplasm's possible fate after its release [14], on-farm management of cassava seems to be extremely conservative vis-à-vis maize. Cassava seed exchange among indigenous Guyanian farmers, for instance, is quite formal and largely restricted to close kin and neighbors [6]. Exchanges outside family, village or ethnic boundaries are reportedly very occasional, suggesting that informal seed systems are surprisingly closed. Nevertheless, occasional exchanges across hundreds of miles have been reported. Moreover, indigenous farmers are known to actively incorporate seedlings from soil seed banks, which presumably facilitates unintentional gene flow across successive occupants' stocks and into wild populations [2]. But to what extent are these practices representative of cassava's center of diversity? We are unaware of systematic analyses of cassava management across any region. In what follows we analyze cassava's management and population dynamics across Cauca, comparing them to those of maize in Mexico.

Open populations and dynamic management

Our results reveal both differences and similarities in the management of cassava and maize. Average replacement rates for cassava in Cauca are only slightly lower than those estimated for maize across Mexico ($1 - p = 0.28$ and 0.32, respectively) [10]. But when the focus is on improved varieties, it is clear that cassava is replaced at much lower rates than maize ($1 - p = 0.35$ and 0.79, respectively) (Table 1A). This is not surprising since improved maize consists mostly of hybrid varieties, whose vigor decreases rapidly after the first cycle, while improved cassava can be maintained indefinitely via cloning. Observed rates suggest nevertheless that improved cassava is replaced every three years in average—i.e., a longer interval than for improved maize but shorter than for cassava landraces. Not surprisingly, improved cassava is introduced at much lower rates than improved maize ($r = 0.35$ vs. 0.76, respectively) (Table 1C).

The distribution of agricultural systems where maize is grown and their reliance on the formal seed system explains considerable differences in the management of improved varieties across regions and elevations in Mexico [10,19]. Significantly, replacement rates for improved cassava vary little across elevations ($1 - p = 0.33$ – 0.38), but large differences in the rate of introduction are observed ($r = 0.13$ – 0.79). But these differences cannot be attributed to a particular agricultural system or its reliance on the formal seed system. Corpoica, Colombia's National Agricultural Research center, had a central role in the improvement and distribution of cassava germplasm in the past, but its presence in Cauca is currently negligible [20]. Most introduced cassava in the study region is acquired from farmers in neighboring localities. This exchange of germplasm presumably is initiated by request from a farmer in need of seed [5]. Not surprisingly, seed-diffusion rates are 7.5 times higher for improved cassava than improved maize ($q = 0.84$ and 0.11, respectively) (Table 2A). In sum, differences in the management of improved germplasm are clearly associated with these crops' biology but also due to institutional factors, e.g., the current absence of a well-developed formal seed system in Cauca.

Comparing the management of landraces is more revealing because farmers breed, maintain and exchange these varieties exclusively through informal systems. Cassava and maize landraces are replaced at nearly the same average rates ($1 - p = 0.25$ and 0.24, respectively). Across elevations, replacement rates range from 0.21 to 0.33 for cassava and 0.23 to 0.31 for maize. The source of this variation has not been sufficiently explained. Some analysts attribute altitudinal differences to environmental gradients: conditions at low altitudes are said to promote a more dynamic management of landrace populations [4]. However, there is no correlation between elevation and replacement rates in cassava (Table 1A) or in maize [19]. Interestingly, intermediate elevations exhibit the lowest replacement rates for cassava but the highest for maize. Similarities in replacement rates are nevertheless surprising given these crops' contrasting biology, but average introduction rates differ more markedly: cassava landraces are introduced at rates three times higher than maize landraces (Table 1C). Average diffusion rates also differ strikingly across crops ($q = 0.92$ for cassava and 0.20 for maize landraces). Across elevations, diffusion rates range from 0.90 to 0.95 for cassava (Table 2A) and from 0.15 to 0.22 for maize [19].

As with improved varieties, it is difficult to attribute particular aspects of landrace management to any single factor; but identifying the causes of farmer practices is beyond our present intent. Our estimates can be used nevertheless to describe cassava's population dynamics. Estimated diffusion and introduction rates suggest, for instance, that cassava populations are remarkably more dynamic (q) and open (r) than maize. A much more detailed exploration of these dynamics is yet possible.

Diversity and non-local seed

Conceptions on non-local germplasm's impact on local diversity differ markedly across crops. Non-local seed is considered an important source of diversity for cassava but a threat to local maize diversity [3,5–6]. Indeed, seed that is introduced into a locality can displace local germplasm when crop populations are of constant size and unstructured, i.e., when all seed is managed indistinctly [10]. The first of these conditions can occur when physical space is limited within individual farms; and its implications for diversity are clearest for an out-crossing species such as maize, whose racial ideotypes must be maintained through cross-pollination of relatively large populations [3]. In contrast, rare cassava genotypes can be preserved through asexual reproduction; and farmers apparently maintain as few as one or two mounds per farm [5]. But given that the number of mounds per farm remains finite, the size of cassava populations also can be considered constant.

Seed introduced into a locality can be compared to immigrants in a population. If "immigrant" seed lots and their descendants are construed as a distinct subpopulation, their abundance in the metapopulation can be stated as a function of seed management [10]. When seed is managed indistinctly after it is introduced into a locality—i.e., when introduced seed is replaced and diffused at the same rates as local seed—the proportion of non-local seed (i.e., immigrants and their descendants) depends entirely on the rate of introduction [10]. When seed is introduced only once, the relative abundance of non-local germplasm remains constant thereafter, but it grows with every cycle when introductions are continuous. If these conditions applied in Cauca, the growth rate of non-local cassava populations (λ_{nl}) would be 0.21 points higher than that of local seed (λ_l) (i.e., $\lambda_{nl} = \lambda_l + 0.21$) or equal to the average rate of introduction (Table 1C). At intermediate elevations, where introduction rates are highest, the differential would be 0.46 points. The implication would be that cassava populations in Cauca consist almost entirely of seed introduced during the last few years. This need not be the case, however, if the second condition described above does not apply to Cauca, e.g., because introduced seed is replaced at higher rates or diffused at lower rates than local seed. In Mexico, introduced maize landraces and improved varieties are diffused at significantly lower rates than local seed ($q = 0.13$ and 0.23, respectively) [10]. Cassava shows a strikingly different pattern. Although differences observed are only marginally significant, introduced cassava tends to diffuse at higher rates than local germplasm. And these rates are extraordinary compared to those of maize. At intermediate elevations, for instance, 95% of introduced seed is diffused after its first cycle (i.e., $q = 0.95$).

The prevalence of non-local germplasm also depends on seed replacement; e.g., non-local populations will not expand if introduced seed is constantly discarded and replaced. Introduced maize landraces, for instance, are replaced as much as improved varieties [10]. This means that introduced maize is both replaced more and diffused less than local maize, thus requiring a constant influx of introduced seed for non-local populations to survive. The high replacement rate of maize suggests that seed is introduced for testing but found wanting and discarded, which could apply to cassava too. We were unable to estimate replacement rates for introduced cassava due to lack of sufficient data. Non-local cassava

landrace populations nevertheless are bound to grow much faster than maize even if they are replaced at equally high rates, given that their introduction rates are three times higher. In fact, estimates for maize suggest that, in contrast to cassava, local and non-local populations grow at virtually the same rate ($\lambda_{nl} - \lambda_l < 0.001$ in regions where traditional agriculture dominates) [10]. It seems certain thus that the share of non-local germplasm is much higher for cassava landraces than for maize.

Introduced cassava is diffused at similar rates whether it is a landrace or an improved variety; but the latter are introduced at higher rates (Table 1C). Whether a landrace or an improved variety, any germplasm should be considered local after it has been planted for more than one generation [3]. The median age of seed lots is higher for cassava than for maize (i.e., 5 and 3 years, respectively). But no seed over 20 years old was recorded in Cauca, while 18% of maize seed lots in southeast Mexico are at least 25 years old—i.e., old enough to be bequeathed across generations. We would need to consider differences in replacement rates across seed types and farms before concluding that most cassava in Cauca has been introduced in the last generation. What our results say unequivocally is that a greater proportion of improved cassava than landraces has been introduced recently. Results also suggest that improved cassava populations are growing faster than landraces, which could mean that the latter are being displaced (see below).

Indeed, when a metapopulation's size is fixed, the subpopulation that grows fastest eventually displaces the rest, assuming that growth rates are constant [10]. In some cases, however, growth might be inversely associated with a subpopulation's abundance—a process that might favor the spread of newly introduced germplasm but stop short of displacing local stocks. Density-dependence could also prevent dwindling populations from disappearing altogether, as reported in Guyana, where farmers rarely discard cassava varieties no longer favored by others [5]. In Mexico, newly introduced maize grows faster than all other subpopulations, but this growth is not long-lasting [10]. Our estimates suggest that Cauca farmers exchange cassava across localities and then distribute this germplasm among neighbors at rates much higher than Mexican maize; yet, local cassava is not replaced at higher rates than maize. It is possible, thus, that density-dependence is constraining the growth of introduced cassava populations and so seed exchange could actually be increasing the crop's diversity within individual localities. Does this mean that cassava farmers are hoarding diversity?

According to the literature, varietal richness is much higher for cassava than other crops, reaching up to 76 varieties per locality [5,17]. This richness could also be associated with cassava's mixed reproductive system [6,23,28]. However, richness estimates at the locality level can be misguiding (see Methods). When the focus is on individual farms, cassava's diversity does not stand out from other crops [17]. Our data shows that the average number of varieties maintained by cassava farmers in Cauca is only slightly higher than the number reported for maize farmers in south-central Mexico (1.62 and 1.44, respectively) [10]. Is it possible then that farmers are not registering the diversity introduced via seed exchange or that they are losing it inadvertently? Guyanian farmers reportedly recall the origin of every variety they acquire; but farmers might not always recognize differences between newly acquired seed and their own, leading them to mix genotypically distinct germplasm [5]. This "confusion" could prevent an increase in varietal richness but simultaneously promote intra-varietal diversity (see below). Indeed, exchanging large amounts of germplasm across localities might be a way of offsetting intra-varietal loss of diversity due to management (e.g., through seed selection) [5], particularly if this loss is greater in clonally-propagated plants than in sexually-reproduced crops such as maize [10].

Improved varieties vs. landraces

Scholars have long associated the spread of improved varieties with the loss of crop diversity; yet, the evidence remains inconclusive [29]. Moreover, improved varieties have not spread across developing areas as widely as expected, and recent surveys suggest that farmers still maintain considerable landrace diversity [17]. Sales records of improved varieties are the most common measure of use, but sales can underestimate the abundance of improved germplasm in the fields because this seed too can be saved and "recycled." Still, it is possible to estimate changes in the abundance of improved varieties by analyzing the growth rate of their populations within the crop's metapopulation.

Sales records suggest that introduced improved maize was more common in Mexico during the mid nineties than at present—a process associated with the expansion of irrigated agriculture outside maize's center of diversity, that is, with an expanding metapopulation. Clearly, this does not imply that improved populations displaced maize landraces, which might have expanded too despite exhibiting lower growth rates. Subpopulations exhibiting subpar growth can expand when the metapopulation increases in size [10]. But the area in maize has decreased gradually across Mexico since then. Growth rate estimates suggest nevertheless that maize landrace populations are stable ($\lambda = 1.03$); but improved varieties would dwindle ($\lambda = 0.33$) due to their high replacement and low diffusion if not infused continuously with new germplasm (i.e., through formal seed systems). In sum, there is no clear evidence that improved varieties have displaced maize landraces in Mexico [10].

Cassava's population dynamics are very different: current infusions of improved germplasm through formal seed systems are noticeably rare; but existent populations are diffused at rates seven times higher than improved maize, and their rate of survival (p) also is twice as high. In the cassava metapopulation, improved germplasm exhibit higher replacement rates than landraces but also higher introduction rates. The first fact reduces a possible growth differential between these populations, but the second one increases it. According to our estimates, the growth rate of improved varieties (λ_{iv}) is 0.10 points higher than that of landraces (λ_{lr}) (i.e., $\lambda_{iv} - \lambda_{lr} = 0.10$). Surprisingly, both populations seem to grow at exceedingly high rates (i.e., 3.82 and 3.72, respectively). Several factors can explain these results.

An expansion of cassava agriculture could be the main reason behind growing populations in Cauca. "Massive exchanges" of large amounts of planting material seem to take place both when new farms are established and when farmers sow large fields [5]. Survey data show that the area sown to cassava in Cauca increased 30% in 2010 after several years of contracting. At the same time, our growth estimates assume that seed from each source becomes a separate seed lot, i.e., that seed lots are grown and maintained as a distinct type. But if planting material is in short supply, farmers may be combining seed from several sources to sow a single field (or form a single seed lot). As discussed earlier, farmers may be mixing genotypically distinct but phenotypically similar seed (i.e., seed of the same named variety) into a single seed lot, inadvertently increasing its diversity. They may also be mixing different varieties on purpose.

Mexican maize farmers are known to mix varieties (particularly landraces and improved varieties) with the intention of hybridizing them, i.e., creolizing improved varieties or improving local varieties [30]. Growing mixed stands of cassava is a common practice in Cauca (Table 1D); but farmers' intentions are not obvious given cassava's clonal propagation. Amerindian farmers are known to incorporate cassava seedlings into their stocks of clones, favoring large-sized, heterozygous individuals—a practice that increases genotypic diversity or might even generate new varieties [7,23]. However, it is uncertain whether this is an intended or completely inadvertent outcome [5,6]. Moreover, there are few indications that the practice is widespread among cassava farmers. Cauca farmers reportedly incorporate volunteers opportunistically, and significantly, 6% of seed lots in the region are considered hybrids. However, there are no reports of farmers purposely hybridizing varieties. Growing mixed stands could also be a strategy to mitigate the risks posed by a complex and changing environment. Significant variability in the response of different varieties to nutrient availability and in resistance to drought and pests suggests that a mixed stand could help farmers stabilize yields and secure a harvest [31].

A clearer understanding of these issues is needed before mixing can be modeled as part of cassava's demography. This gap in our knowledge notwithstanding, several conclusions can be drawn. Cassava populations' surprising growth rates are due to the high introduction and diffusion rates of landraces (vis-à-vis maize) but to high diffusion and survival rates in the case of improved varieties. Far from being the random process implied by current models [9], these differences could reflect the diffusion of technological innovations, the expansion of agriculture, or multiple other factors influencing farmer decisions. To what extent these social process have played a role in Cauca is a complex question, particularly when we consider the interdependence of introduction, replacement and diffusion rates. The higher replacement of improved cassava than of landraces, for instance, might be tied to the frequent introduction of ill-suited germplasm. Alternatively, farmers might be introducing new seed to replace local germplasm that has (or seems to have) decayed [32]. Observed differences in the dynamics of cassava populations across Cauca can shed light on alternative possibilities.

According to our estimates, improved cassava populations exhibit lower growth than landraces at low elevations but higher growth at high elevations. Since landraces at high elevations are replaced and diffused at the same rates as improved varieties, the latter's advantage is due exclusively to introductions, which occur at over twice the rate for landraces (Table 1). This suggests that new improved germplasm is replacing not only older improved varieties but possibly also landraces. Improved cassava also is introduced at much higher rates than landraces at mid elevations, but here the latter are saved and diffused at higher rates, so growth rates are similar for both subpopulations. Thus, at mid elevations, improved cassava might be introduced mostly to replace its own populations. At low elevations, landraces also are saved and diffused at higher rates than improved varieties, but both groups are introduced at the same low rate, suggesting that local landraces could be regaining ground against improved varieties.

Implications for biosafety

Analysts described the introgression of maize transgenes into Mexican landraces—and their presumed disappearance—as both unsurprising and inevitable [33]. In fact, transgenes have not disappeared but dispersed widely, and their sources and mechanisms of dispersal remain controversial [19]. It is now clear that pollen cannot explain transgene dispersal at a geographical level; but neither can farmer practices alone explain the abundance and distribution of transgenes across Mexico. Predictive models will need to consider germplasm's simultaneous flow through formal and informal seed systems, as well as the movement of grain through markets [19].

Cassava's case is different to the extent that its seed is not traded as food or feed, but there are also similarities with maize. Cassava's biology and vegetative propagation might facilitate the confinement of field trials, as expected [27], but simultaneously promote the diffusion of GM germplasm once it is released. The dynamic exchange of seed observed in Cauca could grant local farmers faster access to biotechnological innovations than their counterparts in Mexico despite the lack of a well-developed seed system. Other traditional-farming practices (i.e., mixing seed and incorporating seedlings into seed stocks) could also allow these farmers to transfer useful transgenes into locally-adapted landraces. These same practices might also allow GM varieties to spread unrestrictedly, whether they have been released intentionally or accidentally. Our findings suggest that GM cassava is unlikely to displace landraces or compromise their diversity, but other hazards cannot be ignored. Transgenes that increase cassava's qualities as an industrial crop or as feed, for instance, could compromise food safety if farmers cannot recognize them. This could be the case of transgenes coding for industrial proteins such as pharmaceuticals. Food crops are ideal hosts for the synthesis of industrial proteins in terms of practicality, economy, ease of storage and distribution, but their use also entails poorly known risks [34–36].

We should expect farmers to manage GM cassava like any other improved germplasm. That improved cassava is commonly saved across cycles, unlike hybrid maize, increases the likelihood of gene flow across fields and into seed banks. Given that farmers often prefer growing cassava in rented land, seed banks could facilitate the inadvertent diffusion of transgenes across households. Deliberate exchange of seed is itself surprisingly frequent, and exchanges across localities are much more often for improved cassava than for landraces. This constant introduction would allow GM cassava to spread across localities rapidly, in contrast to maize, whose improved germplasm very rarely spreads. The diffusion and subsequent incorporation of GM germplasm into seed banks could obstruct the tracking and eradication of deleterious transgenes in cassava to a greater extent than in maize. Attempts to regulate these and other traditional farming practices to reduce such risks could compromise the adaptive potential of cassava's populations and at the same time prove ineffectual [37].

Methods

Our methods—i.e., model, data collection and analyses—and definitions follow the literature [10,19]. For expediency, improved varieties and landraces are treated here as clearly delimited and mutually exclusive categories defined by their breeding history [3,38]. Improved varieties, as opposed to landraces, consist of germplasm generated by a formal breeding program [3]. In practice, landraces are vaguely-circumscribed taxonomic units, and hence their alternative designation as "named varieties." More precisely, a landrace is defined as the group of seed lots considered by farmers as belonging to the same type and thus given the same name; a seed lot is the set of propagules of a specific type selected by a farmer and sown during a cropping season to reproduce that particular type [3]. Since the same landrace can be given different names, the number of named varieties registered in an oral survey can overestimate actual varietal richness. At the same time, these estimates as well as those based on sample collections depend on the intensity of sampling. Clearly, none of these problems arise at the level of the individual farm. Farmers manage crop diversity by acting upon individual seed lots: e.g., replacing one variety for another implies discarding one seed lot

and taking up another; introducing a variety into a locality and diffusing it entails acquiring a seed lot from a non-local source and exchanging (i.e., distributing) it among fellow farmers, thereby producing new seed lots. Likewise, mixing varieties means growing two seed lots in a mixed stand.

Model

Consider a closed population N consisting of N_t seed lots at time t. At the end of the period, seed lots are saved with probability p and diffused with probability q from one farmer to C others. These new seed lots become part of the $t+1$ population along surviving lots, such that $N_{t+1} = (p+qC) N_t$. More generally, assuming constant survival and diffusion probabilities over time, population size at t is given by $N_t = (p+qC)^t N_0 = \lambda^t N_0$, where λ is the population's expected growth rate. The population grows (i.e., $\lambda > 1$) if seed diffusion offsets seed loss or replacement. If survival and diffusion probabilities are the same for seed lots in N, λ is the growth rate of both N and every seed line in it. But even if $\lambda > 1$, specific seed lots or seed lines can become extinct unless there is a perfect negative correlation between seed survival and diffusion.

If there is a one-time introduction of non-local seed into the population at $t = \tau$, such that N incorporates $rN_{\tau-1}$ introduced seed lots along with saved and locally diffused seed, then $N_\tau = (p+qC+r)N_{\tau-1}$. The number and proportion of introduced seed lots are, respectively, $N_{I,\tau} = rN_{\tau-1}$ and $n_{I,\tau} = N_{I,\tau} / N_\tau = r/(p+qC+r)$. Assuming that introduced lots are saved and diffused at the same rate as the local lots, the population grows at a rate of $\lambda = p+qC$ after τ, so that $N_t = \lambda^{t-\tau} N_\tau$. Thus, the population at t consists of surviving lines (i.e., original lots plus copies) of the mixed-origin τ population, and the proportion of non-local seed (i.e., introduced lots plus copies) is constant. If introductions are continuous, the rate of introduction (r) becomes part of the population growth rate: $N_t = (p+qC+r)^t N_0 = \lambda^t N_0$. Since the local subpopulation grows at the rate of $\lambda_L = p+qC < \lambda$, the proportion of local lots in the population decreases continuously: $n_{L,t} = N_{L,t} / N_t = (1 - r/\lambda)^t n_{L,0}$. At carrying capacity, $\lambda = 1$ and $\lambda_L < 1$, so the number of local lots drops exponentially until they are completely replaced by introduced seed.

The dynamics of distinct seed types can be analyzed by letting N_I and $N_{\bar{J}}$ represent separate subpopulations of N. If all rates are homogeneous across subpopulations, then both N_I and $N_{\bar{J}}$ grow at the rate of $\lambda = p+qC+r$. If rates differ, $N_{I,t} = (p_I+q_IC+r_I)^t N_{I,0}$ (and likewise for $N_{\bar{J},t}$). Interactions between subpopulations can be made explicit by decomposing diffusion and introduction rates: $q_{I,0} = q_{II}+q_{I\bar{J}} \; n_{\bar{J}I,0}$ and $r_{I,0} = r_{II}+r_{I\bar{J}} \; n_{\bar{J}I,0}$, where q_{II} and $q_{I\bar{J}}$ are, respectively, diffusion rates of seed lots in N_I with respect to itself and w.r.t. $N_{\bar{J}}$ (and likewise for r_{II} and $r_{I\bar{J}}$), and $n_{\bar{J}I,t} = N_{\bar{J},t}/N_{I,t}$ represents relative abundance at time t. Substituting and regrouping terms, $N_{I,t} = \lambda_{I,t-1} N_{I,t-1} = (\lambda_{II}+s_{I\bar{J}} \; n_{\bar{J}I,t-1})N_{I,t-1}$, where $\lambda_{II} = p_I+q_{II}C+r_{II}$ and $s_{I\bar{J}} = q_{I\bar{J}}C+r_{I\bar{J}}$ represent subpopulation N_I's intrinsic growth and its interaction with subpopulation $N_{\bar{J}}$, e.g., seed replacement within N_I and replacement of variety $N_{\bar{J}}$ by N_I. Growth of N_I is thus a function of $n_{\bar{J}I,t}$, whose rate of change is itself the ratio of N_I's and $N_{\bar{J}}$'s growth rates: $n_{\bar{J}I,t} = (\lambda_{\bar{J},t-1}/\lambda_{I,t-1})n_{\bar{J}I,t-1}$.

Inspection of the previous equation reveals two possible stable equilibria $\underline{n}_{\bar{J}I}$: either growth rates balance out and subpopulations coexist, or one subpopulation prevails and the other becomes extinct; i.e., $\underline{n}_{\bar{J}I} = 0$ or ∞. When $\lambda_{\bar{J}\bar{J}} = \lambda_{II}$ and $s_{\bar{J}I} = s_{I\bar{J}}$, $n_{\bar{J}I}$ converges to 1. If rates differ across types, subpopulations coexist as long as there is a strictly positive solution for $\underline{n}_{\bar{J}I}$ in $\lambda_{\bar{J}\bar{J}} - \lambda_{II} = s_{I\bar{J}} \; \underline{n}_{\bar{J}I} - s_{\bar{J}I} \; \underline{n}_{\bar{J}I}^{-1}$; that is, as long as intrinsic growth differences are offset by replacement across populations. When differences are restricted to interaction terms (i.e., $\lambda_{\bar{J}\bar{J}} = \lambda_{II}$ and $s_{I\bar{J}} \neq s_{\bar{J}I}$), there is an analytical solution: $\underline{n}_{\bar{J}I} = (s_{\bar{J}I}/s_{I\bar{J}})^{0.5}$. Subpopulations coexist whenever there is either some cross-replacement ($s_{IJ} > s_J > 0$) or none at all ($s_{IJ} = s_{JI} = 0$) but not when replacement is one-sided ($s_{IJ} > s_{JI} = 0$).

Data analysis

A survey of 275 farms across 14 municipalities in the Department of Cauca, Colombia, conducted in early 2010, provided data on 719 cassava seed lots. The survey was based on a stratified-random sample to ensure that it is representative of cassava farmers in Cauca (but not necessarily of farmers in smaller areas within this region) [20]. Stratification across municipalities (based on the area sown to cassava) does not explain the role played by social and environmental factors within political divisions, but it reduces the variance of descriptive statistics, ensuring 95% confidence for estimates at the household level. Since households are the elementary sampling units, the effects of sample design are restricted to this level. That is, there are no additional effects (due to deviations from simple random sampling) at the seed-lot level since all seed lots owned by sample households were considered in the analysis; i.e. seed lots were censused. We used seed-lot data to estimate rates of seed replacement, diffusion, introduction and mixing. Rate differences across seed types were then determined through the analysis of three-way tables based upon log-linear models [21]. This analysis was not intended as an exhaustive breakdown of the causal

factors involved in seed management but as a way of identifying differences in management across specific seed types (or populations) often addressed in the literature [16]. These include improved varieties and landraces, seed maintained at different elevations, and seed of different geographic origin [3-4,10,19]. The three altitudinal regions considered here are high (>1600 masl), intermediate (1200–1600 masl) and low (<1200 masl) elevations. The influence of cassava breeding programs and the nature of germplasm diffused through formal seed systems differ markedly across these regions [20]. Finally, rate differences observed across seed types were used to describe disparities in these populations' dynamics [10,19].

Acknowledgments

We thank Fernando Calle (Cassava Program - CIAT), Luisa Fory (LACBiosafety Project- CIAT), Roosvelt Escobar (Agrobiodiversity Program - CIAT), Lisimaco Alonso (Clayuca - CIAT), and an anonymous reviewer for their comments and suggestions.

Author Contributions

Conceived and designed the experiments: GAD CG DCL. Performed the experiments: CG DCL. Analyzed the data: GAD CG DCL. Wrote the paper: GAD.

References

1. Duputié A, David P, Debain C, McKey D (2007) Natural hybridization between a clonally propagated crop, cassava (Manihot esculenta Crantz) and a wild relative in French Guiana. Mol Ecol 16: 3025–3038.
2. Pujol B, Renoux F, Elias M, Rival L, McKey D (2007) The unappreciated ecology of landrace populations: conservation consequences of soil seed Banks in casava. Biol Conserv 136: 541–551.
3. Louette D, Charrier A, Berthaud J (1997) In situ conservation of maize in Mexico: Genetic diversity and maize seed management in a traditional community. Econ Bot 51: 20–38.
4. Perales H, Brush SB, Qualset CO (2003) Dynamic management of maize landraces in central Mexico. Econ Bot 57: 21–34.
5. Elias M, Rival L, McKey D (2000) Perception and management of cassava (Manihot esculenta Crantz) diversity among Makushi Amerindians of Guyana (South America). J Ethnobiol 20: 239–265.
6. Elias M, McKey D Panaud O, Anstett MC, Robert T (2001) Traditional management of cassava morphological and genetic diversity by the Makushi Amerindians (Guyana, South America): Perspectives for on-farm conservation of crop genetic resources. Euphytica 120: 143–157.
7. Pujol B, David P, McKey D (2005) Microevolution in agricultural environments: how a traditional Amerindian farming practice favors heterozygosity in cassava (Manihot esculenta Crantz, Euphorbiaceae). Ecol Lett 8: 138–147.
8. Pressoir G, Berthaud J (2004) Patterns of population structure in maize landraces from the Central Valleys of Oaxaca in Mexico. Heredity 92: 88–94.
9. Van Heerwaarden J, van Eeuwijk FA, Ross-Ibarra J (2009) Genetic diversity in a crop metapopulation. Heredity 104: 28–39.
10. Dyer G, Taylor JE (2008) A crop population perspective on maize seed systems in Mexico. Proc Natl Acad Sci U S A 105: 470–475.
11. Andow DA, Zhwalen C (2006) Assessing environmental risks of transgenic plants. Ecol Lett 9: 196–214.
12. Gepts P, Papa R (2003) Possible effects of (trans)gene flow from crops on the genetic diversity from landraces and wild relatives. Environ Biosafety Res 2: 89–103.
13. Celis C, Scurrah M, Cowgill S, Chumbiauca S, Green J, et al. (2004) Environmental biosafety and transgenic potato in a centre of diversity for this crop. Nature 432: 222–225.
14. Palacios JD (2007) Panorama de flujo de genes en yuca en un contexto de liberación de variedades genéticamente modificadas en Colombia. In: Hodson de Jaramillo E, Carrizosa MS, eds. Desarrollo de capacidades para evaluación y gestión de riesgos y monitoreo de organismos genéticamente modificados. Bogotá, Colombia: Instituto de Investigaciones de Recursos Biológicos Alexander von Humboldt. pp 85–118.
15. Parzies HK, Spoor W, Ennos RA (2004) Inferring seed exchange between farmers from population genetic structure of barley landrace Arabi Aswad from Northern Syria. Genet Resour Crop Evol 51: 471–478.
16. Thomas M, Dawson JC, Goldringer I, Bonneuil C (2011) Seed exchanges, a key to analyze crop diversity dynamics in farmer-led on-farm conservation. Genet Resour Crop Evol 58: 321–338.
17. Jarvis DI, Brown AHD, Pham HC, Collado-Panduro L, Latournerie-Moreno L, et al. (2008) A global perspective of the richness and evenness of traditional crop-variety diversity maintained by farming communities. Proc Natl Acad Sci U S A 105: 5326–5331.
18. Lavigne C, Klein EK, Mari J-F, Le Ber F, Adamczyk K, et al. (2008) How do genetically modified (GM) crops contribute to background levels of GM pollen in an agricultural landscape? J Appl Ecol 45: 1104–113.
19. Dyer GA, Serratos-Hernández JA, Perales HR, Gepts P, Pineyro-Nelson A, et al. (2009) Escape and dispersal of transgenes through maize seed systems in Mexico. PLoS ONE 4(5): e5734. doi:10.1371/journal.pone.0005734.
20. Jaramillo G (2008) Diagnostico del cultivo de la yuca y su agroindustria en el Departamento del Cauca. Cali, Colombia: CIAT.
21. Sokal RR, Rohlf FJ (1995) Biometry. San Francisco: Freeman.
22. Ceballos G, Iglesias CA, Pérez JC, Dixon AGO (2004) Cassava breeding: opportunities and challenges. Plant Mol Biol 56: 503–516.
23. Hershey CH (1987) Cassava germplasm resources. In: Hershey CH, ed. Cassava Breeding: A Multidisciplinary Review. Cali, Colombia: CIAT. pp 1–24.
24. Taylor N, Chavarriaga P, Raemakers K, Siritunga D, Azhang P (2004) Development and application of transgenic technologies in cassava. Plant Mol Biol 56: 671–688.
25. Serratos JA, Wilcox MC, Castillo-Gonzalez F (1997) Gene Flow among Maize Landraces, Improved Maize Varieties, and Teosinte: Implications for Transgenic Maize. Mexico: CIMMYT.
26. Quist D Chapela IH (2001) Transgenic DNA introgressed into traditional maize landraces in Oaxaca, Mexico. Nature 414: 541–543.
27. Halsey ME, Olsen KM, Taylor NJ, Chavarriaga-Aguirre P (2008) Reproductive biology of Cassava (Manihot esculenta Crantz) and isolation of experimental field trials. Crop Sci 48: 49–58.
28. Rival L, McKey D (2008) Domestication and diversity in Manioc (Manihot esculenta Crantz ssp. esculenta, Euphorbiaceae). Curr Anthropol 49: 1119–1128.
29. Smale M (1997) The Green Revolution and wheat genetic diversity: some unfounded assumptions. World Dev 25: 1257–1269.
30. Morris ML, Risopoulos J, Beck D (1999) Genetic Change in Farmer-Recycled Maize Seed: A review of the Evidence, CIMMYT Economics Working Paper 99-07. Mexico: CIMMYT.
31. El-Sharkawy MA (1993) Drought-tolerant cassava for Africa, Asia, and Latin America. BioScience 43: 441–451.
32. Zeven AC (1999) The traditional inexplicable replacement of seed and seed ware of landraces and cultivars: A review. Euphytica 110: 181–191.
33. Raven PH (2005) Transgenes in Mexican maize: desirability or inevitability? Proc Natl Acad Sci U S A 102: 13003–13004.
34. Ma JK-C, Chikwamba R, Sparrow P, Fischer R, Mahoney R, et al. (2005) Plant-derived pharmaceuticals—the road forward. Trends Plant Sci 10: 580–585.
35. Twyman RM, Stoger E, Schillberg S, Christou P, Fischer R (2003) Molecular farming in plants: host systems and expression technology. Trends Biotechnol 21: 570–578.
36. Spok A (2006) Molecular farming on the rise—GMO regulators still walking a tightrope. Trends Biotechnol 25: 74–82.
37. McKey D, Elias M, Pujol B, Duputié A (2010) The evolutionary ecology of clonally propagated domesticated plants. New Phytol 186: 318–332.
38. Zeven AC (1998) Landraces: A review of definitions and classifications. Euphytica 104: 127–139.

3

RNA-Seq Analysis of Transcriptome and Glucosinolate Metabolism in Seeds and Sprouts of Broccoli (*Brassica oleracea var. italic*)

Jinjun Gao[1], **Xinxin Yu**[1], **Fengming Ma**[2], **Jing Li**[1]*

1 College of Life Science, Northeast Agricultural University, Harbin, China, **2** Key Laboratory of Breed Improvement and Physioecology of Cold Region Crops, Northeast Agricultural University, Harbin, China

Abstract

Background: Broccoli (*Brassica oleracea var. italica*), a member of Cruciferae, is an important vegetable containing high concentration of various nutritive and functional molecules especially the anticarcinogenic glucosinolates. The sprouts of broccoli contain 10–100 times higher level of glucoraphanin, the main contributor of the anticarcinogenesis, than the edible florets. Despite the broccoli sprouts' functional importance, currently available genetic and genomic tools for their studies are very limited, which greatly restricts the development of this functionally important vegetable.

Results: A total of ~85 million 251 bp reads were obtained. After *de novo* assembly and searching the assembled transcripts against the *Arabidopsis thaliana* and NCBI nr databases, 19,441 top-hit transcripts were clustered as unigenes with an average length of 2,133 bp. These unigenes were classified according to their putative functional categories. Cluster analysis of total unigenes with similar expression patterns and differentially expressed unigenes among different tissues, as well as transcription factor analysis were performed. We identified 25 putative glucosinolate metabolism genes sharing 62.04–89.72% nucleotide sequence identity with the *Arabidopsis* orthologs. This established a broccoli glucosinolate metabolic pathway with high colinearity to *Arabidopsis*. Many of the biosynthetic and degradation genes showed higher expression after germination than in seeds; especially the expression of the myrosinase *TGG2* was 20–130 times higher. These results along with the previous reports about these genes' studies in Arabidopsis and the glucosinolate concentration in broccoli sprouts indicate the breakdown products of glucosinolates may play important roles in the stage of broccoli seed germination and sprout development.

Conclusion: Our study provides the largest genetic resource of broccoli to date. These data will pave the way for further studies and genetic engineering of broccoli sprouts and will also provide new insight into the genomic research of this species and its relatives.

Editor: Sara Amancio, ISA, Portugal

Funding: The national Natural Science Foundation of China 31370334 http://www.nsfc.gov.cn/Portal0/default152.htm; Opening project of Key Laboratory of breed improvement and physiecology of cold region crops, College of Heilongjiang Province http://61.167.33.11/kjgz/Index.asp?page = 2; Heilongjiang Provincial University Science and Technology Innovation Team Building Program 2011TD005; http://61.167.33.11/kjgz/Index.asp?page = 2. The funders had no role in study design, data collection and analysis, decision to publish, or preparation of the manuscript.

Competing Interests: The authors have declared that no competing interests exist.

* E-mail: lijing@neau.edu.cn

Introduction

Consumption of fruits and vegetables has long been associated with better health and lower incidence of a variety of diseases such as coronary heart disease, cancers, etc [1,2]. Notably, a diet rich in cruciferous vegetables especially broccoli (*Brassica oleracea var. italica*) has been recognized as an efficient way to reduce the risk of getting many types of cancers. Epidemiological studies prior to 1996 showed an inverse relationship between cancer risk and cruciferous vegetable intake [3,4]. Some newer studies demonstrate that this inverse relationship is mainly contributed by the breakdown products of glucosinolates [5,6,7]. Glucosinolates are a major group of sulfur-rich secondary metabolites specifically in Cruciferae, which are well-known by their breakdown products to display several bioactivities, including plant defense against

pathogens and insects as well as anticarcinogenesis in mammals [8]. Based on their precursor amino acids, glucosinolates are divided into three major categories: aliphatic, indolic and aromatic glucosinolates [9,10]. Among them, the degradation products of aliphatic glucosinolates are considered to have the higher phase 2 detoxication enzyme inducer ability than the other two groups which is effective in blocking chemical carcinogenesis; therefore, they are thought to be the main contributor to protection against carcinogenesis [11].

Although broccoli heads are generally used as the edible part, the sprouts have been suggested to be a better source for health benefits. A study in 1997 reported the sprouts of eight broccoli cultivars have phase 2 enzyme inducer potency (nearly all arose from glucosinolates) 10–100 times greater than that of mature plants [11]. During the first few days of germination, the inducer activity per unit plant weight declined from the maximum point in seeds in an exponential manner. The declining trend flattened after nine days, and finally approached the value in mature broccoli heads after about 15 days [11]. The most valuable information is that in sprouts the aliphatic glucosinolates are dominant, while in adult plant the indolic ones account for the most [12]. The high content of the aliphatic glucosinolates in broccoli sprouts is mainly attributed to glucoraphanin. Its hydrolytic product, sulforaphane, has been well studied with high anticancer activity. It can not only inhibit phase 1 enzymes but also induce phase 2 enzymes [13]. Besides, sulforaphane has an important ability to target the highly aggressive cancer stem cell population, which is responsible for tumor therapeutics and cannot be eliminated by conventional chemo- or radio-therapy [14,15]. Another interesting fact is that no significant side effects were found in therapy with sulforaphane in the rapeutic concentrations in non-malignant cells or mice [14,16]. In addition, glucoraphanin has an obvious effect on decreasing oxidative stress, hypertension and inflammation in the cardiovascular system of rats [17]. Based on these promising results, the first prospective clinical studies with cancer patients and sulforaphane-enriched broccoli sprouts have now been initiated in the United States. Therefore, broccoli sprouts should have played a more important role in human health than the mature inflorescences.

The currently available genetic and genomic tools for broccoli research are very limited. While most studies of broccoli focused on physiology, few have been done at the genetic level and functional genomic studies are still in the infancy. Conspicuous effects of ESTs have been reported to develop genetic engineering, including gene family expansion [18,19], improving breeding programs by SNP and SSR markers [20,21], facilitating genome annotation [22], and large-scale expression analysis [23,24]. Currently, only 2,324 broccoli ESTs in the national center for biotechnology information (NCBI) database (http://www.ncbi.nlm.nih.gov/) are generated with the aim to identify gene expression profiles in microspore and floret bud. Most of them have no annotations. Despite of the growing demand and high-yield potential, the average yield of both fresh and processed broccoli has remained virtually unchanged in the United States in recent ten years [25]. Thus, very limited genomic resources of broccoli constitute a key limitation to the development of improved crops. The advent of next generation sequencing technologies has triggered a revolution in biological research, for it is cheaper and more rapid in providing genomic and transcriptomic data [26].

Here we performed a high-throughput Illumina Miseq sequencing to characterize the transcriptomes of five samples, including seeds, cotyledons of 3, 7, 11 day sprouts and euphyllas of 11 day sprouts. Since there is no available reference genome for broccoli, abundant short reads are required in order to perform de novo assembly. From the total of five libraries, we generated 557,094,098 raw reads with an average length of 251 bp containing 139,830,744,098 nucleotide bases. Formal research has suggested that to achieve 99% coverage of an mRNA, at least an 8X sequencing depth is required [27]. For this study, the sequencing depth is 50X, enough to get the maximum coverage. Using a de novo assembly method, 19,441 unigenes are obtained with an average length of 2133 bp. These unigenes are used for subsequent annotation analyses to provide a platform of transcriptome information for genes in broccoli sprouts. In this study, we focused our work on identification of glucosinolate metabolism genes in broccoli seed germination and sprout development. This will pave the way for further genetic engineering to improve this species' agronomic traits.

Results and Discussion

Sequencing and Data Analysis

RNA sequencing of the five samples (seeds, 3 day cotyledons, 7 day cotyledons, 11 day cotyledons and 11 day euphyllas) produced a total of ~85million 251 bp paired-end reads with an average of 17million reads for each sample. Cleaning and quality checks were performed to the raw data (cf. Materials and Methods). A total of ~75million trimmed reads were obtained with useful data percentage in five time points ranging from 70.29% to 76.01% and the average length of each read was 207 bp (Table S1). Compared to the reads generated by the formal platforms, the longer length of Illumina Miseq sequencing reads greatly facilitated the accuracy of the subsequent de novo assembly. Using single k-mer assembler Velvet (http://www.ebi.ac.uk/~zerbino/velvet), assembly of reads generated 659,752 contigs with mean sizes of 254 bp and N50 of 222 bp (Table 1). The contigs with length more than 500 bp accounted for about 6.19%. Multiple K-mer assembler OASES (http://www.ebi.ac.uk/~zerbino/oases) was applied to produce 122,345 transcripts for 40,081 loci with average length of 1670 bp. Then, all the transcripts were blasted against the Arabidopsis database. For those "non-BLASTable" transcripts, we searched them against the NCBI non-redundant (nr) database, using BLASTx program with an E-value threshold of 1E-5. A total of 94,255 (77.04%) transcripts were significantly matched to known genes in Arabidopsis and 3,971 (3.25%) transcripts were matched to the nr database. The high percentage of transcripts matched to Arabidopsis database is due to the close relation of these two species. For the transcripts representing the same loci, the top hit ones were clustered as unigenes. Finally, 19,441 unigenes were generated, with the average length of 2,133 bp and ranging in size from 200 bp to 20,580 bp (Table 1). The size distribution of contigs, transcripts and unigenes were compiled (Figure S1). The sequencing data has been deposited into NCBI gene expression omnibus (GEO) and the accession number is GSE53298.

Variable efficiency of matching to sequences in the databases was found in assembled sequences of different lengths, with longer sequences showing higher match proportions (Figure S2). For sequences longer than 1500 bp, the match efficiency was 98.24%. But for sequences between 200–500 bp and 500–1000 bp in length, it was just 53.40% and 79.17%, respectively. E-value distribution of the top hits in the databases had shown 71.17% of matched sequenced with strong homology (<1.0e-50) (Figure 1). 30.54% of the transcripts had a similarity higher than 80%, while 46.47% showed similarity between 60%–80% in identity distribution pattern. The total 77.01% of the transcripts showing identity higher than 60% along with the high quality e-value distribution supported the reliability of the de novo assembly performed in the study.

Annotation and Classification

Since biologists have recognized that there is likely to be a single limited universe of genes and proteins are conserved in most, if not all living cells, the GO (gene ontology) Consortium was created as a joint project of many organism databases to produce a structured, precisely defined, common, controlled vocabulary for describing the functions of genes and gene products in any organisms [28]. To annotate the broccoli transcriptome, GO terms were assigned to broccoli unigenes based on their identity to known protein sequences in the Arabidopsis database and nr database. 19,441 unigenes were assigned to 47 functional groups with 134,938 functional terms using GO assignments (Figure 2). For the three main categories of GO classification scheme, the

Table 1. Statistical summary of cDNA sequences of broccoli generated by the Illumina Miseq platform.

	Total length(bp)	Sequence No.	Max Length(bp)	Average Length(bp)	N50	>N50 Reads No.
Contigs	167802146	659752	7524	254	222	205867
Transcripts	204377219	122345	20680	1670	2527	26568
Unigenes	41474146	19441	20680	2133	2631	5234

assignments to the "biological process" (61,583, 45.64%) made up the majority, followed by the "cellular component" (53,030, 39.30%) and the "molecular function" (20324, 15.06%). Among these GO groups, the high number of unigenes putatively involved in "cellular process" (11,129) and "metabolic process" (10,230) in the biological process category indicated that the broccoli tissues used in this study were undergoing exquisite metabolic activities, which coincided with the samples' status. Interestingly, 5,220 unigenes were assigned to "response to stimulus", showed that during the germination of broccoli seeds and the development of sprouts, there were some protective mechanisms for preparing for potential external and/or internal stresses. Under the category of cellular component, the "cell", "cell part" and "organelle" were prominent groups. It is noteworthy that the unigenes were not gathered into few groups but were generally expressed. This might be due to the widespread requirements during seedling development.

EggNOG (evolutionary genealogy of genes: Non-supervised Orthologous Groups) is a database providing orthologous groups for 943 Bacteria, 69 Archaea and 121 Eukaryotes [29]. According to the previous report, the proteins could be divided into 25 functional categories [30]. Out of 19,441 unigenes with significant identity with *Arabidopsis* database and nr database in this study, 11,242 could be classified into 24 eggNOG categories with only "Nuclear structure" having no annotated unigenes (Figure S3). The categories "function unknown" (2,088, 18.57%) and "general function prediction only" (2,050, 18.24%) were the two largest functional groups of the 25 eggNOG categories. The high percentage of unigenes classified into "general function prediction only" coincided with the transcriptome studies of other species [31,32,33]. But our newly noticed fact that so many unigenes were assigned to unknown functional group might indicate there are some interesting unknown mechanism during germination of broccoli seeds and the development of sprouts. Following the most abundant two groups were "transcription" (929, 8.26%), "replication", "recombination and repair" (802, 7.13%), "signal

transduction mechanisms"(797, 7.09%) and "posttranslational modification, protein turnover, chaperones"(680, 6.05%), whereas the two groups involving "cell motility" and "extracellular structures" consisted of a total of 10 unigenes (0.09%), representing the smallest eggNOG classifications. Noteworthily, 277 unigenes (2.64%) were classified into secondary metabolite biosynthesis group, including glucosinolate biosynthesis in broccoli sprouts.

The KEGG (Kyoto Encyclopedia of Genes and Genomes) is a database linking genomic information with higher order functional information by collecting manually drawn pathway maps representing current knowledge on cellular processes and standardized gene annotations [34]. A total of 9836 genes were classified into six main categories including 38 secondary pathways (Figure S4) in the five tested samples. "Metabolism" is the biggest category (3,624, 36.84%), followed by "human disease" (1,760, 17.89%), "Genetic Information Processing" (1,674, 17.02%), "Organismal Systems" (1,279, 13.00%) and "Cellular Processes" (909, 9.24%), whereas "environmental information processing" (590, 6.00%) containing only 3 sub-units ("membrane transport", "signal transduction and signaling molecules and interaction") was the smallest category. These results indicated that the broccoli sprouts were undertaken active metabolic and genetic processes and the functional classification of KEGG provided a valuable resource for investigating specific processes and pathways in broccoli sprouts.

Gene Expression Pattern

Gene expression patterns can provide important clues as to the roles of unknown genes in biological active processes [35]. While RPKM (reads aligned to gene per kilobase of exon per million mapped reads) was widely used to calculate gene expression values [36], we used a more accurate method called DESeq to estimate gene expression values in this analysis to infer differential expression signals with good statistical power [37]. K-means clustering analysis was performed using the software MeV edition 4.90 (http://www.tm4.prg/mev.html) to group unigenes with

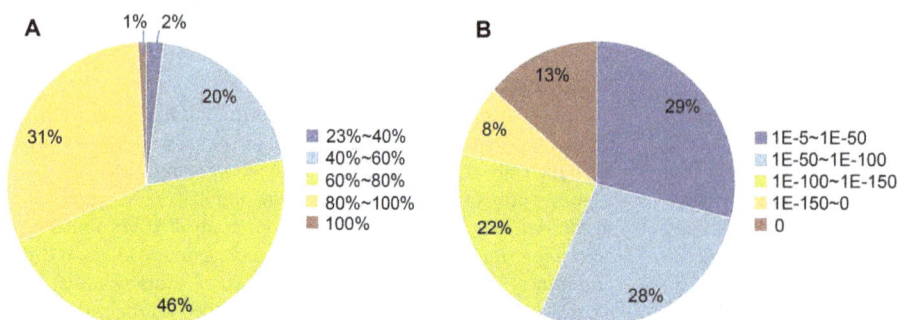

Figure 1. Characterization of broccoli unigenes by searching against public database. A. Identity distribution of unigenes blasted against public databases with E-value cutoff of 1E-5; B. E-value distribution of unigenes blasted against public databases with E-value cutoff of 1E-5.

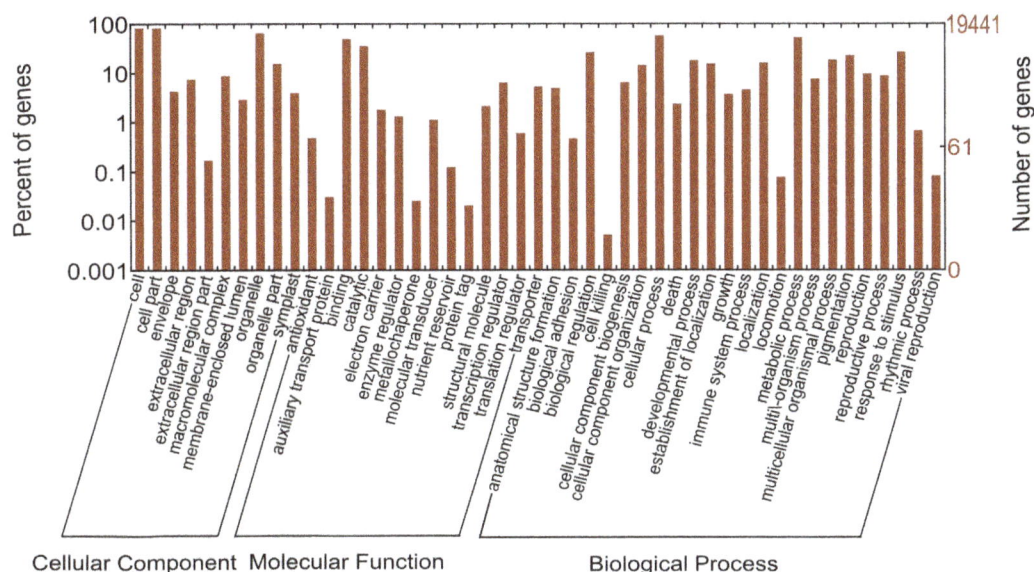

Figure 2. Histogram presentation of Gene Ontology classification of the assembled unigenes.

similar expression patterns under different time points and resulting in 10 different clusters (Figure 3). The most abundant cluster (IX) contained 3,659 genes with highest expression at the very beginning (i.e. seeds) and then their expression levels were down-regulated all through the development of sprouts, and reached the lowest point at the 11D euphyllas. These genes might greatly and specifically contribute to seed germination. Cluster I, II, III, VI and VIII comprised genes whose expression levels were very low in seeds but peaked at any one of the three points in cotyledons. Also, genes in cluster IV showed the highest expression level at 11D euphyllas with relatively low level at other time points.

In order to identify differentially expressed genes between the five types of samples, we compared them with each other and picked out a total of 2675 genes, which were at least 2-fold up- or down-regulated between two samples with p-value smaller than 0.05. Then, hierarchical cluster was generated to gain a global view of the differential expressed genes (Figure 4). Obviously, the 11 day cotyledons showed closer relationship with the 11 day euphyllas than with the cotyledons of other time points. This indicated the similar function between the initial stage of euphyllas and the late stage of cotyledons. As expected, the three time points of cotyledons were more similar to each other than to seeds. Even though the 7th day is the mid-point of 3rd day and 11th day, the expression profile of 7 day cotyledons was more similar to that of 11 day cotyledons than that of 3 day cotyledons. This fact along with the big difference between seeds and 3 day cotyledons illustrated the 3 day might be the special point in broccoli sprout development.

Putative Transcription Factors

Transcription factors (TFs) have been considered as one of the most important functional elements regulating gene expression that leads to developmental and other changes. It has been reported that in response to internal or external environment changes, TF genes exhibit more rapid expression changes than the bulk of the regulated genes [38]. Thus, the expression profile of TF genes may in some way reflect the subsequent transcription activities regulated by them. For their important roles, the key

putative TFs involved in broccoli seed germination and sprout development were analyzed.

A study in 2003 has revealed that in *Arabidopsis*, most (84%) of TFs could be detected in six day old seedlings [38]. Currently, the Plant Transcription Factor Database (PlnTFDB) contains 2451 and 2162 distinct TF sequences from *Arabidopsis* and *Arabidopsis lyrata*, respectively, arranged in 81 families [39]. In the sequenced broccoli seeds and sprouts, 78 TF families including 1,633 putative TF genes had been identified with the five most expressed TF families being AP2-EREBP, bHLH, MYB, HB and C3H (Table S2). A total of 1,581 of the 1,633 genes accounting for 86.82% were annotated with sequences from the close related species *Arabidopsis* and *Arabidopsis lyrata*.

The biggest TF family in our study of broccoli seeds and sprouts was AP2-EREBP with 109 putative family members being detected. AP2-EREBP family is unique to plants and characterized by a conserved AP2 DNA-binding domain of about 60 amino acids [40]. AP2-EREBP genes have been found to play important roles throughout the life cycle including regulating several developmental processes especially leaf epidermal cell identity and forming part of the mechanisms used to respond to stress [41]. Some members of AP2-EREBP, like *AP2*, control seed mass and seed size in *Arabidopsis* which is very important to extended growth of cotyledons [42,43]. The significantly large number of AP2-EREBP family members expressed in broccoli seeds and sprouts indicated the important function of these genes in this period as previously reported in other species [41,42,43].

The basic helix-loop-helix (bHLH) family members are involved in various process of seedling development such as light signaling, brassinosteroid and abscisic acid signaling, flavonoid biosynthesis, axillary meristem formation, stomatal patterning and trichome differentiation [44]. 99 unigenes identified to have bHLH-like sequences formed the second biggest TF family in the tested samples. Light is one of the most important elements in seed germination and seedling development, a subfamily containing 15 members are involved in light signaling. These bHLH proteins are known as PIF (phytochrome interacting factor) or PIL (phytochrome interacting factor-like) [45,46,47,48]. These proteins play important roles in phytochrome signal transduction by interacting

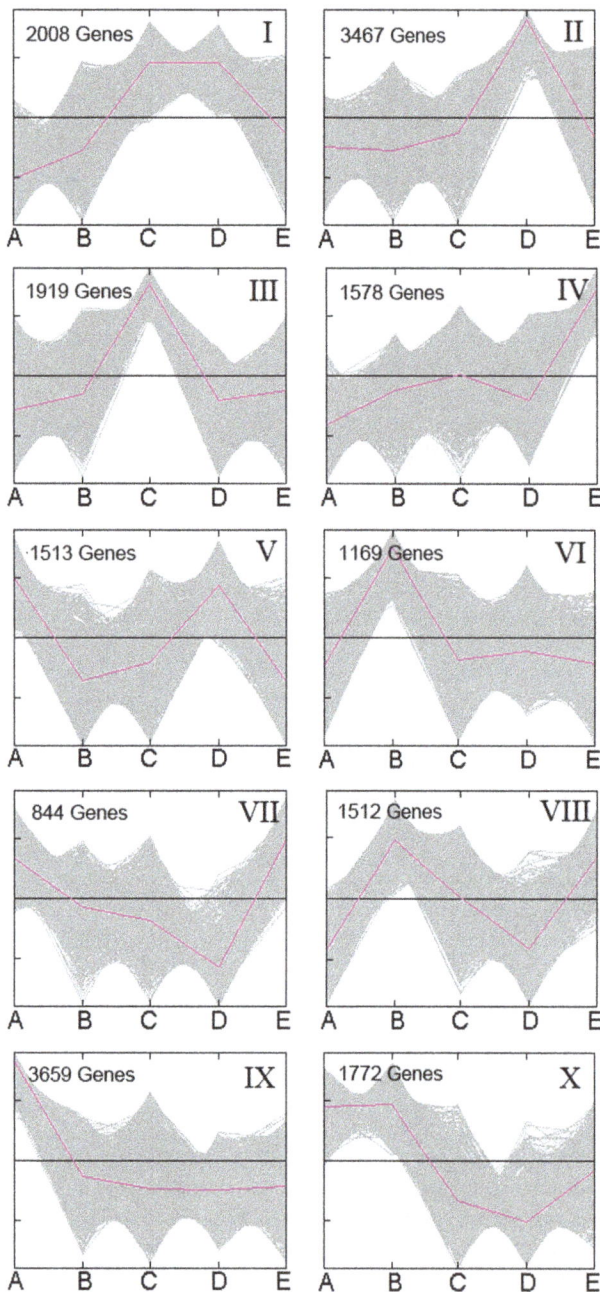

Figure 3. Dynamic expression patterns during broccoli seed germination and sprout development. K-means clustering was performed to identify 10 clusters, each containing various numbers of genes with similar expression pattern during broccoli seed germination and sprout development. The red lines show representative transcriptional regulators. The x-axis represents sequenced samples, and the y-axis represents normalized RNA-seq expression level. A. seeds; B. 3 day cotyledons; C. 7 day cotyledons; D. 11 day cotyledons; E. 11 day euphyllas.

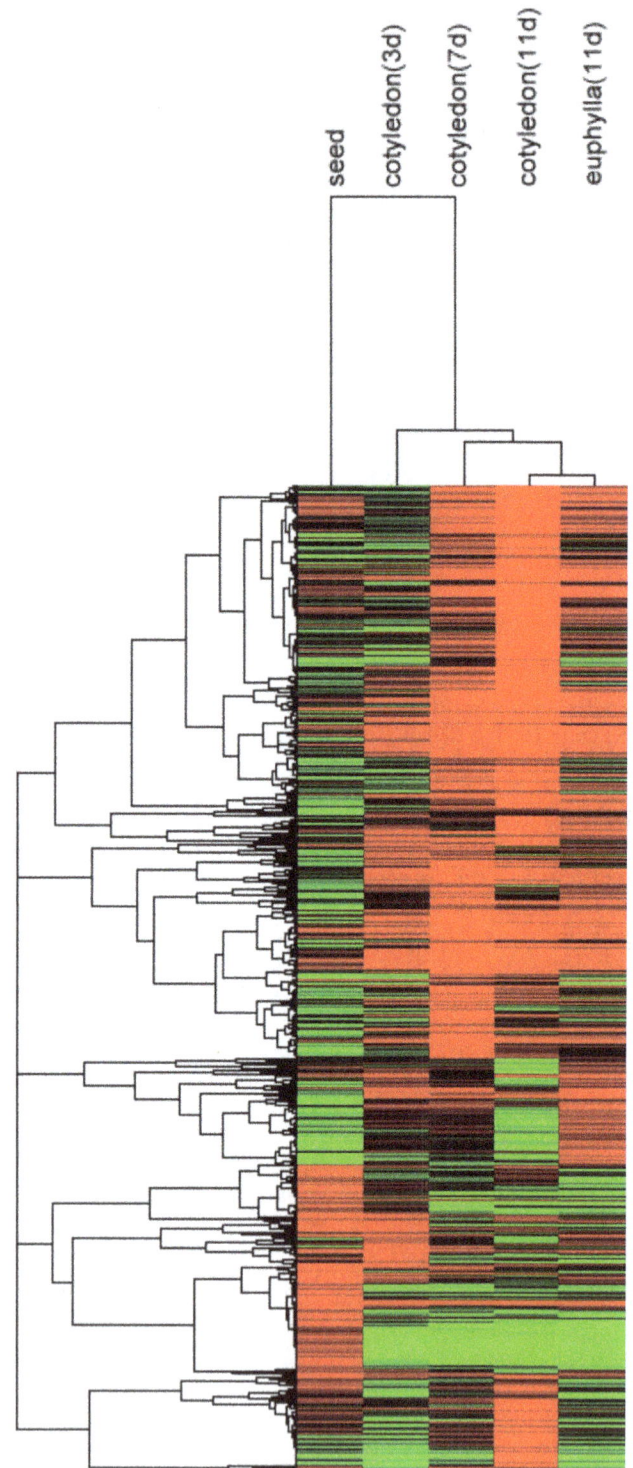

Figure 4. Cluster of differentially expressed genes during broccoli seed germination and sprout development. Expression changes and cluster analysis of 2675 genes that were differentially expressed between any two of the five samples. Each row represents a differentially expressed gene, while each column represents a sample. Changes in expression levels are shown in color scales with saturation at >2.0-fold changes. Green and red color gradients indicate a decrease and increase in transcript abundance, respectively.

with phytochromes. Our study found that putative orthologs of *bHLH56, PIF1, PIF3, PIF4, PIF5, PIL1, PIF7, SPT* were expressed in broccoli seed germination and sprout development, suggesting their similar roles in light signaling during this period. SPATULA (SPT) was found as a leaf size regulator [49]. The SPT ortholog in broccoli sprouts showed highest expression level in the cotyledons of 3 day sprouts, indicating its possible regulation activity on cotyledon size in early broccoli sprouts.

The MYB TFs contain varying numbers of MYB domain repeats to bind DNA. The function of MYB proteins have been well studied in a variety of plant species to be involved in regulatory networks controlling development, metabolism and responses to biotic and abiotic stresses in *Arabidopsis* [50]. In *Arabidopsis* seedlings, MYB115 and MYB118 play important roles in embryogenesis [51]. MYB38 and MYB18 have been proposed to regulate hypocotyls elongation in response to blue [52] and far-red light, respectively [53]. Also, the MYB17 has shown activity in regulating seed germination [54]. Some other MYB proteins are involved in the control of cell wall biosynthesis like MYB58, MYB63, MYB85, MYB68 and MYB46 [55,56,57,58]. In this study, a total of 84 putative MYB genes were detected including those orthologs involved in *Arabidopsis* seedling development. The many putative MYB TF genes expressed in the broccoli seed and sprouts indicated this important family also plays important roles in regulating the biological process during seed germination and sprout development.

Sixty-seven putative NAC TF family members were identified in seed and sprouts of broccoli. The large NAC transcription factor family has been implicated in a variety of plant developmental processes in many species including *Arabidopsis* and soy bean etc [59]. However, the molecular mechanisms of the family members are still unknown even in well studied species. It has been suggested that they have the ability to enable crosstalk between different pathways [60]. Cys2His2 (C2H2)-type zinc finger proteins are a group of widespread eukaryotic TFs. A majority of C2H2 zinc finger proteins are regarded as *trans* regulators of genes playing important roles in development, differentiation and suppression of malignant cell transformation [61]. In the sequenced tissues, 65 putative C2H2 zinc finger genes were identified.

Several other TF families were also found like 61 members in bZIP, 54 members in WRKY, 14 members in ARF, etc. Because of the importance of TFs in regulating the downstream genes in variety of pathways, further investigation of the putative TFs would provide interesting clues to the variety of activities in seed germination and sprout development of broccoli.

Glucosinolate Metabolic Pathways

The high contents of glucosinolates especially the much higher content of aliphatic glucosinolates in broccoli sprouts compared to mature tissues have attracted attention in past decade [11]. The biological basis of this trait especially whether the glucosinolate metabolic genes in *Arabidopsis* or *Brassica rapa* have the same functions in broccoli sprouts, remains an open question. In this study, a total of 36 unigenes were annotated as putative genes involved in aliphatic and indolic glucosinolate biosynthesis, degradation and regulation. By comparing these unigenes with the CDS of *Arabidopsis* ones and setting the identity cutoff of 60%, we abandoned 11 unigenes and finally got 25 putative broccoli glucosinolates metabolic genes sharing 62.04–89.72% nucleotide sequence identity with the *Arabidopsis* orthologs (Figure 5, Table 2).

The glucosinolate biosynthesis proceeds through three independent stages: chain elongation (for aliphatic glucosinolates), core structure formation and side chain modification (Figure 5) [62]. For the 25 selected putative genes, 7 were uniquely involved in the aliphatic pathway including *BoIPMDH3, BoBCAT3, BoCYP83A1, BoGSTF11, BoSOT17, BoFMO$_{GS-OX2}$ and BoFMO$_{GS-OX5}$*. In the chain elongation stage, two genes (*BoIPMDH3, BoBCAT3*) were detected. *BCAT3* encodes a chloroplast branched-chain amino acid aminotransferase [62] and *IPMDH3* is one of the three

isopropylmalate dehydrogenase genes in *Arabidopsis* whose isozyme *IPMDH1* has been characterized as a functional gene involved in aliphatic glucosinolate chain elongation process [63]. Most of the *Arabidopsis* genes involved in biosynthesis of aliphatic glucosinolate core structure have orthologs expressed in the studied tissues except for *CYP79F1, CYP79F2 and UGT74B1, UGT74C1* (Figure 5). It has been reported that in the double-knockout *Arabidopsis* mutant of *CYP79F1* and *CYP79F2*, aliphatic glucosinolate biosynthesis is completely abolished, meaning that *CYP79F1* and *CYP79F2* are necessary for the pathway [64]. The contradiction between the high level of aliphatic glucosinolate content and the missing of both of *CYP79F1* and *CYP79F2* indicates that there may be unknown gene(s) performing the same function in broccoli sprouts. *FMO$_{GS-OX2}$ and FMO$_{GS-OX5}$* are important genes performing S-oxygenation in side chain modification stage in *Arabidopsis* [65,66], they may have the same function in broccoli. However, orthologs of other genes of this stage expressed in *Arabidopsis* have not been identified in our tissues. Notably, glucoraphanin is one of the products produced by *FMO$_{GS-OX2}$* [66]. The missing of the downstream genes of *FMO$_{GS-OX2}$* may explain the accumulation of glucoraphanin in broccoli sprouts.

In broccoli seeds and sprouts, the indolic glucosinolate biosynthetic pathway showed a high colinearity with *Arabidopsis*. Eight genes involved in the indolic pathway were detected with *BoCYP79B2, BoCYP79B3, BoCYP83B1 and BoGSTF10* in core structure formation and *BoCYP81F4, BoCYP81F1, BoCYP81F3 and BoIGMT1* in side chain modification. The enzyme UGT74B1 transforming the Indolylmethyl-thiohydroximate to the Indolylmethyl-desulfoglucosinolate in the indolic glucosinolate pathway was missing. In *Arabidopsis*, when the indolylmethyl-glucosinolates were formed, there were two ways for them to be modified. Part of them would be transformed to 1-methoxy-3indolylmethyl-glucosinolates by CYP81F4 and the others would be transformed to 4-methoxy-3indolylmethyl-glucosinolates by CYP81F1, CYP81F2 or CYP81F3 and then the IGMT1 or IGMT2 would modify them to 4-methoxy-3-indolylmethyl-glucosinolates [67]. The *CYP81F2* and *IGMT2* were not detected in this study.

Beside these genes uniquely expressed in indolic or aliphatic glucosinolate pathway, the three genes involved in both the two pathways including *GGP1, SUR1 and SOT18* were all identified in the studied tissues too. In *Arabidopsis*, *TSB1* is a tryptophan synthesis gene [68] and *GSH1* is a crucial gene to form GSH, which is considered as the sulfur donor to be conjugated with the activated aldoxime [69,70]. The orthologs in broccoli with these two genes not involved in glucosinolate biosynthesis directly but having important roles for the forming of glucosinolates were also identified.

Some transcription factors of MYB family are crucial in regulating glucosinolate biosynthesis pathways of *Arabidopsis*, in which MYB28, MYB29 and MYB76 [71,72,73] are involved in aliphatic glucosinolates biosynthesis whereas MYB51, MYB34 and MYB122 [74,75] could strongly enhance the expression of indolic glucosinolate biosynthesis genes. In broccoli seeds and sprouts, only *BoMYB29* and *BoMYB51* were detected.

In the glucosinolate degradation pathways, the *PEN2, TGG1* and *TGG2* were well studied in *Arabidopsis*. *PEN2* was reported to cleave indolic glucosinolates as a myrosinase [76]. TGG1 and TGG2 were two important myrosinases identified long time ago. The double mutant of these two genes showed nearly no aliphatic glucosinolate degradation but only reduced indolic glucosinolate myrosinase activity [77]. In addition, the *tgg1tgg2* double mutant still exhibited a wild type callose response to fungi simulation which required the degradation products of indolic glucosinolates [78]. These facts indicated the TGG1 and TGG2 mainly degrade

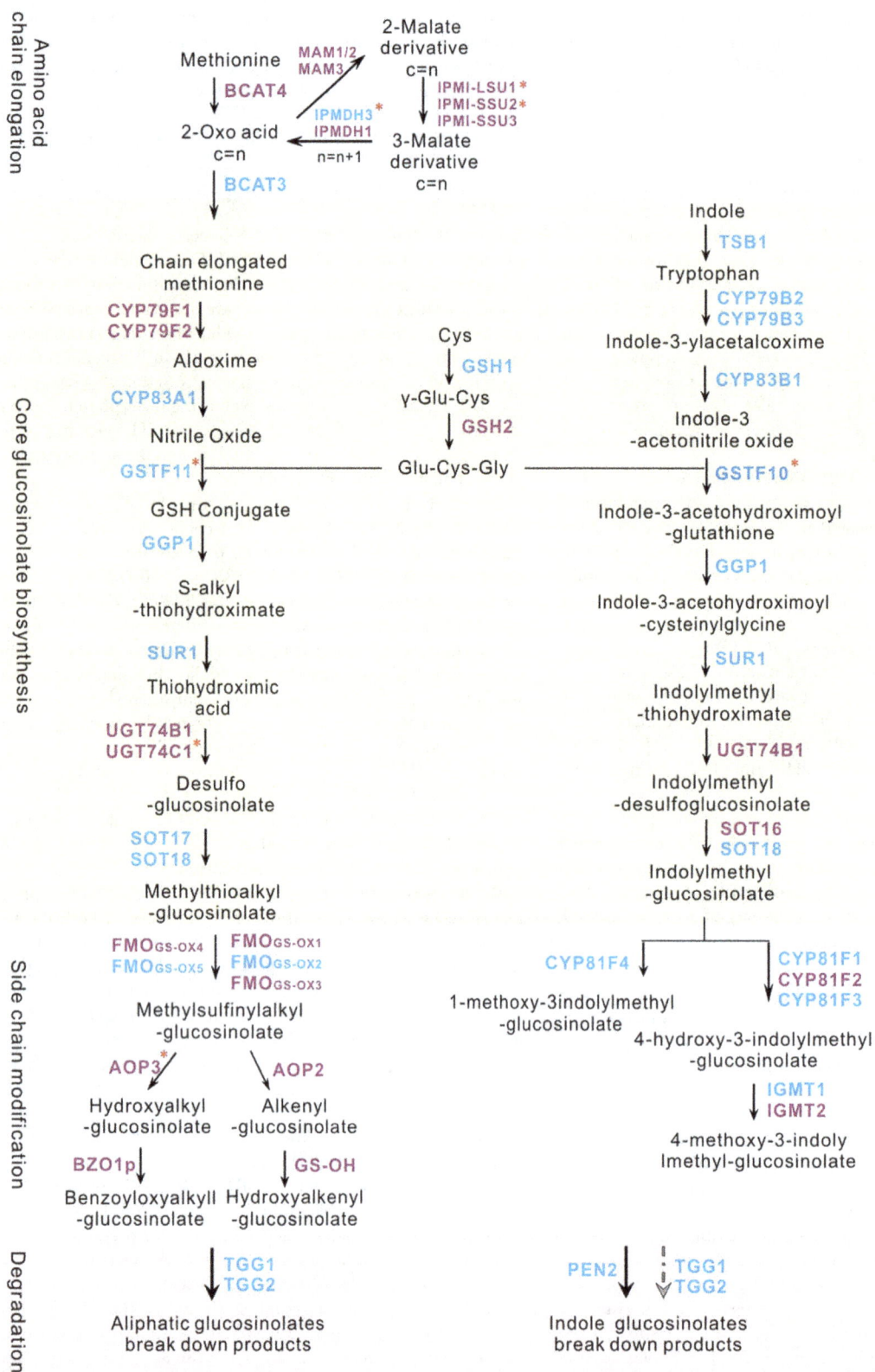

Figure 5. Detected orthologs in the aliphatic and indolic glucosinolate biosynthetic and degradation pathways in broccoli seeds and sprouts. Four stages of the pathways in *Arabidopsis* are showed separately for chain elongation, core structure biosynthesis, side chain modification and degradation. Otrologs identified in Broccoli are marked in blue color. Predicted enzymes are marked by *.

Table 2. Putative genes involved in glucosinolate metabolic pathways in Broccoli.

Name	Arabidopsis orthologs	Basemean of seeds	Basemean of 3 day cotyledons	Basemean of 7 day cotyledons	Basemean of 11 day cotyledons	Basemean of 11 day euphyllas	Identity
Aliphatic glucosinolates							
Glucosinolate synthesis							
BoIPMDH3**	AT1g31180	9.2	2.03	9.45	2.14	8.09	63.28%
BoBCAT3	AT3g49680	274.54	635.68	722.67	482.47	678.31	86.31%
BoCYP83A1	AT4g13770	1.42	54.92	922.24	345.23	427.55	88.20%
BoGSTF11**	AT3g03190	16.27	125.1	32.56	2.14	71.64	83.94%
BoFMO-GSOX2	AT1g62540	9.91	8.14	534.65	11.79	9.24	76.50%
BoFMO-GSOX5	AT1g12140	42.46	15.26	22.06	6.43	18.49	78.61%
BoSOT17	AT1g18590	235.63	49.84	306.71	1627.53	116.71	83.00%
Transcription factor							
BoMyb29	AT5g07690	0	6.1	52.52	0	6.93	83.04%
Indolic glucosinolates							
Tryptophan synthesis							
BoTSB1	AT5g54810	421.02	454.63	560.91	263.75	271.55	79.41%
Glucosinolate synthesis							
BoCYP79B2	AT4g39950	87.03	73.23	829.81	1029.27	145.6	85.41%
BoCYP79B3	AT2g22330	0.71	13.22	281.51	306.64	18.49	89.72%
BoCYP83B1	AT4g31500	961.62	515.66	1543.03	3386.94	479.55	86.67%
BoGSTF10**	AT2g30870	844.16	1554.1	3612.3	3070.66	1061.95	67.23%
BoCYP81F1	AT4G37430	55.19	49.84	3.15	0	0	76.35%
BoCYP81F3	AT4G37400	58.73	6.1	52.52	28.95	0	86.83%
BoCYP81F4	AT4G37410	378.56	12.2	175.42	8.58	16.18	81.70%
BoIGMT1	AT1G21100	89.16	8.14	52.52	700.12	27.73	82.44%
Transcription factor							
BoMyb51	AT1g18570	90.57	18.31	56.72	131.88	33.51	85.46%
Common to all glucosinolates							
BoSOT18	AT1g74090	79.25	39.67	53.57	164.04	55.47	64.22%
BoGGP1*	AT4g30530	780.47	474.98	1406.48	7330.34	746.48	81.14%
BoSUR1	AT2g20610	179.02	58.99	257.35	536.08	104	75.56%
BoGSH1/PAD2	AT4g23100	4680	1962.96	1898.06	5790.72	1453.68	68.61%
Glucosinolate degradation							
BoPEN2	AT2G44490	247.66	157.65	279.4	2437.01	294.66	70.49%
BoTGG1	AT5G26000	55.19	42.72	74.58	111.5	132.89	68.58%
BoTGG2	AT5G25980	38.21	1702.59	1590.29	818.06	4958.45	62.04%

the aliphatic glucosinolates and had slight effects on indolic ones. In the sequenced tissues, the three myrosinase genes' orthologs were all identified.

It is interesting to note that the expression levels of many broccoli glucosinolate related genes were expressed higher in sprouts than in seeds. Some previous studies had indicated the glucosinolate concentration decreased exponentially after germination [11,12]. This contradiction between the decreased concentration level of glucosinolates and the increased level of biosynthesis genes might due to the high consumption of glucosinolates and this dramatic degradation of glucosinolates possibly played an important role in the stage of broccoli seed germination and sprout development.

Besides, the putative genes involved in indolic glucosinolate synthesis have higher expression levels than those involved in aliphatic glucosinolate synthesis in general. This was more obvious in the expression of TFs. The only identified transcription factor BoMYB29 in aliphatic glucosinolate synthesis had no expression in seeds and 11 day cotyledons; in the other three time points, the expression level was also relatively low (Table 2). The expression values of BoMYB51 were much higher compared to the expression of BoMYB29. Furthermore, we noticed that the expression value of BoTGG2, was 45-fold higher in 3 day cotyledons than in seeds (Table 2). The expression value decreased slowly along with the development of sprouts and got to the lowest point in the 11 day cotyledons, which was still about 21-fold higher than that in seeds. Notably, at the time of 11th day, the expression values of BoTGG2 in the new forming euphyllas astonishingly increased to about 130-fold. While the expression values of the indolic glucosinolate degradation gene BoPEN2 were relatively low compared to those of BoTGG2 and not too much different in our sequenced tissues except for the 11 day cotyledons (Table 2). These results demonstrated that in glucosinolate sprouts, the exponentially decreased levels of glucosinolates were mainly due to the degradation of aliphatic glucosinolates especially in the young stage of tissues, while the indolic glucosinolates might be constantly synthesized and stored. The exception was the expression value of BoPEN2 in late stage cotyledons increased to 10-fold higher than that in seeds. This might indicate the degraded indolic glucosinolates had special roles at the old stage of cotyledons.

Actually, previous studies have reported that the expression of TGG1, redundantly functioning with TGG2 in Arabidopsis, is higher in young developing tissues than older tissues [79,80,81] and PEN2 is unlikely to function in glucosinolate turnover during seedling development [82], which are coordinated to our results. Degraded glucosinolates have been proposed to regulate the cellular signaling in response to abiotic stress which is based on the observation that TGG1 is highly enriched in stomatal guard cells and regulate the stomatal opening or closing by affecting the ABA and MeJA signaling [83,84]. So the myrosinases' high level in young seedling may be necessary for proper protection due to lower physical strength barriers. In our study, BoTGG2 was the predominantly expressed myrosinase rather than TGG1 in Arabidopsis. Considering that TGG1 and TGG2 redundantly function in glucosinolate degradation [77], the predominantly functional gene may be BoTGG2 in broccoli sprouts instead of BoTGG1. Also, glucosinolates has been suggested to represent up to 30% of total sulfur content of plant organs and glucosinolate content has been observed to decrease during sulfur deprivation [85]. So we can hypothesis that the enriched degradation products of glucosinolates may contribute greatly to the defense system as a compensation for the lower physical strength barrier in broccoli sprouts and they may also play an important role as potential

sulfur donor during broccoli seed germination and sprout development.

Conclusion

In this study, we performed a transcriptome sequencing of seeds, 3 day, 7 day and 11 day cotyledons and 11day euphyllas to identify the transcripts and quantify their levels of expression in broccoli seed germination and sprout developments. In total, we obtained 19441 unigenes annotated by homologs with sequences in the public databases. 2675 genes differentially expressed between any two of the five samples and 1,633 putative TFs were detected. These genes may be closely related to the seed germination and sprout development in broccoli. Twenty-five unigenes with high identity to Arabidopsis glucosinolate metabolic genes have been investigated. This established a high colinearity between Arabidopsis and broccoli in glucosinolate metabolic pathways for further studies. In addition, expression analysis of these glucosinolate metabolic genes showed contradiction between the increased expression of the candidate synthesis genes and the previously reported decreased concentration of glucosinolate content after germination. The ortholog of TGG2, which mainly degrade aliphatic glucosinolates in Arabidopsis, expressed astonishingly higher after germination. These results indicate the breakdown products of glucosinolates may play important roles in the stage of broccoli seed germination and sprout development. The results here represent the largest genetic resource for broccoli and will provide new insight into the genomic research of this species and its relatives.

Materials and Methods

Sample Preparation

Seeds of broccoli cultivar "Qingxiu" with wide suitability of temperature and soil were germinated and grown in trays containing a soil mixture (peat: vermiculite, 2:3, v/v). Plants were adequately watered with Hoagland's solution and grown in a culture room with the following settings: 24°C, light regime of 16 h light and 8 h dark, 70% relative humidity and a constant illumination of 100 $\mu mol \cdot m^{-2} \cdot s^{-1}$. For the sample of seeds, they were incubated in 5% NaClO with shaking for 8 min and then were washed six times using sterile water with once 30 s. Subsequently, seeds were placed on moist filter paper in petri dishes for one night. Finally, samples of equal weight were harvested for seeds, cotyledons at the 3rd day, 7th day and 11th day and euphyllas at 11th day (Figure 6). To minimize biological variance, each sample was harvested in three independent biological replicates with equal weight and subsequently pooled for sequencing. Samples were immediately frozen in liquid nitrogen and stored at −80°C until RNA was extracted. Total RNA of each sample was isolated using E.Z.N.A. Plant RNA Kit (OMEGA bio-tek, GA) according to the instructions from the manufacturer. RNA quality was characterized on an agarose gels electrophoresis and spectrophotometry. High quality RNA with 28S:18S more than 1.5 and absorbance 260/280 ratios between 1.8 and 2.2 was used for library construction and sequencing.

cDNA Library Construction and Sequencing

Illumina Miseq library construction was performed according to the manufacturer's instructions (Illumina, San Diego, CA). Magnetic beads with poly T oligos attached were used for purifying the mRNA from the total RNA. Fragmentation buffer was added to cleave mRNA into short fragments. The fragments were used to synthesize first-strand cDNA using random hexamerprimers, which was transformed into double stranded

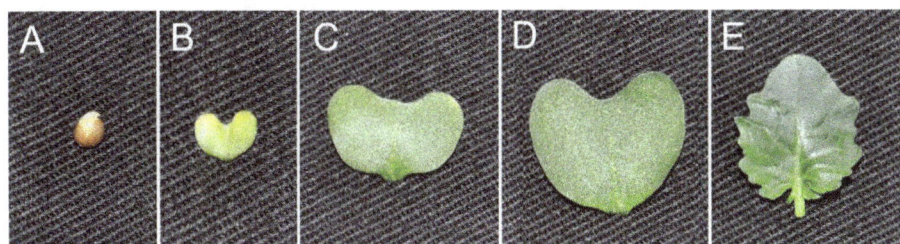

Figure 6. Images of the sampled tissues. A. seed; B. 3 day cotyledon; C. 7 day cotyledon; D. 11 day cotyledon; E. 11 day euphylla.

cDNA with RHase H and DNA polymerase I. A paired-end library was constructed from the cDNA synthesized with Genomic Sample Prep Kit (Illumina). Fragments in desirable lengths were purified with QIAquick PCR (Qiagen) Extraction Kit, end repaired and linked with sequencing adapters (Margulies et al., 2005). AMPureXP beads were used to remove the unsuitable fragments, then the sequencing library was constructed with PCR amplification. After being checked with Pico green staining and fluorospectrophotometry and quantified with Agilent 2100, the multiplexed DNA libraries were mixed by equal volume with normalized 10 nM concentration. The sequencing library was then sequenced with Illumina Miseq platform (Shanghai Personal Biotechnology Cp., Ltd. Shanghai, China).

Data Filtering and *de novo* Assembly

Raw sequencing reads of five samples were mixed together to perform the following filtration using a stringent process and subsequent *de novo* assembly. The adaptor contamination was removed, the reads were screened from the 3′ to 5′ to trim the bases with a quality score of $Q<20$ using 5 bp windows and the reads with final length less than 25 bp were removed. De novo transcriptome assembly was performed by Velvet [86] followed by Oases [87] with default settings except for K-mer value to get contigs and transcripts. Velvet was run using single k-mer length of 69 and OASES was then run with the preliminary Velvet assemblies as input. Because the results after merging with multiple k-mers used in OASES program prompted severe assembly redundancy, we use the same single k-mer in OASES program. High quality reads of every sample were remapped to transcripts to get the abundance of transcripts using Bowtie program [88]. Those transcripts with no reads mapped in all five samples were considered error and removed. All the transcripts were searched against Arabidopsis database, for those with no hits were then searched to NCBI non-redundant (nr) database (ftp://ftp.ncbi.nlm.nih.gov/blast/db/) with BLAST program (E-value <1E-5), and the top-hit transcripts were selected as unigenes. For the unigenes failed to be aligned to the databases, the software GetORF [89] was used to predict their open reading frames (ORFs) and ascertain their sequence directions, with default settings except for the parameter "–find" being set 1.

Gene Annotation and Analysis

To further annotate the unigenes in this study, we used the Blast2GO program [90,91,92,93] to get GO annotation based on GO terms related to the *Arabidopsis* and nr database annotation. EggNOG (evolutionary genealogy of genes: Non-supervised Orthologous Groups) is a database of orthologous groups of genes. To annotate genes with common denominators or functional categories, the unigenes were also aligned to the eggNOG database (http://eggnog.embl.de/version_3.0/). To summarize the pathways information involved in broccoli seeds

and sprouts, the KEGG database were used to perform the pathway annotation (http://www.genome.jp/kegg/). To identify putative TFs presented in this research, we searched the unigenes against the complete TF gene sequences of the Plant Transcription Factor Database (http://plntfdb.bio.uni-potsdam.de/) using BLAST program with an E-value cutoff of 1E-5. To identify the putative sequences related to glucosinolate pathways, the unigenes annotated by putative glucosinolate biosynthetic and regulator genes according to previous studies were chosen [9,66]. Then, CDS of *Arabidopsis* glucosinolate biosynthesis and regulator genes were aligned to broccoli homologs using DNAMAN6.0 (http://www.lynnon.com/) and unigenes with identity larger than 60% were selected.

Comparative Expression Analysis

The R package DESeq was performed to identify differential gene expression [94]. This method represents the widely accepted and accurate analysis approaches of RNA-seq data. We first mapped high-quality reads to unigenes to calculate the number of reads mapped to each unigene in five samples. These raw read counts were then used as the input of DESeq to get the normalized signal for each unigene, and the fold change of unigene expression values with p-values compared to each other of the five samples was used to report differential expression. Those with p-value< 0.05 were considered as significant differential expression. We performed cluster analysis of gene expression patterns with the Cluster [95], MeV [96] and Java treeview software packages [97].

Supporting Information

Figure S1 Length distribution of contigs, transcripts and unigenes.

Figure S2 Matching percentage of broccoli unigenes with different lengths to entries in public databases.

Figure S3 EggNOG classification of the broccoli seed and sprout transcriptome.

Figure S4 Classification of unigenes based on KEGG categorization.

Table S1 Characterization of raw data and trimmed data.

Table S2 Transcription factor members of every family detected in the broccoli seeds and sprouts.

Author Contributions

Conceived and designed the experiments: JJG JL. Performed the experiments: JJG XXY. Analyzed the data: JJG XXY. Contributed

reagents/materials/analysis tools: FMM JL. Wrote the paper: JJG JL. Read and approved the final manuscript: JJG XXY FMM JL.

References

1. US Food and Drug Administration. 2005. Code of Federal Regulations: 21 CFR101.78. Health claims: fruits and vegetables and cancer.
2. US Food and Drug Administration. 2012. Code of Federal Regulations: 21 CFR101.77. Health claims: fruits, vegetables, and grain products that contain fiber, particularly soluble fiber, and risk of coronary heart disease.
3. Van Poppel G, Verhoeven DT, Verhagen H, Goldbohm RA. (1999) Brassica vegetables and cancer prevention. Epidemiology and mechanisms. Adv Exp Med Biol. 472: 159–168.
4. Verhoeven DT, Goldbohm RA, van Poppel G, Verhagen H, van den Brandt PA. (1996) Epidemiological studies on brassica vegetablesand cancer risk, Cancer Epidemiol Biomarkers Prev. 5: 733–748.
5. Michaud DS, Spiegelman D, Clinton SK, Rimm EB, Willett WC et al. (1999) Fruit and vegetable intake and incidence of bladder cancer in a male prospective cohort. J Natl Cancer Inst. 91: 605–613.
6. Rose P, Faulkner K, Williamson G, Mithen R. (2000) 7-Methylsulfinylheptyl and 8-methylsulfinyloctyl isothiocyanates from watercress are potent inducers of phase II enzymes. Carcinogenesis. 21: 1983–1988.
7. Verhoeven DT, Verhagen H, Goldbohm RA, van den Brandt PA, van Poppel G. (1997) A review of mechanisms underlying anticarcinogenicity by brassica vegetables. Chem Biol Interact. 103: 79–129.
8. Chen Y, Yan X, Chen S. (2011) Bioinformatic analysis of molecular network of glucosinolate biosynthesis. Comput Biol Chem. 35: 10–18.
9. Yan X, Chen S. (2007) Regulation of plant glucosinolate metabolism. Planta. 226: 1343–1352.
10. Sønderby IE, Geu-Flores F, Halkier BA. (2010) Biosynthesis of glucosinolates–gene discovery and beyond. Trends Plant Sci. 15: 283–290.
11. Fahey JW, Zhang Y, Talalay P. (1997) Broccoli sprouts: An exceptionally rich source of inducers of enzymes that protect against chemical carcinogens. Proc Natl Acad Sci U S A. 94: 10367–10372.
12. S Pérez-Balibrea, DA Moreno, C García-Viguera. (2008) Influence of light on health-promoting phytochemicals of broccoli sprouts. J Sci Food Agric. 88: 904–910.
13. Fahey JW, Talalay P. (1999) Antioxidant functions of sulforaphane: a potent inducer of Phase II detoxication enzymes. Food Chem Toxicol. 37: 973–979.
14. Kallifatidis G, Rausch V, Baumann B, Apel A, Beckermann BM et al. (2009) Sulforaphane targets pancreatic tumour-initiating cells by NF-kappaBinduced antiapoptotic signalling. Gut. 58: 949–963.
15. Abbott A. (2006) Cancer: the root of the problem. Nature. 442: 742–743.
16. Kallifatidis G, Labsch S, Rausch V, Mattern J, Gladkich J, et al. (2011) Sulforaphane increases drug-mediated cytotoxicity towards cancer stem-like cells of pancreas and prostate. Mol Ther. 19: 188–195.
17. Wu L, Noyan Ashraf MH, Facci M, Wang R, Paterson PG, et al. (2004) Dietary approach to attenuate oxidative stress, hypertension, and inflammation in the cardiovascular system. Proc Natl Acad Sci U S A. 101: 7094–7099.
18. Bourdon V, Naef F, Rao PH, Reuter V, Mok SC, et al. (2002) Genomic and expression analysis of the 12p11-p12 amplicon using EST arrays identifies two novel amplified and overexpressed genes. Cancer Res. 62: 6218–6223.
19. Cheung F, Win J, Lang JM, Hamilton J, Vuong H, et al. (2008) Analysis of the Pythium ultimum transcriptome using Sanger and pyrosequencing approaches. BMC Genomics 9: 542.
20. Ruyter-Spira CP, de Koning DJ, van der Poel JJ, Crooijmans RP, Dijkhof RJ, et al. (1998) Developing microsatellite markers from cDNA: A tool for adding expressed sequence tags to the genetic linkage map of the chicken. Anim Genet 29: 85–90.
21. Gonzalo MJ, Oliver M, Garcia-Mas J, Monfort A, Dolcet-Sanjuan R, et al. (2005) Simple-sequence repeat markers used in merging linkage maps of melon (Cucumis melo L.). Theor Appl Genet 110: 802–811.
22. Seki M, Narusaka M, Kamiya A, Ishida J, Satou M, et al. (2002) Functional annotation of a full-length Arabidopsis cDNA collection. Science 296: 141–145.
23. Fei Z, Tang X, Alba RM, White JA, Ronning CM, et al. (2004) Comprehensive EST analysis of tomato and comparative genomics of fruit ripening. Plant J 40: 47–59.
24. USDA National Agricultural Statistics Service.
25. Adams MD, Kelley JM, Gocayne JD, Dubnick M, Polymeropoulos MH, et al. (1991) Complementary DNA sequencing: expressed sequence tags and human genome project. Science 252: 1651–1656.
26. Metzker ML. (2010) Sequencing technologies - the next generation. Nat Rev Genet 11: 31–46.
27. Jiang L, Schlesinger F, Davis CA, Zhang Y, Li R, et al. (2011) Synthetic spike-in standards for RNA-seq experiments. Genome Res 21: 1543–1551.
28. Ashburner M, Ball CA, Blake JA, Botstein D, Butler H, et al. (2000) Gene ontology: tool for the unification of biology. The Gene Ontology Consortium. 25: 25–29.
29. Powell S, Szklarczyk D, Trachana K, Roth A, Kuhn M, et al. (2012) eggNOG v3.0: orthologous groups covering 1133 organisms at 41 different taxonomic ranges. Nucleic Acids Res 40(Database issue): D284–289.
30. Tatusov RL, Koonin EV, Lipman DJ. (1997) A genomic perspective on protein families. Science 278: 631–637.
31. Fan H, Xiao Y, Yang Y, Xia W, Mason AS, et al. (2013) RNA-Seq analysis of Cocos nucifera: transcriptome sequencing and de novo assembly for subsequent functional genomics approaches. PLoS One 8: e59997.
32. Zhang XM, Zhao L, Larson-Rabin Z, Li DZ, Guo ZH. (2012) De novo sequencing and characterization of the floral transcriptome of Dendrocalamus latiflorus (Poaceae: Bambusoideae). PLoS One 7: e42082.
33. Li D, Deng Z, Qin B, Liu X, Men Z. (2012) De novo assembly and characterization of bark transcriptome using Illumina sequencing and development of EST-SSR markers in rubber tree (Hevea brasiliensis Muell. Arg.). BMC Genomics 13: 192.
34. Kanehisa M, Goto S. (2000) KEGG: kyoto encyclopedia of genes and genomes. Nucleic Acids Res 28: 27–30.
35. Qing DJ, Lu HF, Li N, Dong HT, Dong DF, et al. (2009) Comparative profiles of gene expression in leaves and roots of maize seedlings under conditions of salt stress and the removal of salt stress. Plant Cell Physiol 50: 889–903.
36. Mortazavi A, Williams BA, McCue K, Schaeffer L, Wold B. (2008) Mapping and quantifying mammalian transcriptomes by RNA-Seq. Nat Methods 5(7): 621–628.
37. Anders S, Huber W. (2010) Dierential expression analysis for sequence count data. Genome Biol 11: R106.
38. Jiao Y, Yang H, Ma L, Sun N, (2003) A genome-wide analysis of blue-light regulation of Arabidopsis transcription factor gene expression during seedling development. Plant Physiol 133: 1480–1493.
39. Pérez-Rodríguez P, Riaño-Pachón DM, Corrêa LG, Rensing SA, Kersten B, et al. (2010) PlnTFDB: updated content and new features of the plant transcription factor database. Nucleic Acids Res 38(Database issue): D822–827.
40. Saleh Abdelaty, Pagés Montserrat. (2003) Plant AP2/ERF transcription factors. Genetika 35: 37–50.
41. Riechmann JL, Meyerowitz EM. (1998) The AP2/EREBP family of plant transcription factors. Biol Chem 379: 633–646.
42. Jofuku KD, Omidyar PK, Gee Z, Okamuro JK. (2005) Control of seed mass and seed yield by the floral homeotic gene APETALA2. Proc Natl Acad Sci U S A 102: 3117–3122.
43. Ohto MA, Fischer RL, Goldberg RB, Nakamura K, Harada JJ. (2005) Control of seed mass by APETALA2. Proc Natl Acad Sci U S A 102: 3123–3128.
44. Hongtao Zhao, Xia Li, Ligeng Ma (2012) Basic helix-loop-helix transcription factors andepidermal cell fate determination in Arabidopsis. Plant Signal Behav 7: 1556–1560.
45. Duek PD, Fankhauser C. (2005) bHLH class transcription factors take centre stage in phytochrome signalling. Trends Plant Sci 10: 51–54.
46. Khanna R, Huq E, Kikis EA, Al-Sady B, Lanzatella C, et al. (2004) A novel molecular recognition motif necessary for targeting photoactivated phytochrome signaling to specific basic helix–loop–helix transcription factors. Plant Cell 16: 3033–3044.
47. Ni M, Tepperman JM, Quail PH. (1999) Binding of phytochrome B to its nuclear signalling partner PIF3 is reversibly induced by light. Nature 400: 781–784.
48. Toledo-Ortiz G, Huq E, Quail PH. (2003) The Arabidopsis basic/helix–loop–helix transcription factor family. Plant Cell 15: 1749–1770.
49. Ichihashi Y, Horiguchi G, Gleissberg S, Tsukaya H. (2010) The bHLH transcription factor SPATULA controls final leaf size in Arabidopsis thaliana. Plant Cell Physiol 51: 252–261.
50. Dubos C, Stracke R, Grotewold E, Weisshaar B, Martin C, et al. (2010) MYB transcription factors in Arabidopsis. Trends Plant Sci. 15: 573–581.
51. Wang X, Niu QW, Teng C, Li C, Mu J, et al. (2009) Overexpression of PGA37/MYB118 and MYB115 promotes vegetative-to-embryonic transition in Arabidopsis. Cell Res. 19: 224–235.
52. Hong SH, Kim HJ, Ryu JS, Choi H, Jeong S, et al. (2008) CRY1 inhibits COP1-mediated degradation of BIT1, a MYB transcription factor, to activate blue light-dependent gene expression in Arabidopsis. Plant J 55: 361–371.
53. Yang SW, Jang IC, Henriques R, Chua NH. (2009) FAR-RED ELONGATED HYPOCOTYL1 and FHY1-LIKE associate with the Arabidopsis transcription factors LAF1 and HFR1 to transmit phytochrome A signals for inhibition of hypocotyl elongation. Plant Cell 21: 1341–1359.
54. Zhang Y, Cao G, Qu LJ, Gu H. (2009) Characterization of Arabidopsis MYB transcription factor gene AtMYB17 and its possible regulation by LEAFY and AGL15. J. Genet. Genomics 36: 99–107.
55. Zhou J, Lee C, Zhong R, Ye ZH. (2009) MYB58 and MYB63 are transcriptional activators of the lignin biosynthetic pathway during secondary cell wall formation in Arabidopsis. Plant Cell 21: 248–266.
56. Zhong R, Lee C, Zhou J, McCarthy RL, Ye ZH. (2008) A battery of transcription factors involved in the regulation of secondary cell wall biosynthesis in Arabidopsis. Plant Cell 20: 2763–2782.

57. Feng C, Andreasson E, Maslak A, Mock HP, Mattsson O, et al. (2004) Arabidopsis MYB68 in development and responses to environmental cues. Plant Sci. 167: 1099–1107.

58. Zhong R, Richardson EA, Ye ZH. (2007) The MYB46 transcription factor is a direct target of SND1 and regulates secondary wall biosynthesis in Arabidopsis. Plant Cell 19: 2776–2792.

59. Shamimuzzaman M, Vodkin L. (2013) Genome-wide identification of binding sites for NAC and YABBY transcription factors and co-regulated genes during soybean seedling development by ChIP-Seq and RNA-Seq. BMC Genomics 14: 477.

60. Olsen AN, Ernst HA, Leggio LL, Skriver K. (2005) NAC transcription factors: structurally sidtinct, functionally diverse. Trends Plant Sci 10: 79–87.

61. Razin SV, Borunova VV, Maksimenko OG, Kantidze OL. (2012) Cys2His2 zinc finger protein family: classification, functions, and major members. Biochemistry (Mosc) 77: 217–226.

62. Knill T, Schuster J, Reichelt M, Gershenzon J, Binder S. (2008) Arabidopsis branched-chain aminotransferase 3 functions in both amino acid and glucosinolate biosynthesis. Plant Physiol 146: 1028–1039.

63. Nozawa A, Takano J, Miwa K, Nakagawa Y, Fujiwara T. (2005) Cloning of cDNAs encoding isopropylmalate dehydrogenase from Arabidopsis thaliana and accumulation patterns of their transcripts. Biosci Biotechnol Biochem 69: 806–810.

64. Tantikanjana T, Mikkelsen MD, Hussain M, Halkier BA, Sundaresan V. (2004) Functional analysis of the tandem-duplicated P450 genes SPS/BUS/CYP79F1 and CYP79F2 in glucosinolate biosynthesis and plant development by Ds transposition-generated double mutants. Plant Physiol 134: 840–848.

65. Hansen BG, Kliebenstein DJ, Halkier BA. (2007) Identification of a flavin-monooxygenase as the S-oxygenating enzyme in aliphatic glucosinolate biosynthesis in Arabidopsis. Plant J 50: 902–910.

66. Li J, Hansen BG, Ober JA, Kliebenstein DJ, Halkier BA. (2008) Subclade of flavin-monooxygenases involved in aliphatic glucosinolate biosynthesis. Plant Physiol 148: 1721–1733.

67. Pfalz M, Mikkelsen MD, Bednarek P, Olsen CE, Halkier BA, et al. (2011) Metabolic engineering in *Nicotiana benthamiana* reveals key enzyme functions in Arabidopsis indole glucosinolate modification. Plant Cell 23: 716–729.

68. Zhao Y, Hull AK, Gupta NR, Goss KA, Alonso J, et al. (2002) Trp-dependent auxin biosynthesis in Arabidopsis: involvement of cytochrome P450s CYP79B2 and CYP79B3. Genes Dev 16: 3100–3112.

69. Cobbett CS, May MJ, Howden R, Rolls B. (1998) The glutathione-deficient, cadmiumsensitive mutant, cad2–1, of *Arabidopsis thaliana* is deficient in gamma-glutamylcysteine synthetase. Plant J 16: 73–78.

70. Schlaeppi K, Bodenhausen N, Buchala A, Mauch F, Reymond P. (2008) The glutathione-deficient mutant pad2–1 accumulates lower amounts of glucosinolates and is more susceptible to the insect herbivore Spodoptera littoralis. Plant J 55: 774–786.

71. Gigolashvili T, Yatusevich R, Berger B, Müller C, Flügge UI. (2007) The R2R3-MYB transcription factor HAG1/MYB28 is a regulator of methionine-derived glucosinolate biosynthesis in *Arabidopsis thaliana*. Plant J 51: 247–261.

72. Gigolashvili T, Engqvist M, Yatusevich R, Müller C, Flügge UI. (2008) HAG2/MYB76 and HAG3/MYB29 exert a specific and coordinated control on the regulation of aliphatic glucosinolate biosynthesis in *Arabidopsis thaliana*. New Phytol 177: 627–642.

73. Hirai MY, Sugiyama K, Sawada Y, Tohge T, Obayashi T, et al. (2007) Omics-based identification of Arabidopsis Myb transcription factors regulating aliphatic glucosinolate biosynthesis. Proc Natl Acad Sci U S A 104: 6478–6483.

74. Celenza JL, Quiel JA, Smolen GA, Merrikh H, Silvestro AR, et al. (2005) The Arabidopsis ATR1Myb transcription factor controls indolic glucosinolate homeostasis. Plant Physiol. 2005, 137, 253–262.

75. Gigolashvili T, Berger B, Mock HP, Müller C, Weisshaar B, et al. (2007) The transcription factor HIG1/MYB51 regulates indolic glucosinolate biosynthesis in *Arabidopsis thaliana*. Plant J. 50: 886–901.

76. Bednarek P, Pislewska-Bednarek M, Svatos A, Schneider B, Doubsky J, et al. (2009) A glucosinolate metabolism pathway in living plant cells mediates broad-spectrum antifungal defense. Science 323: 101–106.

77. Barth C, Jander G. (2006) Arabidopsis myrosinases TGG1 and TGG2 have redundant function in glucosinolate breakdown and insect defense. Plant J 46: 549–562.

78. Clay NK, Adio AM, Denoux C, Jander G, Ausubel FM. (2009) Glucosinolate metabolites required for an Arabidopsis innate immune response. Science 323: 95–101.

79. Husebye H, Chadchawan S, Winge P, Thangstad OP, Bones AM. (2002) Guard cell- and phloem idioblast-specific expression of thioglucoside glucohydrolase 1 (myrosinase) in *Arabidopsis*. Plant Physiol 128: 1180–1188.

80. Barth C, Jander G. (2006) Arabidopsis myrosinases TGG1 and TGG2 have redundant function in glucosinolate breakdown and insect defense. Plant J 46: 549–562.

81. Burow M, Rice M, Hause B, Gershenzon J, Wittstock U. (2007) Cell- and tissue-specific localization and regulation of the epithiospecifier protein in Arabidopsis thaliana. Plant Mol Biol 64: 173–185.

82. Wittstock U, Burow M. (2010) Glucosinolate breakdown in Arabidopsis: mechanism, regulation and biological significance. Arabidopsis Book 8: e0134.

83. Zhao Z, Zhang W, Stanley BA, Assmann SM. (2008) Functional proteomics of *Arabidopsis thaliana* guard cells uncovers new stomatal signaling pathways. Plant Cell 20: 3210–3226.

84. Islam MM, Tani C, Watanabe-Sugimoto M. (2009) Myrosinases, TGG1 and TGG2, redundantly function in ABA and MeJA signaling in Arabidopsis guard cells. Plant Cell Physiol 50: 1171–1175.

85. Falk KL, Tokuhisa JG, Gershenzon J. (2007) The effect of sulfur nutrition on plant glucosinolate content: physiology and molecular mechanisms. Plant Biol (Stuttg) 9: 573–581.

86. Zerbino DR, Birney E. (2008) Velvet: algorithms for de novo short read assembly using de Bruijn graphs. D.R. Zerbino and E. Birney. Genome Res 18: 821–829.

87. Schulz MH, Zerbino DR, Vingron M, Birney E. (2012) Oases: Robust de novo RNA-seq assembly across the dynamic range of expression levels. Bioinformatics 28: 1086–1092.

88. Langmead B. (2010) Aligning short sequencing reads with Bowtie. Curr Protoc Bioinformatics Chapter 11:Unit 11.7.

89. Rice P, Longden I, Bleasby A. (2000) EMBOSS: the European Molecular Biology Open Software Suite. Trends Genet 16: 276–277.

90. Conesa A, Götz S, García-Gómez JM, Terol J, Talón M, et al. (2005) Blast2GO: a universal tool for annotation, visualization and analysis in functional genomics research. Bioinformatics 21: 3674–3676.

91. Conesa A, Götz S. (2008) Blast2GO: A Comprehensive Suite for Functional Analysis in Plant Genomics. Int J Plant Genomics 2008: 619832.

92. Götz S, García-Gómez JM, Terol J, Williams TD, Nagaraj SH, et al. (2008) High-throughput functional annotation and data mining with the Blast2GO suite. Nucleic Acids Res 36: 3420–3435.

93. Götz S, Arnold R, Sebastián-León P, Martín-Rodríguez S, Tischler P, et al. (2011) B2G-FAR, a species-centered GO annotation repository. Bioinformatics 27: 919–924.

94. Anders S, Huber W. (2010) Differential expression analysis for sequence count data. Genome Biology 11: R106.

95. Eisen MB, Spellman PT, Brown PO, Botstein D. (1998) Cluster analysis and display of genome-wide expression patterns. Proc Natl Acad Sci U S A 95: 14863–14868.

96. Howe EA, Sinha R, Schlauch D, Quackenbush J. (2011) RNA-Seq analysis in MeV. Bioinformatics 27: 3209–3210.

97. Saldanha AJ. (2004) Java Treeview–extensible visualization of microarray data. Bioinformatics 20: 3246–3248.

4

AtWuschel Promotes Formation of the Embryogenic Callus in *Gossypium hirsutum*

Wu Zheng[1,9], **Xueyan Zhang**[1,9], **Zuoren Yang**[1,9], **Jiahe Wu**[1,2], **Fenglian Li**[1], **Lanling Duan**[1], **Chuanliang Liu**[1], **Lili Lu**[1], **Chaojun Zhang**[1*], **Fuguang Li**[1*]

1 State Key Laboratory of Cotton Biology, Institute of Cotton Research, Chinese Academy of Agricultural Sciences, Anyang, Henan, China, 2 Institute of Microbiology, Chinese Academy of Sciences, Beijing, China

Abstract

Upland cotton (*Gossypium hirsutum*) is one of the most recalcitrant species for *in vitro* plant regeneration through somatic embryogenesis. Callus from only a few cultivars can produce embryogenic callus (EC), but the mechanism is not well elucidated. Here we screened a cultivar, CRI24, with high efficiency of EC produce. The expression of genes relevant to EC production was analyzed between the materials easy to or difficult to produce EC. Quantitative PCR showed that CRI24, which had a 100% EC differentiation rate, had the highest expression of the genes *GhLEC1*, *GhLEC2*, and *GhFUS3*. Three other cultivars, CRI12, CRI41, and Lu28 that formed few ECs expressed these genes only at low levels. Each of the genes involved in auxin transport (*GhPIN7*) and signaling (*GhSHY2*) was most highly expressed in CRI24, with low levels in the other three cultivars. WUSCHEL (WUS) is a homeodomain transcription factor that promotes the vegetative-to-embryogenic transition. We thus obtained the calli that ectopically expressed *Arabidopsis thaliana Wus* (*AtWus*) in *G. hirsutum* cultivar CRI12, with a consequent increase of 47.75% in EC differentiation rate compared with 0.61% for the control. Ectopic expression of *AtWus* in CRI12 resulted in upregulation of *GhPIN7*, *GhSHY2*, *GhLEC1*, *GhLEC2*, and *GhFUS3*. *AtWus* may therefore increase the differentiation potential of cotton callus by triggering the auxin transport and signaling pathways.

Editor: David D. Fang, USDA-ARS-SRRC, United States of America

Funding: This study was supported by National Science Fund for Distinguished Young Scholars (Grant no. 31125020). The funders had no role in study design, data collection and analysis, decision to publish, or preparation of the manuscript.

Competing Interests: The authors have declared that no competing interests exist.

* E-mail: zcj1999@yeah.net (CZ); aylifug@163.com (FL)

9 These authors contributed equally to this work.

Introduction

Somatic embryogenesis (SE) is a principal model for studying the growth and development of zygotic embryos in higher plants. This process includes callus induction, embryogenic callus (EC) formation, embryo development, and plant regeneration. During the past three decades, much effort has attempted to determine important genes controlling SE [1,2].

The gene *WUSCHEL* (*Wus*) is essential for stem cell formation and maintenance in shoot and root apical meristems [3,4]. *Wus* mediates stem cell homeostasis by regulating cell division and differentiation [5–7]. In *wus* mutants, apical meristems are unable to preserve the pool of undifferentiated cells [3]. The maintenance function of WUS can be repressed by inducing *AGAMOUS* (*AG*) expression and floral meristem differentiation [8]. *Wus* was first reported as the key gene promoting SE in *pga6* mutants in *Arabidopsis*, with overexpression of *PGA6/Wus* causing the vegetative-to-embryonic transition [9]. *Wus* is crucial for EC renewal during SE in *Arabidopsis* [10]. Overexpression of *Wus* in *Coffea canephora* can also promote SE [11].

Auxin is necessary for SE [12,13], but the auxin transport and signaling pathways during SE are not well understood. *PIN-FORMED* (*PIN*) genes encoding efflux carrier proteins are involved in auxin transport [14,15]. Genetic analysis indicates that *PIN1* is a major regulatory factor for auxin gradients in EC and embryo

[16]. Auxin regulates auxin-responsive genes via the Aux/IAA (SHY)-ARF module. At sufficiently low auxin concentration, auxin response factors (*ARF*s) are repressed by Aux/IAA. At sufficiently high auxin concentration, Aux/IAA is degraded by SCF[TIR1], and *ARF*s are activated [17–19]. *ARF*s positively or negatively control downstream genes, resulting in responses to auxin signaling. Transcript profiling reveals that the auxin signaling pathway may play a vital role during SE in cotton [13].

LEC1, *LEC2*, and *FUS3* are key genes that control SE progression [20,21]. The capacity for SE is completely repressed in double (*lec1 lec2*, *lec1 fus3*, *lec2 fus3*) or triple (*lec1 lec2 fus3*) mutants of *Arabidopsis thaliana* [21]. *LEC2* expression changes rapidly during auxin responses [22], suggesting that *LEC/FUS* may be downstream genes in the auxin signaling pathway [21,23].

The majority of cotton cultivars are incapable of undergoing SE [24] because of their difficulty in inducing callus differentiation to form EC. Thus, most cultivars are not used for molecular breeding using transgenic technologies with *Agrobacterium*-mediated transformation via SE. Therefore, it is essential to study the mechanism of SE in cotton so as to improve regeneration of various cotton cultivars.

Here, we report the ectopic expression of *A. thaliana Wus* (*AtWus*) in *G. hirsutum* cv. CRI12, a cultivar that shows poor SE ability under established tissue culture methods. *AtWus* promoted differentiation of transgenic callus. Furthermore, ectopic expres-

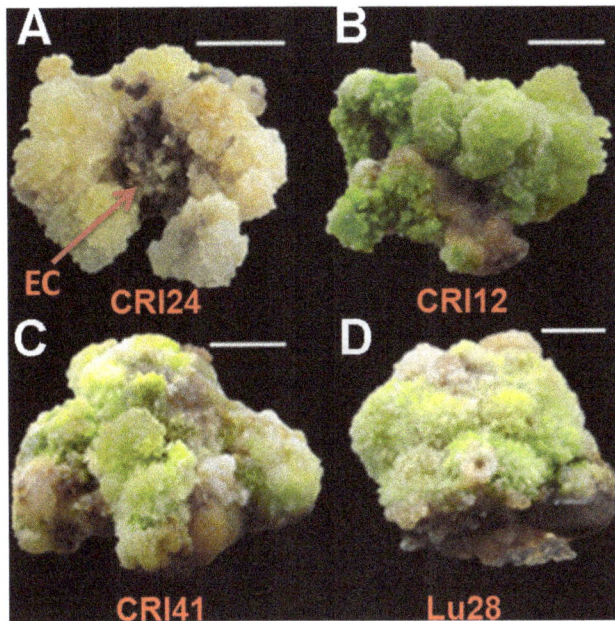

Figure 1. Cotton cultivars have different differentiation rates in EC. After 90 days of culture in NEIM, all calli of CRI24 produced EC, whereas calli of CRI12, CRI41, and Lu28 were dark green and tight and unable to differentiate into EC. Bar, 1 cm.

sion of *AtWus* could upregulate *GhLEC1* (*G. hirsutum LEC1*), *GhLEC2*, and *GhFUS3* expression during SE and alter auxin transport and signaling mechanisms. *AtWus* therefore promotes the efficiency of EC differentiation in cotton callus.

Materials and Methods

Plant Materials and Tissue Culture Conditions

We selected four cotton cultivars, CRI24, CRI12, CRI41 and Lu28, as experiment materials. CRI24 has a 100% EC differentiation rate and is the main transgenic material used for *Agrobacterium*-mediated method in our lab. CRI12 used to be an important basic breeding material because of traits of its relatively high yield and disease resistance, yet it has a low rate of differentiation during SE. CRI41 and Lu28, the main cultivars planted in China, can not undergo SE because of failure in EC induction.

Seeds of the four cotton cultivars were sterilized with 0.1% (w/v) mercuric chloride for 3 min. The seeds were then washed five times with sterilized distilled water and then germinated on

modified Murashige and Skoog (MS) medium (25 g l^{-1} sucrose, 50 ml l^{-1} MSI, 5.6 g l^{-1} agar) for hypocotyl induction. Sterilized seeds were cultured at 28°C with a 14 h/10 h light/dark photoperiod. The hypocotyls from 7-day-old sterile seedlings were cut into 2 cm segments. For transgenic experiments using an *Agrobacterium*-mediated method [25], the hypocotyl cuts of CRI12 were transferred to 250 ml flasks and placed on callus-induction medium (CIM; MS medium plus B5 vitamins, supplemented with 0.05 mg l^{-1} 3-indole acetic acid (IAA), 0.05 mg l^{-1} kinetin, 0.05 mg l^{-1} 2,4-dichlorophexoxyacetic acid, 25 g l^{-1} glucose, 2 g l^{-1} gelrite gellan gum, 50 mg l^{-1} kanamycin, 100 mg l^{-1} cefotaxime, pH 5.8). The medium was changed once per month. After 2 months of culture, all calli were transferred onto EC induction medium (EIM; MSB supplemented with 25 g l^{-1} glucose, 2 g l^{-1} gelrite, 0.5 g l^{-1} MgCl$_2$, 0.16 mg l^{-1} kinetin, 0.08 mg l^{-1} IAA, 50 mg l^{-1} kanamycin, 100 mg l^{-1} cefotaxime, pH 6.5). The medium was changed monthly. After 4 months of culture, ECs were transferred to somatic embryo induction medium (SIM; MSB supplemented with 25 g l^{-1} glucose, 2 g l^{-1} gelrite, 0.5 g l^{-1} MgCl$_2$, 0.08 g l^{-1} kinetin, 0.12 mg l^{-1} 6-benzylaminopurine, 50 mg l^{-1} kanamycin, 100 mg l^{-1} cefotaxime, pH 6.8), and the medium was refreshed monthly. For non-transgenic experiments, the hypocotyl cut explants of CRI24, CRI12, CRI41, and Lu28 were transferred onto NCIM (CIM lacking kanamycin and cefotaxime), and the medium was changed once per month. After 30 days of culture, all calli were transferred onto NEIM (EIM lacking kanamycin and cefotaxime), and the medium was changed monthly. After 90 days of culture, all calli of CRI24 differentiated into EC, most calli of CRI12 and all calli of CRI41 and Lu28 did not differentiate into ECs. To confirm the lack of capacity for EC differentiation among the latter three cultivars, the calli which can not differentiate in ECs were cultured in NEIM for another 120 days, with the medium being refreshed monthly. Indeed, those calli did not differentiate into ECs, and thus no further experiments were conducted with these non-transgenic cultivars.

Gene Cloning and Vector Construction

The nucleotide sequences of *GhLEC1*, *GhLEC2*, and *GhFUS3* were obtained from the D subgenome database of *Gossypium. raimondii* by comparing with amino acid sequences of *AtLEC1*, *AtLEC2*, and *AtFUS3* using the tblastn tool. The three genes were then amplified from a full-length cDNA library of CRI24 with specific primers (Table S1). For ectopic expression of *AtWus*, the full-length coding regions (CDS) of the gene was cloned from wild-type *Arabidopsis* (Columbia ecotype) (Table S1). The full-length CDS of *AtWus* was amplified via PCR with specific primers (Table S1) and ligated into vector pMD18-T. After verifying the sequence, each of the *AtWus* fragment and Vector pBI121 was digested with *Bam*H I and *Sac* I. and the *AtWus* fragment was

Table 1. EC induction in four cotton cultivars.

Replication	CRI12 (WT)			CRI24 (WT)			CRI41 (WT)			Lu28 (WT)		
	C[a]	EC[b]	EC rate (%)	C[a]	EC[b]	EC rate (%)	C[a]	EC[b]	EC rate (%)	C[a]	EC[b]	EC rate (%)
1	175	1	0.57	371	371	100	177	0	0.00	209	0	0.00
2	221	3	1.36	204	204	100	303	0	0.00	242	0	0.00
3	286	0	0.00	261	261	100	319	0	0.00	138	0	0.00
Average	682	4	0.59	836	836	100	799	0	0.00	589	0	0.00

[a]Number of explants forming a callus.
[b]Number of EC forming from a callus. WT, wild type.

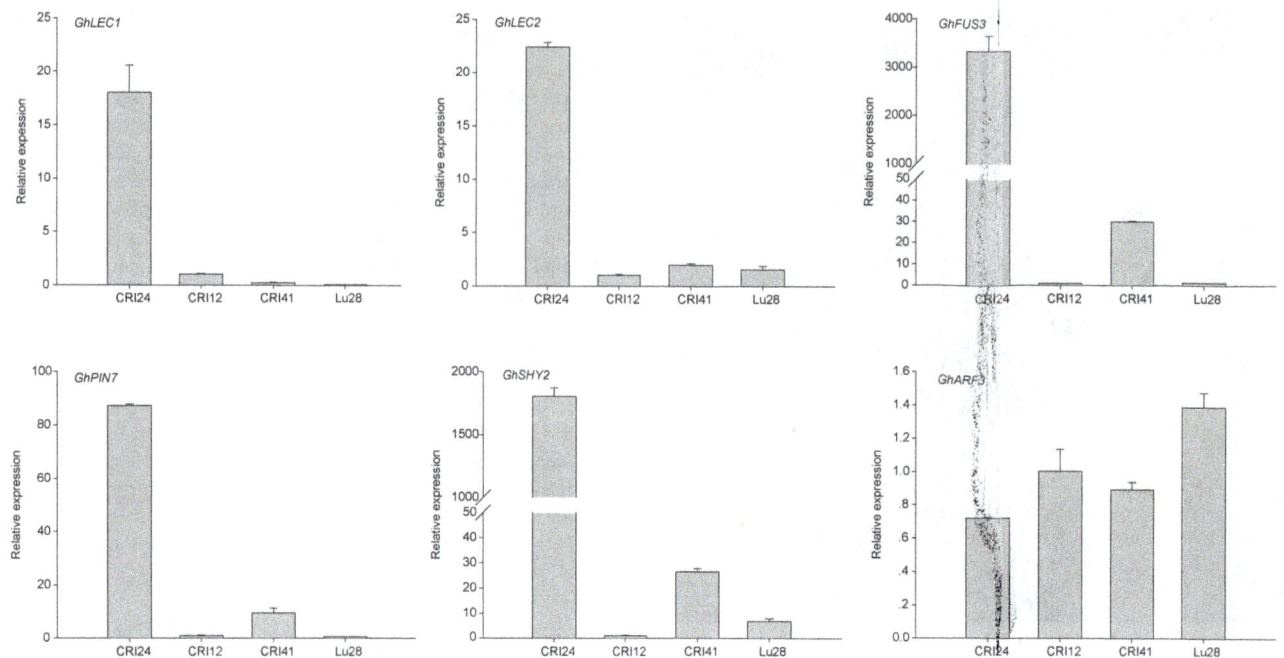

Figure 2. Analysis of gene expression in the non-transgenic calli of the four cultivars.

inserted into pBI121. The nucleotide sequences of *GhPIN7*, *GhSHY2*, and *GhARF3* were obtained from the D subgenome database of *G. raimondii* by comparing with amino acid sequences of *AtPIN1*, *AtSHY2*, and *AtARF1* using the tblastn tool.

RNA Extraction

All calli of CRI24, CRI12, CRI41 and Lu28 cultured for 90 days in NEIM and of 35S:WUS and CK lines cultured for 4 months in EIM were stored at −80°C. We extracted RNA of the above samples using a modified CTAB method [26]. RNA samples with A260/A280 ratios between 1.8 and 2.0 and A260/A230 ratios >1.5 were considered acceptable.

Quantitative Real Time PCR (QPCR)

Approximately 1 μg total RNA samples were reverse transcribed using the PrimeScript RT reagent kit with gDNA Eraser (Takara). The cDNA templates were diluted three times prior to amplification. The QPCR experiment was conducted according to the guidelines of SYBR Premix Ex Taq^{TM} kit (Takara). QPCR was performed in 96-well plates with a total volume of 20 μL containing 10 μL 2× SYBR Premix Ex Taq^{TM}, 6.8 μL PCR-grade water, 2 μL cDNA template, 0.4 μL 50× ROX reference dye I, and 0.4 μL each of forward and reverse primers (10 μM). All QPCRs were run with three technical replicates on an ABI 7900 Real-Time PCR system (Applied Biosystems). The thermal cycling conditions were as follows: an initial denaturation step of 30 s at 95°C, followed by 40 cycles of 95°C annealing for 5 s and 60°C extension for 30 s. The primers used for QPCR are shown in Table S2.

Scanning Electron Microscopy

Scanning electron microscopy was performed on somatic embryos obtained after 1.5 months culture on SIM. Samples were prefixed at room temperature for 12 h in 2.5% (v/v) glutaraldehyde (phosphate buffer, pH 7.2). After dehydration using a graded ethanol series, samples were dried with a CO_2

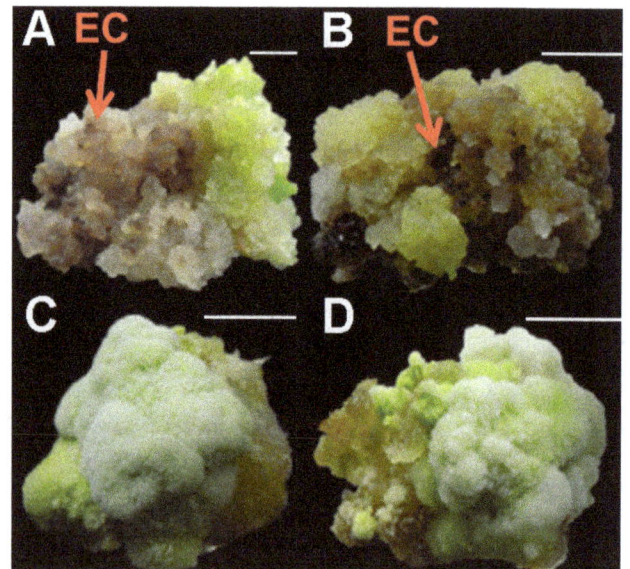

Figure 3. The callus of 35S:WUS and CK cultured for 1.5 months in EIM. A, **B:** Calli of 35S:Wus lines at the beginning of EC formation, **C, D:** Calli of CK lines were unable to differentiate. Bar, 1 cm.

critical-point drying system (HITACHI HCP-2). Subsequently, samples were sputtered with gold dust and observed under a HITACHI S-530 scanning electron microscopy.

Statistics

The rate of EC = number of EC/number of calli. After 45 days of culture in SIM, we determined the weight of the abnormal embryos in 35S:WUS lines and normal cotyledonary embryos in CK lines with three technical replicates. The average weight of each individual somatic embryo = weight of 10 somatic embryos/

Table 2. EC induction in CRI12.

Replication	35S:WUS			Empty vector (CK)		
	C[a]	EC[b]	EC rate (%)	C[a]	EC[b]	EC rate (%)
1	173	88	50.66	151	2	1.32
2	201	94	46.77	114	0	0.00
3	93	41	44.09	223	1	0.45
Average	467	223	47.75**	488	3	0.61

[a]Number of explants forming callus.
[b]Number of EC forming from a callus. **$p < 0.01$.

10. We conducted t-tests to determine significant differences ($p < 0.05$ or $p < 0.01$, depending on the experiment).

Results

Expression of *GhLEC* and *GhFUS3* in Cultivars with Diverse Differentiation Rates

We used four cotton cultivars (*G. hirsutum*) for tissue culture. The EC differentiation rate of CRI24 was 100% but only 0.59% for CRI12, whereas the calli of Lu28 and CRI41 could not form EC (Table 1). After culturing hypocotyl segments for 30 days on NCIM (callus induction medium lacking kanamycin and cefotaxime), the induced calli were transferred onto NEIM (EC induction medium lackling kanamycin and cefotaxime). After 90 days in NEIM, all the CRI24 calli were non-compacted and had differentiated into EC (Figure 1**A**). However, most calli of CRI12 and all calli of CRI41 and Lu28 were compacted, dark green, and did not differentiate (Figure 1**B–D**).

LEC1, *LEC2*, and *FUS3* are essential for SE [21], but there are few studies on the roles of *LEC* and *FUS3* in SE in plants other than *Arabidopsis*. Hence, we first isolated the homologs *GhLEC1*, *GhLEC2*, *GhFUS3* in CRI24 [27]. At the amino acid level, these homologs share high sequence similarity with those of in *Arabidopsis* (Figure S1). The results implied that *GhLEC1*, *GhLEC2*, and *GhFUS3* may have functions similar to those of the *Arabidopsis* homologs that control the capacity for SE [21].

To investigate the expression of *GhLEC1*, *GhLEC2*, and *GhFUS3* in calli of the four cultivars, we carried out QPCR on calli cultured for 90 days on NEIM. The expression levels of *GhLEC1*, *GhLEC2*, and *GhFUS3* were significantly higher in the calli of CRI24 than in CRI12, Lu28, and CRI41, with barely detectable expression of *GhFUS3* in CRI12 and Lu28 (Figure 2).

Expression of Genes Involved in Auxin Transport or Signaling Pathways

Auxin plays an important role in SE of cotton [13,28], but genes involved in auxin transport (*PLNs*) and auxin signaling are not well studied in cotton SE. QPCR analysis of expression levels of these genes in SE used the same samples as for expression patterns of *LECs* and *FUS3*. *GhPLN7* and *GhSHY2* (*AUX/IAA2*) levels in calli of CRI24 were much higher than in the other three cultivars (Figure 2). In contrast, *GhARF3* expression was lower in CRI24 than in other cultivars (Figure 2), suggesting that *GhARF3* expression may be inhibited by the *GhSHY2* gene product. These results indicated that the auxin transport and signaling pathways were more active in calli with high EC differentiation rates than those with low EC differentiation rates.

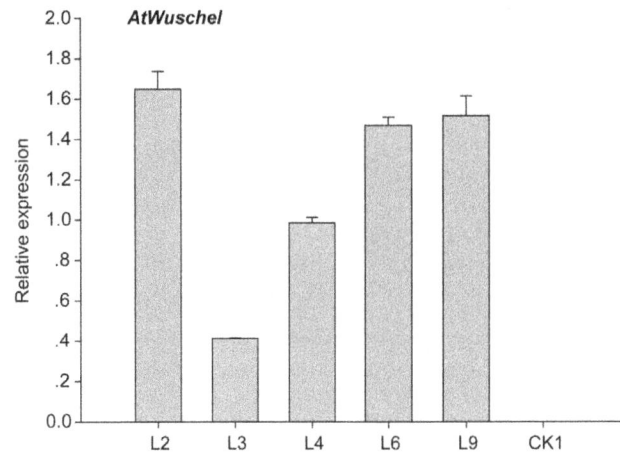

Figure 4. *AtWus* **expression in calli of CRI12.**

Ectopic Expression of *AtWus* Improves EC Induction

CRI12 is one of the most recalcitrant cotton cultivars for plant regeneration via SE. To study the function of *AtWus* in SE, we overexpressed *AtWus* in CRI12 (35S:WUS). An empty vector was used as the control (CK) for parallel transformation. Hypocotyl segments transformed with 35S:WUS or the control CK were cultured on CIM containing kanamycin and cefotaxime to induce resistant calli. One month later, these segments were transferred to fresh CIM. There were no apparent differences between 35S:WUS and CK within the first 2 months. Then, calli that formed on segments were transferred to EIM containing kanamycin and cefotaxime for inducing EC. Calli were transferred to fresh EIM monthly. After culture for 1.5 months on EIM, most 35S:WUS calli were non-compacted and light green, and some began to differentiate into EC (Figure 3**A, B**). However, the CK calli were compact and dark green and did not undergo EC differentiation (Figure 3**C, D**). After culturing for 4 months on EIM, the differentiation rate of 35S:WUS transformants was 47.75% compared with 0.61% for CK (Table 2).

QPCR was used to determine *AtWus* expression level in transformed calli. The calli of transgenic cultures (L2, L3, L4, L6, and L9) cultured for 4 months in EIM were selected for analysis, whereas the calli of CK that did not differentiate into EC served as the control (CK1). The data revealed *AtWus* was overexpressed in 35S:WUS cultures (Figure 4).

Ectopic Expression of *AtWus* Regulates Auxin Transport and Signal Transduction

AtWus induction by IAA can regulate *AtPIN* expression during SE in *Arabidopsis* [10]. Hence, the role of *AtWus* in auxin transport and signal transduction in cotton was studied. Transgenic calli of L2, L3, L4, L6, and L9 cultured for 4 months in EIM were analyzed by QPCR, with CK1 calli serving as a control. *GhPIN7* expression was higher in 35S:WUS transformed callus lines than in CK (Figure 5).

Transcript profiling during SE in cotton has been used to establish the association between the auxin signaling pathway and callus differentiation [13]. The above-mentioned data also demonstrated an interaction between the auxin signaling pathway and EC induction whereby *GhSHY2* transcripts were increased in 35S:WUS lines compared with the CK1 line. However, *GhARF3* transcripts were reduced in 35S:WUS, possibly owing to suppression by *GhSHY2* (Figure 5).

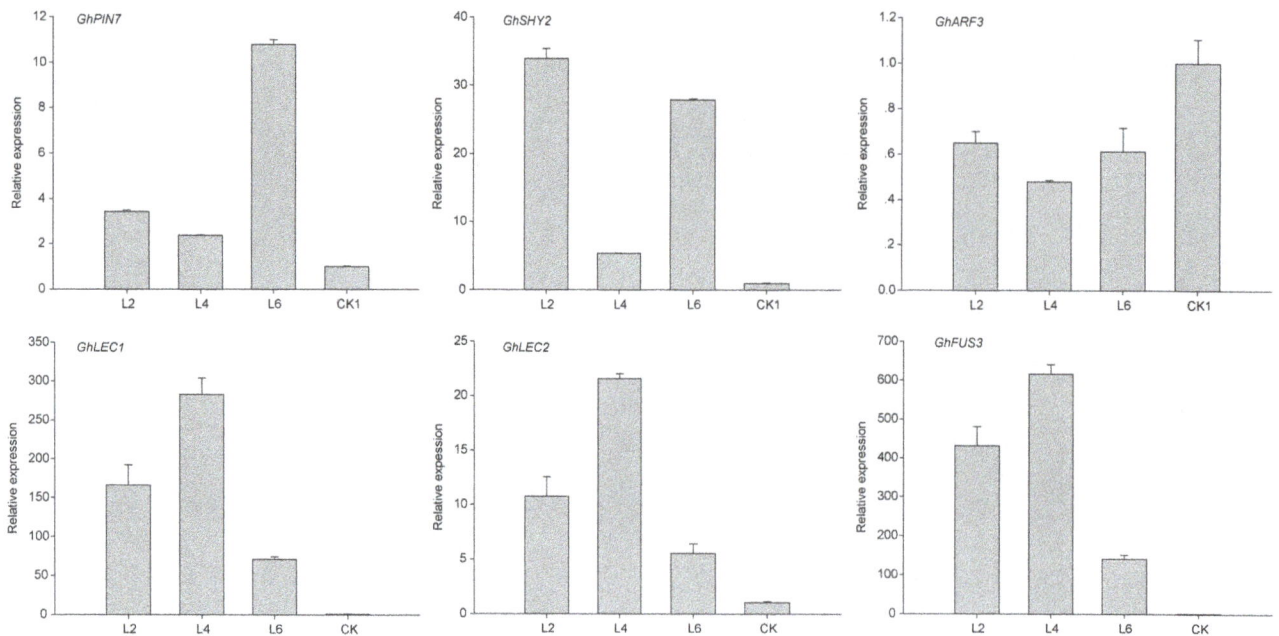

Figure 5. Analysis of gene expression in the transgenic calli carrying *AtWus*.

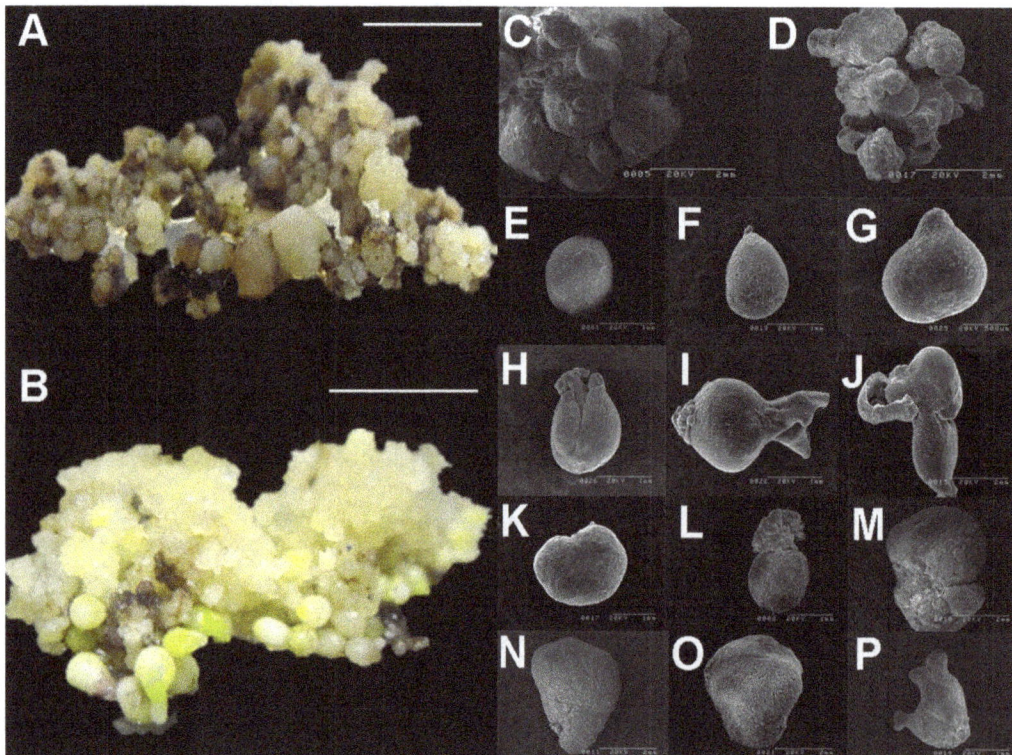

Figure 6. AtWus overexpression results in abnormal development of somatic embryos. A: Many abnormal somatic embryos were produced in 35S:WUS lines, and the somatic embryos were inflated and lacked cotyledons. **B:** Formation of normal somatic embryos in CK lines at different stages. Scanning electron microscopy: Holistic perspective of somatic embryos in 35S:WUS lines (**C**) and CK lines (**D**). **E–J:** Normal somatic embryos at different stages. **E, F:** globular embryo. **G:** heart-shape embryo. **H–J:** cotyledonary embryo. **K–P:** Abnormal somatic embryos having various appearance. **O:** leaf-like embryo. **P:** multiple-cotyledon embryo. Bar in **A** or **B,** 1 cm.

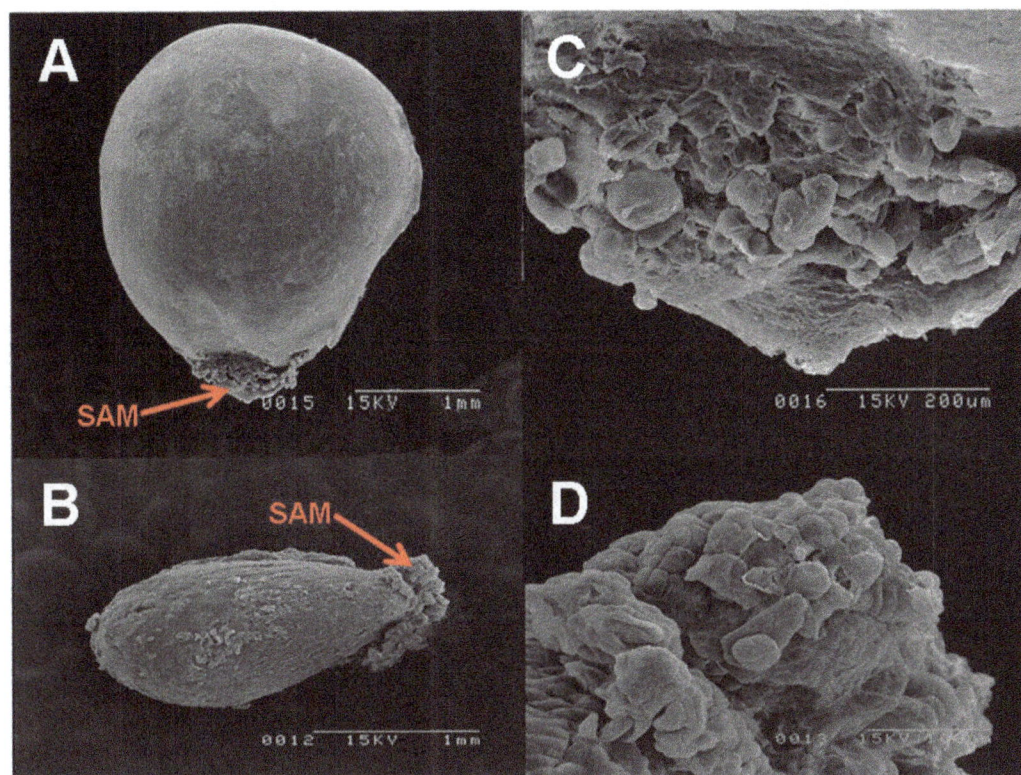

Figure 7. SAM structure of somatic embryos in transgenic lines observed with scanning electron microscopy. A: Abnormal somatic embryo of 35S:WUS lines. **B:** Normal somatic embryo of CK lines. **C:** Enlarged image of SAM in **A**. **D:** Enlarged image of SAM in **B**.

AtWus Activates the Expression of *LEC* and *FUS3* during SE in Cotton

GhLEC1, *GhLEC2*, and *GhFUS3* were expressed in calli of CRI24 but only barely detected in CRI12. To characterize the regulatory relation between *AtWus* and *LEC/FUS3*, we selected L2, L4, and L6 callus lines cultured for 4 months in EIM for further study. QPCR analysis revealed that *GhLEC1*, *GhLEC2*, and *GhFUS3* transcript levels were low in calli of CK1. In calli of L2, L4 and L6, however, higher expression levels of the three genes were detected owing to upregulation by *AtWus* (Figure 5). Hence, *AtWus* promoted the expression of *GhLEC1*, *GhLEC2*, and *GhFUS3* during SE in cotton, similar to *Arabidopsis* [10].

AtWus Overexpression Results in Abnormal Development of Somatic Embryos

For somatic embryo induction, 4-month-old EC of 35S:WUS and CK lines were transferred to SIM containing kanamycin and cefotaxime for somatic embryo induction. ECs were transferred

Table 3. Somatic embryo weight in CRI12.

Replication	35S:WUS (g)	CK (g)
1	0.0138	0.00764
2	0.0124	0.00445
3	0.0181	0.00530
Average	0.0148	0.00580

*p<0.05.

onto a fresh SIM once per month and cultured for 50 days on SIM. EC formed from calli of CK lines grew into normal-looking globular, heart-shaped, and cotyledonary embryos (Figure 6 **B**). However, ECs that developed from 35S:WUS callus lines grew into various abnormal translucent embryos (Figure 6**A**) that could not form the normal-looking cotyledonary embryos except for some leaf-like or multi-cotyledon embryos, although they produced more somatic embryos than CK (Figure 6**C, D**). Scanning electron microscopy was used to investigate the structural abnormalities of these embryos. Somatic embryos from 35S:WUS lines were much larger and heavier than those from CK lines (Figure 6**C**, Table 3). In CK lines, globular, heart-shaped and cotyledonary embryos were clearly observed (Figure 6**E–J**). However, somatic embryos of 35S:WUS lines exhibited several abnormal morphologies, such as leaf-like or multi-cotyledon embryos (Figure 6**K–P**). Cotyledons are derived from the shoot apical meristem (SAM) region. SAM cells were arranged in an organized manner in CK somatic embryos (Figure 7**B, D**), but the SAM cells of 35S:WUS lines were disorganized and rather unstructured (Figure 7**A, C**).

Discussion

Wus is a key gene that controls renewal of stem cells in the apical meristem [29]. However, because *AtWus* promotes the vegetative-to-embryonic transition during SE in *Arabidopsis*, numerous studies have elucidated its functions during SE in several plant species [30]. During preparation of the current report, *AtWus* was shown to promote SE differentiation in *G. hirsutum* Coker cotton [31], a cultivar that easily undergoes SE. However, it is unknown if these results apply to other cotton genotypes. Furthermore, little work has been done on the regulatory role of *Wus* in the auxin signaling pathway during SE.

Here, we examined the role and molecular mechanism of *Wus*-promoted SE in *G. hirsutum*. Cultivar CRI12, a genotype that is difficult to regenerate. We constitutively overexpressed *AtWus* in CRI12 using an *Agrobacterium*-mediated transformation and found that *AtWus* can promote the formation of non-compact light green calli that induce EC easily. The EC differentiation rate was higher in 35S:WUS of CRI12 (47.75%) in comparison with <1% in CK. Hence, *AtWus* is a possible candidate that could promote the nonembryonic-to-embryonic transition during SE in cotton.

Transcript analysis has revealed that the auxin transport and signaling pathways may play a substantive role in EC induction [13]. Our QPCR results showed that *GhPINs* and *GhSHY/GhARF* may regulate the efficiency of callus differentiation into EC. *PINs* play a fundamental role in embryonic auxin distribution in plant embryos [16], and *AtPIN1* is associated with the establishment of auxin gradients during SE in *Arabidopsis*. Previous studies revealed diverse changes in the endogenous auxin levels when EC was induced [13]. The antisense cDNA of *AtWus* suppresses *PINs* during SE in *Arabidopsis* [10]. When auxin levels are low, *SHY* (*AUX/IAA*) expression is induced and certain *ARF* activity is repressed, but at high auxin levels SHY is degraded by SCF$^{TIR1/AFBs}$ and the *ARF* inhibitory action is abolished [17,19]. Our results revealed enhanced *GhSHY2* expression and repressed *GhARF3* expression in CRI12 calli overexpressing *AtWus*. However, *AtWus* is unable to modify the amount of auxin in cotton calli [31]. Hence, *AtWus* may play an important role in upregulating *GhPINs* to redistribute auxin gradients, which may alter expression patterns of *SHY-ARF* at low levels of auxin.

LEC1, *LEC2*, and *FUS3* are crucial for SE. The capacity of SE is almost completely repressed in double and triple mutants of the three genes, indicating that *LEC* genes may function downstream of endogenous auxin-induced SE in *Arabidopsis* [21,23]. In our present study, *GhLEC1*, *GhLEC2*, and *GhFUS3* transcript levels were extremely low in callus that was unable to differentiate into EC, but levels were high in callus producing EC. Hence, *AtWus* positively regulated *LECs* and *FUS3*. In *Arabidopsis*, WUSCHEL and PGA37/MYB118 promote SE and activate the expression of *LEC1*, *LEC2*, and *FUS3* [32]. In our study, *GhLEC1*, *GhLEC2*, and *GhFUS3* were also upregulated in callus of 35S:WUS lines. These results suggest that *AtWus* may alter *PIN* expression, which leads to the establishment of new auxin gradients in the callus. Subsequently, a new auxin response was formed and stimulated *GhLEC1*, *GhLEC2*, and *GhFUS3* in the callus of CRI12. *AtWus* may provoke the ability of differentiation in the callus by reactivating *GhLECs* and *GhFUS3* expression through auxin transport and signaling mechanisms.

Although *AtWus* improved EC induction, the observed abnormal somatic embryos were an unexpected consequence of *AtWus* overexpression and prevented seedling generation. *AtWus* expression is limited to the SAM because auxin accumulates in the cotyledon primordial cells during somatic embryo development in *Arabidopsis* [10]. Therefore, ectopic expression of *Wus* could cause loss of expression specificity in somatic embryos, leading to asymmetric growth. In *Arabidopsis*, *LEC1* and *FUS3* may control multiple aspects of seed development [33]. Constitutive expression of *LEC1* leads to occasional formation of somatic embryo like structures [34]. Therefore, constitutive expression of *AtWus* in CRI12 may have led to constitutive expression of *GhLEC1*, *GhLEC2* and *GhFUS3* in embryos, and then the expression of

GhLECs and *GhFUS3* may have resulted in the observed abnormal embryos and failure of seedling regeneration. Using inducible promoters such as the estradiol-inducible promoter [35] rather than the 35S promoter during SE in cotton may avoid abnormal embryo formation. Estradiol could be added into the CIM and EIM for EC induction with subsequent transfer of ECs onto SIM without estradiol for normal embryo development. This promoter has been successfully applied for SE in several species [30].

Cotton is an important source of textile fiber and edible oil, but cotton yield is adversely affected by abiotic or biotic stresses [36]. Therefore, efforts to improve cotton resistance against such stresses by genetic modification may play a vital role in efforts to increase production. *Agrobacterium*-mediated transformation via SE has been the most popular transgenic technology in cotton. Most genotypes cannot undergo EC induction or have low rates of differentiation, although many of those recalcitrant genotypes have certain positive agronomic characters [37]. Thus, the difficulty of EC induction always restricts the application of transgenic breeding and *in vitro* regeneration in additional cultivars. For example, CRI12 used to be an important elite cultivar widely cultured in China for its disease resistance, high yeild and superior fiber quality. However, SE production in this cultivar is not easy, making it difficult to improve traits using transgenic technology. In our study, the introduction of *AtWus* into the recalcitrant cotton cultivars enhanced their somatic embryogenesis. With this foundation established, we may now construct a vector to overexpress *AtWus* with an estradiol-inducible promoter and transfer it into CRI12 or other cultivars. This will require a simple addition of estradiol to CIM and EIM to ensure EC induction with high frequency and production of transgenic seedlings. This will then enable us to improve the rate of cotton transformation for many foreign genes or cotton genes. Such transgenic plants can be used directly as germplasm for cotton breeding. This protocol will achieve our goals of creating more germplasm resources that facilitate SE and expand the scope of transgenic breeding in more cultivars.

Supporting Information

Figure S1   Characterization of *GhLEC1*, *GhLEC2* and *GhFUS3*.

Table S1   Specific primers for cloning *GhLEC1*, *GhLEC2*, *GhFUS3* and *AtWuschel*.

Table S2   Primers used for quantitative real time PCR.

Acknowledgments

We are grateful to Lihua Ma and Tianping Suo (Cotton Research Institute, Chinese Academy of Agricultural Sciences) for their technical assistance in scanning electronic microscopy. We thank Dr. Jianru Zuo (State Key Laboratory of Plant Genomics, Institute of Genetics and Developmental Biology, Chinese Academy of Sciences) for helpful suggestions and discussions.

Author Contributions

Conceived and designed the experiments: Fuguang Li CZ WZ. Performed the experiments: WZ. Analyzed the data: WZ Fuguang Li CZ JW XZ ZY. Contributed reagents/materials/analysis tools: CZ LD Fuguang Li CL Fenglian Li ZY. Wrote the paper: WZ. Check writing: LL JW.

References

1. Schmidt ED, Guzzo F, Toonen MA, de Vries SC (1997) A leucine-rich repeat containing receptor-like kinase marks somatic plant cells competent to form embryos. Development 124: 2049–2062.

2. Thomas TL (1993) Gene expression during plant embryogenesis and germination: an overview. Plant Cell 5: 1401.

3. Laux T, Mayer KF, Berger J, Jurgens G (1996) The WUSCHEL gene is required for shoot and floral meristem integrity in Arabidopsis. Development 122: 87–96.

4. Kamiya N, Nagasaki H, Morikami A, Sato Y, Matsuoka M (2003) Isolation and characterization of a rice WUSCHEL-type homeobox gene that is specifically expressed in the central cells of a quiescent center in the root apical meristem. Plant J 35: 429–441.

5. Yadav RK, Reddy GV (2011) WUSCHEL-mediated cellular feedback network imparts robustness to stem cell homeostasis. Plant Signal Behav 6: 544–546.

6. Yadav RK, Perales M, Gruel J, Girke T, Jonsson H, et al. (2011) WUSCHEL protein movement mediates stem cell homeostasis in the Arabidopsis shoot apex. Genes Dev 25: 2025–2030.

7. Yadav RK, Tavakkoli M, Reddy GV (2010) WUSCHEL mediates stem cell homeostasis by regulating stem cell number and patterns of cell division and differentiation of stem cell progenitors. Development 137: 3581–3589.

8. Lenhard M, Bohnert A, Jurgens G, Laux T (2001) Termination of stem cell maintenance in Arabidopsis floral meristems by interactions between WUSCHEL and AGAMOUS. Cell 105: 805–814.

9. Zuo J, Niu QW, Frugis G, Chua NH (2002) The WUSCHEL gene promotes vegetative-to-embryonic transition in Arabidopsis. Plant J 30: 349–359.

10. Su YH, Zhao XY, Liu YB, Zhang CL, O'Neill SD, et al. (2009) Auxin-induced WUS expression is essential for embryonic stem cell renewal during somatic embryogenesis in Arabidopsis. The Plant Journal 59: 448–460.

11. Arroyo-Herrera A, Gonzalez AK, Moo RC, Quiroz-Figueroa FR, Loyola-Vargas VM, et al. (2008) Expression of WUSCHEL in Coffea canephora causes ectopic morphogenesis and increases somatic embryogenesis. Plant Cell Tissue and Organ Culture 94: 171–180.

12. Ikeda-Iwai M, Satoh S, Kamada H (2002) Establishment of a reproducible tissue culture system for the induction of Arabidopsis somatic embryos. J Exp Bot 53: 1575–1580.

13. Yang X, Zhang X, Yuan D, Jin F, Zhang Y, et al. (2012) Transcript profiling reveals complex auxin signalling pathway and transcription regulation involved in dedifferentiation and redifferentiation during somatic embryogenesis in cotton. BMC Plant Biol 12: 110.

14. Wisniewska J, Xu J, Seifertova D, Brewer PB, Ruzicka K, et al. (2006) Polar PIN localization directs auxin flow in plants. Science 312: 883.

15. Zazimalova E, Krecek P, Skupa P, Hoyerova K, Petrasek J (2007) Polar transport of the plant hormone auxin - the role of PIN-FORMED (PIN) proteins. Cell Mol Life Sci 64: 1621–1637.

16. Weijers D, Sauer M, Meurette O, Friml J, Ljung K, et al. (2005) Maintenance of embryonic auxin distribution for apical-basal patterning by PIN-FORMED-dependent auxin transport in Arabidopsis. Plant Cell 17: 2517–2526.

17. Goh T, Kasahara H, Mimura T, Kamiya Y, Fukaki H (2012) Multiple AUX/IAA-ARF modules regulate lateral root formation: the role of Arabidopsis SHY2/IAA3-mediated auxin signalling. Philos Trans R Soc Lond B Biol Sci 367: 1461–1468.

18. Sauer M, Balla J, Luschnig C, Wisniewska J, Reinohl V, et al. (2006) Canalization of auxin flow by Aux/IAA-ARF-dependent feedback regulation of PIN polarity. Genes Dev 20: 2902–2911.

19. Weijers D, Benkova E, Jager KE, Schlereth A, Hamann T, et al. (2005) Developmental specificity of auxin response by pairs of ARF and Aux/IAA transcriptional regulators. EMBO J 24: 1874–1885.

20. Fambrini M, Durante C, Cionini G, Geri C, Giorgetti L, et al. (2006) Characterization of LEAFY COTYLEDON1-LIKE gene in Helianthus annuus

and its relationship with zygotic and somatic embryogenesis. Development genes and evolution 216: 253–264.

21. Gaj MD, Zhang S, Harada JJ, Lemaux PG (2005) Leafy cotyledon genes are essential for induction of somatic embryogenesis of Arabidopsis. Planta 222: 977–988.

22. Stone SL, Braybrook SA, Paula SL, Kwong LW, Meuser J, et al. (2008) Arabidopsis LEAFY COTYLEDON2 induces maturation traits and auxin activity: implications for somatic embryogenesis. Proceedings of the National Academy of Sciences 105: 3151–3156.

23. Stone SL, Kwong LW, Yee KM, Pelletier J, Lepiniec L, et al. (2001) LEAFY COTYLEDON2 encodes a B3 domain transcription factor that induces embryo development. Proceedings of the National Academy of Sciences 98: 11806–11811.

24. Hu L, Yang X, Yuan D, Zeng F, Zhang X (2011) GhHmgB3 deficiency deregulates proliferation and differentiation of cells during somatic embryogenesis in cotton. Plant Biotechnol J 9: 1038–1048.

25. Jin S, Zhang X, Nie Y, Guo X, Huang C (2005) Factors affecting transformation efficiency of embryogenic callus of upland cotton (Gossypium hirsutum) with Agrobacterium tumefaciens. Plant cell, tissue and organ culture 81: 229–237.

26. Wan C-Y, Wilkins TA (1994) A Modified Hot Borate Method Significantly Enhances the Yield of High-Quality RNA from Cotton (Gossypium hirsutum L.). Analytical biochemistry 223: 7–12.

27. Wang K, Wang Z, Li F, Ye W, Wang J, et al. (2012) The draft genome of a diploid cotton Gossypium raimondii. Nat Genet 44: 1098–1103.

28. Xu Z, Zhang C, Zhang X, Liu C, Wu Z, et al. (2013) Transcriptome Profiling Reveals Auxin and Cytokinin Regulating Somatic Embryogenesis in Different Sister Lines of Cotton Cultivar CCRI24. J Integr Plant Biol.

29. Mayer KFX, Schoof H, Haecker A, Lenhard M, Jurgens G, et al. (1998) Role of WUSCHEL in regulating stem cell fate in the Arabidopsis shoot meristem. Cell 95: 805–815.

30. Solís-Ramos LY, González-Estrada T, Nahuath-Dzib S, Zapata-Rodriguez LC, Castaño E (2009) Overexpression of WUSCHEL in C. chinense causes ectopic morphogenesis. Plant Cell, Tissue and Organ Culture (PCTOC) 96: 279–287.

31. Bouchabke-Coussa O, Obellianne M, Linderme D, Montes E, Maia-Grondard A, et al. (2013) Wuschel overexpression promotes somatic embryogenesis and induces organogenesis in cotton (Gossypium hirsutum L.) tissues cultured in vitro. Plant Cell Rep 32: 675–686.

32. Wang X, Niu QW, Teng C, Li C, Mu J, et al. (2009) Overexpression of PGA37/MYB118 and MYB115 promotes vegetative-to-embryonic transition in Arabidopsis. Cell Res 19: 224–235.

33. Parcy F, Valon C, Kohara A, Miséra S, Giraudat J (1997) The ABSCISIC ACID-INSENSITIVE3, FUSCA3, and LEAFY COTYLEDON1 loci act in concert to control multiple aspects of Arabidopsis seed development. The Plant Cell Online 9: 1265–1277.

34. Lotan T, Ohto M-a, Yee KM, West MA, Lo R, et al. (1998) Arabidopsis LEAFY COTYLEDON1 Is Sufficient to Induce Embryo Development in Vegetative Cells. Cell 93: 1195–1205.

35. Zuo J, Niu QW, Chua NH (2000) Technical advance: An estrogen receptor-based transactivator XVE mediates highly inducible gene expression in transgenic plants. Plant J 24: 265–273.

36. Leelavathi S, Sunnichan V, Kumria R, Vijaykanth G, Bhatnagar R, et al. (2004) A simple and rapid Agrobacterium-mediated transformation protocol for cotton (Gossypium hirsutum L.): embryogenic calli as a source to generate large numbers of transgenic plants. Plant Cell Rep 22: 465–470.

37. Wu J, Zhang X, Nie Y, Jin S, Liang S (2004) Factors affecting somatic embryogenesis and plant regeneration from a range of recalcitrant genotypes of Chinese cottons (Gossypium hirsutum L.). In Vitro Cellular & Developmental Biology-Plant 40: 371–375.

5

Comparative Proteomic Analysis of Embryos between a Maize Hybrid and Its Parental Lines during Early Stages of Seed Germination

Baojian Guo[1,2,9], **Yanhong Chen**[1,2,9], **Guiping Zhang**[1,2], **Jiewen Xing**[1,2], **Zhaorong Hu**[1,2], **Wanjun Feng**[1,2], **Yingyin Yao**[1,2], **Huiru Peng**[1,2], **Jinkun Du**[1,2], **Yirong Zhang**[1,2], **Zhongfu Ni**[1,2]*, **Qixin Sun**[1,2]*

1 State Key Laboratory for Agrobiotechnology and Key Laboratory of Crop Heterosis and Utilization (MOE), Beijing Key Laboratory of Crop Genetic Improvement, China Agricultural University, Beijing, China, 2 National Plant Gene Research Centre (Beijing), Beijing, China

Abstract

In spite of commercial use of heterosis in agriculture, the molecular basis of heterosis is poorly understood. It was observed that maize hybrid Zong3/87-1 exhibited an earlier onset or heterosis in radicle emergence. To get insights into the underlying mechanism of heterosis in radicle emergence, differential proteomic analysis between hybrid and its parental lines was performed. In total, the number of differentially expressed protein spots between hybrid and its parental lines in dry and 24 h imbibed seed embryos were 134 and 191, respectively, among which 47.01% (63/134) and 34.55% (66/191) protein spots displayed nonadditively expressed pattern. Remarkably, 54.55% of nonadditively accumulated proteins in 24 h imbibed seed embryos displayed above or equal to the level of the higher parent patterns. Moreover, 155 differentially expressed protein spots were identified, which were grouped into eight functional classes, including transcription & translation, energy & metabolism, signal transduction, disease & defense, storage protein, transposable element, cell growth & division and unclassified proteins. In addition, one of the upregulated proteins in F_1 hybrids was ZmACT2, a homolog of *Arabidopsis thaliana* ACT7 (AtACT7). Expressing *ZmACT2* driven by the *AtACT7* promoter partially complemented the low germination phenotype in the *Atact7* mutant. These results indicated that hybridization between two parental lines can cause changes in the expression of a variety of proteins, and it is concluded that the altered pattern of gene expression at translational level in the hybrid may be responsible for the observed heterosis.

Editor: Jinfa Zhang, New Mexico State University, United States of America

Funding: This work was financially supported by the National Basic Research Program of China (973 Program) (2012CB910900), the National Natural Science Foundation of China (31230054 and 30925023) and 863 Project of China (2012AA10A305) (http://www.sciencedirect.com/science/article/pii/S037811912004131#gts0005). The funders had no role in study design, data collection and analysis, decision to publish, or preparation of the manuscript.

Competing Interests: The authors have declared that no competing interests exist.

* E-mail: nzhf2002@aliyun.com (ZN); qxsun@cau.edu.cn (QS)

9 These authors contributed equally to this work.

Introduction

Heterosis or hybrid vigor was defined as the advantage of hybrid performance over its parents in terms of biomass, size, yield, speeds of development, fertility, resistance to biotic and abiotic stresses. However, the genetic and molecular basis of heterosis remains enigmatic [1]. At the genomic level, a significant loss of colinearity at many loci between different inbred lines of maize was observed [2]. At the level of gene expression, complex transcriptional networks specific for different developmental stages and tissues in maize, wheat, rice and *Arabidopsis* were monitored, and these results indicated that hybridization between two parental lines could cause expression changes of different genes, which might be responsible for the observed heterosis [3–10]. Although transcriptomic analysis have contributed greatly to our understanding of the heterosis, changes on the level of mRNA do not necessarily indicate changes on the protein level, thus studies are needed to determine the differential proteomes between hybrids and its parental lines, and understand their functional roles in heterosis.

In fact, as early as 1970s, several investigators had estimated the correlations between isozyme allelic diversity and grain yield of single-cross maize hybrid [11–13]. Lately, two-dimensional gel electrophoresis was employed to determine correlations between polymorphism of individual protein amounts indices and hybrid vigor for agronomic traits [14–18]. For example, comparative proteomic analysis of 25 and 35 DAP (days after pollination) seed embryos of maize reciprocal F_1-hybrids and their parental inbred lines revealed that 141 proteins exhibited nonadditive accumulation in at least one hybrid and approximately 44% of differently expressed proteins displayed low-dominant in hybrid [17]. Taken together, these observations at translational level add circumstantial evidence that expression differences between hybrid and its parental lines exist not only at mRNA levels, but also at protein abundances.

In maize, F_1 hybrid seeds have a superior germination capacity as compared to their parental inbred lines, and hybrid vigor in maize is detectable at early stages of germination [19]. Studies have also indicated that vigorous growth of the embryonic axis in germinating F_1 seed is related to a higher rate of RNA and protein synthesis [20]. Recently, proteomic analysis of heterosis during maize seed germination was analyzed and 257, 363, 351, 242, and 244 nonadditively expressed proteins were identified in hybrids Zhengdan 958, Nongda 108, Yuyu 22, Xundan 20, and Xundan 18 corresponding to their parents, respectively. Additionally, 54 different proteins were identified and the most interesting were those involved in germination-related hormone signal transduction, including abscisic acid and gibberellin regulation networks [21]. In the present study, high-throughput two-dimensional gel electrophoresis (2-DE) was used to establish the differentially expressed protein profiles in dry and 24 h imbibed embryos of maize hybrid Zong3/87-1 and its parental lines and differentially expressed proteins were further identified by MALDI TOF MS, with the purpose to gain insights into the molecular basis of maize heterosis of radicle emergence.

Materials and Methods

Plant Materials and Total Protein Extraction from Embryo

One highly heterotic hybrid Zong3/87-1 and its female parent Zong3 and male parent 87-1 were selected for this study. Germination efficiency was determined in three replicates. For each replicate, 100 seeds were placed embryo side down on two pieces of Whatman Grade No. 1 filter paper and placed in a plastic Petri dish, which were incubated at 28°C in 24 h dark. Seeds were considered germinated when radicle protrusion was visible. We recorded the germination rate at 12, 24, 36, 48 and 60 h after imbibitions in terms of radicle emergence, and the changes of 100 seeds weight were recorded during the germination process. Statistical analysis of the differences in traits was performed by using t-test.

The dry and 24 h imbibed embryos were detached from seeds, respectively. Remarkably, since germination of a batch of maize seeds was not strictly synchronous, 24 h imbibed embryos of each genotype with no radicle emergence were dissected and mixed for protein extraction and three replicates were performed. Twenty embryos were pooled as one biological replicate for protein extraction, and then frozen in liquid nitrogen stored at -80°C before use. Total protein was isolated from embryos using Invitrogen's TRIZOL® reagent according to the manufacturer's instruction. Protein concentration was determined by Ramagli assay [22].

2-DE and Image Analysis

The seed embryo proteins in the dried powder were solubilized in 7 M urea, 2 M thiourea, 2% CHAPS (powder to solution, w/v), 0.5% IPG buffer (v/v) (pH 4–7) (GE Healthcare, USA) and 36 mM DTT (5.6 mg/mL) via incubation at room temperature for 1 h, vortexing every 10 min, the mixture was then centrifuged (15 000 rpm) for 15 min, and the supernatant was collected. Total protein extract (500 μg) was loaded onto GE Healthcare 24 cm IPG gel strips (pH 4–7) during strip rehydration overnight. IEF of the acidic range IPG strips (pH 4–7) were conducted using IPGPhor II (GE Healthcare, USA) at 20°C for a total of 65 kVh. The IPG strips were equilibrated according to the manufacturer (GE Healthcare, USA). The second dimension SDS-PAGE gels (12.5% linear gradient) were run on an Ettan Daltsix (GE Healthcare, USA), 0.5 h at 2.5 W per gel, then at 15 W per gel until the dye front reached the gel bottom. Upon electrophoresis, the protein spots were stained with silver nitrate according to the instruction of protein PlusOne™ Silver Staining Kit (GE Healthcare, USA), which offered improved compatibility with subsequent mass spectrometric analysis. Briefly, gels were fixed in 40% ethanol and 10% acetic acid for 30 min, and then sensitized

with 30% ethanol, 0.2% sodium thiosulfate (w/v) and 6.8% sodium acetate (w/v) for 30 min. Then gels were rinsed with distilled water three times, five minutes for each time, then incubated in silver nitrate (2.5 g/L) for 20 min. Incubated gels were rinsed with distilled water and developed in a solution of sodium carbonate (25 g/L) with formaldehyde (37%, w/v) added (300 μL/L) before use. Development was stopped with 1.46% EDTA-Na$_2$·2H$_2$O (w/v), and gels were stored in distilled water until they could be processed and the reproducible spots removed from them. Gel images were acquired using Labscan (GE Healthcare, USA) at 300 dpi resolution. Image was analyzed with Imagemaster 2D Platinum Software Version 5.0 (GE Healthcare, USA). Spot detection was performed with the parameters smooth, minimum area and saliency set to 2, 15 and 8, respectively, and was done automated by the software used, followed by manual spot editing, such as artificial spot deletion, spot splitting and merging. The experiments were carried out in triplicate. Only those protein spots that could be detected in all three replicated gels were considered for further analysis.

Protein Expression Patterns Analysis

All the gels were matched to the reference gel in automated mode, combined with manual pair correction. The volume of each spot from three replicate gels was normalized and quantified against total spot volume. In this analysis the parental lines midparent value (the average value of the parental inbred lines) was calculated for all three replicates and compared with the hybrid expression for each of the three replicates. Only those protein spots with the fold changes more than 1.5 and significant at $p < 0.05$ were considered as differentially expressed protein between hybrid and its parental lines. Among which any deviation from the midparent value was considered nonadditive expressed proteins, others were consider as additive protein. According to the system suggested by Hoecker et al., the nonadditive proteins were further grouped into the following possible modes [16], including (i) above high parent expression (++), the expression in the hybrid significantly exceeded both parental inbred lines; (ii) below low parent expression (–), the expression in the hybrid was significantly lower than in both parents; (iii) high parent expression (+), expression in hybrid is equal to the highly expressed parent, but significantly difference from low expressed parent; (iv) low parent (-), expression in hybrid is equal to the lowly expressed parent, but significantly distinct from the highly expressed parent; (v) partial dominance (+/−), the expression abundance of the hybrid protein is significantly higher than the lower performing parent and significantly lower than the higher performing parent; (vi) different (D), displayed an expression pattern that was not fit in any of the above expression pattern.

In-gel Digestion

Spots of varied intensities were excised manually and transferred to 1.5 mL microcentrifuge tubes. The protein spots of lower abundance were removed from all the replicate gels, pooled and digested in a single tube. Protein spots were destained twice with 30 mM potassium ferricyanide and 100 mM sodium thiosulfate, and then were rinsed with 25 mM ammonium bicarbonate in 50% acetonitrile. Gel pieces were dehydrated with 100% acetonitrile, dried under vacuum and incubated for 16 h at 37°C with 10 μL of 10 ng/μL trypsin in 25 mM ammonium bicarbonate. The resulting tryptic fragments were eluted by diffusion into 50% v/v acetonitrile and 0.5% v/v trifluoroacetic acid.

Mass Spectrometry

Protein MS was conducted using AUTOFLEX II TOF-TOF (Bruker Daltonics, Germany). Digested protein samples (70% v/v acetonitrile and 0.5% v/v trifluoroacetic) were spotted on an AnchorChipTM plate (1.0 μL) and recrystallized CHCA matrix (Bruker Daltonics, Germany) dissolved in 0.1% TFA/70% ACN (0.5 μL). External standards from the manufacturer dissolved in the same matrix solution and spotted on the fixed positions labeled on the plate. Each sample spot was desalted with 0.01% TFA, and dried completely. The peptide ions generated by autolysis of trypsin (with m/z 2163.333 and 2273.434) were used as internal standards for calibration. The list of peptide masses from each PMF was saved for database analysis.

Data Analysis

Monoisotopic peptide masses generated from the PMFs were analyzed with Auto Flexanalysis (Bruker Daltonics, Germany) and were searched using using MASCOT distiller 2.2 software (http://www.matrixscience.com/). Matches to protein sequences from the Viridiplantae taxon in NCBInr or MSDB database were considered acceptable if : 1) A MOWSE score was obtained from MASCOT, which rates scores as significant if they are above the 95% significance threshold ($p < 0.05$); 2) At least four different predicted peptide masses matched the observed masses for an identification to be considered valid; 3) The coverage of protein sequences by the matching peptides should be higher than 10%; 4) a peptide mass tolerance of 25 ppm; 5) a parent ion mass tolerance of 0.2 Da [23,24]. In addition, some proteins successfully identified have substantial discrepancies between the experimental and calculated pI and Mr, which could be caused by numerous factors such as posttranslational modification (PTM), polymeric forms of proteins, proteolytic degradation of proteins, matches to proteins from different organisms, or genomic sequence, which could contain segments that are spliced out of the functional protein. Such protein identifications were deemed acceptable as long as the other statistical criteria were met.

Western Blot Analysis

Total protein extracts from maize dry/24 h imbibed embryos and 48 h germinating *Arabidopsis* seed. The protein was resolved by SDS-PAGE imprinted on membranes, and incubated with monoclonal actin-antibodies MabGpa (Sigma) that recognize different subsets of actin isovariants. The anti-actin monoclonal antibody covalently binds with appropriate horseradish peroxidase-conjugated secondary antibodies (Sigma). Quantification of actin proteins was conducted using ECL kit (Sigma) and X-ray films with short exposure times. Experiments repeated at least twice, coomassie brilliant blue stained gels demonstrate approximate equal loading of protein samples as loading control in maize. In addition, histone H3 was loading control in *Arabidopsis*.

Isolation of Total RNA and Reverse Transcription

Total RNA was isolated from 24 h imbibed embryos of maize inbred line Zong3 and using a standard Trizol RNA isolation protocol (Life Technologies, USA), and treated with DNase (Promega Corporation, USA) following the manufacturer's instructions. The amount and quality of the total RNA was checked through electrophoresis in 1% agarose gel. The concentration of RNA was measured by spectrophotometer NanoDrop, ND1000 (Nano Drop Technologies). Equal amounts of 2 μg total RNA was reverse transcribed to cDNA in 20 μL reaction using M-MLV reverse transcriptase (Promega Corporation, USA). Reverse transcription was performed for 60 min at 37°C with a final denaturation step at 95°C for 5 min. Aliquots of 2 μL of the obtained cDNA was subjected to RT-PCR analysis.

Generation of *Arabidopsis* Transgenic Plants

The full-length coding sequence of *ZmACT2* (GRMZM2G006765) was cloned into the plant transformation vector pEGAD between *BsrGI* and *EcoRI* sites. After that, a 1992 bp promoter sequence of *Arabidopsis AtACT7* gene was ligated into the *pEGAD-ZmACT2* between *PacI* and *AgeI* sites, in which *ZmACT2* was drived by *AtACT7P* promoter. The primer sequences were as follows: ZmACT2F: 5'-ACGAGCTGTACAATGGCT-GACGGCGAGGACAT-3', ZmACT2R: 5'-GGAAT -TCTTA-GAAGCACTTCCGATGAA-3' and AtACT7PF: 5'-CCTTAATTAAGATTAT -TTAAGTTGCCAACCAAGC-3', AtACT7PR: 5'-AGCGCTACCGGTCACTAAAA -AAAAAG-TAAAATGAAACCG-3'. Plasmid *AtACT7P::ZmACT2* was introduced into the Agrobacterium tumefaciens strain *GV3101* and used to transform *Atact7* (*Salk_131610*) *Arabidopsis* mutant plants using a floral dip method. Transgenic plants were selected by herbicide, and two representative T3 homozygous lines were obtained for seed germination analysis. Seeds were sterilized and plated on GM medium containing 0.9% agarose but no sucrose. After 3 days of stratification at 4°C in the dark, plates were incubated at 20°C in the darkness. We recorded the germination rate at 12, 24, 36, 48 and 60 h after germination. Experiments were done in triplicate with 70 seeds for each experiment and genotype. Statistical analyses of the differences in germination rate were performed using *t*-test.

Results

Heterosis of Radicle Emergence in Maize Hybrid Zong3/87-1

Growth patterns of germinating seeds in a maize hybrid and its parental lines were shown in Figure 1 and Table S1. It can be noted that, after 24 h of imbibitions at 28°C, 27% of hybrid seeds showed visible radicle emergence, but the percentage of germinated seeds for parental lines Zong3 and 87-1 were only 4% and 0%, respectively. This superiority of hybrid Zong3/87-1 was maintained throughout 48 h when all the seeds were fully germinated, indicating that hybrid Zong3/87-1 exhibited an earlier onset or heterosis in radicle emergence.

Previous study indicated that small kernel weight exhibited a faster and a more efficient germination in maize [25]. In this study, one hundred dry grains weight (HGW) per genotype were record also. The HGW of hybrid Zong3/87-1 (32.15±0.41 g) did not show any significant difference with respect to its female parent Zong3 (32.05±0.36 g), but lower than that of male parent 87-1 (36.93±0.24 g) ($p < 0.05$). Therefore, it can be concluded that kernel weight was not related to our observed heterosis in terms of germination rates.

Two-Dimensional Electrophoresis Analysis

In present study, dry and 24 h imbibed seed embryos of maize hybrid and its parental lines were used for high-throughput two-dimensional gel electrophoresis analysis. In total, 1140 and 1443 protein spots were reproducibly detected in dry and 24 h imbibed seed embryos of the three genotypes, respectively, among which 134 (11.75%) and 191 (13.24%) spots were found to be differentially expressed between hybrids and its parental lines (Table 1, Figure S1).

When comparing the patterns of differentially expressed protein spots between hybrid and midparent value (the average value of the parental inbred lines), it was found that a large proportion of

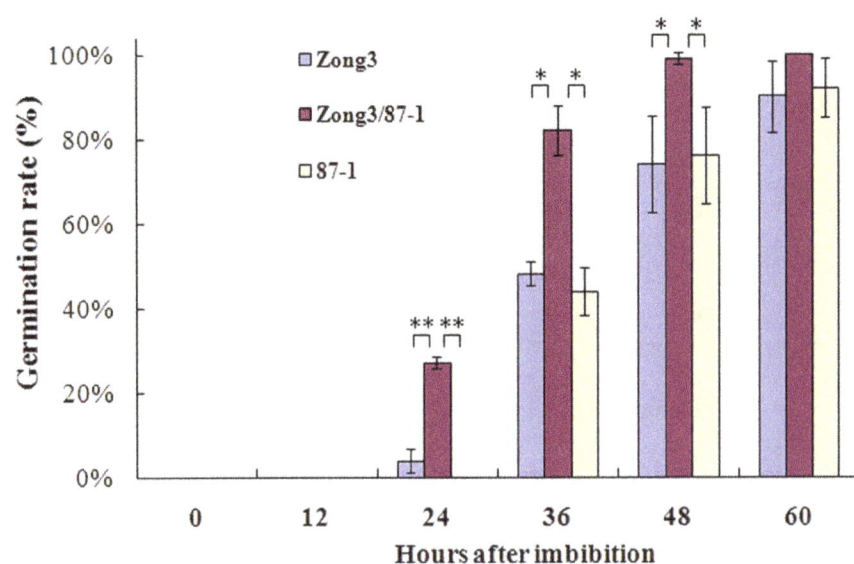

Figure 1. Seed germination rate of maize hybrid Zong3/87-1 and its parents. HAI, hours after imbibitions *, represent Significance level $p<0.05$; ** represent Significance level $p<0.01$.

proteins in dry embryos (71/134, 52.99%) and 24 h imbibed embryos (125/191, 65.45%) exhibited expression patterns that are not statistically distinguishable from additivity. Remarkably, although the number of nonadditive expressed protein spots between hybrid and its parental lines are very close in dry and 24 h imbibed embryos (63 and 66, respectively), significant differences were observed in patterns of nonadditively expressed proteins (Table 1). For example, the number of protein spots expressed above or equal to the level of the higher parent (++ and +) in 24 h imbibed embryos (36) was much higher than in dry embryos (18). On the contrary, 42 protein spots were found to be expressed in hybrid below or equal to the level of the lower parent proteins (− and −), which was twice more than in 24 h imbibed embryos (Table 1, Figure 2A).

For each genotype, comparison of dry and 24 h imbibed embryos 2-DE maps was also performed. It can be seen that the many proteins are differentially expressed between dry and 24 h imbibed embryos, and the number of differentially expressed protens in hybrid, Zong3 and 87-1 were 124, 115 and 110, respectively. Further analysis revealed that 68 protein spots existed

in 24 h imbibed embryos of hybrid, but they could not be detected in dry embryos of hybrid, which was much higher than that of its two parental lines (40 and 40) (Table S2).

Identification of Differentially Expressed Protein Spots

All differentially expressed protein spots between maize hybrid and its parental lines were eluted from representative 2-DE gels for identification and 155 protein spots were successfully identified, among which 69 and 86 were derived from dry and 24 h imbibed embryos, respectively (Table S3 and S4). Differential expression patterns of some identified protein spots between hybrids and its parents were shown in Figure 2A. According to the criteria used previously [26], these identified differentially expressed protein spots were classified into eight functional classes, including transcription & translation, energy & metabolism, signal transduction, disease & defense, storage protein, transposable element, cell growth & division, and unclassified proteins (Figure S2). Except for a high portion of unclassified proteins, the largest category in dry and 24 h imbibed embryos is transcription & translation (22% and 30%), followed by energy & metabolism

Table 1. The differentially expressed protein spots of additive and nonadditive between maize hybrid and its parental lines.

Stages[a]	No. of displayed protein spots	Differentially expressed pattern							Total
		Additive	Nonadditive						
			++[b]	+[c]	+/−[d]	D[e]	−[f]	− −[g]	Sum	
DS	1140	71	6	12	2	1	8	34	63	134
24 HAI	1443	125	18	18	4	5	9	12	66	191

[a]DS, dry seed; 24 HAI, 24 hours after imbibitions;
[b]++: above high parent expression;
[c]+: high parent expression;
[d]+/−: partial dominance expression;
[e]D: different from additivity (midpoint value), not belonging to any of the other classes;
[f]−: low parent expression;
[g]− −: Below low parent expression.

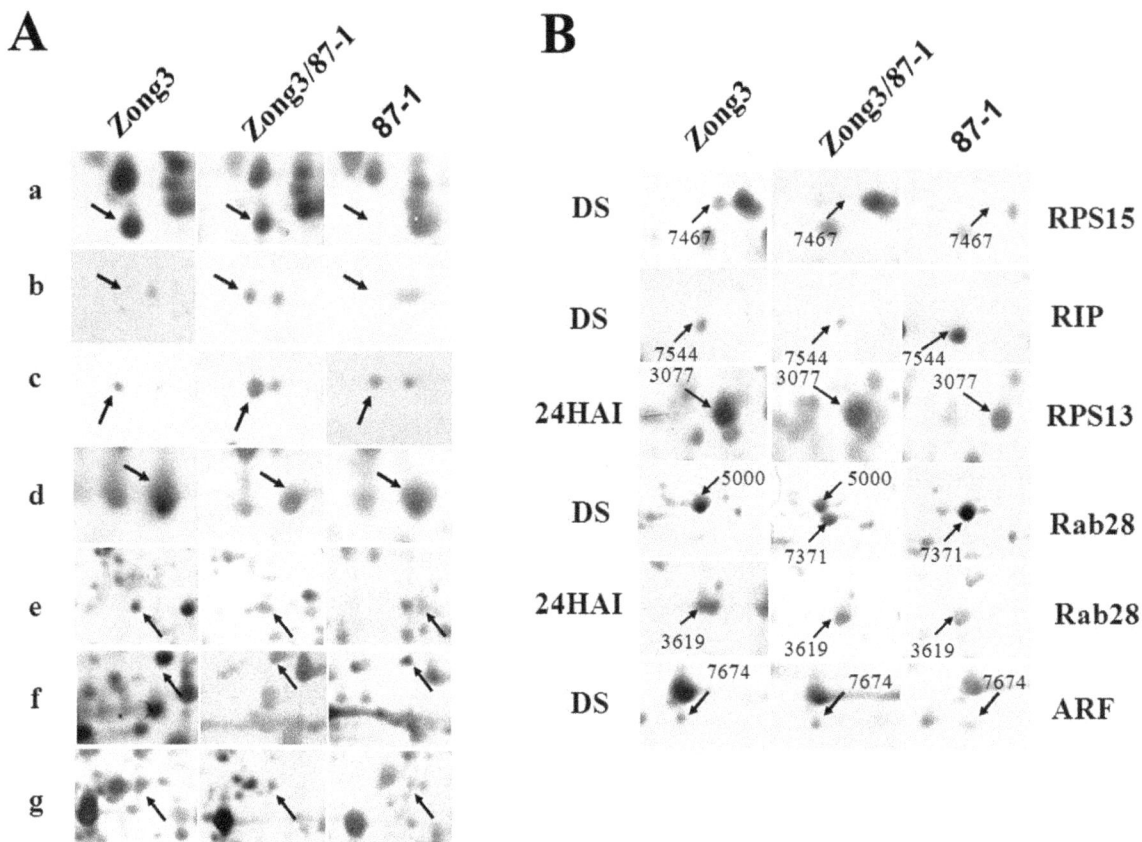

Figure 2. Differential protein expression patterns and some some identified protein spots between hybrids and its parents. A Differential protein expression patterns between hybrids and its parents in maize seed embryos. a) additive expression (spot 7262); b)++: above high parent expression (spot 7228); c)+: high parent expression (spot 7554); d)-: below low parent expression (spot 6686); e) –: low parent expression (spot 2992); f) +/−: partial dominance expression (spot 6547); g)D: different from additivity (midparent value), not belonging to any of the other classes (spot 2728); B. Differential expression patterns of some identified protein spots between hybrids and its parents. DS, dry seed; 24 HAI, 24 hours after imbibitions.

(16% and 19%), disease & defense (15% and 13%) and storage protein (10% and 17%). It was to be noted that a large proportion (9%) of the identified differentially expressed proteins in dry embryos of hybrid and its parental inbred lines was involved in signal transductin, whereas low proportion (2%) proteins in this category was identified in 24 h imbibed embryos.

Further analysis revealed that 155 protein spots identified in this study derived from 118 different genes or gene families, among which 20 proteins representing 39 protein isoforms were found in multiple spots. Additive expressed protein spots identified as protein disulfide isomerase (spot 2174, 2226, 2227, 3257, 6544 and 6562) and Rab28 (spot 5000 and 7371) were only expressed in one parent and hybrid, indicating dominant inheritance from Zong3 and 87-1, which might be a complementation in the hybrid. Some proteins exhibited nonadditive expression patterns, such as UDP-glucose pyrophosphorylase (spots 6675 and 6686). Moreover, globulin2 storage proteins (spots 2551, 3239, 3470, 3473, 7288, 7507, 7510 and 7533) cover two different expression patterns (additive and nonadditive), which made them most complicated. These protein spots could represent different modifications of the same gene product. Alternatively, they might also represent closely related members of gene families, which could not unambiguously identified. Strikingly, there were 17 differentially expressed protein spots were overlapped in dry and

24 h imbibed embryos, among which 4 protein spots displayed the same expression patterns (Table 2).

Complementation of *Arabidopsis Atact7* Allele with Maize Hybrid Upregulated Gene *ZmACT2*

In *Arabidopsis*, it has been reported that *Atact7* mutant display delayed and less efficient in germination [27]. In present study, we also found that one actin protein (designated as ZmACT2 hereafter) was upregulated in dry embryos of maize hybrid (Spot 7044, Figure 3A), which was selected for further analysis. Firstly, western blot analysis was performed on protein samples from dry and 24 h imbibed embryos of maize hybrid Zong3/87-1 and its parental lines. Equal protein loading was confirmed by staining replicate gels with Coomasie Brilliant Blue (Figure 3B and C). Although MAbGPa could recognize all actins, it clearly showed that the abundance of actin protein in dry embryos of maize hybrid was much higher than in its two parental lines, which confirmed the result of 2-DE; Secondly, in order to obtain the full-length cDNA corresponding to hybrid upregulated ZmACT2 protein, RT-PCR was performed with primer pair ZmACT2F/ ZmACT2R and open reading frame (ORF) was obtained, which encoded 377 amino acids. Comparison with amino acid sequences in GenBank revealed that the ZmACT2 protein was high similarity (97%) to *Arabidopsis* actin protein 7 (AtACT7) (Data

Table 2. Differentially expressed proteins overlapped in dry and 24 h imbibed embryos.

Spot Number (DS/24 HAI)[a]	Differential Expression Pattern[b] (DS/24 HAI)	Functions	Ac. Number
7543/3462	+/D	Predicted protein	gi|168056539
7544/3453	−/additive	Ribosome-inactivating protein	gi|18149181
7551/3479	additive/additive	Putative polyprotein	gi|57863895
7529/3460	+/additive	Vicilin-like embryo storage protein	gi|22284
7228/3008	++/- -	Unnamed protein product	gi|116058145
7163/3093	additive/++	Histone deacetylase superfamily (ISS)	gi|116055532
7458/3412	additive/+	Glutathione transferase30	gi|162458953
7510/3470	+/additive	Globulin 2	gi|228310
7455/3405	+/additive	Hypothetical protein At1g03160	Q67Z21_ARATH
7421/3448	+/additive	FZL	Q1KPV0_ARATH
7234/3179	additive/additive	Cyclo-DOPA 5-O-glucosyltransferase	Q59J81_MIRJA
7110/2992	additive/−	Hypothetical protein CHLREDRAFT_118990	gi|159474678
6674/2614	- -/additive	ATPase subunit 1	gi|94502565
7377/3257	additive/additive	Protein disulfide isomerase	gi|145666464
7473/3423	additive/−	Hypothetical protein CHLREDRAFT_167615	gi|159485762
7278/3223	additive/additive	Cinful1 polyprotein	Q7XBD4_MAIZE
7533/3546	+/additive	Globulin 2	gi|228310

[a]: DS, dry seed; 24 HAI, 24 hours after imbibitions.
[b]: ++: above high parent expression; +: high parent expression; D: different from additivity (midpoint value), not belonging to any of the other classes; −: low parent expression; - -: below low parent expression.

not shown); Finally, to take an insight into the possible function of *ZmACT2* gene in maize, cDNA of *ZmACT2* driven by the promoter of *Arabidopsis AtACT7* gene was transformed into *Atact7* mutant. Western blot analysis indicated that the total actin increased at least one fold in seedling of two T3 homozygous transgenic lines (*AtACT7P::ZmACT2#4* and *6*), as compared to *Atact7* mutant (Figure 4A). As shown in Figure 4B and C, after 36 h imbibitions at 20°C under dark condition, the percentage of germinated seeds for *AtACT7P::ZmACT2* transgenic lines was about 60%, which was lower than that of *Columbia* (80%), while only 20% of *Atact7* seeds showed visible radicle. Collectively, it can be concluded that *ZmACT2* was able to partially rescue the phenotype of the *Atact7* mutant in *Arabidopsis*.

Discussions

Significant Alterations in Protein Expression Occurred in Embryos of Hybrids and it Parental Lines During Seed Germination

In most flowering plants, seed germination is the first and may be the foremost growth stage in the plant's life cycle. Previous studies and experimental data demonstrated that maize hybrid F_1 seeds have a superior germination capacity in comparison with their parental lines, but the molecular basis for the heterosis in terms of radical emergence is unknown [20]. Recently, a proteomic study on mature embryos of rice hybrid and parents detected 150 differentially expressed protein spots [28]. In maize seed embryos of 24 h imbibitions, 257, 363, 351, 242, and 244 nonadditively expressed proteins were identified in hybrids Zhengdan 958, Nongda 108, Yuyu 22, Xundan 20, and Xundan 18 corresponding to their parents, respectively [21]. In the present study, the number of differentially expressed protein spots between hybrids and its parental lines in dry and 24 h imbibed seed

embryos were 134 and 191, respectively, among which 47.01% (63/134) and 34.55% (66/191) protein spots displayed nonadditively expressed pattern. These results indicated that significant alterations in protein level occurred in hybrids as compared to their parents, which might correlate with the observed heterosis.

During the last decades, with the advent of quantitative trait locus (QTL) mapping, three types of QTL interactions were proposed to contribute to heterosis in different studies, i.e., overdominance in maize [29], dominance in rice [30], and epistasis in rice [31]. More recently, however, Hua *et al.* proposed that all kinds of genetic effects, including partial dominance, complete dominance, and overdominance, as well as epistasis contributed to heterosis in a rice "immortalized F_2" population [32], indicating that these genetic effects were not mutually exclusive in the genetic basis of heterosis. At transcriptional level, all possible modes of gene action, including additivity, high- and low-parent dominance, underdominance, and overdominance, are observed in a global comparison of gene expression in F_1 hybrid and its inbred parents of maize, rice, and wheat [3,4,6,8]. In this study, 47.01% (63/134) and 34.55% (66/191) differentially expressed proteins were accumulated in a nonadditive fashion in the hybrid compared to the average of their parental inbred lines in dry and 24 h imbibed embryos, respectively. These nonadditively accumulated proteins can be grouped into six differential expression patterns, which were equivalent to overdominance (above high parent), underdominance (below low parent), dominance (high, low parent dominance and partial dominance). These findings are consistent with the hypothesis that multiple molecular models contribute to heterosis [3].

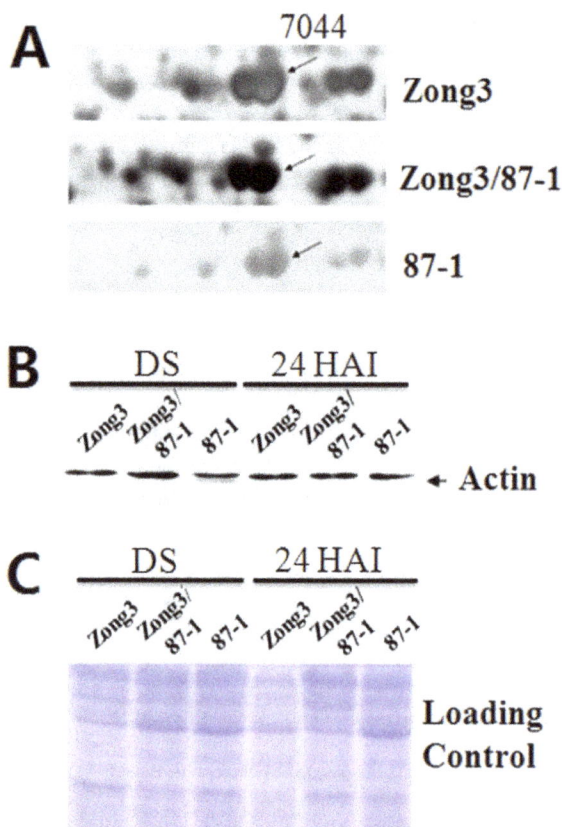

Figure 3. Analysis of actin expression during germination of maize dry and 24 h imbibed seeds embryos at the protein level shown by western immunoblotting with monoclonal antibodies MabGpa. A, The presence of actin isoforms in the maize embryo was analysed by 2-DE. B, Western immunoblotting detected the 42 kDa actin. Equivalent amounts of protein (20 µg) were loaded. C, Protein profile obtained on a 12.5% polyacrylamide gel stained with Coomassie Brilliant Blue. DS, dry seed; 24 HAI, 24 hours after imbibitions.

Expression Patterns between Maize Hybrid and its Parental Lines were Different in Dry and 24 h Imbibed Seed Embryos

Up to date, a number of studies have analyzed heterosis associated gene expression between maize hybrids and their parental inbred lines in different organs and developmental stages [3–5]. On the protein level, it was reported that 33% of the proteins detected via 2-DE in the hybrid have been more abundant or newly translated in comparison to the parental lines in maize primary root tip [20]. Recently, in a study of 3.5-day-old primary roots, 49% of the proteins detected in the hybrid accumulated in a nonadditive fashion compared to the average of its parental lines [16]. Similarly, in the reciprocal hybrids and their parental inbred lines 25 and 35 days after pollination, 141 of 597 detected proteins (24%) exhibited nonadditive accumulation in at least one hybrid [17]. In present study, 47.01% and 34.55% differentially expressed proteins were accumulated in a nonadditive fashion in dry and 24 h imbibed hybrid embryos, respectively. Moreover, significant differences were also observed in patterns of nonadditively expressed proteins. For instance, in 24 h imbibed embryos, 36 protein spots displayed above or equal to the level of the higher parent patterns, which is much higher than in dry embryos (18). On the contrary, more protein spots displayed below or equal to the level of the lower parent patterns were observed in

dry embryos (42) as compare to that in 24 h imbibed embryos (21). In addition, only 17 of 155 identified protein spots were overlapped in dry and 24 h imbibed embryos, among which 4 displayed same expression patterns. Collectively, our present data provide further evidence for the notion that nonadditive protein accumulation in maize depends on the analyzed developmental stage and plant organ [16].

Seed germination is a complicated physiological process, which starts with the uptake of water by the seed (imbibition) and ends when the embryonic axis starts to elongate and the radicle emerges [33]. Upon imbibition, the quiescent dry seed rapidly resumes metabolic activity, and respiration, enzymatic activity, RNA and protein synthesis are fundamental cellular activities reestablished during germination and are a prerequisite for seedling growth [34]. Based on our present data, it can be seen that 54.55% of nonadditively accumulated proteins in 24 h imbibed embryos displayed above or equal to the level of the higher parent patterns (++ and +). Considering the phenomenon that hybrids germinate earlier than its parental lines, it seems reasonable that more proteins are highly expressed in hybrid embryos during seed germination, but the underlying molecular mechanism responsible for these nonadditively expressed proteins is still an area for further investigated.

Proteins Involved in Transcription and Translation Overrepresented among Differentially Expressed Proteins

To further get an insight into the possible role of each individual differentially accumulated protein spots in heterosis, 155 protein spots were successfully identified from dry and 24 h imbibed embryos, which can be grouped into eight functional classes. Further analysis revealed that, except for unclassified proteins, the largest category is transcription & translation. Recently, in a study of the close *Arabidopsis* relative *Lepidium sativum*, it was found that both transcription and translation play an important role in seed germination [35]. Moreover, in our present study, it was found that significant alterations in protein expression occurred in dry and 24 h imbibed embryos of hybrid as compared to its parental lines. Thus, it was not surprising that a number of identified nonadditive protein spots were involved in gene transcription and protein translation. Although the precise roles of these identified proteins in heterosis are unknown, some have been demonstrated to be related to seed germination in maize and other plant species.

Firstly, the major plant hormones regulating seed germination in diverse plant species are GA and ABA, which promote and inhibit germination, respectively [36]. Abscisic acid (ABA) responsive protein Rab28 has been shown to be ABA-inducible in embryos. During normal germination, the amount of rab28 proteins decreased after 1 day and became undetectable after 2 days imbibed in water [37]. In the present study, three proteins (spots 3619, 5000 and 7371 (Figure 2B)) were identified as Rab28, especially spot 3619 was only highly expressed in 24 h imbibed embryos of parental line Zong3, but not in F_1. More recently, it has been shown that auxin also plays a critical role in seed germination, at least in part by mechanisms involving interactions between *ARF10*-dependent auxin and ABA pathways [36]. Interestingly, one auxin response factor protein (spot 7674, Figure 2B) was found to be highly expressed in dry seed embryos of hybrid F_1. However, at present it would be premature to determine how expression changes in genes involved in hormonal signaling transduction in hybrid might affect heterosis, these expression changes might be important to regulate down-stream gene expression in hybrids that affect heterosis in their turn. Consistent with this hypothesis, a comparative proteomic analysis between imbibed embryos of maize hybrids and their parental

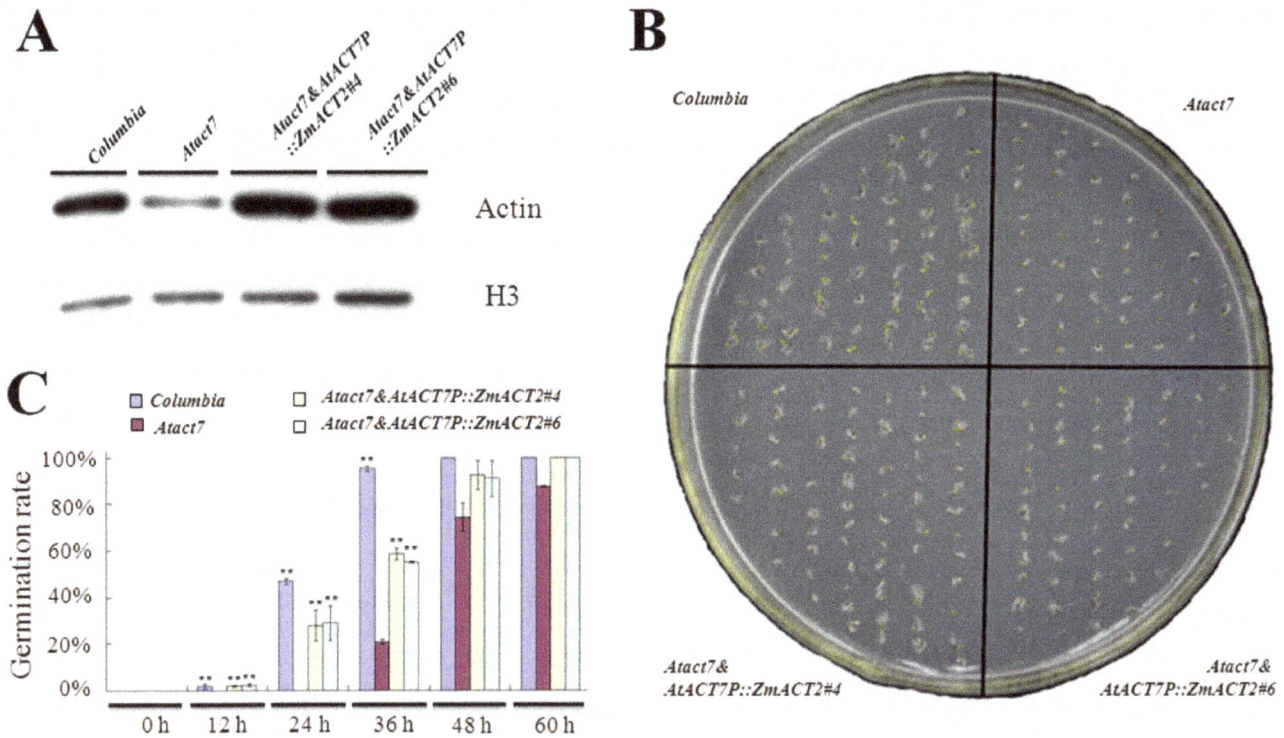

Figure 4. The germination phenotypes of the *Atact7* mutants and complementation of the *Atact7* allele with the maize gene *ZmACT2*.
A, Actin protein expression in the *Columbia*, *Atact7* mutant and complementation of the *Atact7* allele with the maize *ZmACT2*. The delay in germination is restored in *Atact7* mutants by the presence of *ZmACT2* as shown for two complemented lines (*Atact7&AtACT7P::ZmACT2#4* and *#6*). Total protein extracts from 48 h germinating seeds was resolved by SDS-PAGE and western immunoblotting detected the actin, hisone H3 as loading control. B, Seeds were stratified at 4°C in the dark for 48 h and incubated at 20°C with MS media for seed germination. C, Seed germination of *Arabidopsis thaliana Columbia*, *Atact7* mutants and complementation of the *Atact7* mutant with the maize gene *ZmACT2*. Germination percentage was counted with time course. Asterisks indicate statistically significant differences compared with the *Atact7* mutants. (**Represent Significance level $P<0.01$).

lines also exhibited that some differentially expressed proteins were involved with germination-related hormone signal transduction [21].

Secondly, ribosomal proteins are major components of ribosomes, and some appeared to be associated with plant development [38]. For example, knock-out of the plastid ribosomal protein S21 causes impaired photosynthesis and sugar-response during germination and seedling development in *Arabidopsis thaliana* [39]. Ribosome -inactivating proteins (RIPs) are a group of toxic proteins that can irreversibly inactivate ribosome for protein synthesis [40]. Previous studies suggested that early synthesized proteins were essential for germination and were programmed by a conserved polyadenylate-containing mRNA, preserved in dry embryos [41]. Therefore, it was not surprising that three ribosomal related proteins were identified to be differentially accumulated between hybrid and its parental lines. The abundance of one protein, which represented ribosomal proteins S13 (RPS13, Spot 3077, Figure 2B) was equal to that of the highly expressed parent line Zong3 in 24 h imbibed embryos, while ribosome-inactivating protein was detected to be down-regulated in dry embryos of hybrid F_1. Thus, the change in the expression profiles of these proteins may contribute to the observed vigorous growth of hybrid embryos. It should also be noted that one ribosomal proteins S15 (RPS15, Spot 7467, Figure 2B) were detected to be lowly expressed in dry embryos of hybrid F_1, the causal reason need further investigation.

Hybrid Upregluated ZmACT2 Partially Rescue the Delayed Germination Phenotype of the *Atact7* Mutant in *Arabidopsis*

As a major component of the plant cytoskeleton, actins are involved in several basic plant developmental processes including the establishment of cell polarity, cell division plane determination, cell wall deposition, and cell elongation [42], and play an essential role in processes like cytoplasmic streaming, organelle orientation and tip growth certain cells [43]. Most notably, in *Arabidopsis*, *Atact7* mutant displayed delayed and less efficient in germination [27]. In maize, it has been reported that maize actin isoforms expression at these early germination growth stages is a highly regulated event [44]. In present study, we found that the abundance of one protein, which represented ZmACT2 protein, was upregulated in dry embryos of maize hybrid. Further analysis revealed that genes encoding ZmACT2 showed high similarity to *Arabidopsis* AtACT7 proteins. Moreover, transgenic analysis exhibited that maize *ZmACT2* was able to partially rescue the phenotype of the *Atact7* mutant in *Arabidopsis*. Taken together, it can be concluded that the altered pattern of *ZmACT2* gene expression at translational level in the hybrid may be responsible for the observed heterosis. However, the underlying mechanism for the upregulation of ZmACT2 protein in maize hybrid is still an area to be further elucidated.

Conclusion

In this study, we demonstrated that maize hybrid Zong3/87-1 exhibited an earlier onset or heterosis of radicle emergence. At translational level, 11.75% and 191 13.24% were differentially expressed between hybrid Zong3/87-1 and its parental lines in dry and 24 h imbibed embryos, respectively, among which 47.01% (63/134) and 34.55% (66/191) differentially expressed proteins were accumulated in a nonadditive fashion. Remarkably, significant differences were observed for nonadditively expressed proteins between dry and 24 h imbibed embryos, which provided further evidence for the notion that nonadditive protein accumulation in maize depends on the analyzed developmental stage and plant organ. In addition, a total of 155 differentially expressed proteins were successfully identified, among which some have been demonstrated to be related to seed germination in maize and other plant species, including ABA responsive proteins, ribosomal related proteins and actin protein. Moreover, one of the upregulated proteins in F_1 hybrids was ZmACT2, a homolog of *Arabidopsis thaliana* ACT7 (AtACT7). Expressing *ZmACT2* driven by the *AtACT7* promoter partially complemented the low germination phenotype in the *Atact7* mutant. These results indicated that hybridization between two parental lines can cause changes in the expression of a variety of proteins, and it can be concluded that the altered pattern of gene expression at translational level in the hybrid may be responsible for the observed heterosis.

Supporting Information

Figure S1 Seed embryo 2-DE gel reference maps of maize hybrid Zong3/87-1 and inbred lines Zong3, 87-1 at dry and 24 h after imbibitions. Dry and 24 HAI indicated that dry and 24 h after imbibed embryo, respectively. Spot number displayed the location of differentially expressed proteins between hybrid Zong3/87-1 and its parental lines in dry and 24 h

after imbibitions seed embryo on the 2-DE gel. Protein spots exhibiting changes in intensity are indicated, along with their spot number.

Figure S2 Functional classification of the identified proteins and protein isoforms differentially expressed in dry and 24 h imbibed seed embryos between hybrid and its parental lines.

Table S1 The results of *t*-test for the germination rate differences between maize hybrid and its parental lines.

Table S2 Changes in the number and intensity of protein spots in dry and 24 h imbibed seed embryo.

Table S3 The results of *t*-test for the protein spots accumulation differences between maize embryo hybrid and its parental lines.

Table S4 The 155 identified differentially expressed proteins between hybrid and its parental lines.

Acknowledgments

The authors wish to thank Mr. Jidong Feng for his assistance in mass spectrometry analysis.

Author Contributions

Conceived and designed the experiments: ZN QS. Performed the experiments: BG YC GZ. Analyzed the data: BG JX ZH WF YY. Contributed reagents/materials/analysis tools: HP JD YZ. Wrote the paper: BG ZN QS.

References

1. Birchler JA, Auger DL, Riddle NC (2003) In search of the molecular basis of heterosis. Plant Cell 15: 2236–2239.
2. Fu H, Dooner HK (2002) Intraspecific violation of genetic colinearity and its implications in maize. Proc. Natl. Acad. Sci. 99: 9573–9578.
3. Swanson-Wagner RA, Jia Y, DeCook R, Borsuk LA, Schnable PS (2006) All possible modes of gene action are observed in a global comparison of gene expression in a maize F_1 hybrid and its inbred parents. Proc. Natl. Acad. Sci. 103: 6805–6810.
4. Stupar RM, Hermanson PJ, Springer NM (2007) Nonadditive expression and parent-of-origin effects identified by microarray and allele-specific expression profiling of maize endosperm. Plant Physiol 145: 411–425.
5. Guo M, Rupe MA, Yang X, Crasta O, Zinselmeier C, et al. (2006) Genome-wide transcript analysis of maize hybrids: allelic additive gene expression and yield heterosis. Theor Appl Genet 113: 831–845.
6. Meyer S, Pospisil H, Scholten S (2007) Heterosis associated gene expression in maize embryos 6 days after fertilization exhibits additive, dominant and overdominant pattern. Plant Mol Biol 63: 381–391.
7. Song SH, Qu HZ, Chen C, Hu SN, Yu J (2007) Differential gene expression in an elite hybrid rice cultivar (*Oryza sativa*, L) and its parental lines based on SAGE data. BMC Plant Biology 7: 49.
8. Yao YY, Ni ZF, Zhang YH, Chen Y, Han ZF, et al. (2005) Identification of differentially expressed genes in leaf and root between wheat hybrid and its parental inbreds using PCR-based cDNA subtraction. Plant Mol Biol 58: 367–384.
9. Andorf S, Selbig J, Altmann T, Poos K, Witucka-Wall H, et al. (2010) Enriched partial correlations in genome-wide gene expression profiles of hybrids (*A. thaliana*): a systems biological approach towards the molecular basis of heterosis. Theor Appl Genet 120: 249–259.
10. Huang Y, Li LH, Chen Y, Li XH, Xu C, et al. (2006) Comparative analysis of gene expression at early seedling stage between a rice hybrid and its parental lines using a cDNA microarray of 9198 uni-sequences. Sci China 49: 519–529.
11. Hunter RB, Kannenberg LW (1971) Isozyme characterization of corn (*Zea mays*) inbreds and its relationship to single cross hybrid performance. Can J Genet Cytol 13: 649–655.
12. Heidrich-Sobrinho E, Cordeiro AR (1975) Codominant isoenzymic allele as markers of genetic diversity correlated with heterosis in maize (*Zea mays* L.). Theor Appl Genet 46: 197–199.
13. Gonella JA, Peterson PA (1978) Isozyme relatedness of inbred lines of maize and performance of their hybrids. Maydica 23: 55–61.
14. Song X, Ni ZF, Yao YY, Xie CJ, Li ZX, et al. (2007) Wheat (*Triticum aestivum* L.) root proteome and differentially expressed root proteins between hybrid and parents. Proteomics 7: 3538–3557.
15. Song X, Ni ZF, Yao YY, Zhang YH, Sun QX (2009) Identification of differentially expressed proteins between hybrid and parents in wheat (*Triticum aestivum* L.) seedling leaves. Theor Appl Genet 118: 213–225.
16. Hoecker N, Lamkemeyer T, Sarholz B, Paschold A, Fladerer C, et al. (2008) Analysis of nonadditive protein accumulation in young primary roots of a maize (*Zea mays* L.) F (1)-hybrid compared to its parental inbred lines. Proteomics 8: 3882–3894.
17. Marcon C, Schützenmeister A, Schütz W, Madlung J, Piepho HP, et al. (2010) Nonadditive protein accumulation patterns in maize (*Zea mays* L.) hybrids during embryo development. J Proteome Res 9: 6511–6522.
18. Dahal D, Mooney BP, Newton KJ (2012) Specific changes in total and mitochondrial proteomes are associated with higher levels of heterosis in maize hybrids. Plant J 72: 70–83.
19. Sarkissian IV, Kessinger MA, Harris W (1964) Differential rates of development of heterotic and nonheterotic young maize seedlings, I. Correlation of differential morphological development with physiological differences in germinating seeds. Proc Natl Acad Sci 51: 212–218.
20. Romagnoli S, Maddaloni M, Livini C, Motto M (1990) Relationship between gene expression and hybrid vigor in primary root tips of young maize. Theor Appl Genet 80: 769–775.
21. Fu ZY, Jin XN, Ding D, Li YL, Fu ZJ, et al. (2011) Proteomic analysis of heterosis during maize seed germination. Proteomics 11: 1462–1472.
22. Ramagli LS (1999) Quantifying protein in 2-D PAGE solubilization buffers. Methods Mol Biol 112: 99–103.
23. Donnelly BE, Madden RD, Ayoubi P, Porter DR, Dillwith JW (2005) The wheat (*Triticum aestivum* L.) leaf proteome. Proteomics 5: 1624–1633.

24. Porubleva L, Velden KV, Kothari S, Oliver DJ, Chitnis PR (2001) The proteome of maize leaves: use of gene sequences and expressed sequence tag data for identification of proteins with peptide mass fingerprints. Electrophoresis 22: 1724–1738.

25. Limami AM, Rouillon C, Glevarec G, Gallais A, Hirel B (2002) Genetic and physiological analysis of germination efficiency in maize in relation to nitrogen metabolism reveals the importance of cytosolic glutamine synthetase. Plant Physiol 130: 1860–1870.

26. Bevan M, Bancroft I, Bent E, Love K, Goodman H, et al. (1998) Analysis of 1.9 Mb of contiguous sequence from chromosome 4 of the Arabidopsis thaliana. Nature 391: 485–488.

27. Gilliland LU, Pawloski LC, Kandasamy MK (2003) Arabidopsis actin gene ACT7 plays an essential role in germination and root growth. Plant J 33: 319–328.

28. Wang WW, Meng B, Ge XM, Song SH, Yang Y, et al. (2008) Proteomic profiling of rice embryos from a hybrid rice cultivar and its parental lines. Proteomics 8: 4808–4821.

29. Stuber CW, Lincoln SE, Wolff DW, Helentjaris T, Lander ES (1992) Identification of genetic factors contributing to heterosis in a hybrid from two elite maize inbred lines using molecular markers. Genetics 132: 823–839.

30. Xiao J, Li J, Yuan L, Tanksley SD (1995) Dominance is the major genetic basis of heterosis in rice as revealed by QTL analysis using molecular markers. Genetics 140: 745–754.

31. Yu S, Li JX, Xu CG, Tan YF, Gao YJ, et al. (1997) Importance of epistasis as the genetic basis of heterosis in an elite rice hybrid. Proc Natl Acad Sci 94: 9226–9231.

32. Hua JP, Xing YZ, Wu WR, Xu CG, Sun XL, et al. (2003) Single-locus heterotic effects and dominance by dominance interactions can adequately explain the genetic basis of heterosis in an elite rice hybrid. Proc Natl Acad Sci 100: 2574–2579.

33. Bewley JD (1997) Seed Germination and Dormancy. Plant Cell 9(7): 1055–1066.

34. Potokina E, Sreenivasulu N, Altschmied L, Michalek W, Graner A (2002) Differential gene expression during seed germination in barley (Hordeum vulgare L.). Funct Integr Genomics 2(1–2): 28–39.

35. Morris K, Linkies A, Müller K, Oracz K, Wang X, et al. (2011) Regulation of seed germination in the close Arabidopsis relative Lepidium sativum: a global tissue-specific transcript analysis. Plant Physiol 155: 1851–1870.

36. Liu PP, Montgomery TA, Fahlgren N, Kasschau KD, Nonogaki H, et al. (2007) Repression of AUXIN RESPONSE FACTOR10 by microRNA160 is critical for seed germination and post-germination stages. Plant J 52: 133–146.

37. Niogret MF, Culiáñez-Macià FA, Goday A, Albà MM, Pagès M (2002) Expression and cellular localization of rab28 mRNA and Rab28 protein during maize embryogenesis. Plant J 9: 549–557.

38. Yao YY, Ni ZF, Du JK, Wang XL, Wu HY, et al. (2006) Isolation and characterization of 15 genes encoding ribosomal proteins in wheat (Triticum aestivum L.). Plant Sci 170: 579–586.

39. Morita-Yamamuro C, Tsutsui T, Tanaka A, Yamaguchi J (2004) Knock-out of the plastid ribosomal protein S21 causes impaired photosynthesis and sugar-response during germination and seedling development in Arabidopsis thaliana. Plant and Cell Physiol 45: 781–788.

40. Xu H, Liu WY (2004) Cinnamomina versatile type II ribosome-inactivating protein. Acta. Biochim. Biophys. Sin. 36: 169–176.

41. Sopory SK, Puri-Avinashi M, Deka N, Datta A (1980) Early protein synthesis during germination of barley embryos and its relationship to RNA synthesis. Plant and Cell Physiol 21: 649–657.

42. Meagher RB, McKinney EC, Kandasamy MK (2000) The significance of diversity in the plant actin gene family. In Staiger CJ, Baluska F, Volkmann D. and Barlow P., editors. Actin: a dynamic framework for multiple plant cell functions. Amsterdam, the Netherlands: Kluwer Academic Publishers. 3–27.

43. Meagher RB, McKinney EC, Vitale AV (1999) The evolution of new structures: clues from plant cytoskeletal genes. Trends Genet 15: 278–284.

44. Diaz-Camino C, Conde R, Ovsenek N (2005) Actin expression is induced and three isoforms are differentially expressed during germination in Zea mays. Journal of Experimental Botany 56: 557–565.

6

Characterization of a Soluble Phosphatidic Acid Phosphatase in Bitter Melon (*Momordica charantia*)

Heping Cao*, Kandan Sethumadhavan, Casey C. Grimm, Abul H. J. Ullah

U.S. Department of Agriculture, Agricultural Research Service, Southern Regional Research Center, New Orleans, Louisiana, United States of America

Abstract

Momordica charantia is often called bitter melon, bitter gourd or bitter squash because its fruit has a bitter taste. The fruit has been widely used as vegetable and herbal medicine. Alpha-eleostearic acid is the major fatty acid in the seeds, but little is known about its biosynthesis. As an initial step towards understanding the biochemical mechanism of fatty acid accumulation in bitter melon seeds, this study focused on a soluble phosphatidic acid phosphatase (PAP, 3-sn-phosphatidate phosphohydrolase, EC 3.1.3.4) that hydrolyzes the phosphomonoester bond in phosphatidate yielding diacylglycerol and P_i. PAPs are typically categorized into two subfamilies: Mg^{2+}-dependent soluble PAP and Mg^{2+}-independent membrane-associated PAP. We report here the partial purification and characterization of an Mg^{2+}-independent PAP activity from developing cotyledons of bitter melon. PAP protein was partially purified by successive centrifugation and UNOsphere Q and S columns from the soluble extract. PAP activity was optimized at pH 6.5 and 53–60°C and unaffected by up to 0.3 mM $MgCl_2$. The K_m and V_{max} values for dioleoyl-phosphatidic acid were 595.4 µM and 104.9 ηkat/mg of protein, respectively. PAP activity was inhibited by NaF, Na_3VO_4, Triton X-100, $FeSO_4$ and $CuSO_4$, but stimulated by $MnSO_4$, $ZnSO_4$ and $Co(NO_3)_2$. In-gel activity assay and mass spectrometry showed that PAP activity was copurified with a number of other proteins. This study suggests that PAP protein is probably associated with other proteins in bitter melon seeds and that a new class of PAP exists as a soluble and Mg^{2+}-independent enzyme in plants.

Editor: Shrikant Anant, University of Kansas School of Medicine, United States of America

Funding: This work was supported by the USDA-Agricultural Research Service Quality and Utilization of Agricultural Products Research Program 306 through CRIS 6435-41000-102-00D and 6435-41000-102-10N. Mention of trade names or commercial products in this publication is solely for the purpose of providing specific information and does not imply recommendation or endorsement by the U.S. Department of Agriculture. USDA is an equal opportunity provider and employer. The funders had no role in study design, data collection and analysis, decision to publish, or preparation of the manuscript.

Competing Interests: The authors have declared that no competing interests exist.

* Email: Heping.Cao@ars.usda.gov

Introduction

Momordica charantia is often called bitter melon, bitter gourd or bitter squash because its fruit has a bitter taste. It is a tropical and subtropical vine of the *Cucurbitaceae* family and widely grown in Asia, Africa and the Caribbean. The plant grows as herbaceous, tendril-bearing vine up to 5 m long. Bitter melon flowering occurs during June-July and fruit develops during September-November in the Northern Hemisphere. The fruit has a distinct warty exterior and an oblong shape. It is hollow in cross-section with a relatively thin layer of flesh surrounding a central seed cavity filled with large, flat seeds and pith. The fruit is generally consumed in the green or early yellowing stage. The fruit's flesh is crunchy and watery in texture and tasted bitter at these stages. The skin is tender and edible. Seeds and pith appear white in unripe fruits, are not intensely bitter and can be removed before cooking. Bitter melon is often used in Chinese cooking for its bitter flavor, typically in stir-fries, soups and herbal teas. It has also been used as the bitter ingredient in some Chinese and Okinawan beers. Bitter melon seeds are rich in fatty acids and minerals including iron, beta carotene, calcium, potassium and many vitamins. The fatty acid compositions of bitter melon oil include 37% of saturated fatty acids mainly stearic acid; 3% of monounsaturated fatty acid dominantly linoleic acid, and 60% of polyunsaturated fatty acid

predominately alpha-eleostearic acid (α-ESA, 9*cis*, 11*trans*, 13*trans* octadecatrienoic acid) which counts for 54% of the total fatty acids [1].

Bitter melon has been used as herbal medicine in Asia and Africa for a long time. It has been used as an appetite stimulant, a treatment for gastrointestinal infection, and to lower blood sugar in diabetics in traditional Chinese medicine. Recent studies have demonstrated the potential uses of bitter melon oil with a wide range of nutritional and medicinal applications because of its anti-cancer effect [2–10], anti-diabetic activity [11–19], anti-inflammatory effect [20], antioxidant activity [21–23], anti-ulcerogenic effect [24–26] and wound healing effect [27]. Alpha-ESA, a conjugated linolenic acid, may be the key bioactive compound in the seed oil. Alpha-ESA from bitter melon seeds has cytotoxic effect on tumor cells [6], induces apoptosis and upregulates GADD45, p53 and PPARγ in human colon cancer Caco-2 cells [3], blocks breast cancer cell proliferation and induces apoptosis through a mechanism that may be oxidation dependent [2], protects plasma, low density lipoprotein and erythrocyte membrane from oxidation which may be effective in reducing the risk of coronary heart disease in diabetes mellitus [28] and unregulates mRNA expression of PPARα, PPARγ and their target genes in C57BL/6J mice [29]. These studies suggest that α-ESA has anti-cancer, anti-diabetic, and anti-inflammatory activities, inhibits tumor cell proliferation, lowers blood fat and prevents cardiovascular diseases.

Currently, little is known about the enzymatic mechanism for the biosynthesis of α-ESA in bitter melon seeds. In general, acyltransferases including diacylglycerol transferases [30,31], add fatty acyl groups sequentially to the sn-1, sn-2 and sn-3 positions of glycerol-3-phosphate (G3P) to form triacylglycerol (TAG). This pathway is commonly referred to the Kennedy or G3P pathway [32]. A key step in TAG biosynthesis is the dephosphorylation of the sn-3 position of phosphatidate (PtdOH) catalyzed by phosphatidic acid phosphatase (PAP or lipins) to produce diacylglycerol (DAG) and inorganic phosphate (P_i) (Figure 1) [33]. PtdOH is synthesized by the actions of glycerophosphate acyltransferase (GPAT) and lysophosphatidic acid acyltransferase (LPAAT). DAG formation is believed to be the penultimate key step in Kennedy pathway because it is a critical metabolite for the synthesis of TAG, phosphatidylethanolamine (PtdEtn), and phosphatidylcholine (PtdCho) [33–35].

As an initial step towards understanding the biochemical mechanism of fatty acid accumulation in bitter melon seeds, we focused our studies on PAP (3-sn-phosphatidate phosphohydrolase, EC 3.1.3.4) that dephosphorylates phosphatidic acid (PA, also called PtdOH) to generate DAG and P_i. PAP family enzymes are currently classified as either soluble PAP [36] or membrane-bound PAP [37]. Based on the requirement of Mg^{2+} for activity, the enzyme could also be divided into 2 classes: Mg^{2+}-dependent and Mg^{2+}-independent PAP [38]. Typically, soluble PAP is Mg^{2+}-dependent; whereas membrane-bound PAP is Mg^{2+}-independent. We recently identified a soluble PAP in bitter melon cotyledons [39]. We report here the partial purification and characterization of PAP from developing cotyledons of bitter melon as a soluble and Mg^{2+}-independent enzyme.

Results and Discussion

Subcellular distribution of PAP activity in bitter melon cotyledons

PAP family enzymes are currently classified as either soluble PAP or membrane-bound PAP [36]. Differential centrifugation was used to separate the cytosol and microsomal membrane fractions for determining the subcellular localization of PAP activity in Momordica charantia (bitter melon, bitter gourd or bitter squash). The cotyledon extract was successively centrifuged at 3,000 g, 18,000 g and 105,000 g. The final pellet and supernatant after ultracentrifugation are generally regarded as the microsomal membranes and the cytosol, respectively [40]. PAP activity in the 105,000 g pellet (P3-pellet) was only 11% of

the total PAP activity (Figure 2). Following dialysis and centrifugation of the supernatant, the S3-pellet contained only 1.8% whereas the S3-supernatant contained 87.2% of the total activity (Figure 2). These subcellular distributions of PAP activity clearly demonstrated that the great majority of PAP activity in bitter melon seeds was soluble and localized in the cytosol.

Purification of PAP

PAP protein was partially purified from bitter melon cotyledons by a combination of Q and S columns. Enzyme activity was determined by two assays: PAP and phosphatase activity assays. By PAP activity assay, separation of the soluble PAP with Q column resulted in 3.2-fold of purification with a recovery of 90.3% of total PAP activity (Table 1). The PAP protein did not bind to S column effectively and 83.3% of PAP activity was recovered in the flow-through. Separation of the proteins from the flow-through by a second Q column slightly increased the purity with 1.2-fold of specific activity compared to the load and ¾ of the total activity was recovered (Table 1). After changing the binding condition, some PAP protein was bound to S column which resulted in better purification (3-fold) but this step reduced the yield of PAP in the elution to 20% of the load (Table 1). This scheme of purification generated PAP proteins with an overall purification factor of 12.5-fold and a yield of 11.4% using PAP activity assay (Table 1). A higher purification factor (16-fold) and higher yield (14.5%) of PAP purification was obtained from the same protein samples when using phosphatase activity assay (Table 1). Further purification with Affigel Blue resin did not result in any improvement of purity and actually reduced the total activity of PAP (data not shown). The yield and specific activity of PAP from bitter melon is much higher than those of PAP purified from Lagenaria siceraria (bottle gourd, opo squash or long melon) [41]. Additional purification steps resulted in significant loss of activity and the protein yield was low, probably due to PAP-associated with other proteins (see below).

Linearity of PAP assays

In the initial characterization of the partially purified PAP enzyme, linearity between the incubation time and PAP activity was observed in 50 mM imidazole buffer containing 0.3 mM $MgCl_2$, pH 6.5 (Figure 3A). Similar linearity of enzymatic reaction was observed in responding to the amount of enzymes used in the assay (Figure 3B). Data fitting showed that the coefficient of determination (R^2) was over 0.99 in both analyses, suggesting a great correlation between PAP activity and reaction time or amount of enzyme used. These assay results suggest that

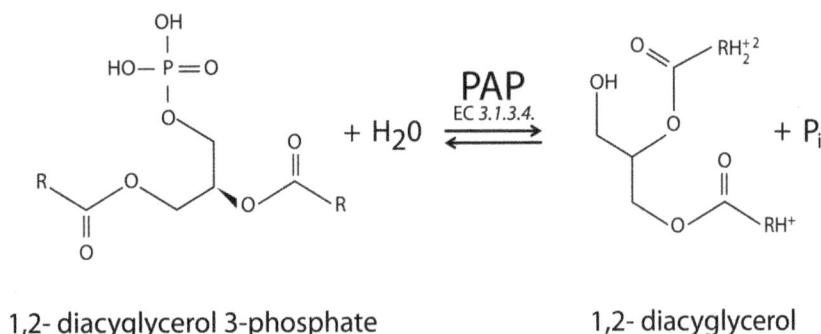

1,2- diacyglycerol 3-phosphate   1,2- diacyglycerol

Figure 1. The schematic representation of enzymatic reaction catalyzed by PAP. The enzyme hydrolyzes the phosphoester linkage of PtdOH and generates DAG and P_i.

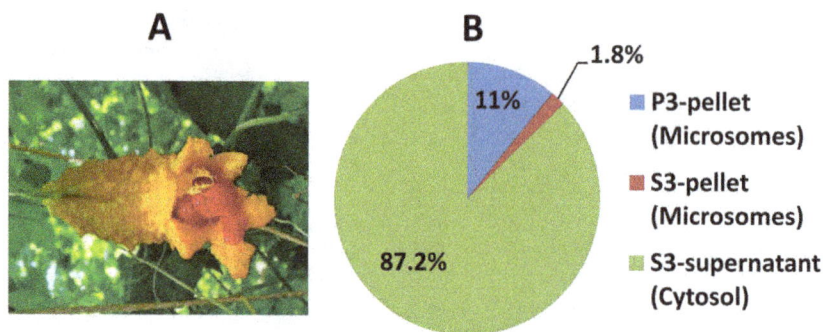

Figure 2. Subcellualr distribution of PAP activity in bitter melon cotyledons. (A) bitter melon of Indian origin was used in the study. (B) Distribution of PAP activity in bitter mellow cotyledons. The cotyledon extract was successively centrifuged at 3,000 g, 18,000 g and 105,000 g. The final pellet (P3-pellet) after ultracentrifugation is generally regarded as the microsomal membranes. Following dialysis and centrifugation of the 105,000 g supernatant, the S3-pellet (contaminated microsomes) and the S3-supernatant (cytosol) were obtained. All three fractions were analyzed for PAP activity. The data presented are the mean of 2 assays for each sample.

the established assays are suitable for characterization of the partially purified PAP from bitter melon.

The pH and temperature optima of PAP

The pH optima of PAP was observed at pH 6.5 under the assay conditions using 5 μL of the enzyme and 500 μM dioleoyl-phosphatidic acid (DPA) in 50 mM imidazole buffer containing 0.3 mM $MgCl_2$ (Figure 4A). The temperature optima were ranged from 53 to 60°C (Figure 4B). These optimal pH and temperature values are slightly higher than those obtained from the crude extract [39]. It is not clear from this study why the temperature optimum is unusually higher than other metabolic enzymes such as the soluble starch synthases from maize kernels [42].

Mg^{2+}-independent activity of PAP

Based on the requirement of Mg^{2+} for activity, PAP family enzymes including lipins and lipid phosphate phosphatases (LPPs) are divided into 2 classes: Mg^{2+}-dependent PAP and Mg^{2+}-independent PAP [38]. It is known that yeast and invertebrates have a single lipin ortholog, but plants have two PAP/lipin genes and mammals have three lipin genes [33]. The lipin family members of PAP are soluble enzymes and require Mg^{2+} for their activity. LPPs also exhibit PAP activity but they are structurally unrelated to lipin proteins. LPPs are localized to the plasma membrane and do not require Mg^{2+} for their activity. We determined the effect of Mg^{2+} on PAP activity using the purified PAP. Our results showed that PAP activity was not affected or minimally affected by up to 0.3 mM $MgCl_2$ in the assay mixtures (Figure 5). The effects of ion chelators EDTA and EGTA on PAP activity were measured. However, both chelators did not have significant effects on PAP activity (data not shown). This result further confirmed that the soluble PAP activity is Mg^{2+}-independent in bitter melon extract. These assay results confirm the previous observations from crude extract of bitter melon and partially purified bottle gourd PAP that these PAPs are Mg^{2+}-independent enzyme [39,41]. The overall results suggest that a new class of PAP exists in bitter melon and bottle gourd which is a soluble and Mg^{2+}-independent enzyme.

Kinetic parameters of PAP

The kinetic parameters of PAP were determined using the purified protein and DPA as the substrate under the optimized assay conditions (pH 6.5, 53°C and 0.3 mM $MgCl_2$). The enzyme gave a typical sigmoidal curve for the substrate (Figure 6). The K_m

and V_{max} values for DPA were 595.4 μM and 104.9 ηkat/mg of protein, respectively. These K_m and V_{max} values of the purified PAP from bitter melon were approximately 4-fold and 56-fold, respectively, of those obtained from crude bitter melon extract [39]. These K_m and V_{max} values of the purified PAP from bitter melon were approximately 3-fold of those obtained from partially purified bottle gourd PAP [41]. The differences in PAP kinetic parameters between the partially purified PAP and crude PAP preparations are in agreement with some observations from previous publications. It is expected that the V_{max} values of purified or partially purified enzymes are much higher than those of the crude enzyme preparations because the V_{max} values are calculated based on the amount of proteins used in the assays. It is also observed that the K_m values of purified enzymes are higher than those of the crude enzyme preparation. For example, we observed previously that the K_m value of partially purified starch synthase II is approximately 2.5-fold of those of the crude extracts [42,43].

Effect of phosphatase inhibitors, additives and cations on PAP activity

Three phosphatase inhibitors were used to test their effects on PAP activity. NaF partially inhibited PAP activity. Sodium orthovanadate inhibited PAP activity up to 90%. However, sodium tartrate did not affect PAP activity under the assay conditions (Figure 7A). Triton X-100 significantly reduced PAP activity but other tested additives did not affect its activity (Figure 7B). The effects of cations in the assay buffers significantly altered PAP activity. Particularly, $MnSO_4$, $ZnSO_4$ and $Co(NO_3)_2$ increased PAP activity, but $FeSO_4$ and $CuSO_4$ decreased its activity (Figure 7C).

Native gel analysis of PAP activity

The purified PAP fractions from Affigel Blue column contained a number of copurified proteins on SDS-PAGE (Figure 8A), but no clear protein bands were observed on the native gel (Figure 8C). PAP activity was analyzed by the in-gel phosphatase activity assay. Activity gel showed that the positive control band of B-phycoerythrin was sharp, but the activity band of PAP displayed a wide range of smear bands with much large size on the native gel (Figure 8B). These results suggest that PAP protein is probably multimerlized and/or associated with other proteins directly and/or indirectly, which is probably one of the reasons for the difficulties in further purification of PAP to homogeneity.

Table 1. Purification of PAP from bitter melon cotyledons.

Purification step	Protein (mg)	Volume (mL)	PAP activity				Phosphtase activity			
			Total activity (ηKat)	Specific activity (ηKat/mg/min)	Purification factor	Yield (%)	Total activity (ηKat)	Specific activity (ηKat/mg/min)	Purification factor	Yield (%)
1. Supernatant	276	67	411	1.5	1	100	2928	10.6	1	100
2. 1st Q column flow-through	192	107	13	0.1	0.07	3.2	1220	6.4	0.6	41.7
3. 1st Q column elute	79	33	371	4.7	3.2	90.3	1716	21.7	2.0	58.6
4. 1st S column flow-through	57	47.5	309	5.4	3.6	75.2	1183	20.8	2.0	40.4
5. 2nd Q column elute	35	42	235	6.7	4.5	57.2	936	26.7	2.5	31.96
6. 2nd S column elute	2.5	1.5	47	18.7	12.5	11.4	425	170.0	16.0	14.5

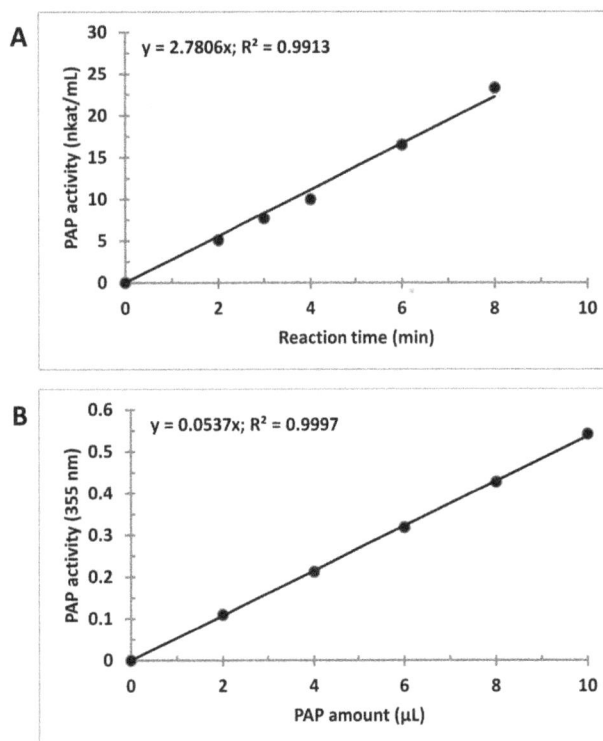

Figure 3. Linearity of PAP assays. PAP activity was assayed using aliquots of the proteins from the second S column fraction duing PAP purification. (A) PAP activity vs. reaction time. The assay was performed at 53°C for various times with 500 μM DPA and 0.3 mM $MgCl_2$. Aliquots of the enzymatic reactions were withdrawn for measurement at the indicated time points. (B) PAP activity vs. amount of enzyme. The assay was performed at 53°C using various amounts of the PAP preparation with 500 μM DPA and 0.3 mM $MgCl_2$. The data presented are the mean of 2 assays for each sample.

Mass spectral identification of PAP-associated proteins

We attempted to identify PAP-associated proteins in the purified fractions by mass spectrometry. The PAP fractions from the second Q column and the Affigel Blue column were pooled and concentrated for tryptic and endoproteinase GluC digestions, respectively. The digested peptides were separated by LC and identified by MS/MS. MS-generated ms ions were searched against available plant sequence databases in the NCBI non-redundant protein library. LC/MS/MS identified 10 and 24 PAP-associated proteins in the trypsin and GluC digests, respectively (Table 2). MS/MS identified four of the proteins in both trypsin and GluC digests corresponding to ribosome-inactivating protein momordin I, elastase inhibitor 4, malate dehydrogenase and trypsin inhibitor 2 (Table 2). However, none of the ms corresponded to PAP sequences in other species. These MS/MS results suggest that PAP protein is probably associated with other proteins and that PAP itself is a minor component in the purified fraction.

Conclusions

Phosphatidic acid phosphatases (PAPs) catalyze the dephosphorylation of phosphatidic acid to diacylglycerol, the penultimate step in TAG synthesis. PAPs are widely present in plants, animals, microbes and human. PAPs are typically categorized into two subfamilies: Mg^{2+}-dependent soluble PAP and Mg^{2+}-independent

A

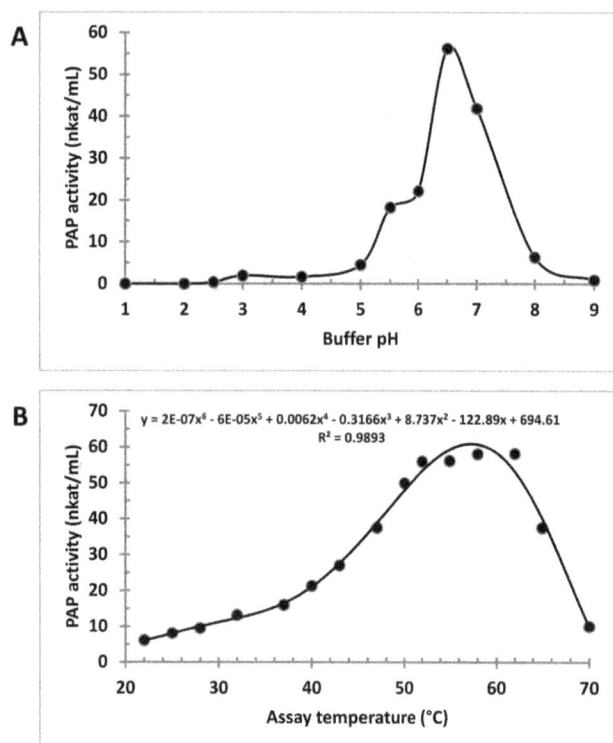

B

$$y = 2E-07x^6 - 6E-05x^5 + 0.0062x^4 - 0.3166x^3 + 8.737x^2 - 122.89x + 694.61$$
$$R^2 = 0.9893$$

Figure 4. Effect of buffer pH and assay temperature on PAP activity. The assay was performed at 53°C for 30 min using 5 µL of the PAP preparation with 500 µM DPA and 0.3 mM $MgCl_2$. (A) pH profile of PAP enzyme catalyzing PtdOH. (B) Temperature profile of PAP enzyme catalyzing PtdOH. The data presented are the mean of 2 assays for each sample.

membrane-associated PAP. In this study, we provided evidence for the existence of a new class of PAP enzyme in bitter melon (*Momordica charantia*). This class of PAP is soluble and Mg^{2+}-independent. PAP protein is probably associated with other proteins in the oilseeds. Bitter melon has been used as herbal medicine for a long time. The molecular basis of these uses is supported by recent studies showing the potential of bitter melon oil being used in a wide range of nutritional and medicinal applications. Therefore, understanding and regulating PAP activities may lead to increased yield of bitter melon oil. The elucidation of PAP functions may lead to novel approaches to modulate cellular lipid storage and metabolic diseases.

Materials and Methods

Plant material

Momordica charantia (bitter melon) fruits (Indian origin) were purchased at an oriental grocery store in New Orleans, LA, USA. The fruits were collected at approximately 6 inches in length (mid-level maturity) which is at an ideal developmental stage for human consumption. The seeds were removed from the seed cavity of the fruit and washed with 0.9% ice cold saline solution. The outer coverings of the seeds were removed at room temperature using scalpel and the cotyledons removed manually.

Preparation of tissue extract

All operations were carried out at 4°C. The seed cotyledons weighing 91 g were homogenized in 100 mL extraction buffer (50 mM NaOAc, pH 5.0, 150 mM NaCl and 10 mM $MgCl_2$)

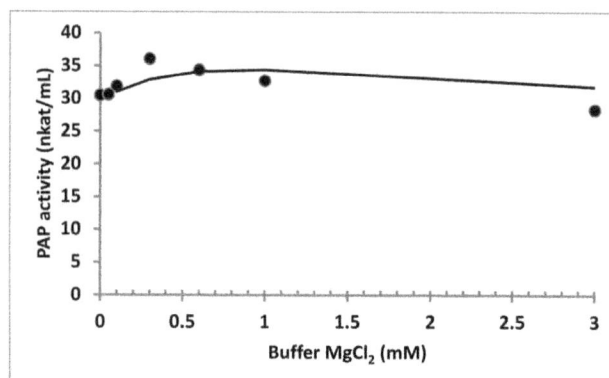

Figure 5. Effect of Mg^{2+} on PAP activity catalyzing PtdOH. The assay was performed at 53°C, pH 6.5 for 30 min using 5 µL of the PAP preparation with 500 µM DPA and various concentrations of $MgCl_2$. The data presented are the mean of 2 assays for each sample.

using polytron tissuemizer (Tekmar Tissumizer MarkII, Cincinnati, OH) at low, medium and high speed for 30 s each. The homogenate was cooling down on ice for 1 min between bursts. The homogenate was centrifuged at 3,000 *g* for 15 min and the resulting supernatant (S1) was centrifuged at 18,000 *g* for 30 min at 4°C (Sorvall RC2B, Miami, FL). This supernatant (S2) was ultracentrifuged at 105,000 *g* for 60 min at 4°C (Sorvall Discovery 100SE, Hitachi Ltd, Tokyo, Japan) and the resulting supernatant (S3, cytosol) and the pellet (P3, microsomal membranes) were collected. The S3 supernatant was dialyzed to remove Pi from the seed extract against imidazole buffer (25 mM imidazole buffer, pH 6.5, 1 mM $MgCl_2$) with three 500 mL buffer changes. The dialyzed supernatant became cloudy after dialysis, which was removed by centrifugation at 18,000 *g* for 30 min at 4°C. The final supernatant (S3-supernatant) and pellet (S3-pellet) as well as P3 were used for determining subcellular distribution of PAP activity and S3-supernatant was used for PAP purification. The protein content of the supernatant was determined by Bicinchoninic acid (BCA) method (Thermo Scientific, Rockford, IL).

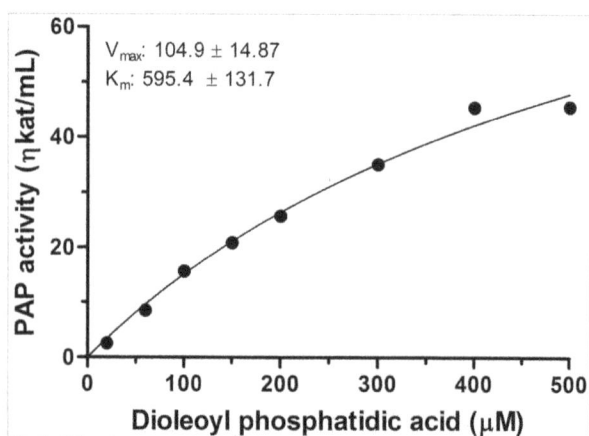

Figure 6. Kinetcis of PAP enzyme activity. PAP activity vs. substrate concentration. The assay was performed at 53°C, pH 6.5 for 10 min using various concentrations of DPA. The data presented are the mean of 2 assays for each sample.

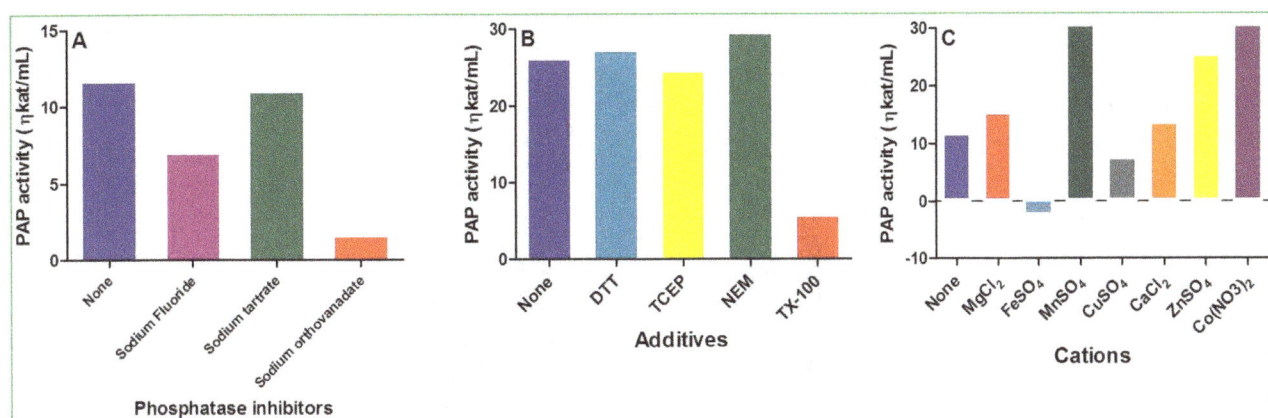

Figure 7. Effects of phosphatase inhibitors, additives and cations on PAP activity. The assays were performed at 53°C, pH 6.5 for 10 min using 5 µL of the PAP preparation with 500 µM DPA. (A) Effect of phosphatase inhibitors on PAP activity. The phosphatase inhibitors used were NaF (1 mM), sodium tartrate (4 mM) and sodium orthovanadate (2 mM). (B) Effect of additives on PAP activity. The final concentration of each additive in the assay buffer was 100 mM, except TX-100 was used at 10% concentration. (C) Effect of cations on PAP activity. The final concentration of each cation in the assay buffer was 0.3 mM. The data presented are the mean of 2 assays for each sample.

Purification of PAP

PAP purification protocol included four steps of chromatography consisting of a repeat of UNOsphere Q and S columns (Bio-Rad Laboratories, Hercules, CA). The column chromatography was performed at ambient 25°C using BioLogic LP chromatographic workstation (Bio-Rad Laboratories). During various steps of purification, active fractions were identified by PAP assay. Parallel phosphatase assay using p-nitrophenol was also carried out to confirm the purity because it was observed that PAP enzyme has phosphatase activity whereas not all phosphatases had PAP activity. All the buffers used during various stages of purification contained 1 mM MgCl₂. Briefly, the dialyzed S3-supernatant from bitter melon cotyledon extract was applied to a 20 mL UNOsphere Q column (2.5×4.0 cm) equilibrated with the 25 mM imidazole buffer, pH 6.5 at the flow rate of 3.0 mL/min. The column was then washed with the same buffer followed by a stepwise elution with NaCl gradient in imidazole buffer (25 mM, pH 6.5) ranging from 0.2 to 0.4 M at 0.05 M NaCl increment and then with 0.5 and 1.0 M NaCl. The eluted fractions with PAP activities were combined, dialyzed against (25 mM, pH 6.5) and applied on to a 20 mL UNOsphere S column (2.5×4.0 cm) equilibrated with the same buffer at the flow rate of 3.0 mL per min. PAP activity was found in the unbound fractions. Further purification involved sequential UNOsphere Q and S columns (1 mL). The buffer and elution conditions were the same except the S column was done at pH 4.5 (25 mM acetate buffer). The PAP activity was found in the eluted fractions. The final protein with PAP activity were concentrated by ultrafiltration and used for enzyme activity assays. The concentrated proteins were further purified by Affigel Blue column (Bio-Rad Laboratories) in 25 mM imidazole, pH 6.5 and 1 mM MnSO₄. The great majority of PAP activity was not bound to the column. The active fractions were pooled and concentrated by Amicon Ultra-0.5 mL Centrifugal Filters (Millipore Corporation, Billerica, MA) for SDS-PAGE, in-gel phosphatase activity assay and protease digestion.

PAP activity assay

PAP activity was determined by Pᵢ-release assay and phosphatase assay. These activity assays for bitter melon PAP were described previously [39]. The Pi-release assay followed the ammonium molybdate-acetone-acid (AMA) method [44]. For standard assay except otherwise noted below, a 50 µL of PAP enzyme was added to 900 µL of 50 mM imidazole buffer, pH 6.5 in a 53°C water bath. The enzymatic reaction was initiated by the addition of 50 µL of PtdOH/DPA (dioleoyl-phosphatidic acid or 1,2-dioleoyl-sn-glycero-3-phosphate, sodium salt, Avanti Polar Lipids, Inc., Alabaster, Alabama), incubated for 30 min and terminated by 2 mL of AMA reagent. Citric acid (0.1 mL, 1.0 M) was added to each tube 30 s later to fix the color followed by centrifugation at 13,000 g for 7 min (Eppendorf 5415C, Westbury, NY). The absorbance at 355 nm was measured after blanking the spectrophotometer with the appropriate control, which was stopped at zero time. The PAP activity was expressed as nanokatals per milliliter (ηkat/mL, ηmoles orthophosphate released per sec). One International Unit (IU) is equivalent to 16.67 nkat.

To determine the pH optima of bitter melon PAP, we used 25 mM glycine-HCl (pH 1.5–3.0), 50 mM sodium acetate (pH 3.5–5.5), and 25 mM imidazole (pH 6–9). To measure the optimum temperature, the samples were incubated with substrate between 20 and 70°C in 25 mM imidazole, pH 6.5. The Mg²⁺ optima were determined by the phosphatase assay with up to 3 mM MgCl₂ in the assay mixtures. The general procedures for PAP characterization were similar to those used for soluble starch synthases [42].

Phosphatase activity

For phosphatase assay, PAP enzyme (2 to 10 µL) was added to a volume of 950 µL of buffer (50 mM imidazole, pH 6.5) and incubated at 53°C for 2 to 5 min with 1.25 mmole of p-nitrophenylphosphate (pNPP) in a final volume of 1.0 mL. The reaction was terminated with 0.1 mL of 1.0 N NaOH and the released p-nitrophenol was measured spectrophotometrically at 400 nm [45].

Kinetics of PAP

The K_m and V_{max} for bitter melon PAP using the Pᵢ-release assay were determined at 53°C and pH 6.5 as mentioned above. The concentration of DPA ranged from 0 to 500 µM. Window-Chem's software Enzyme Kinetics version 1.1 (Fairfield, CA) was used to compute the K_m and V_{max} values.

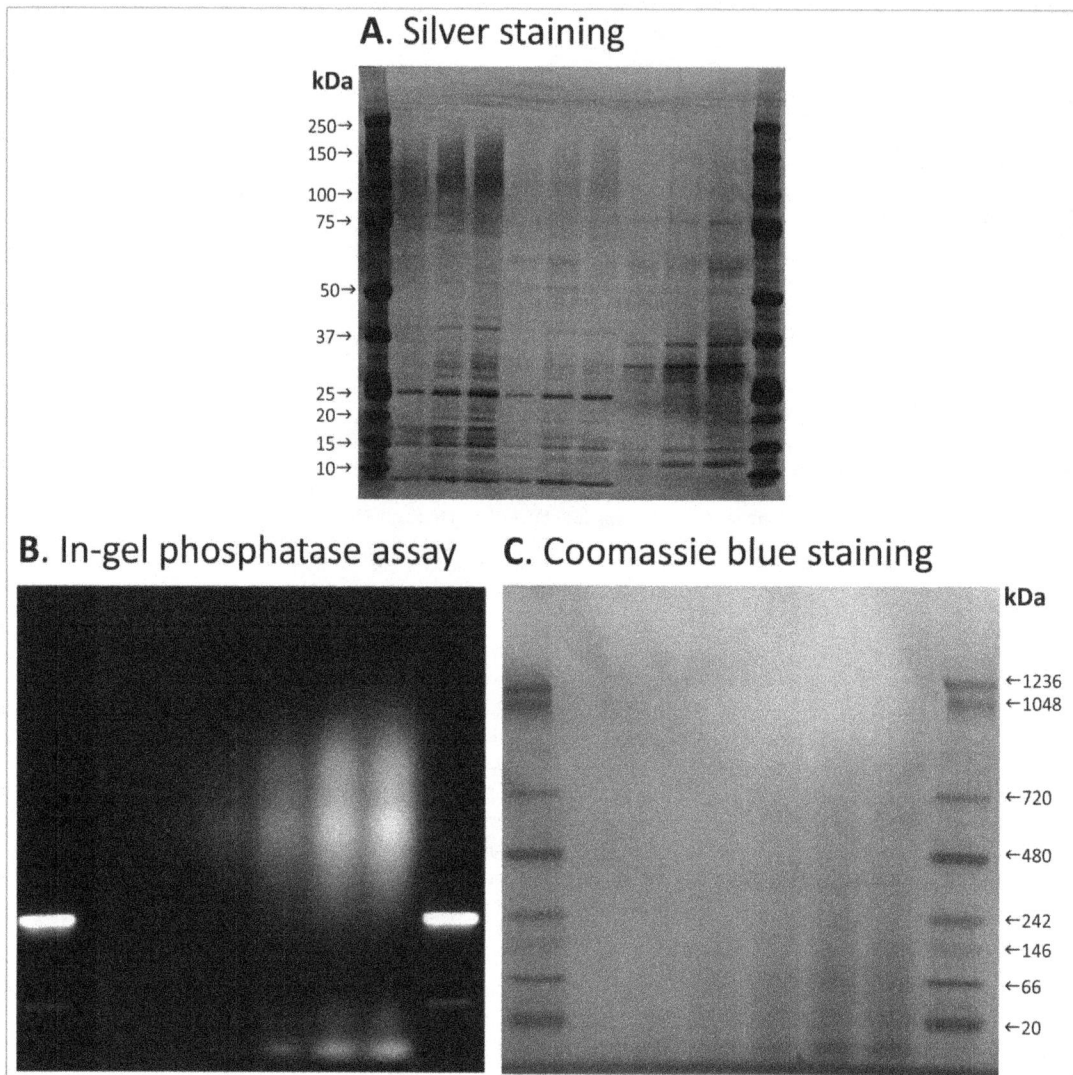

Figure 8. SDS-PAGE and native gel analysis of protein composition and PAP activity. (A) Denaturing gel analysis of the purified PAP fractions from Affigel Blue column by SDS-PAGE and stained by silver nitrate. Lanes 1 and 11, Bio-Rad Precision plus protein standards; lanes 2–4, 0.28, 0.55 and 0.83 μg of fraction #4; lanes 5–7, 0.19, 0.37 and 0.56 μg of fraction #5; lanes 8–10, 0.30, 0.60 and 0.90 μg of the pooled fractions 11–17. (B) PAP activity determined by in-gel phosphatase assay with DiFMUP. Lanes 1 and 8, positive control B-phycoerythrin; lanes 2–7, 0.5, 1, 3, 6, 10 and 12 μg of PAP from the pooled fraction. (C) Native gel analysis of the purified PAP fractions stained with Coomassie Blue. Lanes 1 and 8, native marker unstained protein standards; lanes 2–7, 0.5, 1, 3, 6, 10 and 12 μg of PAP from the pooled fraction.

Denaturing gel electrophoresis

The purified PAP enzyme preparation from Affigel Blue column was analyzed by SDS-PAGE (200 V and 70 min) using Xcell II, Mini-Cell and 4–12% Novex NuPage Bis-Tris gels with MOPS as running buffer (Life Technologies, Grand Island, NY). The separated protein bands were visualized with Pierce silver staining kit using prestained and multicolored molecular weight markers (4 to 250 kDa) as size standards (Thermo Scientific).

Native gel electrophoresis and in-gel phosphatase activity assay

The purified PAP enzyme preparation was also analyzed by native gel for in-gel PAP activity assay using Xcell II, Mini-Cell and 3–12% Novex NuPage Bis-Tris gels with NuPAGE native running buffer (Life Technologies). The native gel was loaded with 0.5–12 μg of proteins purified from the Affigel Blue column. The

protein bands on the gel were visualized with Coomassie Blue staining (Simple Blue SafeStain, Life Technologies) using native mark unstained protein standards (20 to 1236 kDa) as size standards (Thermo Scientific). PAP activity on the unstained native gel was performed using 6,8-difluoro-4-methylumbelliferyl phosphate (DiFMUP) as the substrate and B-phycoerythrin as the positive control (Life Technologies). The activity assay was performed by incubating the native gel into 3-mL reaction mixture containing 0.5 mM DiFMUP, 50 mM imidazole, pH 6.5, 3 mM $MnCl_2$ and 100 mM DTT at 37°C for 5 min.

Trypsin digestion of the purified PAP enzyme

Enzyme fraction from Affigel Blue column was concentrated in a Centricon concentrator (Amicon, centrifugal filter devices, Millipore Corporation) and during the course the imidazole buffer was exchanged to protease digestion buffer (50 mM ammonium carbonate, pH 8.0). The tryptic digests were prepared following

Table 2. LC/MS/MS analysis of PAP-copurified proteins from bitter melon cotyledons.

No.	Co-purified proteins	Accession	Sequence	Protease
1	Ribosome-inactivating protein momordin I (Alpha-momorcharin)	P16094.2	(K)ITLPYSGNYER(L)	Trypsin
			(K)VVTSNIQLLLNTR(N)	Trypsin
			(K)YIEQQIQER(A)	Trypsin
			(K)IPIGLPALDSAISTLLHYDSTAAAGALLVLIQTTAEAAR(F)	Trypsin
			(K)VVTSNIQLLLNTR(N)	Endoproteinase Glu-C
2	Elastase inhibitor 4 (MCEI-IV)	P10296.2	(R)DSDCLAQCICVDGHCG(−)	Trypsin
			(K)RDSDCLAQCICVDGHCG(−)	Trypsin
			(R)DSDCLAQCICVDGHCG(−)	Endoproteinase Glu-C
3	Malate dehydrogenase, mitochondrial	P17783.1	(K)LALYDIAGTPGVAADVGHVNTR(S)	Trypsin
			(R)FVESSLR(A)	Trypsin
			(K)LFGVTTLDVVR(A)	Trypsin
			(D)IAGTPGVAADVGHVNTRSE(V)	Endoproteinase Glu-C
4	Trypsin inhibitor 2 (MCTI-A)	P10295.1	(R)DSDCMAQCICVDGHCG(−)	Trypsin
			(R)DSDCMAQCICVDGHCG(−)	Endoproteinase Glu-C
5	Superoxide dismutase [Cu-Zn] 1	Q42611.3	(R)AVVVHADPDDLGK(G)	Trypsin
			(R)AVVVHAEPDDLGRGGHELSKTTGNAGGR(V)	Trypsin
			(K)GGHELSLTTGNAGGR(V)	Trypsin
			(R)VACGIIGLQG(−)	Trypsin
			(K)GGHELSLTTGNAGGRVACGIIGLQG(−)	Trypsin
			(R)LACGVVGLTPV(−)	Trypsin
6	Allergen Ara h 1 (allergen Ara h I)	P43238.1	(R)IFLAGDKDNVIDQIEK(Q)	Trypsin
7	Probable lactoylglutathione, chloroplast	Q8W593.1	(R)GPTPEPLCQVMLR(V)	Trypsin
8	Peroxidase C1B	P15232.1	(R)MGNITPLTGTQGEIR(L)	Trypsin
9	Cell division protein ftsB homolog	B8GQ76.1	(K)TGLDAIEER(A)	Trypsin
10	Putative F-box/kelch-repeat protein At3g43710	Q9M2B5.1	(R)LFTLCR(R)	Trypsin
11	Enolase 2 (2-phospho-D-glycerate hydro-lyase 2)	Q9LEI9.1	(V)KVQIVGDDLLVTNPKRVE(K)	Glu-C
			(E)LRDGGSDYLGKGVSKAVE(N)	Glu-C
			(L)ELRDGGSDYLGKGVSKAVE(N)	Glu-C
12	Malate dehydrogenase, chloroplastic (pNAD-MDH)	Q9SN86.1	(D)AKAGAGSATLSMAYAAARFVE(S)	Glu-C
			(D)LPFFASR(V)	Glu-C
			(N)AFIHIISNPVNSTVPIAAE(V)	Glu-C
13	Heat shock 70 kDA protein	P26791.1	(E)TAGGVMTVLIPRNTTIPTKKE(Q)	Glu-C
14	Malate dehydrogenase, glyoxysomal	P46488.1	(E)LPFFATKVRLGRNGID(E)	Glu-C
15	Malate dehydrogenase, glyoxysomal	P19446.1	(E)LPFFASKVRLGRNGIE(E)	Glu-C
16	Fructose-biphosphate aldolase, cytoplasmic isozyme	P29356.1	(D)GGVLPGIKVDKGTVE(L)	Glu-C
17	Embryonic abundant protein 1	P46520.1	(Q)TVVPGGTGGKSLE(A)	Glu-C
18	DnaJ protein homolog (DNAJ-1)	Q04960.1	(E)ILGVSKNASQDD(L)	Glu-C
19	Nucleoside diphosphate kinase 4, chloroplastic	Q8RXA8.1	(E)IKLWFKPEE(L)	Glu-C
20	Nascent polypeptide-associated complex subunit alpha-like protein	Q9M612.1	(D)TGVEPKDIE(L)	Glu-C
21	Glutathione reductase, chloroplastic, GR	Q43154.1	(A)QFDSTVGIHPSAAEE(F)	Glu-C
22	Ubiquitin-activating enzyme E1 (AtUBA1)	P93028.1	(V)KGGIVTQVKQPK(L)	Glu-C
23	Tetratricopeptide repeat protein 18 (TPR repeat protein)	Q5T0N1.2	(Q)VVLGDSAKITVSPE(G)	Glu-C
24	Ubiquitin-activating enzyme E11	P20973.1	(E)FQDGDLVVFSE(V)	Glu-C
25	Ferredoxin -NADP reductase, root isozyme, chloroplastic (FNR)	Q41014.2	(E)KLSQLKKNKQWHVE(V)	Glu-C

Table 2. Cont.

No.	Co-purified proteins	Accession	Sequence	Protease
26	Peptide deformylase (PDF)	A5D1C0.1	(A)VYKIVE(L)	Glu-C
27	DNA polymerase zeta catalytic subunit (REV3-like)	Q61493.2	(K)ATSSSRSELEGRK(G)	Glu-C
28	DNA mismatch repair protein mutL	Q87VJ2.1	(V)HDFLYGTLHRALGDVRPE(N)	Glu-C
29	Trigger factor (TF)	Q55511.1	(M)AVDETKLIPVTFPE(D)	Glu-C
30	Nascent polypeptide-associated complex subunit alpha-like protein 1 (NAC-alpha-like protein 1)	Q9LHG9.1	(A)LKAADGDIVSAIME(L)	Glu-C

the instructions of the in-solution tryptic digestion and guanidination kit (Thermo Scientific). Briefly, PAP-containing proteins (5–10 µg) were added to 15 µL digestion buffer containing 5.6 mM DTT in a total volume of 27 µL in a microcentrifuge tube. The digestion mixture was incubated at 95°C for 5 min. To alkylate the proteins, 3 µL of 100 mM iodoacetamide were added to the mixture and kept at room temperature in the dark. A 2 µL aliquot of activated trypsin (100 µg/µL) was added to each tube and incubated at 37°C for 3 h followed by addition of 1 µL of trypsin to the digestion mixture and incubated for another 2 h for better digestion. The digests were guanidilated to convert lysine to homoarginine by adding 10 µL of 30% NH$_4$OH and 6 µL of guanidination reagent (50 mg O-methylisourea hemisulfate in 51 µL H$_2$O) and incubated at 65°C for 12 min. The reaction was stopped by the addition of 3 µL of trifluoroacetic acid (TFA) and stored at −20°C before LC/MS/MS analysis.

Endoproteinase GluC digestion of the purified PAP enzyme

Enzyme fraction from the Affigel Blue column was concentrated in a Centricon concentrator (Amicon, centrifugal filter devices, Millipore Corporation) and at the same time the buffer was exchanged to protease digestion buffer (50 mM ammonium carbonate, pH 7.8). The endoproteinase GluC digest was prepared following the manufacturer's instructions (New England Biolabs, Boston, MA). Briefly, PAP-containing proteins (10–20 µg) were mixed with 0.5 µg endoproteinase GluC and incubated at 37°C for 16 h. The reaction was stopped by the addition of 6 µL of TFA and stored at −20°C before LC/MS/MS analysis.

Peptide separation, mass spectral sequencing and database search

The protease digests were analyzed by LC/MS/MS consisting of an Agilent 1200 LC system, an Agilent Chip-cube interface, and an Agilent 6520 Q-TOF tandem mass spectrometer (Agilent Technologies, Santa Clara, CA). The peptides were separated using a Chip consisting of a 40 nL enrichment column and a 43 mm analytical column packed with C$_{18}$ (5 µm beads with 300 Å pores). One-µL aliquot of the sample was transferred to the

enrichment column via a capillary pump operating at a flow rate of 4 µL/min. The nano pump was operated at a flow rate of 600 nL/min. An initial gradient of 97% Solvent A (0.1% formic acid in H$_2$O) and 3% Solvent B (90% acetonitrile/0.1% formic acid in H$_2$O) was changed to 60% Solvent A at 12 min, 20% at 13 min, and held till 15 min. A post run time of 3 min was employed for column equilibration. The MS source was operated at 300°C with 5 L/min N$_2$ flow and a fragmentor voltage of 175 V. N$_2$ was used as the collision gas with collision energy varied as a function of mass and charge using a slope of 3.7 V/100 Da and an offset of 2.5 V. Both quad and Time-of-Flight (TOF) were operated in positive ion mode. Reference compounds of 322.048121 Da and 1221.990637 Da were continually leaked into the source for mass calibration. An initial MS scan was performed from m/z 300 to 1600 and up to three multiply charged ions were automatically selected for MS/MS analysis. Following the initial run, a second injection was made excluding ions previously targeted for MS/MS analysis. LC chromatograms and mass spectra were examined using Mass-Hunter software (Version B.0301; Agilent Technologies). Data files were transferred to an Agilent workstation equipped with Spectrum Mill software (Agilent Technologies). The raw MS/MS data files were extracted, sequenced and searched against the National Center for Biotechnology Information (NCBI) non-redundant protein library.

Acknowledgments

This paper is dedicated to the memory of Dr. Abul Hasnat Jaffor Ullah who worked at the USDA-ARS Southern Regional Research Center for 28 years and initiated this study but suddenly passed away with a heart attack on August 21, 2013 at the age of 65. The authors thank Dr. K. Thomas Klasson for encouragement and Drs. Si-Yin Chung, Thach-Mien D Nguyen and Dunhua Zhang for helpful comments on the manuscript.

Author Contributions

Conceived and designed the experiments: AHJU KS HC. Performed the experiments: KS CCG. Analyzed the data: HC KS. Contributed reagents/materials/analysis tools: HC CCG. Contributed to the writing of the manuscript: HC.

References

1. Liu XR, Deng ZY, Fan YW, Li J, Liu ZH (2010) Mineral elements analysis of Momordica charantiap seeds by ICP-AES and fatty acid profile identification of seed oil by GC-MS. Guang Pu Xue Yu Guang Pu Fen Xi 30: 2265–2268.

2. Grossmann ME, Mizuno NK, Dammen ML, Schuster T, Ray A et al. (2009) Eleostearic Acid inhibits breast cancer proliferation by means of an oxidation-dependent mechanism. Cancer Prev Res (Phila) 2: 879–886.

3. Yasui Y, Hosokawa M, Sahara T, Suzuki R, Ohgiya S et al. (2005) Bitter gourd seed fatty acid rich in 9c,11t,13t-conjugated linolenic acid induces apoptosis and up-regulates the GADD45, p53 and PPARgamma in human colon cancer Caco-2 cells. Prostaglandins Leukot Essent Fatty Acids 73: 113–119.

4. Kohno H, Suzuki R, Yasui Y, Hosokawa M, Miyashita K et al. (2004) Pomegranate seed oil rich in conjugated linolenic acid suppresses chemically induced colon carcinogenesis in rats. Cancer Sci 95: 481–486.

5. Kohno H, Yasui Y, Suzuki R, Hosokawa M, Miyashita K et al. (2004) Dietary seed oil rich in conjugated linolenic acid from bitter melon inhibits azoxymethane-induced rat colon carcinogenesis through elevation of colonic PPARgamma expression and alteration of lipid composition. Int J Cancer 110: 896–901.

6. Suzuki R, Noguchi R, Ota T, Abe M, Miyashita K et al. (2001) Cytotoxic effect of conjugated trienoic fatty acids on mouse tumor and human monocytic leukemia cells. Lipids 36: 477–482.

7. Kwatra D, Subramaniam D, Ramamoorthy P, Standing D, Moran E et al. (2013) Methanolic extracts of bitter melon inhibit colon cancer stem cells by affecting energy homeostasis and autophagy. Evid Based Complement Alternat Med 2013: 702869.

8. Rajamoorthi A, Shrivastava S, Steele R, Nerurkar P, Gonzalez JG et al. (2013) Bitter melon reduces head and neck squamous cell carcinoma growth by targeting c-Met signaling. PLoS ONE 8: e78006.

9. Zeng YW, Yang JZ, Pu XY, Du J, Yang T et al. (2013) Strategies of functional food for cancer prevention in human beings. Asian Pac J Cancer Prev 14: 1585–1592.

10. Brennan VC, Wang CM, Yang WH (2012) Bitter melon (Momordica charantia) extract suppresses adrenocortical cancer cell proliferation through modulation of the apoptotic pathway, steroidogenesis, and insulin-like growth factor type 1 receptor/RAC-alpha serine/threonine-protein kinase signaling. J Med Food 15: 325–334.

11. Ahmad Z, Zamhuri KF, Yaacob A, Siong CH, Selvarajah M et al. (2012) In vitro anti-diabetic activities and chemical analysis of polypeptide-k and oil isolated from seeds of Momordica charantia (bitter gourd). Molecules 17: 9631–9640.

12. Dhar P, Chattopadhyay K, Bhattacharyya D, Roychoudhury A, Biswas A et al. (2006) Antioxidative effect of conjugated linolenic acid in diabetic and non-diabetic blood: an in vitro study. J Oleo Sci 56: 19–24.

13. Oishi Y, Sakamoto T, Udagawa H, Taniguchi H, Kobayashi-Hattori K et al. (2007) Inhibition of increases in blood glucose and serum neutral fat by Momordica charantia saponin fraction. Biosci Biotechnol Biochem 71: 735–740.

14. Saha SS, Ghosh M (2012) Antioxidant and anti-inflammatory effect of conjugated linolenic acid isomers against streptozotocin-induced diabetes. Br J Nutr 108: 974–983.

15. Efird JT, Choi YM, Davies SW, Mehra S, Anderson EJ et al. (2014) Potential for improved glycemic control with dietary Momordica charantia in patients with insulin resistance and pre-diabetes. Int J Environ Res Public Health 11: 2328–2345.

16. Shih CC, Shlau MT, Lin CH, Wu JB (2014) Momordica charantia ameliorates insulin resistance and dyslipidemia with altered hepatic glucose production and fatty acid synthesis and AMPK phosphorylation in high-fat-fed mice. Phytother Res 28: 363–371.

17. Wang HY, Kan WC, Cheng TJ, Yu SH, Chang LH et al. (2014) Differential anti-diabetic effects and mechanism of action of charantin-rich extract of Taiwanese Momordica charantia between type 1 and type 2 diabetic mice. Food Chem Toxicol 69: 347–356.

18. Bao B, Chen YG, Zhang L, Na Xu YL, Wang X et al. (2013) Momordica charantia (Bitter Melon) reduces obesity-associated macrophage and mast cell infiltration as well as inflammatory cytokine expression in adipose tissues. PLoS ONE 8: e84075.

19. Hsieh CH, Chen GC, Chen PH, Wu TF, Chao PM (2013) Altered white adipose tissue protein profile in C57BL/6J mice displaying delipidative, inflammatory, and browning characteristics after bitter melon seed oil treatment. PLoS ONE 8: e72917.

20. Saha SS, Ghosh M (2012) Antioxidant and anti-inflammatory effect of conjugated linolenic acid isomers against streptozotocin-induced diabetes. Br J Nutr 108: 974–983.

21. Dhar P, Ghosh S, Bhattacharyya DK (1999) Dietary effects of conjugated octadecatrienoic fatty acid (9 cis, 11 trans, 13 trans) levels on blood lipids and nonenzymatic in vitro lipid peroxidation in rats. Lipids 34: 109–114.

22. Padmashree A, Sharma GK, Semwal AD, Bawa AS (2011) Studies on the antioxygenic activity of bitter gourd (Momordica charantia) and its fractions using various in vitro models. J Sci Food Agric 91: 776–782.

23. Kim HY, Sin SM, Lee S, Cho KM, Cho EJ (2013) The Butanol fraction of bitter melon (Momordica charantia) scavenges free radicals and attenuates oxidative stress. Prev Nutr Food Sci 18: 18–22.

24. Gurbuz I, Akyuz C, Yesilada E, Sener B (2000) Anti-ulcerogenic effect of Momordica charantia L. fruits on various ulcer models in rats. J Ethnopharmacol 71: 77–82.

25. Ozbakis DG, Gursan N (2005) Effects of Momordica charantia L. (Cucurbitaceae) on indomethacin-induced ulcer model in rats. Turk J Gastroenterol 16: 85–88.

26. Mardani S, Nasri H, Hajian S, Ahmadi A, Kazemi R et al. (2014) Impact of Momordica charantia extract on kidney function and structure in mice. J Nephropathol 3: 35–40.

27. Piskin A, Altunkaynak BZ, Tumentemur G, Kaplan S, Yazici OB et al. (2014) The beneficial effects of Momordica charantia (bitter gourd) on wound healing of rabbit skin. J Dermatolog Treat 25: 350–357.

28. Dhar P, Chattopadhyay K, Bhattacharyya D, Roychoudhury A, Biswas A et al. (2006) Antioxidative effect of conjugated linolenic acid in diabetic and non-diabetic blood: an in vitro study. J Oleo Sci 56: 19–24.

29. Chao CY, Yin MC, Huang CJ (2011) Wild bitter gourd extract up-regulates mRNA expression of PPARalpha, PPARgamma and their target genes in C57BL/6J mice. J Ethnopharmacol 135: 156–161.

30. Cao H (2011) Structure-function analysis of diacylglycerol acyltransferase sequences from 70 organisms. BMC Res Notes 4: 249.

31. Cao H, Shockey JM, Klasson KT, Chapital DC, Mason CB et al. (2013) Developmental regulation of diacylglycerol acyltransferase family gene expression in tung tree tissues. PLoS ONE 8: e76946.

32. Smith SW, Weiss SB, Kennedy EP (1957) The enzymatic dephosphorylation of phosphatidic acids. J Biol Chem 228: 915–922.

33. Csaki LS, Dwyer JR, Fong LG, Tontonoz P, Young SG et al. (2013) Lipins, lipinopathies, and the modulation of cellular lipid storage and signaling. Prog Lipid Res 52: 305–316.

34. Kennedy EP, Weiss SB (1956) The function of cytidine coenzymes in the biosynthesis of phospholipides. J Biol Chem 222: 193–214.

35. Kocsis MG, Weselake RJ (1996) Phosphatidate phosphatases of mammals, yeast, and higher plants. Lipids 31: 785–802.

36. Butterwith SC, Hopewell R, Brindley DN (1984) Partial purification and characterization of the soluble phosphatidate phosphohydrolase of rat liver. Biochem J 220: 825–833.

37. Lin YP, Carman GM (1989) Purification and characterization of phosphatidate phosphatase from Saccharomyces cerevisiae. J Biol Chem 264: 8641–8645.

38. Han GS, Wu WI, Carman GM (2006) The Saccharomyces cerevisiae Lipin homolog is a Mg2+-dependent phosphatidate phosphatase enzyme. J Biol Chem 281: 9210–9218.

39. Ullah AHJ, Sethumadhavan K (2013) Identification of a soluble phosphatidate phosphohydrolase in the developing cotyledons of Momordica charantia. Advances in Biological Chemistry 3: 11–17.

40. Cao H, Chapital DC, Howard OD Jr, Deterding LJ, Mason CB et al. (2012) Expression and purification of recombinant tung tree diacylglycerol acyltransferase 2. Appl Microbiol Biotechnol 96: 711–727.

41. Ullah AHJ, Sethumadhavan K, Grimm C, Shockey J (2012) Purification, characterization, and bioinformatics studies of phosphatidic acid phosphohydrolase from Lagenaria siceraria. Advances in Biological Chemistry 2: 403–410.

42. Cao H, James MG, Myers AM (2000) Purification and characterization of soluble starch synthases from maize endosperm. Arch Biochem Biophys 373: 135–146.

43. Ozbun JL, Hawker JS, Preiss J (1971) Adenosine diphosphoglucose-starch glucosyltransferases from developing kernels of waxy maize. Plant Physiol 48: 765–769.

44. Heinonen JK, Lahti RJ (1981) A new and convenient colorimetric determination of inorganic orthophosphate and its application to the assay of inorganic pyrophosphatase. Anal Biochem 113: 313–317.

45. Ullah AH, Cummins BJ (1987) Purification, N-terminal amino acid sequence and characterization of pH 2.5 optimum acid phosphatase (E.C. 3.1.3.2) from Aspergillus ficuum. Prep Biochem 17: 397–422.

7

Novel SSR Markers from BAC-End Sequences, DArT Arrays and a Comprehensive Genetic Map with 1,291 Marker Loci for Chickpea (*Cicer arietinum* L.)

Mahendar Thudi[1,9], **Abhishek Bohra**[1,2,9], **Spurthi N. Nayak**[1,2], **Nicy Varghese**[1], **Trushar M. Shah**[1], **R. Varma Penmetsa**[3], **Nepolean Thirunavukkarasu**[4], **Srivani Gudipati**[1], **Pooran M. Gaur**[1], **Pawan L. Kulwal**[5], **Hari D. Upadhyaya**[1], **Polavarapu B. KaviKishor**[2], **Peter Winter**[6], **Günter Kahl**[7], **Christopher D. Town**[8], **Andrzej Kilian**[9], **Douglas R. Cook**[3], **Rajeev K. Varshney**[1,10]*

1 Grain Legumes Research Program, International Crops Research Institute for the Semi-Arid Tropics (ICRISAT), Hyderabad, India, 2 Department of Genetics, Osmania University, Hyderabad, India, 3 Department of Plant Pathology, University of California Davis, Davis, California, United States of America, 4 Division of Genetics, Indian Agricultural Research Institute, New Delhi, India, 5 State Level Biotechnology Centre, Mahatma Phule Agricultural University, Ahmednagar, India, 6 GenXPro GmbH, Frankfurt am Main, Germany, 7 Molecular BioSciences, University of Frankfurt, Frankfurt am Main, Germany, 8 J. Craig Venter Institute (JCVI), Rockville, Maryland, United States of America, 9 DArT Pty. Ltd., Yarralumla, Australia, 10 CGIAR Generation Challenge Programme (GCP), CIMMYT, Mexico DF, Mexico

Abstract

Chickpea (*Cicer arietinum* L.) is the third most important cool season food legume, cultivated in arid and semi-arid regions of the world. The goal of this study was to develop novel molecular markers such as microsatellite or simple sequence repeat (SSR) markers from bacterial artificial chromosome (BAC)-end sequences (BESs) and diversity arrays technology (DArT) markers, and to construct a high-density genetic map based on recombinant inbred line (RIL) population ICC 4958 (*C. arietinum*)×PI 489777 (*C. reticulatum*). A BAC-library comprising 55,680 clones was constructed and 46,270 BESs were generated. Mining of these BESs provided 6,845 SSRs, and primer pairs were designed for 1,344 SSRs. In parallel, DArT arrays with ca. 15,000 clones were developed, and 5,397 clones were found polymorphic among 94 genotypes tested. Screening of newly developed BES-SSR markers and DArT arrays on the parental genotypes of the RIL mapping population showed polymorphism with 253 BES-SSR markers and 675 DArT markers. Segregation data obtained for these polymorphic markers and 494 markers data compiled from published reports or collaborators were used for constructing the genetic map. As a result, a comprehensive genetic map comprising 1,291 markers on eight linkage groups (LGs) spanning a total of 845.56 cM distance was developed (http://cmap.icrisat.ac.in/cmap/sm/cp/thudi/). The number of markers per linkage group ranged from 68 (LG 8) to 218 (LG 3) with an average inter-marker distance of 0.65 cM. While the developed resource of molecular markers will be useful for genetic diversity, genetic mapping and molecular breeding applications, the comprehensive genetic map with integrated BES-SSR markers will facilitate its anchoring to the physical map (under construction) to accelerate map-based cloning of genes in chickpea and comparative genome evolution studies in legumes.

Editor: Bengt Hansson, Lund University, Sweden

Funding: Financial support from Tropical Legumes I of Generation Challenge Program (GCP; http://www.generationcp.org/) of CGIAR, Bill and Melinda Gates Foundation (BMGF) and National Fund for Basic and Strategic Research in Agriculture (NFBSRA; http://www.icar.org.in/nfbsfara/index.html) of Indian Council of Agricultural Research (ICAR) and Department of Biotechnology (DBT), Government of India, are greatly acknowledged. The funders had no role in study design, data collection and analysis, decision to publish, or preparation of the manuscript.

Competing Interests: The authors have read the journal's policy and have the following conflicts: Dr. Andrzej Kilian is the founder and Director of DArT P/L. Dr. Peter Winter is the Managing Director for GenXPro.

* E-mail: r.k.varshney@cgiar.org

9 These authors contributed equally to this work.

Introduction

Chickpea (*Cicer arietinum* L.) is a self-pollinated, diploid ($2n = 2x = 16$), grain legume crop with a genome size of 740 Mb [1]. It is the third most important legume crop of the world and the first most important pulse crop of India (http://www.icrisat.org/crop-chickpea.htm). The *kabuli* types are generally grown in the Mediterranean region including Southern Europe, Western Asia and Northern Africa and the *desi* types are grown mainly in Ethiopia and Indian subcontinent. It is cultivated mostly on low-input and residual moisture from monsoon rains on the Indian subcontinent and semi-arid regions of Sub-Saharan Africa. Chickpeas are high in protein (23%), dietary fiber, carbohydrates (64% of total carbohydrates), and minerals like calcium, magnesium, potassium, phosphorus, iron, zinc and manganese, hence it is considered a neutraceutical crop. Besides terminal drought, *Helicoverpa armigera* (pod borer) is the most devastating pest of chickpea, amounting to annual yield losses to the tune of US$ 400 million per annum in India, and over US$ 2 billion in the semi-arid tropics [2]. Hence, despite the growing demands and high yield potential, chickpea yields are stable and productivity has remained almost stagnant at unacceptably low levels [3,4].

In spite of tireless efforts of the chickpea breeding community at a global scale, not much progress has been made to overcome these production obstacles. Nevertheless, recent advances in crop genomics offer a great potential for improving crop productivity by deploying marker-assisted selection (MAS) for production constraints in chickpea breeding. Simple sequence repeats (SSRs) or sequence tagged microsatellites (STMS) markers have proven as molecular markers of choice for plant genetics and breeding [5]. In case of chickpea, a few hundred SSR markers were isolated from genomic DNA libraries [6–8] or mined from expressed sequence tags (ESTs) [9,10], and some of them were integrated into genetic maps of chickpea [7,8,10]. Similarly, a set of 233 SSR markers were developed after screening a bacterial artificial chromosome (BAC)-library with synthetic oligonucleotides complementary to SSRs [11]. However, only 52 SSR markers were integrated into genetic map [8]. Another method of SSR marker development is the sequencing of BAC-end sequences (BESs), and the resulting SSR markers are referred as BAC-end derived SSR (BES-SSR) markers [12,13]. Mapping of BES-SSR markers facilitates alignment of genetic and physical maps for applications in map-based cloning and genome sequencing [14,15].

Diversity arrays technology (DArT), developed by Jaccoud et al. [16], is another approach for screening a large number of marker loci in parallel. DArT markers have been employed for developing high-density genetic maps and assessing genetic diversity at a large scale in several crops, e.g. barley [17], wheat [18], pearl millet [19], to name some. Among legumes, so far DArT markers have only been mapped for pigeonpea [20].

In addition to SSR and DArT marker systems, single nucleotide polymorphism (SNP) markers, because of their higher abundance and amenability for high-throughput approaches are becoming popular as well in crop genetics and breeding [21]. By using allele-specific sequencing for candidate genes and mining the ESTs derived from several genotypes, SNPs have been identified in chickpea [8,10,22]. Some of these SNPs have been integrated into genetic maps of chickpea [10].

By using different marker systems, high-density genetic maps have been developed for several crop species including legumes like soybean [23], cowpea [24] and common bean [25]. However, this has not been the case for chickpea, mainly because of the narrow genetic basis of the cultivated gene pool of chickpea. Therefore, the chickpea community has used *C. reticulatum*, a closely related wild species, to develop an inter-specific mapping population for genetic mapping of a maximum number of marker loci. The recombinant inbred line (RIL) mapping population, namely *C. arietinum* (ICC 4958) × *C. reticulatum* (PI 489777), has been extensively used and considered as the reference mapping population for genome mapping [7,8,10,26]. Even based on this mapping population, the most advanced genetic map reported so far, provides the order of maximally 521 markers including SSR and SNP marker loci [8]. Nevertheless, Millán et al. [26] has developed a consensus map based on five inter-specific maps and integrated 555 marker loci including 251 random amplified polymorphic DNAs (RAPDs), 149 STMSs, 47 amplified fragment length polymorphisms (AFLPs), 33 cross-genome markers, 28 gene-specific markers, 10 isozyme markers, 10 inter-simple sequence repeats (ISSRs) and 7 resistance gene analogue (RGA) loci.

With an objective of developing a high-density genetic map based on a single mapping population with maximum genome coverage and precise marker order, the present study reports: (i) construction of a new bacterial artificial chromosome (BAC) library and generation of BAC-end sequences (BESs), (ii) development of novel BES-SSR markers, (iii) development of

DArT arrays, and (iv) construction of a dense genetic map based on the BES-SSR and DArT markers (developed in this study) and legacy markers. Genetic mapping data from this study as well as their comparison with two other maps [8,26] are available in the CMap database at http://cmap.icrisat.ac.in/cmap/sm/cp/thudi/.

Results

Construction of BAC-library and generation of BAC-end sequences

The bacterial artificial chromosome (BAC) library (CAA1Ba) was developed from chickpea accession ICC 4958. The library consisted of 55,680 clones, with most inserts ranging from 100 to 130 kbp. A set of 25,000 BAC clones, randomly selected, were sequenced from both ends. Terminal vector sequences were then trimmed and BESs shorter than 100 bp were discarded. As a result, a total of 46,270 high quality BESs were generated. These sequence data are available in the form of genome survey sequences (GSS) at National Center for Biotechnology Information (NCBI) with GenBank accession numbers EI846478.1 to GS878115.1 and GenBank gi numbers 14645554 to 270242271.

Identification and distribution of BES-SSRs

With an aim of increasing the molecular marker repertoire for chickpea, 46,270 BESs representing 33.22 Mbp of the genome were surveyed for the presence of SSRs by means of the MIcroSAtellite (*MISA*) search module ([27], http://pgrc.ipk-gatersleben.de/misa/). In total 6,845 SSRs were identified in 5,123 BESs, scanning one SSR per every 4.85 kb. The SSRs were either perfect (i.e., containing a single repeat motif such as 'ATA') or compound SSRs (i.e., composed of two or more SSRs separated by ≤100 bp). About 1,245 BESs contained more than one SSR motif, while 913 SSRs identified were in compound form. Perfect SSRs were further subdivided according to the length of SSR tracts [28,29]: Class I SSRs (≥20 nucleotides in length) and Class II SSRs (≥10 but <20 nucleotides in length). Among Class I repeats, di-nucleotide repeats (42.7%) were most abundant, followed by tri-nucleotide repeats (26%), while Class II repeats consisted mostly of penta-nucleotides (65.30%), followed by hexa-nucleotide repeats (26.10%; Figure 1). Among the SSR repeats, mono-nucleotide (51.35% of total) and di-nucleotide repeats (37.03% of total) were dominating. Excluding mono-nucleotide repeats, which were almost exclusively poly-A motifs, A/T-rich repeats accounted for 49.84% of all SSRs. The frequency of AT-rich repeats increased as motif length increased, from a low of 71.18% in di-nucleotide repeats to a high of 94.75% in hexa-nucleotide repeats. Majority of the SSR motifs occurred in the range of <10 to 20 repeat units category (Figure 2).

Development of novel genetic markers

Out of 6,845 SSRs identified in 5,123 BESs, the primer pairs were designed for 2,189 non-redundant BES-SSRs (Table S1). The markers based on these primer sequences have been referred as *C*icer *a*rietinum *M*icrosatellite (CaM) markers. However, based on criteria mentioned in our earlier study [8] for getting higher proportion of polymorphic markers, only 1,344 primer pairs were synthesized and tested for amplification and polymorphism potential. Primer sequence information, repeat motifs, amplicon sizes, and polymorphism features for all 1,344 primers are provided in Table S1. In addition, primer sequence information is also provided in Table S1 for 845 primers pairs that were not characterized in the present study so that the chickpea community can utilize the developed resource.

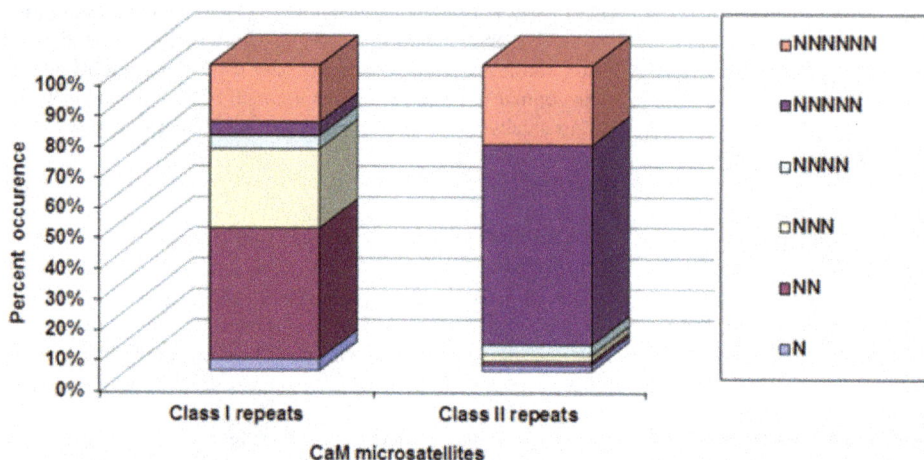

Figure 1. Distribution of Class I and Class II repeats in newly isolated chickpea microsatellites. Class I microsatellites contain >20 nucleotides, Class II repeats perfect SSRs with >12 but <20 nucleotides. Among Class I repeats, tri-nucleotide repeats were most abundant, followed by di-nucleotide repeats, while in Class II repeats, penta-nucleotide repeats were most prevalent, followed by hexa-repeats. N, mono-nucleotide repeats; NN, di-nucleotide repeats; NNN, tri-nucleotide repeats; NNNN, tetra-nucleotide repeats; NNNNN, penta-nucleotide repeats, NNNNNN, hexa-nucleotide repeats.

Of 1,344 primer pairs tested on the two genotypes ICC 4958 and ICC 1882, scorable amplification was observed with 1,063 primer pairs. Furthermore, 737 (69.33%) primer pairs or markers showed polymorphism on a panel of 48 genotypes including 33 genotypes from cultivated species (*C. arietinum*) and 15 genotypes from eight wild species including *C. echinospermum*, *C. bijugum*, *C. cuneatum*, *C. judaicum*, *C. microphyllum*, *C. pinnatifidum*, *C. reticulatum* and *C. yamashitae* (Table S2). In terms of polymorphism detection, markers derived from hexa-nucleotide repeats were highly polymorphic followed by tetra-, penta-, tri- and di-nucleotide repeats. In brief, 69.33% (737) markers were polymorphic and detected a total of 3,144 alleles ranging from 2–25 with an average of 4.26 alleles per marker locus. The PIC value of these polymorphic markers ranged from 0.04 to 0.94 with an average of 0.30. Of 737 polymorphic markers, 602 markers had a PIC value of ≤0.50 and a set of 86 (11.66%)

highly informative SSR markers with PIC values >0.60 was identified (Figure 3).

Among 737 polymorphic markers, 517 were polymorphic in 15 genotypes of eight wild species and 329 markers were polymorphic across 33 genotypes of the cultivated species. As expected, a higher level of polymorphism was detected in inter-specific crosses as compared to intra-specific crosses. For instance, 126 – 253 markers showed polymorphism between parents of inter-specific mapping populations, while 99 – 171 markers displayed polymorphism between parents of intra-specific mapping populations (Table 1).

Development of DArT markers

A DArT array with 15,360 DArT clones was developed from a *Pst*I/*Taq*I representation generated from a mixture of DNA of 94 diverse chickpea genotypes as well as some other chickpea

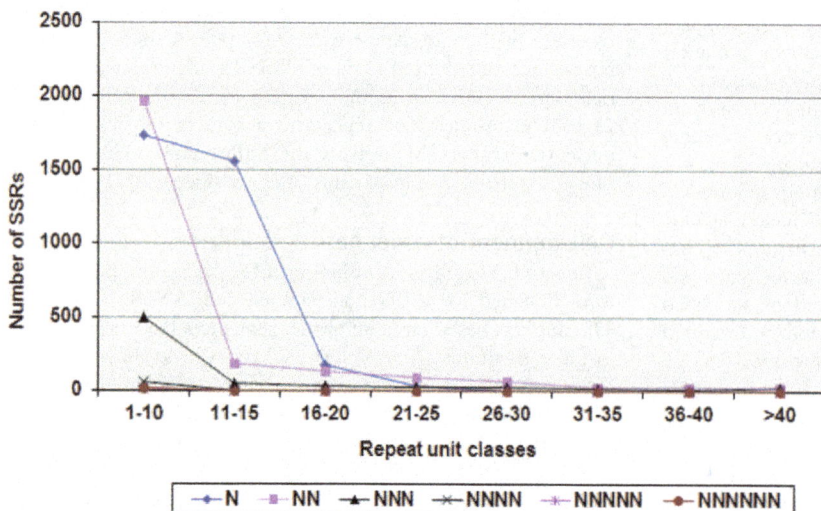

Figure 2. Distribution of microsatellites with varying repeat units in BAC-end sequences. N: mono-nucleotide repeats; NN: di-nucleotide repeats; NNN: tri-nucleotide repeats; NNNN: tetra-nucleotide repeats; NNNNN: penta-nucleotide repeats and NNNNNN: hexa-nucleotide repeats.

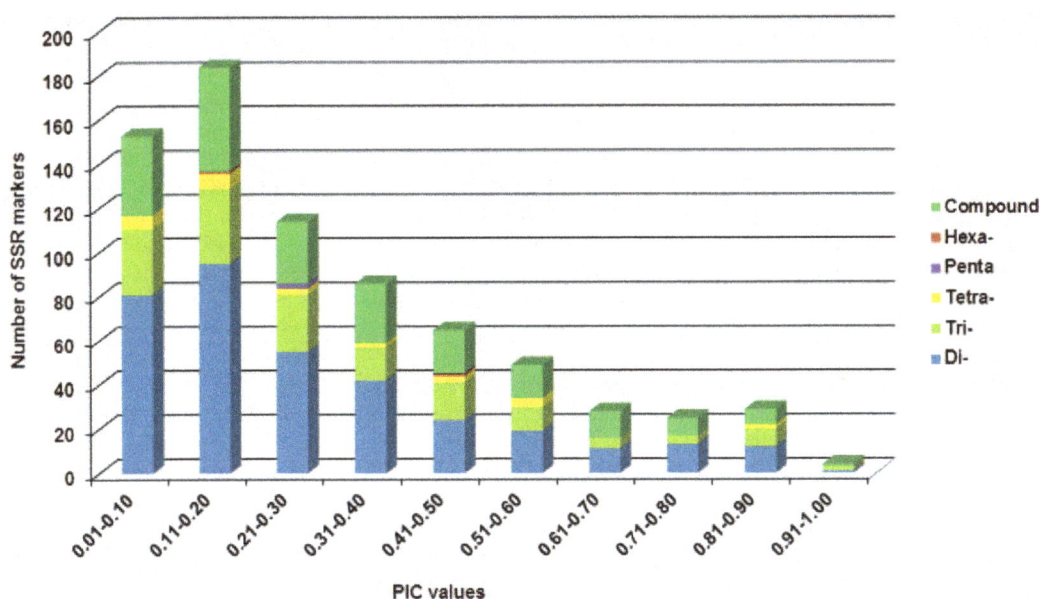

Figure 3. Number of BES-SSR markers in different PIC value classes. The number of di-, tri-, tetra-, penta-, hexa- nucleotide repeats and compound SSRs in different PIC value classes are in blue, light blue, yellow, purple, dark red and green respectively.

genotypes of Australian origin. After scanning the developed DArT arrays on the set of 94 genotypes, a total of 5,397 DArT markers exhibited polymorphism. The number of polymorphic markers among parents of different intra- and inter-specific mapping populations ranged from 35 to 496 and 210 to 906, respectively (Table 1). Only 675 DArT markers showed polymorphism between the parental genotypes of the inter-specific mapping population (ICC 4958×PI 489777).

The PIC values for DArT markers were relatively low, with only 11.72% of DArTs having PIC values of 0.30–0.50, whereas 81.7% DArTs exhibited PIC values of <0.20 (Table S3). The average

mean PIC value was 0.13. Further, when the quality of the DArT markers was analyzed against their performance, which is determined by call rate and PIC values, 34.64% of the polymorphic DArT markers ($n = 1,870$) were in the 80–100% quality category with an average PIC value of 0.18 and a call rate of 99.36%, respectively (Table S3). The average PIC value decreased with the average quality value. The PIC values for 108 markers possessing marker quality of <50% ranged from 0.02 – 0.14. Of 1,870 markers with a quality of more than 80%, only 328 markers had a PIC value of >0.30 (Table S3). Out of 5,397 polymorphic DArT markers in the germplasm analyzed, there are

Table 1. Polymorphism survey of novel SSR and DArT markers between parental genotype combinations of different intra- and inter-specific mapping populations.

Crosses	BES-SSR markers (total 1,063 used)			DArT markers (total 15,360 clones used)		
	Markers amplified	Number of polymorphic markers	Polymorphism (%)	Number of DArT clones giving signals	Number of polymorphic markers	Polymorphism (%)
Intra-specific (*C. arietinum* ×*C. arietinum*)						
ICC 4958×ICC 1882	931	100	10.74	5,285	496	9.38
ICC 283×ICC 8261	864	159	18.40	5,308	327	6.16
ICCV 2×JG 11	909	99	10.89	5,368	35	0.65
ICCV 2×JG 62	882	171	19.39	5,380	36	0.66
ICC 506EB×Vijay	949	117	12.33	5,303	99	1.86
ICC 6263×ICC 1431	913	128	14.02	5,352	447	8.35
Inter-specific (*C. arietinum* ×*C. reticulatum*)						
ICC 4958×PI 489777	990	253	25.55	5,262	675	12.82
ICC 3137×IG 72953	931	129	13.86	5,299	680	12.83
ICC 3137×IG 72933	863	126	14.60	5,266	210	3.98
ICC 8261×ICC 17160	848	229	27	5,248	845	16.10
ICCV 2×ICC 17160	823	248	30.13	5,262	906	17.21

266 and 270 markers in the PIC value range of 0.30–0.40 and 0.40–0.50, respectively (Figure 4).

Construction of a high-density inter-specific map

With an objective to construct a high-density genetic map of chickpea, newly developed BES-SSR markers and DArT arrays were screened on the parental genotypes, i.e. ICC 4958 and PI 489777 of the reference mapping population. As a result, 253 BES-SSR and 675 DArT markers were polymorphic between the parental genotypes. Subsequently, segregation data were obtained for all 928 polymorphic markers on 131 RILs of the mapping population. In addition, genotyping data were collected for 192 genic molecular markers (GMMs) including 83 conserved orthologous sequences (COS)-based SNPs (COS-SNPs), 54 cleaved amplified polymorphic sequences (CAPS), 35 conserved intron spanning region (CISR) and 20 EST-derived SSR (EST-SSR, with the name ICCeM) marker loci published in Gujaria et al. [10] and 494 first generation DNA markers that have been used in construction of genetic maps in several studies [8,10,30], referred as legacy markers. In summary, genotyping data obtained for all 1,614 markers were compiled and used for developing the genetic map. In the first instance, all the markers showing good segregation were used for the construction of the genetic map. Subsequently, with an objective of not losing the genetic information of other published markers, the markers showing segregation distortion were also tried to integrate into the map (Table S4). Finally, a total of 1,291 (79.99%) marker loci were mapped onto eight linkage groups (LGs) spanning a distance of 845.56 cM (Table 2; Figure 5; http://cmap.icrisat.ac.in/cgi-bin/cmap_public/viewer?data_source=CMAP_PUBLIC;saved_link_id=5). The linkage groups have been numbered LG 1 - LG 8 following the nomenclature style of our earlier study [8].

In summary, the developed genetic map in this study comprises 157 novel SSR loci, 621 novel DArT loci, 145 GMM, and 368 legacy marker loci (Table 2). The number of markers per linkage group varied from 68 (LG 8) to 218 (LG 5). The Figure 5 shows distribution of all the marker loci, as mentioned above, on 8 linkage groups (LGs). The length of individual linkage groups ranged from 79.06 (LG 8) to 133.97 cM (LG 7). LG 5 had the highest number of marker loci (218) while the highest map length was recorded for LG 7 (133.97 cM). On the other hand, LG 8 exhibited the lowest number of mapped markers (68) as well as the

shortest map distance (79.06 cM). On average, one marker is present for every 0.65 cM per each linkage group.

Of 621 DArT loci mapped, 355 (57.2%) loci were mapped on only three linkage groups (LG 4, LG 5 and LG 6). Maximum number of novel CaM marker loci was mapped on LG 3 (34) followed by LG 7 (33) and LG 5 (32). None of the EST-SSR (ICCeM) markers was mapped onto LG 2 and LG 7, and similarly none of the CISR markers was mapped to LG 4 and LG 7. For making the map more informative for selecting the markers for future genetic mapping and diversity analysis studies in chickpea, each LG was divided into 10 cM long bins (Figure 5). The PIC value and number of alleles, wherever possible, were calculated for all the mapped markers. The average PIC value of the mapped SSR markers on individual LGs varied from 0.29 (LG 6) to 0.51 (LG 1), while the average number of alleles ranged from 4 (LG 2) to 6.33 (LG 8) (Table S5). The information on PIC values and number of alleles associated with the SSR markers in different bins will help selection of highly informative SSR markers from each bin in a systematic way that will represent the genome as well as enhance the probability of displaying high polymorphism in the germplasm to be analyzed.

Uneven distribution was observed for the mapped markers across all linkage groups (Figure 5). A total of 59 major clusters (≥5 loci/cM) were identified on all eight linkage groups (Table 3). The largest cluster included 25 loci within 1 cM interval on LG 3. Furthermore, at least one cluster of DArT loci was found on each linkage group in the current map. A maximum of 13 clusters comprising 97 marker loci was observed on LG 5. Uneven distribution of markers was also evident from the occurrence of gaps in different linkage groups. A total of 16 minor gaps (5 – 10 cM) between adjacent markers were spread across seven linkage groups (Table 3), except for LG 3. A large gap between adjacent markers (>20 cM) was observed on LG 7, and a gap >10 cM on LG 4. Nevertheless, 16 gaps between 5 and 10 cM on all LGs exist, except for LG 5 (Figure 5; and Table 3).

To assess the congruency of marker order and map position, the present comprehensive genetic map was compared with four earlier genetic maps [7,8,26,31]. On comparison, the linkage group position of different markers remained conserved in case of LG 2, LG 3, LG 4, LG 5 and LG 8 with Nayak et al. [8]. However, markers on LG 1, LG 6 and LG 7 of the present map exhibited some discrepancies in their position. For instance, of 218

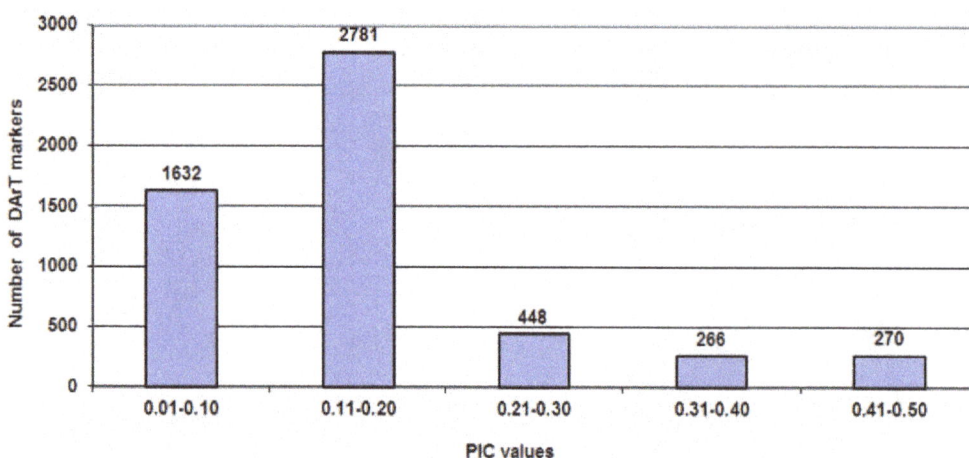

Figure 4. Number of DArT markers in different PIC value classes. Polymorphic markers have been grouped into five classes of PIC values namely 0.01– 0.10, 0.11–0.20, 0.21– 0.30, 0.31– 0.40 and 0.41–0.50.

Table 2. Distribution of different type of markers on eight chickpea linkage groups (LGs).

Marker series	NovelSSR markers (CaM)	Genic molecular markers (GMMs)				DArT markers	Legacy markers	Total
		EST-SSR	CISR	CAPS	COS-SNP			
Markers used	253	20	35	54	83	675	494	1,614
Total markers mapped	157	11	18	35	81	621	368	1,291
Percent mapped	62.06	55	51.43	64.81	97.59	92	74.49	79.99
Markers unlinked	96	9	17	19	2	54	126	323
Percent unlinked	37.94	45	48.57	35.19	2.41	8	25.51	20.01
Markers mapped on different linkage groups (LGs)								
LG 1	6	2	3	4	16	77	48	156
LG 2	3	-	1	8	8	26	52	98
LG 3	34	2	1	3	10	91	52	193
LG 4	17	2	-	7	11	122	53	212
LG 5	32	1	4	3	8	114	56	218
LG 6	25	3	5	4	8	119	45	209
LG 7	33	-	-	4	13	47	40	137
LG 8	7	1	4	2	7	25	22	68
Total	**157**	**11**	**18**	**35**	**81**	**621**	**368**	**1,291**

markers mapped on LG 5, in the current study, 35 were present on LG 5 and 10 markers on LG 2 of Nayak et al. [8]. LG 1, LG 2 and LG 4b of the consensus map of Millán et al. [26] based on narrow (intra-specific) crosses correspond to LG 1, LG 2 and LG 4 of the current map (Table 4). Similarly, LG 1, LG 3, LG 4, LG 5 and LG 6 of the consensus map based on wide (inter-specific) crosses correspond to LG 1, LG 3, LG 4, LG 5 and LG 6 of the present map. The linkage groups LG 4 and LG 11 of Winter et al. [7] correspond to LG 4 of present map. The linkage group wise correspondence among current map and the maps developed by Winter et al. [7] and Millán et al. [26] have been shown via CMap (http://cmap.icrisat.ac.in/cmap/sm/cp/thudi/).

Discussion

Novel SSR markers from BESs

A new 10X BAC library and 46,270 BESs have been generated for the reference genotype ICC 4958 in the present study. Although BAC libraries have been targeted for isolation of SSRs in chickpea earlier [11,32], this is the first time that SSR markers have been developed after mining the BESs. This study adds a new set of 1,063 BES-SSR markers of which 737 markers showed polymorphism in the set of 48 tested genotypes, of which 58 markers with a PIC value of >0.70 were highly informative. In terms of mapping newly developed BES-SSR markers to the genetic map, success was obtained only in the case of 157 (11.68%) markers. This reduction in number of markers from designing the primer pairs to mapping is termed as "SSR attritions" [33]. Higher attrition rates have also been reported earlier in the case of BES-SSR markers e.g. rye [34]. Nevertheless, one of the most important advantage of the developed BES-SSR markers over genomic or EST-SSR markers is that they serve as anchor points between genetic and physical maps [13,25,35]. Screening of these markers on a set of parental genotypes of 11 mapping populations provided 99 to 253 polymorphic markers in different intra- and inter-specific mapping populations (Table 1). These markers can

be used for map construction and trait mapping in the respective populations.

In the total set of 6,845 SSRs identified in 5,123 BESs, the Class I SSRs (≥20 nucleotides in length) include a higher proportion of di-nucleotide repeats (42.7%), followed by tri-nucleotide repeats (26%), while Class II SSRs were mostly derived from penta-nucleotides (65.3%), and followed by hexa-nucleotides (26.1%). Availability of information on this aspect of SSRs is important for the selection of potential polymorphic SSR markers. In case of ICCM markers, the average PIC value of Class I SSRs was higher (0.38) than that of Class II SSRs (PIC = 0.22), thus demonstrating the potential of Class I SSRs over Class II SSRs [8]. Similarly, in the case of CaM markers, average PIC value of Class I SSRs was higher (0.21) compared to Class II SSRs (0.11). The majority of Class I SSRs contains tri-nucleotide repeats, indicating the importance of tri-nucleotide repeat motifs over others.

SSR frequency in the present study was found to be one SSR in every 4.85 kb. The frequency and distribution of SSRs, however, depends on various factors such as size of sequence dataset, tools and criteria used [36]. As a result, in the same species, a varied level of frequency of SSRs has been reported in different studies [36]. Similar is the case of chickpea where SSR frequencies have been reported as 1/707 bp in coding regions [9], 1/1.3 kb in transcriptome assembly [37] and 1/4.85 kb in BESs in the present study.

In general, tri-nucleotide repeats were considered the most polymorphic sites [36]. In addition to tri-nucleotide repeats, compound SSRs constituted the majority of polymorphic markers during the present study. Contrary to majority of the other plant species where di-nucleotide repeats showed high polymorphism [27,38], hexa-nucleotide repeats were highly polymorphic in the present study. Similar results have been reported in the case of pigeonpea [39] and common bean [40]. PIC values of compound SSRs (average PIC values of ICCM = 0.29 [8] and CaM = 0.27) were comparable to those of tri-nucleotide repeats. This can be attributed to the fact that the markers with compound SSRs have

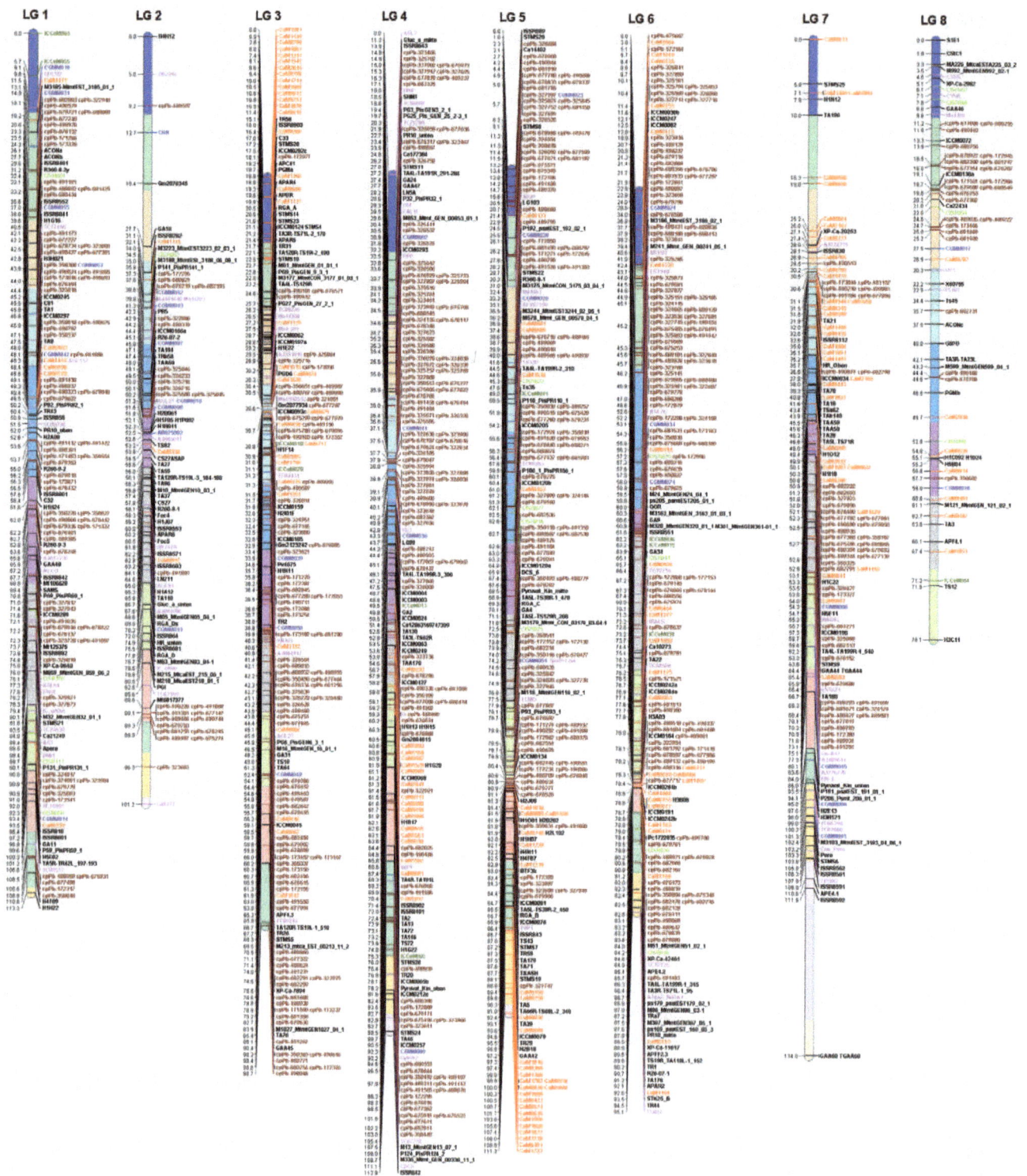

Figure 5. Interspecific reference genetic map with 1,291 loci, spanning 845.56 cM. The map distance is indicated on the left and the marker names on the right side of each linkage group. Each linkage group is divided into 10 cM bins. Marker series are colour coded: CaM (red), DArT (brown), ICCeM (green), CISR (light green), COS-SNP (pink), CAPS (blue) and legacy markers (black).

more than one SSR motif, which increases their chances to be polymorphic [8]. The present study demonstrated a positive correlation between number of alleles and PIC values. For instance, CaM0713 produced the highest number of alleles (25) with highest PIC values (0.94) followed by CaM0836 with 21 alleles and PIC value of 0.93.

Table 3. Features of the inter-specific reference genetic map.

Linkage group (LG)	Number of markers	Length (cM)	Density (markers/cM)	Number of intervals	Number of gaps (>5 cM)	Number of clusters	Genetic mapping position and number of markers (in parenthesis) in clusters observed
LG 1	156	113.32	0.73	121	2	9	18 cM (6), 44 cM (8), 47 cM (5), 8 cM (7), 49 cM (6), 54 cM (6), 2 cM (8), 72 cM (7), 91 cM (6)
LG 2	98	101.19	1.03	84	4	3	50 cM (5), 52 cM (5), 89 cM (8)
LG 3	193	98.66	0.51	153	1	8	21 cM (7), 30 cM (12), 31 cM (25), 32 cM (6), 36 cM (5), 42 cM (7), 60 cM (7), 79 cM (5)
LG 4	213	112.91	0.53	160	1	10	15 cM (6), 34 cM (8), 35 cM (12), 36 cM (15), 38 cM (9), 39 cM (6), 40 cM (5), 59 cM (9), 61 cM (11), 98 cM (9)
LG 5	218	111.29	0.51	160		14	5 cM (5), 6 cM (9), 11 cM (8), 26 cM (5), 52 cM (6), 56 cM (9), 63 cM (9), 74 cM (9), 79 cM (8), 80 cM (6), 81 cM (8), 82 cM (10), 87 cM (5), 100 cM (6)
LG 6	208	95.12	0.46	149	2	11	10 cM (9), 22 cM (5), 38 cM (5), 45 cM (16), 46 cM (6), 47 cM (6), 67 cM (6) 78 cM (24), 79 cM (10), 82 cM (6), 87 cM (5)
LG 7	137	133.97	0.98	109	5	4	31 cM (9), 52 cM (6), 53 cM (17), 68 cM (6)
LG 8	68	79.06	1.16	59	1	1	20 cM (6)
Total	1,291	845.56	-	995	16	60	
Average	161.38	105.70	0.74	124.38	2	7.5	

DArT marker system for chickpea

DArT markers are typically developed from a representation that is generated from a pool of DNA samples from a number of accessions, cultivars or breeding lines which as a group represent the genetic diversity within a species [16]. In the current study, high-density DArT arrays comprising of 15,360 clones were generated from genomic representations of 94 diverse genotypes (used as parents of mapping populations), genotypes from the reference set and wild genotypes exploited for introgression studies. A total of 5,397 (35.13%) markers were found polymorphic on the panel of 94 genotypes. Thus it is very evident that, compared to other marker technologies, DArT markers can be developed and typed quickly and cheaply [20]. The PIC values of DArT markers for chickpea germplasm are comparable to other germplasms such as sorghum [41] and cassava [42]. Ten percent of DArT markers had a PIC value in the range of 0.30 to 0.50, and these markers, therefore, may be considered useful or informative.

The most comprehensive genetic map of the chickpea genome

Despite the availability of a few hundred SSR markers in chickpea, putting them on the genetic map has been a challenging task due to the low level of polymorphism in cultivated chickpea germplasm [43]. MAS is most effective when markers are tightly linked to the gene of interest so that the probability of crossing-over between the gene and markers decreases. Moreover, map-based cloning requires very fine resolution mapping in the target interval, since the highest marker density can shorten chromosome walking. Hence, an inter-specific mapping population derived from ICC 4958 (*C. arietinum*) and PI 489777 (*C. reticulatum*) was used to integrate the novel markers developed in this study together with earlier published or some unpublished markers. This mapping population has been widely used by the chickpea community to incorporate several hundred microsatellite [7,8] and gene based markers [10,44]. The diverse genetic background of the parents showed higher degree of polymorphism not only at

Table 4. Comparison of the linkage groups (LGs) of the present reference genetic map with some key genetic maps.

Map developed in this study	Winter et al. [7]	Radhika et al. [31]	Nayak et al. [8]	Millán et al. [26]	
				Narrow crosses	Wide crosses
LG 1	LG 1	LG 2	LG 1, LG 3	LG 1	LG 1
LG 2	LG 2	LG 2, LG 3	LG 2	LG 2	LG 1, LG 2
LG 3	LG 3	LG 1	LG 3	-	LG 3
LG 4	LG 4, LG 11	LG 1, LG 2	LG 4	LG 4b	LG 4
LG 5	LG 7	LG 1	LG 5	-	LG 5
LG 6	LG 6	LG 1, LG 4	LG 2, LG 6	-	LG 6
LG 7	LG 5, LG 9, LG 15	LG 5	LG 5, LG 7	-	-
LG 8	LG 8	LG 3, LG 6	LG 8	-	-

the genetic level but also at phenotypic levels such as resistance to *Fusarium* wilt [7] and *Ascochyta* blight [45], thus, facilitating trait mapping. Consequently, this population could generally be considered as the "international reference" mapping population [8].

The integrated genetic map developed in this study comprises 157 novel SSR (CaM loci), 621 novel DArT marker loci, 145 GMM (81 COS-SNP, 35 CAPS, 18 CISR and 11 ICCeM) loci [10] and 368 legacy marker loci [7,8,11,46]. The current map with 1,291 loci is the most comprehensive genetic map ever reported for chickpea. Although the direct comparison of this map with the other published maps is not possible as different studies use different mapping programmes and criteria, the smaller map distance (845.56 cM) of the current map as compared to other published maps so far on this population indicates that this map is probably the densest genetic map for chickpea. Higher marker density can be attributed to: (i) the integration of large number of markers including both developed in this study and from other studies, and (ii) the use of JoinMap v 4 for calculating the map distance. In general, the maps constructed with JoinMap are shorter than those generated with a multi-locus likelihood package such as MAPMAKER [47]. The multi-locus likelihood method used by MAPMAKER assumes an absence of crossover interference, and, JoinMap allows interference and correctly produces shorter maps, even though both programs use the Kosambi mapping function [48]. The marker density of each individual linkage group ranged from 1 marker/0.46 cM (LG 6) to 1 marker/1.16 cM (LG 8). No correlation was found between the number of mapped markers and the length of linkage groups. For instance, LG 3 spans a distance of 98.66 cM with 193 markers, but LG 2 with only 98 loci spans a distance of 101.19 cM (Table 3). Similar observations were recorded in earlier chickpea mapping studies [7,8,26].

It is important to mention that the developed map has 94 marker loci that showed segregation distortion (p≤0.05). These markers have been retained in the map intentionally so that genetic information associated with such markers mapped in past (33 legacy markers and 8 GMM markers) or future (11 CaM and 42 DArT markers) may not be lost (Table S4). The legacy markers showed segregation distortion in earlier genetic mapping studies [7,8,26] and DArT markers have also shown segregation distortion in several species like triticale [49], wheat [50]. Comparison of this map with other key genetic maps [7,8,26,31] that contained majority of legacy markers showed a good congruency in terms of both marker orders as well as nomenclature of linkage groups was observed. These observations reconfirm the quality of the map and rule out the possibility of being concerned with the markers showing segregation distortion on the genetic map. As this map is the densest genetic map and includes the majority of the mapped loci available on the genetic map, this map could be considered and used as the reference genetic map of chickpea for developing and comparing new genetic maps in future.

The utility of our dense map could be demonstrated by fine mapping of locus CS27, the resistance locus for *Fusarium* wilt race 1 (*Foc1*), that was mapped onto LG 2 by Winter et al. [7] and the same LG of the present map. This locus is also linked to *Fusarium* race 4 (*Foc4*) and 5 (*Foc5*) at distances of 0.57 and 2.44 cM, respectively, on the current map. However, race 4 and 5 were 3.7 and 21.5 cM, respectively, away from locus CS27 in the map of Winter et al. [7]. Clustering of resistance genes for different races of pathogens and also different pathogens has been demonstrated in different crop plants including legumes [46]. The SSR markers flanking CS27, *Foc4*, *Foc5*, for instance TA37, H1J07 and

CaM0955 and other markers on either side of these loci can be employed in marker-assisted breeding programs. Of five resistance gene analogs (RGAs; RGA-D, RGA-Ds, RGA-A, RGA-C, RGA-B and RGA-G) mapped onto four linkage groups of chickpea [46], namely LG 2, 3, 5 and 6, only four RGAs (RGA-D, RGA-Ds, RGA-A, RGA-C, RGA-B) could be mapped onto three respective linkage groups (LG 2, LG 3 and LG 5).

With an objective of enhancing the utility of the reference genetic map for genetics research and breeding applications, the reference genetic map developed here has been divided into bins of 10 cM length. For several marker loci from these bins, we also have the information on PIC values or number of alleles. Information on distribution of marker loci into different bins along with the polymorphism features is an added-value. This will help geneticists and breeders to select an informative set of markers in appropriate numbers that represent the genome as well as display a high degree of polymorphism for developing new genetic maps, trait mapping and diversity analysis.

Uneven distribution of recombination in the chickpea genome

The present map indicates that recombination in chickpea, like some other plant species, is unevenly distributed with "hot-spots" and "cold-spots" across chromosomes. Clustering around centromeres is a well-known phenomenon with all types of markers, resulting from centromeric recombination suppression [51]. A set of 11 DArT markers were clustered near the centromeric region of LG 1. A remarkable clustering of DArT and BES-SSR markers was found in telomeric regions of LG 3 and LG 5. Although such clustering of markers was not reported in earlier mapping studies in chickpea, a stronger tendency of DArT markers towards clustering, as compared to SSR markers, in particular in gene-rich telomeric regions was shown in some crop species like wheat [52] and barley [53]. Markers sometimes tend to cluster, either as a consequence of an uneven distribution of recombination events along chromosomes, or because markers preferentially survey DNA polymorphism that is unevenly distributed along chromosomes [54,55]. For instance, clustering of *Pst*I-based DArT markers may reflect the abundance of *Pst*I restriction sites in hypomethylated telomeric chromosome regions [56].

In summary, this study reports the development of a set of 1,063 novel BES-SSR markers, of which 737 were polymorphic in the surveyed germplasm, and 157 could be integrated into the genetic map. Similarly, a DArT array with 15,360 clones was developed of which 5,397 were polymorphic in the surveyed germplasm, and 621 DArT loci were mapped. Using the above mentioned BES-SSR and DArT marker loci together with other marker datasets, a comprehensive genetic map with 1,291 marker loci has been developed. It is anticipated that the new markers and the dense genetic map will be useful for genetic analysis and breeding of chickpea and for the comparative study of genome evolution in legumes.

Materials and Methods

Plant material and DNA extraction

The reference chickpea genotype ICC 4958 was used for the construction of the BAC library. The developed set of BES-SSR markers were screened on ICC 4958 and ICC 1882, the parental genotypes of an intra-specific mapping population, for the amplification of SSR loci. Subsequently, a set of forty eight chickpea genotypes listed 1–48 in Table S2, were used for identification and characterizing of an informative set of the BES-SSR markers.

For the development of DArT arrays, a set of 94 genotypes (19-112 genotypes listed in Table S2) including parental genotypes of several mapping populations, diverse accessions from the reference set [57] and 19 accessions of wild *Cicer* species were used.

An F_{10} population comprising 131 recombinant inbred lines (RILs), derived from the inter-specific cross of ICC 4958 (*Cicer arietinum*) and PI 489777 (*C. reticulatum*), was used for screening and genotyping with the newly developed set of BES-SSR and DArT markers in this study and with the H-series markers [11].

DNA was extracted from the two weeks old seedlings of above mentioned genotypes using a high-throughput mini-DNA extraction as mentioned in Cuc et al. [58].

Construction of BAC library and generation of BAC-end sequences

The accession ICC 4958 was grown under greenhouse conditions for 6 weeks and transferred to continuous darkness for 2 days prior to use. Nuclei were isolated and embedded in low melting agarose, restriction digested with *Hin*dIII and size selected by two rounds of pulsed field gel electrophoresis (PFGE). Large size DNA fragments were ligated into vector pCCBAC1H and transformed into Epicenter's *E. coli* EPI300-T1R cells by electroporation.

A set of 25,000 BAC clones from the above library was prepared for end-sequencing at the J. Craig Venter Institute (JCVI), USA. Base calling and sequence trimming were performed as described in Bohra et al. [59].

Mining of SSRs in BESs and primer design

BESs were used for mining the SSRs with Perl based *MIcroSAtellite* (*MISA*) ([27], http://pgrc.ipk-gatersleben.de/misa/) search module which is capable of identifying perfect as well as compound SSRs. All BESs with a minimum size of 100 bp were arranged in a single FASTA format text file, and this file was used as an input for *MISA*. True and compound SSRs were classified through criteria defined by Nayak et al. [8].

In general, one SSR-containing BES was selected from each cluster for the design of the primer pairs, employing standalone Primer3 (http://frodo.wi.mit.edu/) program using *MISA* generated Primer3 input file [27].

SSR amplification and analysis

All BES-SSR markers and the H-series SSR markers were used for screening polymorphisms between the parents of the inter-specific mapping population. Subsequently the polymorphic SSR markers were applied to genotype all RILs. PCR amplification conditions and size separation procedures were the same as described in our earlier studies [8,59].

Development and genotyping of DArT arrays

A 15,360-clone DArT genomic library ('diversity array'- forty 384-well plates) was developed from a mixture of DNA samples of 94 chickpea genotypes included in the study (Table S2). Genomic representations for the diversity panel and genotyping were prepared by the complexity reduction method described by Yang et al. [60]. Briefly, ca. 100 ng of DNA were digested with restriction enzymes *Pst*I and *Hae*III (New England Biolabs, USA) and the *Pst*I adapter was simultaneously ligated. One µl of restriction/ligation reaction served as a template in a 50 µl amplification reaction with *Pst*I+0 primer. Adaptor and primer sequences and cycling conditions are given in the earlier study [60]. Arrays were hybridized with fluorescently labeled targets from all genotypes used for the array development [20,60].

For mapping the DArT markers, genomic representations were generated for all 131 RILs employing the same complexity reduction method (*Pst*I/*Hae*III) mentioned above. After overnight hybridization at 62°C, the slides were washed and scanned with a Tecan LS300 confocal laser scanner (Grödig, Salzburg, Austria). Individual samples were processed identically to samples for marker discovery and with similar marker quality thresholds in DArTsoft analysis [61].

Polymorphism information content (PIC) value

The PIC values for the SSR and DArT markers were calculated as mentioned in our earlier studies [8,41].

Linkage mapping

The genotyping data generated in this study as well as from other published studies [7,8,30] and collaborators were used for map construction with JoinMap v 4.0 ([62], www.kyazma.nl/index.php/mc.JoinMap). Prior to map construction, segregation ratios for both alleles (1:1) were tested for goodness of fit to assess deviations from the expected Mendelian segregation for all markers. Initially, markers showing goodness of fit were used for map construction, but later on markers showing segregation distortion were also attempted to be integrated into the map however always on >LOD 3.0. Linkage groups were determined based on "Independence test LOD score". Placement of markers into different linkage groups was done with "LOD groupings" and "Create group using the mapping tree" commands. Map calculations were performed with parameters like LOD value ≥2.0, recombination frequency ≤0.40 and a chi-square jump threshold for removal of loci = 5. Addition of a new locus may influence the optimum map order; hence a "Ripple" (enables to identify "the best marker order" by computing goodness-of-fit among three adjacent markers, for each order of the map) was performed after adding each marker into map. Map distances were calculated by the Kosambi mapping function [49], and the third round was set to allow mapping of an optimum number of loci into the genetic map. Mean chi-square contributions or average contributions to the goodness of fit of each locus were also checked to determine the best fitting position for markers in the genetic map. The markers showing negative map distances and large jumps in mean chi-square values were discarded from mapping. The final map was drawn with the help of MapChart v 2.2 [63]. The marker order of the current map was compared with already published maps using CMap v 1.01 (http://cmap.icrisat.ac.in/cmap/sm/cp/thudi/).

Supporting Information

Table S1   Details on novel simple sequence repeat (SSR) markers developed after mining bacterial artificial chromosome (BAC)- end sequences.

Table S2   List of 112 chickpea genotypes used for SSR and DArT analysis.

Table S3   Features of 5,397 polymorphic DArT loci based on marker analysis in 94 genotypes.

Table S4   Details on mapped markers showing segregation distortion (p<0.05).

Table S5 Bin wise description of the inter-specific reference genetic map of chickpea.

Acknowledgments

Authors are thankful to A BhanuPrakash, Ashish Kumar and Neha Gujaria for their help in various ways while preparing this MS. Thanks are also due to several collaborators for sharing their published/unpublished mapping data.

Author Contributions

Conceived and designed the experiments: RKV. Performed the experiments: MT AB SNN NV SG AK TMS. Analyzed the data: MT AB SNN NT PLK TMS. Contributed reagents/materials/analysis tools: PMG HDU PBKK PW GK RVP CDT AK DRC. Wrote the paper: RKV MT AB.

References

1. Arumuganathan K, Earle ED (1991) Nuclear DNA content of some important plant species. Plant Mol Biol Rep 9: 208–218.

2. Sharma HC, Varshney RK, Gaur PM, Gowda CLL (2008) Potential for using morphological, biochemical, and molecular markers for resistance to insect pests in grain legumes. J Food Legumes 21: 211–217.

3. Varshney RK, Thudi M, May GD, Jackson SA (2010) Legume genomics and breeding. Plant Breed Rev 33: 257–304.

4. Gaur PM, Gowda CLL (2005) Trends in world chickpea production, research and development. In: Knights EJ, Merril R, eds. Focus 2005: Chickpea in the farming systems. Queensland, Australia: Pulse Australia Publishers. pp 8–15.

5. Gupta PK, Varshney RK (2000) The development and use of microsatellite markers for genetics and plant breeding with emphasis on bread wheat. Euphytica 113: 163–185.

6. Winter P, Pfaff T, Udupa SM, Sharma PC, Sahi S, et al. (1999) Characterization and mapping of sequence-tagged microsatellite sites in the chickpea (C. arietinum L.) genome. Mol Gen Genet 262: 90–101.

7. Winter P, Benko-Iseppon AM, Hüttel B, Ratnaparkhe M, Tullu A, et al. (2000) A linkage map of the chickpea (Cicer arietinum L.) genome based on the recombinant inbred lines from a C. arietinum × C. reticulatum cross: localization of resistance genes for Fusarium races 4 and 5. Theor Appl Genet 101: 1155–1163.

8. Nayak SN, Zhu H, Varghese N, Datta S, Choi H-K, et al. (2010) Integration of novel SSR and gene-based SNP marker loci in the chickpea genetic map and establishment of new anchor points with Medicago truncatula genome. Theor Appl Genet 120: 1415–1441.

9. Varshney RK, Hiremath P, Lekha P, Kashiwagi J, Balaji J, et al. (2009) A comprehensive resource of drought- and salinity- responsive ESTs for gene discovery and marker development in chickpea (Cicer arietinum L.). BMC Genomics 10: 523.

10. Gujaria----- N, Kumar A, Dauthal P, Hiremath P, Bhanu Prakash A, et al. (2011) Development and use of genic molecular markers (GMMs) for construction of a transcript map of chickpea (Cicer arietinum L.). Theor Appl Genet 122: 1577–1589.

11. Lichtenzveig J, Scheuring C, Dodge J, Abbo S, Zhang H-B, et al. (2005) Construction of BAC and BIBAC libraries and their applications for generation of SSR markers for genome analysis of chickpea, Cicer arietinum L. Theor Appl Genet 110: 492–510.

12. Mun JH, Kim D-J, Choi H-K, Gish J, Debellé, et al. (2006) Distribution of microsatellites in the genome of Medicago truncatula: a resource of genetic markers that integrate genetic and physical maps. Genetics 172: 2541–2555.

13. Shultz JL, Samreen K, Rabia B, Jawaad AA, Lightfoot DA (2007) The development of BAC-end sequence-based microsatellite markers and placement in the physical and genetic maps of soybean. Theor Appl Genet 114: 1081–1090.

14. Schlueter JA, Lin JY, Schlueter SD, Vasylenko-Sanders IF, Deshpande S, et al. (2007) Gene duplication and paleopolyploidy in soybean and the implications for whole genome sequencing. BMC Genomics 8: 330.

15. Jiao Y, Wang Y, Xue D, Wang J, Yan M, et al. (2010) Regulation of OsSPL14 by OsmiR156 defines ideal plant architecture in rice. Nat Genet 42: 541–544.

16. Jaccoud D, Peng K, Feinstein D, Kilian A (2001) Diversity arrays: a solid state technology for sequence information independent genotyping. Nucleic Acids Res 29: E25.

17. Hearnden PR, Eckermann PJ, McMichael GL, Hayden MJ, Eglinton JK, et al. (2007) A genetic map of 1,000 SSR and DArT markers in a wide barley cross. Theor Appl Genet 115: 383–391.

18. Peleg Z, Saranga Y, Suprunova T, Ronin Y, Röder M (2008) High-density genetic map of durum wheat × wild emmer wheat based on SSR and DArT markers. Theor Appl Genet 117: 103–115.

19. Supriya A, Senthilvel S, Nepolean T, Eshwar K, Rajaram V, et al. (2011) Development of a molecular linkage map of pearl millet integrating DArT and SSR markers. Theor Appl Genet 123: 239–250.

20. Yang SY, Saxena RK, Kulwal PL, Ash GJ, Dubey A, et al. (2011) The first genetic map of pigeonpea based on diversity arrays technology (DArT) markers. J Genet 90: 103–109.

21. Varshney RK (2010) Gene-based marker systems in plants: high-throughput approaches for discovery and genotyping. In: Jain SM, Brar DS, eds. Molecular techniques in crop improvement. Dordrecht, The Netherlands: Springer. pp 119–142.

22. Rajesh PN, Muehlbauer FJ (2008) Discovery and detection of single nucleotide polymorphism (SNP) in coding and genomic sequences in chickpea (Cicer arietinum L.). Euphytica 162: 291–300.

23. Hwang TY, Sayama T, Takahashi M, Takada Y, Nakomoto Y, et al. (2009) High-density integrated linkage map based on SSR markers in soybean. DNA Res 6: 213–225.

24. Muchero W, Diop NN, Bhat PR, Fenton RD, Wanamaker S, et al. (2009) A consensus genetic map of cowpea [Vigna unguiculata (L.) Walp.] and synteny based on EST-derived SNPs. Proc Natl Acad Sci USA 106: 18159–18164.

25. Cordoba JM, Chavarro C, Schlueter JA, Jackson SA, Blair MW, et al. (2010) Integration of physical and genetic maps of common bean through BAC-derived microsatellite markers. BMC Genomics 11: 436.

26. Millán T, Winter P, Jungling R, Gil J, Rubio J, et al. (2010) A consensus genetic map of chickpea (Cicer arietinum L.) based on 10 mapping populations. Euphytica 175: 175–189.

27. Thiel T, Michalek W, Varshney RK, Graner A (2003) Exploiting EST databases for the development and characterization of gene-derived SSR-markers in barley (Hordeum vulgare L.). Theor Appl Genet 106: 411–422.

28. Varshney RK, Thiel T, Stein N, Langridge P, Graner A (2002) In silico analysis on frequency and distribution of microsatellites in ESTs of some cereal species. Cell Mol Biol Lett 7: 537–546.

29. Temnykh S, Declerck G, Lukashova A, Lipovich L, Cartinhour S, et al. (2001) Computational and experimental analysis of microsatellites in rice (Oryza sativa L.): frequency, length variation, transposon associations, and genetic marker potential. Genome Res 11: 1441–1452.

30. Hüttel B, Winter P, Weising K, Choumane W, Weigand F, et al. (1999) Sequence-tagged microsatellite site markers for chickpea (Cicer arietinum L.). Genome 42: 210–217.

31. Radhika P, Gowda SJM, Kadoo NY, Mhase LB, Jamadagni BM, et al. (2007) Development of an integrated intra-specific map of chickpea (Cicer arietinum L.) using two recombinant inbred line populations. Theor Appl Genet 115: 209–216.

32. Rajesh PN, Coyne C, Meksem K, Dev Sharma K, Gupta V, et al. (2004) Construction of a HindIII bacterial artificial chromosome library and its use in identification of clones associated with disease resistance in chickpea. Theor Appl Genet 108: 663–669.

33. Squirrell J, Hollingsworth PM, Woodhead M, Lowe A, Gibby M, et al. (2003) How much effort is required to isolate nuclear microsatellites from plants? Mol Ecol 12: 1339–1348.

34. Kofler R, Bartoš J, Gong L, Stift G, Suchankova P, et al. (2008) Development of microsatellite markers specific for the short arm of rye (Secale cereale L.) chromosome 1. Theor Appl Genet 117: 915–926.

35. Zhang X, Scheuring CF, Zhang M, Dong JJ, Zhang Y, et al. (2010) A BAC/BIBAC-based physical map of chickpea, Cicer arietinum L. BMC Genomics 11: 501.

36. Varshney RK, Graner A, Sorrells ME (2005) Genic microsatellite markers in plants: features and applications. Trends Biotechnol 23: 48–55.

37. Hiremath PJ, Farmer A, Cannon SB, Woodward J, Kudapa H, et al. (2011) Large-scale transcriptome analysis in chickpea (Cicer arietinum L.), an orphan legume crop of the semi-arid tropics of Asia and Africa. Plant Biotech Jour 9: 922–931.

38. Han ZG, Guo WZ, Song XL, Guo W, Guo J, et al. (2004) Genetic mapping of EST-derived microsatellites from the diploid Gossypium arboretum in allotetraploid cotton. Mol Genet Genomics 272: 308–327.

39. Dutta S, Kumawat G, Singh BP, Gupta DK, Singh S, et al. (2011) Development of genic-SSR markers by deep transcriptome sequencing in pigeonpea [Cajanus cajan (L.) Millspaugh]. BMC Plant Biology 11: 17.

40. Yu K, Park SJ, Poysa V, Gepts P (2000) Integration of simple sequence repeat (SSR) markers into a molecular linkage map of common bean (Phaseolus vulgaris L.). J Hered 91: 429–434.

41. Mace ES, Xia L, Jordan DR, Halloran K, Parh DK, et al. (2008) DArT markers: diversity analyses and mapping in Sorghum bicolor. BMC Genomics 9: 26.

42. Xia L, Peng K, Yang S, Wenzl P, de Vicente MC, et al. (2005) DArT for high-throughput genotyping of cassava (Manihot esculenta) and its wild relatives. Theor Appl Genet 110: 1092–1098.

43. Varshney RK, Hoisington DA, Upadhyaya HD, Gaur PM, Nigam SN, et al. (2007) Molecular genetics and breeding of grain legume crops for the semi-arid

tropics. In: Varshney RK, Tuberosa R, eds. Genomics assisted crop improvement, Vol II: Genomics applications in crops. Dordrecht, The Netherlands: Springer. pp 207–242.

44. Gaur R, Sethy NK, Choudhary S, Shokeen B, Gupta V, et al. (2011) Advancing the STMS genomic resources for defining new locations on the intra-specific genetic linkage map of chickpea (*Cicer arietinum* L.). BMC Genomics 12: 117.

45. Rakshit S, Winter P, Tekeoglu M, Muñoz J, Pfaff T, et al. (2003) DAF marker tightly linked to a major locus for *Ascochyta* blight resistance in chickpea (*Cicer arietinum* L.). Euphytica 132: 23–30.

46. Hüttel B, Santra D, Muehlbauer FJ, Khal G (2002) Resistance gene analogues of chickpea (*Cicer arietinum* L.): Isolation, genetic mapping and association with a *Fusarium* resistance gene cluster. Theor Appl Genet 105: 479–490.

47. Sewell MM, Sherman BK, Neale DB (1999) A consensus map for loblolly (*Pinus taeda* L.). I. Construction and integration of individual linkage maps from two outbred three-generation pedigree. Genetics 151: 321–330.

48. Kosambi DD (1944) The estimation of map distance from recombination values. Ann Eugen 12: 172–175.

49. Alheit KV, Reif JC, Maurer HP, Hahn V, Weissmann EA, et al. (2011) Detection of segregation distortion loci in triticale (x *Triticosecale* Wittmack) based on a high-density DArT marker consensus genetic linkage map. BMC Genomics 12: 380.

50. Semagn K, Bjørnstad Å, Skinnes H, Maroy AG, Tarkegne Y, et al. (2006) Distribution of DArT, AFLP, and SSR markers in a genetic linkage map of a doubled-haploid hexaploid wheat population. Genome 49: 545–555.

51. Korol, AB, Preygel IA, Preygel SI (1994) Recombination variability and evolution: algorithms of estimation and population-distribution study. Heredity 70: 254–265.

52. Akbari M, Wenzl P, Caig V, Carling J, Xia L, et al. (2006) Diversity arrays technology (DArT) for high-throughput profiling of the hexaploid wheat genome. Theor Appl Genet 113: 1409–1420.

53. Wenzl P, Li H, Carling J, Zhou M, Raman H, et al. (2006) A high-density consensus map of barley linking DArT markers to SSR, RFLP and STS loci and agricultural traits. BMC Genomics 7: 206.

54. Tanksley SD, Ganal MW, Prince JP, de Vicente MC (1992) High density molecular linkage maps of the tomato and potato genomes: biological inferences and practical applications. Genetics 132: 1141–1160.

55. Ramsay L, Macaulay M, Ivanissevich SD, MacLean K, Cardle L, et al. (2000) A simple sequence repeat-based linkage map of barley. Genetics 156: 1997–2005.

56. Moore G (2000) Cereal chromosome structure, evolution, and pairing. Annu Rev Plant Physiol Plant Mol Biol 51: 195–222.

57. Upadhyaya HD, Dwivedi SL, Baum M, Varshney RK, Udupa SM, et al. (2008) Genetic structure, diversity, and allelic richness in composite collection and reference set in chickpea (*Cicer arietinum* L.). BMC Plant Biol 8: 106.

58. Cuc LM, Mace E, Crouch J, Long TD, Varshney RK, et al. (2008) Isolation and characterization of novel microsatellite markers and their application for diversity assessment in cultivated groundnut (*Arachis hypogaea*). BMC Plant Biol 8: 55.

59. Bohra A, Dubey A, Saxena RK, Penmetsa RV, Poornima KN, et al. (2011) Analysis of BAC-end sequences (BESs) and development of BES-SSR markers for genetic mapping and hybrid purity assessment development in pigeonpea (*Cajanus* spp.). BMC Plant Biol 11: 56.

60. Yang S, Pang W, Ash G, Harper J, Carling J, et al. (2006) Low level of genetic diversity in cultivated pigeonpea compared to its wild relatives is revealed by diversity arrays technology (DArT). Theor Appl Genet 113: 585–595.

61. Kopecký D, Bartoš J, Lukaszewski AJ, Baird JH, Černoch V, et al. (2009) Development and mapping of DArT markers within the *Festuca - Lolium* complex. BMC Genomics 10: 473.

62. Van Ooijen (2006) JoinMap® 4, Software for the calculation of genetic linkage maps in experimental populations. Wageningen, Netherlands.

63. Voorips RE (2002) MapChart: software for the graphical presentation of linkage maps and QTLs. J Hered 93: 77–78.

8

Proper PIN1 Distribution Is Needed for Root Negative Phototropism in *Arabidopsis*

Kun-Xiao Zhang[1], Heng-Hao Xu[1], Wen Gong[1], Yan Jin[1], Ya-Ya Shi[2], Ting-Ting Yuan[1], Juan Li[1], Ying-Tang Lu[1]*

1 State Key Laboratory of Hybrid Rice, College of Life Sciences, Wuhan University, Wuhan, China, **2** Institute of Fruit and Tea, Hubei Academy of Agricultural Sciences, Wuhan, China

Abstract

Plants can be adapted to the changing environments through tropic responses, such as light and gravity. One of them is root negative phototropism, which is needed for root growth and nutrient absorption. Here, we show that the auxin efflux carrier PIN-FORMED (PIN) 1 is involved in asymmetric auxin distribution and root negative phototropism. In darkness, PIN1 is internalized and localized to intracellular compartments; upon blue light illumination, PIN1 relocalize to basal plasma membrane in root stele cells. The shift of PIN1 localization induced by blue light is involved in asymmetric auxin distribution and root negative phototropic response. Both blue-light-induced PIN1 redistribution and root negative phototropism is mediated by a BFA-sensitive trafficking pathway and the activity of PID/PP2A. Our results demonstrate that blue-light-induced PIN1 redistribution participate in asymmetric auxin distribution and root negative phototropism.

Editor: Malcolm Bennett, University of Nottingham, United Kingdom

Funding: This work was supported by the National Natural Science Foundation of China (#90917001) and Key Project of Chinese Ministry of Education (#311026) to YT Lu. The funders had no role in study design, data collection and analysis, decision to publish, or preparation of the manuscript.

Competing Interests: The authors have declared that no competing interests exist.

* E-mail: yingtlu@whu.edu.cn

Introduction

Plants are sessile by nature, and can be adapted to the changing environments through tropic responses, such as hypocotyl phototropism and root negative phototropism [1–3]. Whereas plant shoots can maximize capture of light source by hypocotyl phototropism, plant roots bend away from light source as root negative phototropic response to avoid the damage of light and other stressful stimulus from the upper layers of soil, and to facilitate water and nutrient absorption from the soil [2].

For the mechanism of tropic responses, a role for differential distribution of auxin was proposed in classical Cholodny-Went theory [4], in which asymmetric auxin distribution leads to unequal growth of two sides of a bending organ. In recent years, it has been reported that an increased *DR5* activity in the shaded side of hypocotyl is required for hypocotyl phototropism [5–8]. In contrast, higher *DR5* activity was demonstrated in the illuminated side of roots exposed to unilateral blue light in root negative phototropic response [9]. Notably, the asymmetric auxin distribution during tropic response is mediated by auxin transporters of the AUXIN RESISTANT/LIKE AUXIN RESISTANT, P-GLYCOPROTEIN, and PIN families [10–13].

Root negative phototropism and hypocotyl phototropism are specially regulated by blue light receptor PHOT1 [5,6,8,9,14–20], which perceives the blue light signals and translates into the auxin signaling pathway. Recently, it has been reported that the auxin efflux carriers PIN2 and PIN3 are necessary for asymmetric auxin transport and root negative phototropic response [9,19]. Upon unilateral blue light illumination, the shift of PIN2 localization that is controlled by blue light and BFA sensitive recycling can change

the auxin distribution in roots and result in root negative tropic response [19]. However, the polar localization of PIN2 only controls the basipetal flow of auxin to the elongation zone [21], which implies that the lateral auxin flow in the root tip underlying root negative phototropism also needs other auxin transporters. Recently, it has been indicated that unilateral blue light illumination polarizes PIN3 to the outer lateral membrane of columella cells at the illuminated root side, and increase auxin activity at the illuminated side of roots, where auxin promotes growth and causes roots bending away from the light source [9]. Moreover, the blue-light-induced PIN3 polarization in root negative phototropism is mediated by a BFA-sensitive, GNOM-dependent trafficking pathway and the activity of PID/PP2A. Interestingly, the polar distribution of PIN3 for hypocotyl phototropism, hypocotyl gravitropism and root gravitropism is also regulated by a BFA-sensitive trafficking pathway and the activity of PID [6,8,22–24].

Previous reports show that PIN1 is required for hypocotyl phototropic response [8,25] and the unilateral blue light illumination can result in PIN1 relocalization in hypocotyl cells in this process [25]. In gain-of-function *PID* mutants, which exhibit a collapsed root phenotype, PIN1 is relocated to the apical plasma membrane (PM) and the auxin gradient is disrupted in the roots [24,26]. Notably, several different *PP2A* mutants have a similar phenotype as the *PID* gain-of-function mutants [27], indicating the antagonistic regulation of PIN1 polarization by PID and PP2A. Recently, several PID-dependent Ser/Thr phosphorylation sites in PIN1 were identified that are involved in the basal-to-apical PIN1 polarity shift [28,29]. The shift in PIN1 localization also requires a BFA-sensitive trafficking pathway. BFA, a fungal toxin that inhibits GNOM, causes PIN1 to accumulate in endosomes called BFA compartments [30–32]. GNOM, a member of the ARF-GEF (exchange factors for ARF-GTPases) family, is needed for PIN1 recycling from endosomes to the plasma membrane [33].

In this study, we investigate the role of *PIN1*-regulated auxin distribution during root negative phototropic response. Our results show that blue light illumination can shift the PIN1 localization from intracellular compartments to the basal plasma membrane in root stele cells, which result in asymmetric auxin distribution and root negative phototropism. Moreover, the BFA-sensitive vesicle trafficking pathway and the activity of PID/PP2A are also needed for blue-light-induced PIN1 distribution and root negative phototropic response.

Results

PIN1 is Needed for Root Negative Phototropism and Asymmetric Auxin Distribution

Recently, it has been reported that auxin efflux carriers are involved in plant tropic response, such as hypocotyl phototropism, hypocotyl gravitropism, root gravitropism and root negative phototropism [5,6,8,9,19,22,25]. The reduced root negative phototropic response in *pin3-4* mutant implies the redundant function of auxin transporters in this process [9]. PIN1 as a key factor in hypocotyl phototropism may also participate in root negative phototropism. In order to test the contribution of PIN1 to the root negative phototropism, we used the loss-of-function *pin1* in the following experiments. Because *pin1* null mutants are completely sterile [34], *PIN1/pin1* heterozygous seedlings were used for the physiological experiments and *pin1* homozygous plants were identified by genotyping after the experiments. As expected, the detailed kinetics of root negative phototropic bending in *pin1* homozygous mutants confirmed that PIN1 is involved in root negative phototropic response (Fig. 1A–1C). Furthermore, the effects of polar auxin transport inhibitor N-1-naphthylphthalamic acid (NPA) in wild-type seedlings (Fig. 1B) indicated that polar auxin transport is needed for root negative phototropic response.

Next, we investigated whether PIN1 is needed for generating the asymmetric auxin distribution in root negative phototropic response. We used the auxin responsive *DR5$_{REV}$::GFP* line, which reliably reveals the pattern of auxin distribution in roots [35]. Consistent with previous report [9], an asymmetric *DR5* activity is detected in unilateral blue light illuminated roots of wild-type seedlings (Fig. 2A). In contrast, reduced asymmetry in *DR5* activity is observed in *pin1* homozygous mutants as compared with the wild-type plants (Fig. 2A–2C). These results reveal a role of PIN1 in root negative phototropism, and indicate that PIN1 activity is needed for the generation of asymmetric auxin distribution during root negative phototropic response.

The Shift of PIN1 Localization is Regulated by Blue Light

The directional flow of auxin is mediated by auxin polar transporters of the AUX and PIN families [10,11,13], and PIN1 is key factor in the asymmetric distribution of auxin during hypocotyl phototropism. The polarity of the subcellular localization of PIN1 has been shown to determine the acropetal flow of auxin and thereby regulate auxin redistribution in roots [36]. Thus, we analyzed the subcellular localization of PIN1 under the blue light illumination using the *PIN1::PIN1-GFP* marker line. In darkness, PIN1-GFP was internalized and lost from the plasma membrane (PM) in the root stele cells (Fig. 3A, and 3B). Endoplasmic reticulum (ER) and Golgi tracker dye staining indicated that PIN1-GFP localized to the ER and Golgi (Fig. 3A

and 3B), in addition to the vacuole in the dark [37,38]. However, upon unilateral blue light illumination, PIN1-GFP relocalized to the basal plasma membrane, as determined by co-localization with the FM 4–64 membrane stain (Fig. 3C). We also assayed the effect of blue light on PIN1 distribution in the roots of wild-type plants. After 15 minutes of blue light illumination, some PIN1-GFP protein began to localize to the basal PM of root stele cells, and PIN1-GFP continued to relocate to the basal PM for up to 2 h of light treatment, suggesting that blue light induces the redistribution of PIN1 (Fig. S1A–1D).

Both phototropin and cryptochrome are blue-light receptor families in *Arabidopsis*. To test whether these blue light receptors are involved in PIN1 redistribution, *cry1*, *phot2*, and *phot1* were crossed with the *PIN1::PIN1-GFP* line and the subcellular localization of PIN1-GFP was observed in the resulting progeny. When these progenies were treated with unilateral blue light, the roots of *cry1* and *phot2* plants, exhibit a normal root negative phototropic response [9] and normal distribution of PIN1-GFP at the basal PM of stele cells (Fig. 4A–4C, and 4E–4G). However, in *phot1* mutant, PIN1-GFP was still internalized and no visible effect on blue-light-induced PIN1 redistribution was detected in the roots, even after 2 h of unilateral blue light illumination (Fig. 4D, and 4H). Therefore, the blue-light-induced PIN1 redistribution is regulated by the blue-light receptor PHOT1.

Blue-Light-Induced PIN1 Redistribution is Regulated by a BFA-sensitive, GNOM-dependent Trafficking Pathway

The blue-light-induced PIN1 redistribution in root negative phototropism can result from *de novo* protein synthesis or degradation. To investigate whether protein synthesis is involved in blue-light-induced PIN1 redistribution in root negative phototropism, we used the protein synthesis inhibitor cycloheximide (CHX). Four-day-old etiolated wild-type seedlings were pretreated with CHX for 1 h in the dark, and then exposed to blue light illumination for 2 h. Our results showed that the blue-light-induced PIN1 redistribution occurred normally in *PIN1::PIN1-GFP* plants treated with CHX (Fig. S2A). These results suggest that *de novo* protein synthesis is not involved in the blue-light-induced PIN1 redistribution. In addition, the possible role of proteolytic protein degradation in this process was analyzed using MG132, an inhibitor of the 26S proteasome. No visible effect on blue-light-induced PIN1 redistribution was observed by using MG132 treatment (Fig. S2B). Furthermore, *snx1* mutants, defective in PIN2 degradation [39,40], showed normal PIN1 localization and root negative phototropic response under blue light illumination (Fig. S2C, and 2D). These results indicated that protein degradation does not participate in blue-light-induced PIN1 redistribution in root negative phototropic response.

PIN proteins are recycled constitutively between endosomes and the plasma membrane, and the recycling is sensitive to the vesicle trafficking inhibitor BFA [30,31]. Thus, the BFA-sensitive, vesicle trafficking pathway may be involved in blue light-induced PIN1 redistribution in root negative phototropism. To explore this possibility, BFA was used to test whether the BFA-sensitive trafficking pathway is involved in blue-light-induced PIN1 redistribution in root negative phototropism. BFA treatment strongly inhibited the shift localization of PIN1 from intracellular compartments to basal plasma membrane (Fig. 5A, and 5B). Furthermore, previous reports also demonstrated that the root negative phototropic response is affected by BFA treatment [9,19]. These results demonstrate that the BFA-sensitive trafficking pathway is involved in blue-light-induced PIN1 redistribution and root negative phototropism.

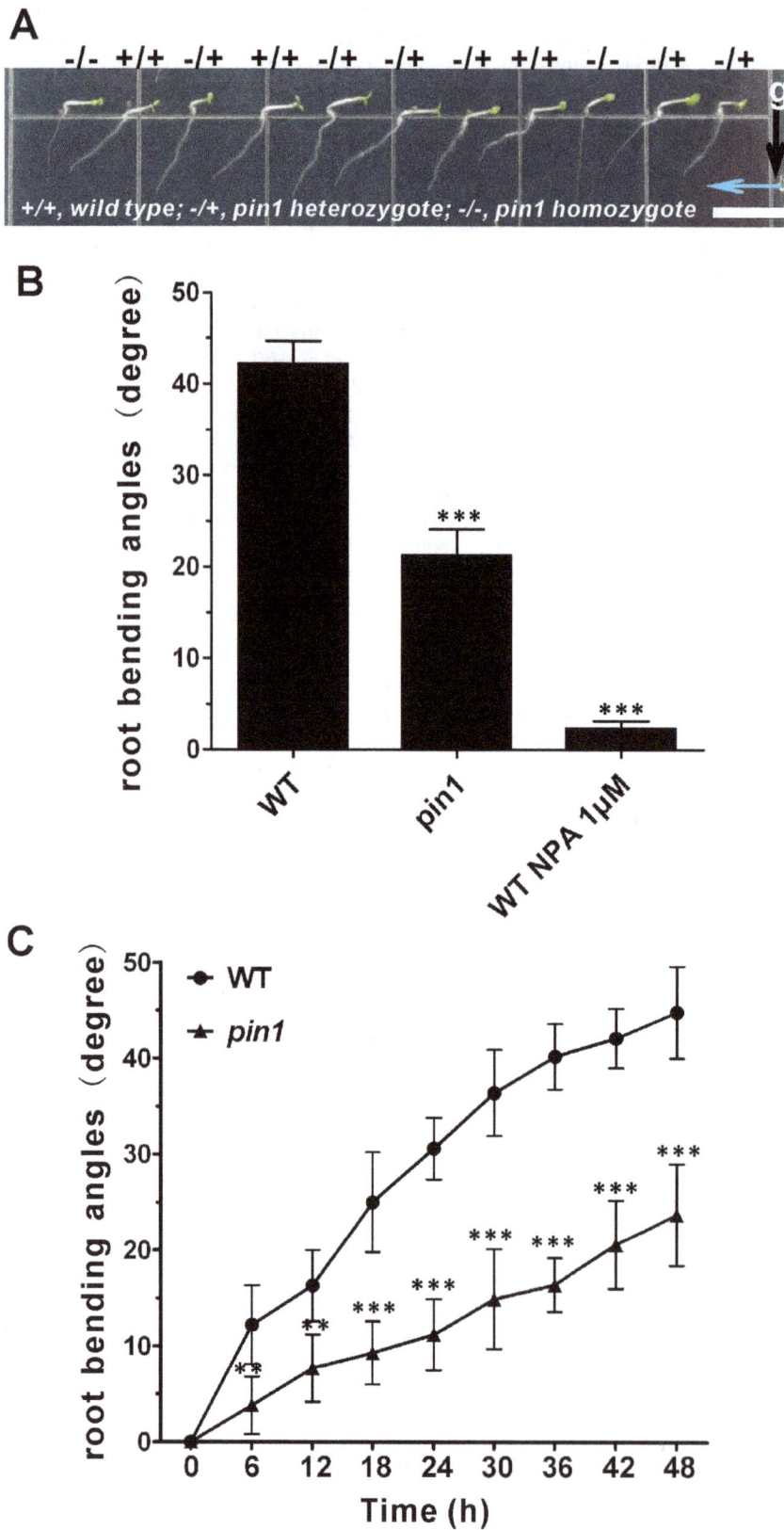

Figure 1. Root negative phototropic response of *pin1* mutant seedlings. (A) Images of 2-day-old etiolated seedlings of the *pin1* homozygous mutants grown on vertical plates, and then exposed to unilateral blue light (10 μmol m^{-2}sec^{-1}) for another 48 h. For the experiments with *pin1* mutant, the seeds from *pin1* heterozygote plants were used because *pin1* homozygote is infertile. After root bending assays were performed, the seedlings were identified to be wild-type, *pin1* homozygote or heterozygote by PCR. Only data for root bending from the seedlings of *pin1* homozygote were used for statistical analysis. The arrows indicate the direction of blue light (blue) and gravity (black). +/+, wild type; −/+, *pin1*

heterozygote; *1−/−*, *pin1* homozygote. Bar = 1 cm. (B) Root bending angles of *pin1* homozygous, WT and NPA-treated WT plants (1 µM). The bending angles of the roots away from the vertical direction were measured after 48 h unilateral blue light illumination and average curvatures were calculated. Values are the average of three biological replicates (n >10 per time point on each replicate). (C) Root bending kinetics of WT, and *pin1* homozygous seedlings. Root curvatures were measured every 6 hours under unilateral blue light illumination and average curvatures were calculated. Values are the average of three biological replicates (n >10 per time point on each replicate). Error bars represent SE and the symbols ** and *** indicate significant difference at P<0.01 or P<0.001 between WT and *pin1* or NPA treated WT in (B) or between WT and *pin1* at each time point in (C), as determined by Student's *t*-test.

GNOM has been reported to mediate PIN proteins recycling to the plasma membrane and is inhibited by BFA [33]. To test whether GNOM is involved in BFA-sensitive vesicle trafficking pathway for blue-light-induced PIN1 redistribution, the *GNOM^{M696L}* lines that express a genetically engineered BFA-resistant version of GNOM were used [33]. In *GNOM^{M696L}* roots, no visible differences on blue-light-induced PIN1 redistribution and root negative phototropic responses were detected in both the presence and absence of BFA (Fig. 5C, and 5D) [9]. In addition, it also has been shown that the partial loss-of-function *gnom^{R5}* mutants exhibit the reduced root negative phototropic response [9]. These results suggested that blue-light-induced PIN1 redistribution is regulated by BFA-sensitive, GNOM-dependent trafficking pathway.

Blue-Light-Induced PIN1 Distribution and Root Negative Phototropism are Mediated by *PID/PP2A*

Given that the shift in PIN1 polarity is mediated by the antagonistic PID/PP2A phosphorylation pathway [27], and that PID/PP2A-dependent PIN3 polarization is involved in root negative phototropism [9], the polar distribution of PIN1 in blue-light-induced root negative phototropic response may be also modulated by this pathway.

To test this hypothesis, we first examined the effect of PID on blue-light-induced PIN1 redistribution in root negative phototropism. Thus, *Pro35S:PID* seedlings constitutively expressing PID [26] were used. In the dark, the localization of PIN1-GFP was the same as in the wild type (Fig. 6A, and 6B). Upon unilateral blue light illumination, in the illuminated roots of *Pro35S:PID* plants,

which had severe defects in root negative phototropism [9], most of the PIN1-GFP localized to the apical plasma membrane (Fig. 6E). Given that WAG1 and WAG2 are the closest homologues of PID, the triple mutant *wag1 wag2 pid* was used for phenotype analyses [41,42]. This triple mutant *wag1 wag2 pid* showed the reduced root negative phototropic response (Fig. 6G, and 6H), consistent with a previous report [9]. These results suggest that the PID-mediated pathway is involved in blue-light-induced PIN1 redistribution during root negative phototropism.

PP2A phosphatase is also an important regulator of PIN1 apical-basal targeting and auxin distribution [27]. To explore the possible role of *PP2A* in blue-light-induced PIN1 redistribution, the *pp2aa1* (*rcn1*) mutant that lacks the phosphatase activity of PP2AA1 was used. In the dark, PIN1-GFP localization was the same in the *pp2aa1* mutant and wild type (Fig. 6A, and 6C). However, upon illumination, only some of the PIN1-GFP became polarized to the basal PM in *pp2aa1* seedlings (Fig. 6D, and 6F). Combined with our previous observation that *pp2aa1* had a reduced root negative phototropic response [9], these results demonstrate that PP2A activity is involved in blue-light-induced PIN1 redistribution during root negative phototropism.

Discussion

The classical Cholodny-Went theory [4] states that tropic responses are due to the asymmetric distribution of the growth regulator auxin. Recently, studies using the auxin response reporter *DR5* demonstrate that increased *DR5* activity on the shaded side of the hypocotyl is required for hypocotyl phototro-

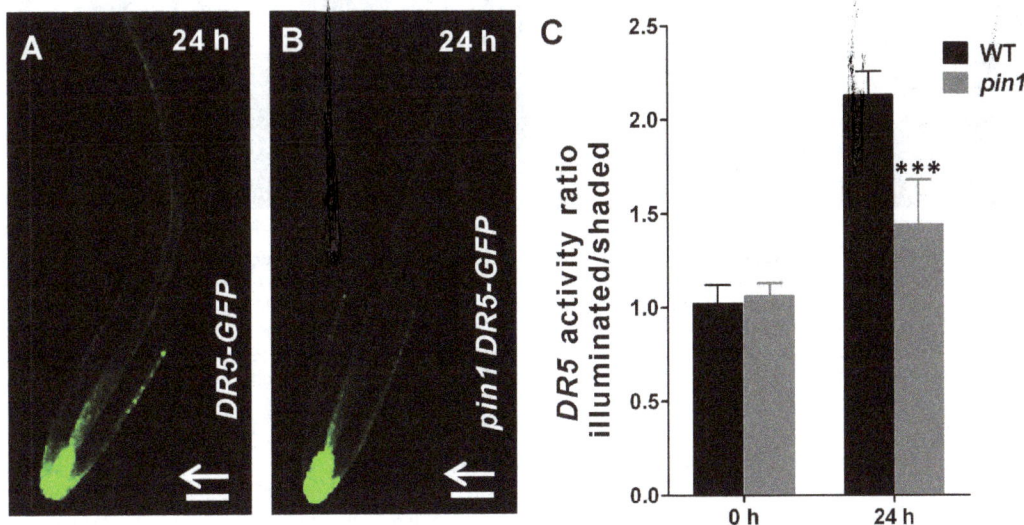

Figure 2. PIN1 activity is needed for asymmetric auxin distribution in root negative phototropism. (A–B) *DR5* activity was monitored in the *DR5_{REV}::GFP* (A) and *pin1 DR5_{REV}::GFP* (B) seedlings exposed to unilateral blue light illumination (10 µmol m^{-2}sec^{-1}) for 24 h. Arrows indicate blue light direction. Bars = 50 µm. (C) GFP signal intensities in (A–B) were quantified and their ratios at the illuminated side versus the shaded side are presented in (C). At least twelve seedlings were imaged per line for each of three replicates. Error bars represent standard deviation and *** indicate significant difference at P<0.001, as determined by Student's *t*-test.

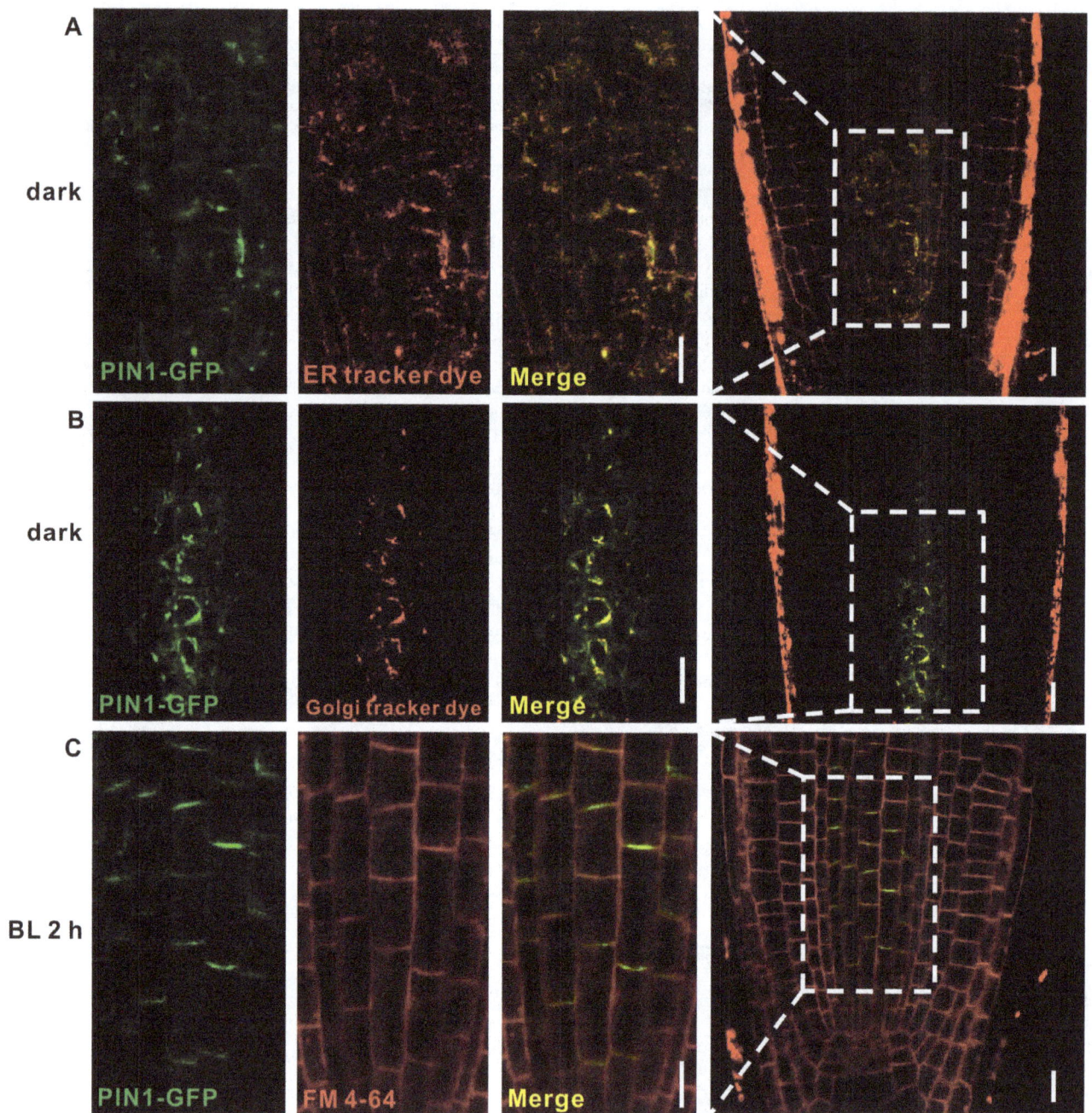

Figure 3. Subcellular localization of PIN1 in the roots of plants grown in the dark or exposed to blue light. (A–B) Co-localization (yellow) of PIN1-GFP (green) with ER or Golgi tracker dye (red) in the dark. Four-day-old etiolated seedlings of the *PIN1::PIN1-GFP* (green) marker line were pretreated with either ER tracker dye (A) or Golgi tracker dye (B) for 30 minutes in the dark before imaging. (C) Co-localization of PIN1-GFP (green) with FM 4-64 (red) after 2 h of blue light illumination. Four-day-old etiolated seedlings of the *PIN1::PIN1-GFP* marker line were exposed to blue light (10 μmol m^{-2}sec^{-1}) for 2 h, and treated with FM 4-64 for 10 minutes before imaging. (A–C) At least twelve seedlings were imaged per line for each of three replicates. Bars = 10 μm. Panels in the left are enlargements of the boxed regions shown in the rightmost column.

pism [5–7]. In this study, we showed that *DR5* activity is increased at the illuminated side of roots exposed to unilateral blue light (Fig. 2A) [9]. Moreover, we found that PIN1 is necessary for the generation of asymmetric auxin distribution and root negative phototropic response. Upon blue light illumination, blue light receptor PHOT1 modulates the expression of *PID* and *PP2A*. Furthermore, PID and PP2A antagonistically regulate the polar targeting of PIN1. The basal plasma membrane localization of PIN1 in root stele directs the acropetal flow of auxin to root tip

[36]. In addition, blue light induces the asymmetric distribution of PIN3 at the outer lateral membrane of columella cells illuminated with unilateral blue light, resulting in the flow of auxin to the illuminated side of roots [9]. Then, blue-light-induced apical PM localization of PIN2 in the epidermis cells of the root [19,40,43] directs the basipetal flow of auxin to the elongation zone [21]. The resultant asymmetric distribution of auxin promotes differential growth between the shaded and illuminated side of roots, resulting in root negative phototropism (Fig. S3).

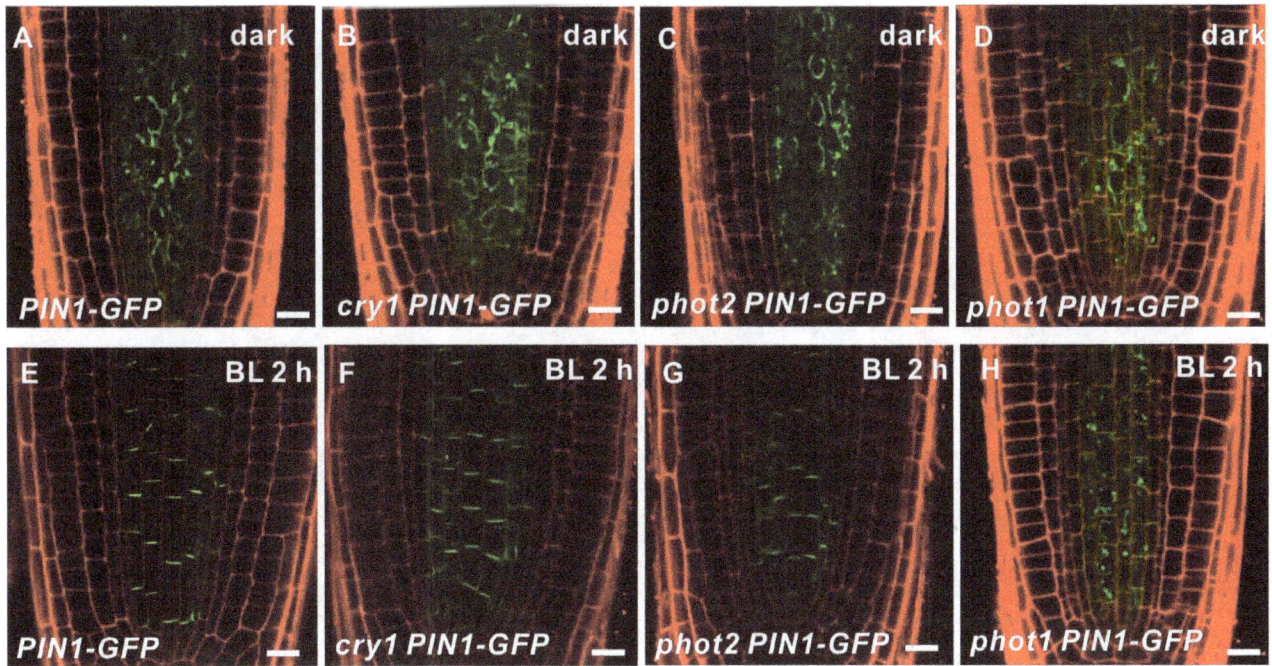

Figure 4. Blue-light-induced PIN1 is regulated by phot1 during root negative phototropism. (A–H) PIN1 localization, as revealed by GFP fluorescence, in the stele of *PIN1::PIN1-GFP* (A, E), *PIN1::PIN1-GFP cry1* (B, F), *PIN1::PIN1-GFP phot2* (C, G), and *PIN1::PIN1-GFP phot1* (D, H) seedlings grown in darkness (A–D) or under blue light (E–H) (10 μmol m^{-2}sec^{-1}) for 2 h. At least twelve seedlings were imaged per line for each of three replicates. Bars = 10 μm.

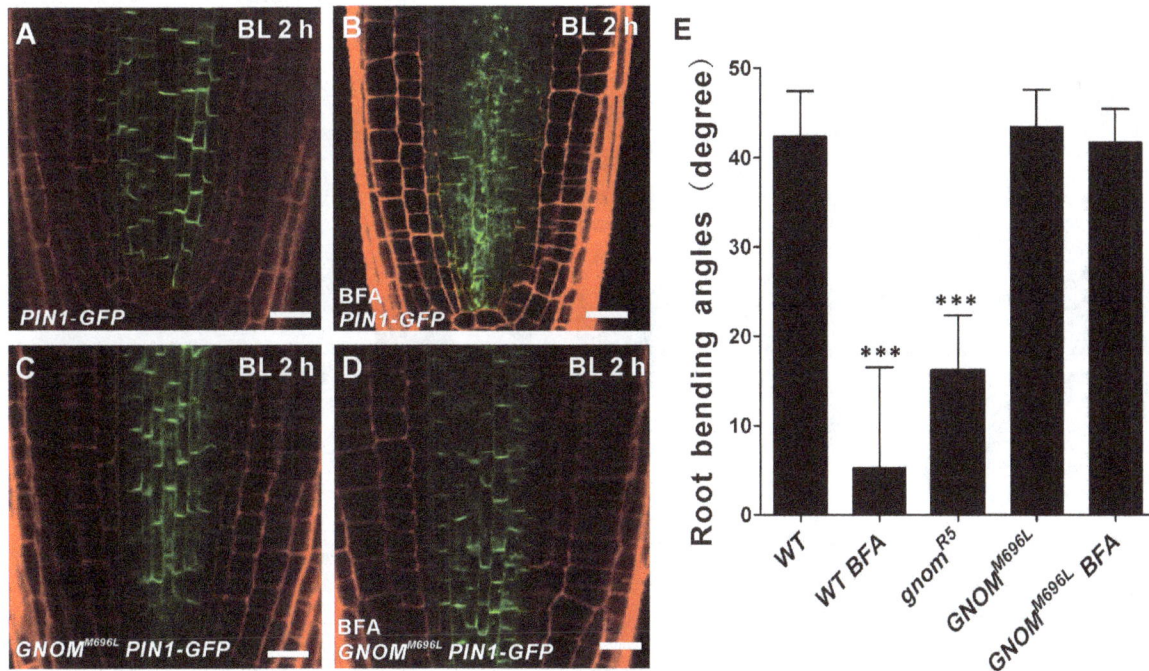

Figure 5. Involvement of the BFA-sensitive, GNOM-dependent trafficking pathway in blue-light-induced PIN1 distribution. (A–D) PIN1 localization, as revealed by GFP fluorescence, was examined in the steles of *PIN1::PIN1-GFP* (A or C) and *PIN1::PIN1-GFP GNOM^{M696L}* (B or D) seedlings treated or not with BFA and exposed to unilateral blue light for 2 h. Four-day-old etiolated seedlings were pretreated with DMSO as a control or BFA (50 μM) in the dark for 1 h, and subsequently exposed to unilateral blue light illumination (10 μmol m^{-2}sec^{-1}) for 2 h. At least twelve seedlings were imaged per line for each of three replicates. Arrows indicate blue light direction. Bars = 10 μm. (E) Root bending angles under BFA treatment. The bending angles of the roots away from the vertical direction were measured after 48 h unilateral blue light illumination and average curvatures were calculated. Values are the average of three biological replicates (n >10 per time point on each replicate). Error bars represent SE and *** indicate significant difference at P<0.001, as determined by Student's *t*-test.

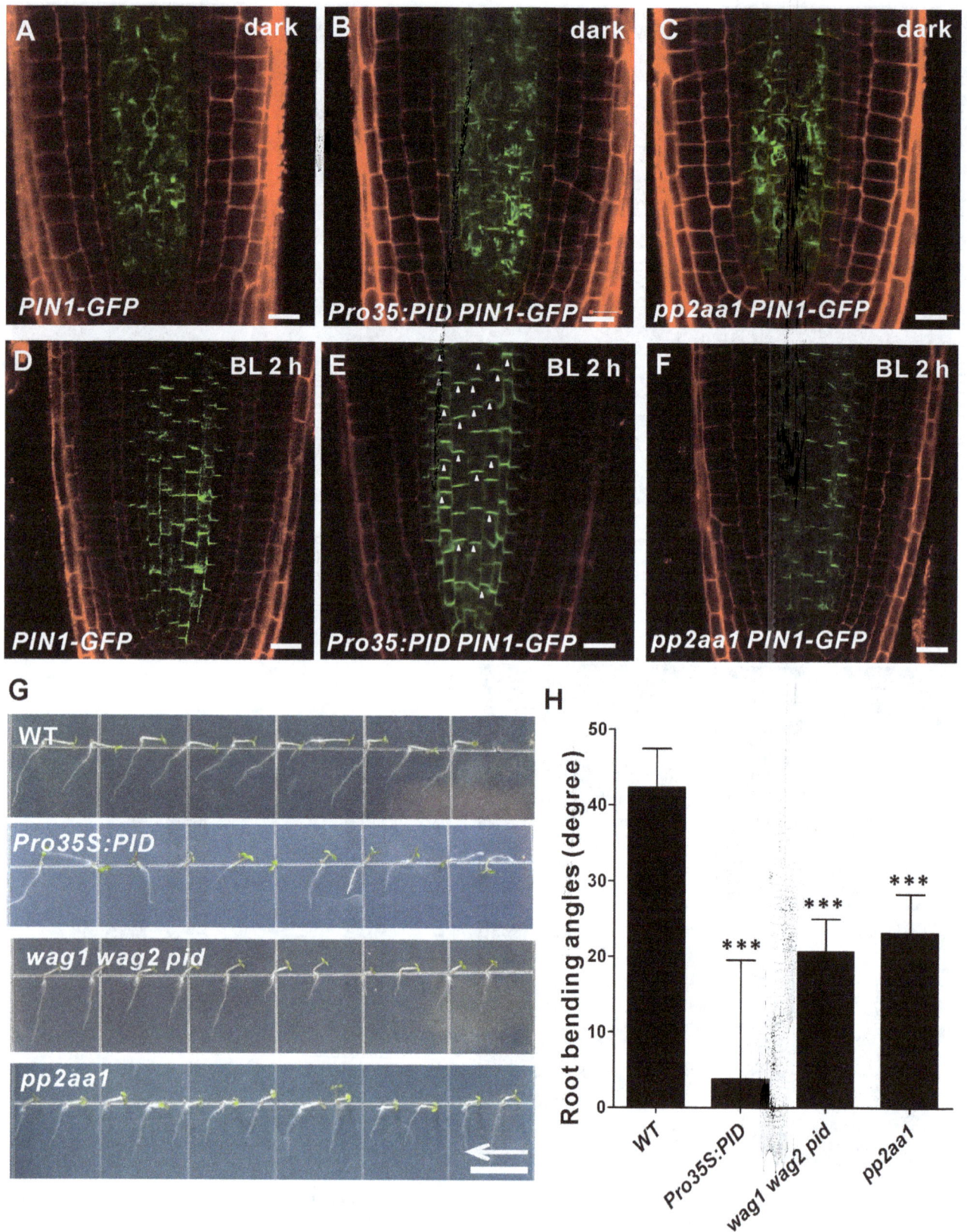

Figure 6. Blue-light-induced PIN1 distribution is mediated by PID/PP2A. (A–F) PIN1 localization, as revealed by GFP fluorescence, in the steles of 4-day-old etiolated *PIN1::PIN1-GFP* (A, D), *PIN1::PIN1-GFP Pro35S:PID* (B, D), and *PIN1::PIN1-GFP pp2aa1* (C, F) seedlings grown in darkness (A–C) or under blue light (D–F) (10 µmol m⁻²sec⁻¹) for 2 h. At least twelve seedlings were imaged per line for each of three replicates. Arrowheads show the apical plasma membrane localization of PIN1 in *Pro35S:PID* seedlings under blue light illumination. Arrows indicate blue light direction. Bars = 20 µm. (G) Images of 2-day-old etiolated seedlings of the wild-type, *Pro35S:PID*, *wag1 wag2 pid* and *pp2aa1* mutants grown on vertical plates,

and then exposed to unilateral blue light (10 µmol m^{-2}sec^{-1}) for another 48 h. The arrows indicate the direction of blue light. Bar = 1 cm. (H) Root bending angles of mutants. The bending angles of the roots away from the vertical direction were measured after 48 h unilateral blue light illumination and average curvatures were calculated. Values are the average of three biological replicates (n >10 per time point on each replicate). Error bars represent SE and *** indicate significant difference at P<0.001, as determined by Student's t-test.

Our data indicate that the blue-light-induced PIN1 distribution, which is regulated by the activity of PID and PP2AA1, is essential for root negative phototropism. It has been reported that PID/ PP2A antagonistically mediate PIN1 localization [24,27–29,32,44]. In these reports, the phosphorylation status of PIN1 have been shown to alter its polar localization in response to different environmental and endogenous cues [24,27–29,32,44]. While PID kinase can directly phosphorylate PIN1 to promote the apical localization of PIN1, PP2A phosphatase operates antagonistically to promote the basal localization of PIN1 in the embryo and the root [27]. The PID-dependent PIN1 polarization is regulated by a BFA-sensitive, GNOM-dependent trafficking pathway during organogenesis and development [32,42,45]. Furthermore, blue-light-induced PIN2 distribution is regulated by a BFA-sensitive trafficking pathway during root negative phototropism (Wan et al. 2012), and blue-light-induced PIN3 polarization is regulated by a BFA-sensitive, GNOM-dependent trafficking pathway during root negative phototropic response [9]. Thus, these data suggest that PID/PP2A-mediated PIN1 polarization via a BFA-sensitive, GNOM-dependent trafficking pathway is a universal mechanism to direct polar auxin transport in response to environmental and endogenous cues.

It has been reported that blue light can enhance the expression of DR5 activity in root stele (Wan et al. 2012). The increased auxin accumulation in root stele should be transported to the root tip, where the blue-light-induced PIN3 polarization can control the auxin flow to the illuminated side of roots [9]. Based on the analyses of the pin1 mutant, our results also indicated that PIN1 is involved in root negative phototropism. Notably, we also showed that blue light illumination can regulate the shift of PIN1 from intracellular compartments to basal plasma membrane in root stele cells, which is essential for the acropetal transport of auxin to increae auxin accumulation in the root tip. Thus, the blue-light-induced PIN1 redistribution in root stele cells is needed for the asymmetric auxin distribution during root negative phototropic response.

In this report, our mutant analyses revealed that PID and PP2A are essential for root negative phototropic response. Both the PID over-expression line and the pp2aa1 mutant can disturb PIN1 polar targeting in root stele cells with abnormal root negative phototropic response. Also, other reports indicate that PID and PP2A can antagonistically regulate PIN1 phosphorylation to direct auxin flux and that PID-dependent phosphorylation pathway is needed for PIN3 polarization in hypocotyl phototropism [6,27]. However, it has been reported that PID, as well as WAG1 and WAG2, is not expressed in root stele [42], implying that PID cannot regulate PIN1 polar targeting through physical interaction. Thus, how these kinases modulate PIN1 distribution in root stele remains unknown in root negative phototropic response.

Increased auxin activity at the illuminated side of roots is expected to induce the expression of auxin response factors that promote asymmetric growth and root bending away from the light source. ARF7/NPH4, an auxin response factor, has been reported to be involved in hypocotyl phototropism [46,47]. As expected, we found that the arf7-1 mutant exhibits a reduced root negative phototropic response (Fig. S4A and B), indicating the involvement of ARF7 in this process.

Materials and Methods

Plant Material

The following published transgenic and mutant lines were used in this study: DR5$_{REV}$:GFP [35]; PIN1:PIN1-GFP [36]; GNOM^{M696L} [33]; pin1 [8]; Pro35S:PID [26]; pp2aa1 [48]; cry1 (SALK_069292); phot1 (SALK_146058); phot2 (SALK_142275); wag1 wag2 pid [42]; and arf7-1 (SALK_040394). The double or triple mutants and/or transgenic lines were obtained by crossing the respective lines above, were confirmed by PCR and are available upon request. All PCR primers used for genotyping are listed in Table S1.

Plant Growth and Light Conditions

Arabidopsis thaliana seeds were surface sterilized with 5% bleach for 5 min, washed three times with sterile water, and plated on agar medium containing half-strength Murashige and Skoog medium (1962) and 0.8% agar (w/v). Seedlings were grown in darkness for 2–4 days at 22°C and stimulated by blue light for further analysis.

The blue light (nm = 475) was provided by an R30 LED Light (enLux) and LH-100SP-LED (NK system). Light fluence rates were measured by a Li250 quantum photometer (Li-Cor, www.licor.com). All experiments in darkness were carried out under a dim green safe light.

Measurement of Root Negative Phototropism

Arabidopsis thaliana seeds were grown in the dark for 2 days and then transferred to unilateral blue light illumination for 2 days as described previously [14]. Root bending angles were analyzed using Image J software (http://rsb.info.nih.gov/ij/) and plotted using Prism 5.0 software (GraphPad, www.graphpad.com). At least three independent experiments were carried out.

Pharmacological Treatments

Four-day-old etiolated seedlings on half-strength Murashige and Skoog medium were treated with brefeldin A (BFA; 50 µM), cycloheximide (CHX; 50 µM), or carbobenzoxy-Leucyl-Leucyl-leucinal (MG132; 50 µM) for 1 h in the dark. The seedlings were then exposed to unilateral blue light and imaged. In experiments involving the phototropic response, 2-day-old etiolated seedlings grown on plates without drug treatment were transferred to solid half-strength Murashige and Skoog medium containing NPA (1 µM) and exposed to unilateral blue light for 2 days. In control experiments, seedlings were treated with an equal amount of solvent (DMSO). Propidium iodide (PI; 0.05%) was dissolved in distilled water. ER tracker dye (1 µM), Golgi tracker dye (330 µg/ml), and FM 4–64 (5 µM) were used to define the localization of PIN1-GFP. Each experiment was performed at least three times.

Confocal Microscopy

An Olympus (www.olympus.com) FV1000 ASW confocal scanning microscope was used. Emission wavelengths were as follows: PI, 600 to 640 nm; FM 4–64, 600 to 700 nm and GFP, 500 to 540 nm. Four-day-old etiolated seedlings were grown in the dark and stimulated with unilateral blue light. Etiolated seedlings were placed on slides and shoots were removed with a blade, leaving only the roots for quick confocal observations. The signal intensity was measured using Photoshop CS4 and Image J software (http://rsb.info.nih.gov/ij/). The fluorescence intensity ratios were obtained by comparing DR5-GFP fluorescence intensities between the illuminated side and shaded side of the root in the responsive part. At least twelve seedlings were imaged per line for each of three replicates.

Supporting Information

Figure S1   The effect of blue light on PIN1 distribution over time. (A–D) PIN1 localization, as revealed by GFP fluorescence, in the root stele cells of *PIN1::PIN1-GFP* plants grown in darkness (A) and then exposed to unilateral blue light (10 μmol m^{-2}sec^{-1}) for 15 min (B), 30 min (C) or 2 h (D).

Figure S2   The *de novo* protein synthesis and degradation are not involved in root negative phototropism. (A–C) PIN1 localization, as revealed by GFP fluorescence, in the stele of *PIN1::PIN1-GFP* plants treated with CHX (A), MG132 (B) or *snx1 PIN1::PIN1-GFP* mutants (C). Four-day-old etiolated seedlings of the *ProPIN1:PIN1-GFP* marker line were pretreated with CHX (50 μM) or MG132 (50 μM) in the dark for 1 h, and then subsequently exposed to unilateral blue light illumination (10 μmol m^{-2}sec^{-1}) for 2 h. Bars = 10 μm. (D) Root bending angles of wild-type and *snx1* mutants. The bending angles of the roots away from the vertical direction were measured after 48 h unilateral blue light illumination (10 μmol m^{-2}sec^{-1}) and average curvatures were calculated. Values are the average of three biological replicates (n >10 per time point on each replicate). Error bars represent SE.

Figure S3   Model for root negative phototropism. Based on the model for hypocotyl phototropism [6], the regulation pathway of root negative phototropism is summarized in Fig. S3. Upon blue light illumination, blue light receptor PHOT1 modulates the expression of *PID* and *PP2A*. Furthermore, PID and PP2A antagonistically regulate the polar targeting of PIN1.

References

1. Estelle M (1996) Plant tropisms: the ins and outs of auxin. Curr Biol 6: 1589–1591.
2. Esmon CA, Pedmale UV, Liscum E (2005) Plant tropisms: providing the power of movement to a sessile organism. Int J Dev Biol 49: 665–674.
3. Holland JJ, Roberts D, Liscum E (2009) Understanding phototropism: from Darwin to today. J Exp Bot 60: 1969–1978.
4. Went FW (1974) Reflections and speculations. Annu Rev Plant Physiol 25: 1–26.
5. Christie JM, Yang H, Richter GL, Sullivan S, Thomson CE, et al. (2011) phot1 inhibition of ABCB19 primes lateral auxin fluxes in the shoot apex required for phototropism. PLoS Biol 9: e1001076.
6. Ding Z, Galvan-Ampudia CS, Demarsy E, Langowski L, Kleine-Vehn J, et al. (2011) Light-mediated polarization of the PIN3 auxin transporter for the phototropic response in Arabidopsis. Nat Cell Biol 13: 447–452.
7. Friml J, Wisniewska J, Benkova E, Mendgen K, Palme K (2002) Lateral relocation of auxin efflux regulator PIN3 mediates tropism in Arabidopsis. Nature 415: 806–809.
8. Haga K, Sakai T (2012) PIN Auxin Efflux Carriers Are Necessary for Pulse-Induced But Not Continuous Light-Induced Phototropism in Arabidopsis. Plant Physiol 160: 763–776.
9. Zhang KX, Xu HH, Yuan TT, Zhang L, Lu YT (2013) Blue light-induced PIN3 polarization for root negative phototropic response in Arabidopsis. Plant J.
10. Petrasek J, Mravec J, Bouchard R, Blakeslee JJ, Abas M, et al. (2006) PIN proteins perform a rate-limiting function in cellular auxin efflux. Science 312: 914–918.
11. Geisler M, Murphy AS (2006) The ABC of auxin transport: the role of p-glycoproteins in plant development. FEBS Lett 580: 1094–1102.
12. Swarup R, Kramer EM, Perry P, Knox K, Leyser HM, et al. (2005) Root gravitropism requires lateral root cap and epidermal cells for transport and response to a mobile auxin signal. Nat Cell Biol 7: 1057–1065.
13. Yang Y, Hammes UZ, Taylor CG, Schachtman DP, Nielsen E (2006) High-affinity auxin transport by the AUX1 influx carrier protein. Curr Biol 16: 1123–1127.

The basal plasma membrane localization of PIN1 in root stele directs the acropetal flow of auxin to root tip. In addition, blue light induces the asymmetric distribution of PIN3 at the outer lateral membrane of columella cells illuminated with unilateral blue light, resulting in the flow of auxin to the illuminated side of roots. Then, blue-light-induced apical PM localization of PIN2 in the epidermis cells of the root directs the basipetal flow of auxin to the elongation zone. The resultant asymmetric distribution of auxin promotes differential growth between the shaded and illuminated side of roots, resulting in root negative phototropism.

Figure S4   NPH4/ARF7 is involved in root negative phototropism. (A–B) The phenotypes (A) and root bending angles (B) in the wild-type and *arf7-1* mutant. Two-day-old etiolated seedlings of the wild-type and *arf7-1* mutant were exposed to unilateral blue light illumination (10 μmol m^{-2}sec^{-1}) for 2 days. Values are the average of three biological replicates (n >10 per time point on each replicate). The arrows indicate the direction of blue light (blue) and gravity (black). Error bars represent SE and *** indicate significant difference at p<0.001, as determined by Student's *t*-test. Bar = 0.5 cm.

Table S1   Primers for genotyping analysis.

Acknowledgments

We thank J. Friml, G. Jrgens, and Jian Xu for sharing their published materials. We thank for Dr. D. Patrick Bastedo (University of Toronto) critical reading.

Author Contributions

Conceived and designed the experiments: KXZ HHX YTL. Performed the experiments: KXZ HHX WG YJ YYS TTY JL. Wrote the paper: KXZ YTL.

14. Sakai T, Wada T, Ishiguro S, Okada K (2000) RPT2. A signal transducer of the phototropic response in Arabidopsis. Plant Cell 12: 225–236.
15. Sakai T, Kagawa T, Kasahara M, Swartz TE, Christie JM, et al. (2001) Arabidopsis nph1 and npl1: blue light receptors that mediate both phototropism and chloroplast relocation. Proc Natl Acad Sci U S A 98: 6969–6974.
16. Christie JM, Reymond P, Powell GK, Bernasconi P, Raibekas AA, et al. (1998) Arabidopsis NPH1: a flavoprotein with the properties of a photoreceptor for phototropism. Science 282: 1698–1701.
17. Boccalandro HE, De Simone SN, Bergmann-Honsberger A, Schepens I, Fankhauser C, et al. (2008) PHYTOCHROME KINASE SUBSTRATE1 regulates root phototropism and gravitropism. Plant Physiol 146: 108–115.
18. Liscum E, Briggs WR (1995) Mutations in the NPH1 locus of Arabidopsis disrupt the perception of phototropic stimuli. Plant Cell 7: 473–485.
19. Wan Y, Jasik J, Wang L, Hao H, Volkmann D, et al. (2012) The Signal Transducer NPH3 Integrates the Phototropin1 Photosensor with PIN2-Based Polar Auxin Transport in Arabidopsis Root Phototropism. Plant Cell 24: 551–565.
20. Kutschera U, Briggs WR (2012) Root phototropism: from dogma to the mechanism of blue light perception. Planta 235: 443–452.
21. Muller A, Guan C, Galweiler L, Tanzler P, Huijser P, et al. (1998) AtPIN2 defines a locus of Arabidopsis for root gravitropism control. EMBO J 17: 6903–6911.
22. Rakusova H, Gallego-Bartolome J, Vanstraelen M, Robert HS, Alabadi D, et al. (2011) Polarization of PIN3-dependent auxin transport for hypocotyl gravitropic response in Arabidopsis thaliana. Plant J 67: 817–826.
23. Sukumar P, Edwards KS, Rahman A, Delong A, Muday GK (2009) PINOID kinase regulates root gravitropism through modulation of PIN2-dependent basipetal auxin transport in Arabidopsis. Plant Physiol 150: 722–735.
24. Friml J, Yang X, Michniewicz M, Weijers D, Quint A, et al. (2004) A PINOID-dependent binary switch in apical-basal PIN polar targeting directs auxin efflux. Science 306: 862–865.

25. Blakeslee JJ, Bandyopadhyay A, Peer WA, Makam SN, Murphy AS (2004) Relocalization of the PIN1 auxin efflux facilitator plays a role in phototropic responses. Plant Physiol 134: 28–31.

26. Benjamins R, Quint A, Weijers D, Hooykaas P, Offringa R (2001) The PINOID protein kinase regulates organ development in Arabidopsis by enhancing polar auxin transport. Development 128: 4057–4067.

27. Michniewicz M, Zago MK, Abas L, Weijers D, Schweighofer A, et al. (2007) Antagonistic regulation of PIN phosphorylation by PP2A and PINOID directs auxin flux. Cell 130: 1044–1056.

28. Huang F, Zago MK, Abas L, van Marion A, Galvan-Ampudia CS, et al. (2010) Phosphorylation of conserved PIN motifs directs Arabidopsis PIN1 polarity and auxin transport. Plant Cell 22: 1129–1142.

29. Zhang J, Nodzynski T, Pencik A, Rolcik J, Friml J (2010) PIN phosphorylation is sufficient to mediate PIN polarity and direct auxin transport. Proc Natl Acad Sci U S A 107: 918–922.

30. Geldner N, Friml J, Stierhof YD, Jurgens G, Palme K (2001) Auxin transport inhibitors block PIN1 cycling and vesicle trafficking. Nature 413: 425–428.

31. Dhonukshe P, Aniento F, Hwang I, Robinson DG, Mravec J, et al. (2007) Clathrin-mediated constitutive endocytosis of PIN auxin efflux carriers in Arabidopsis. Curr Biol 17: 520–527.

32. Kleine-Vehn J, Huang F, Naramoto S, Zhang J, Michniewicz M, et al. (2009) PIN auxin efflux carrier polarity is regulated by PINOID kinase-mediated recruitment into GNOM-independent trafficking in Arabidopsis. Plant Cell 21: 3839–3849.

33. Geldner N, Anders N, Wolters H, Keicher J, Kornberger W, et al. (2003) The Arabidopsis GNOM ARF-GEF mediates endosomal recycling, auxin transport, and auxin-dependent plant growth. Cell 112: 219–230.

34. Okada K, Ueda J, Komaki MK, Bell CJ, Shimura Y (1991) Requirement of the Auxin Polar Transport System in Early Stages of Arabidopsis Floral Bud Formation. Plant Cell 3: 677–684.

35. Friml J, Vieten A, Sauer M, Weijers D, Schwarz H, et al. (2003) Efflux-dependent auxin gradients establish the apical-basal axis of Arabidopsis. Nature 426: 147–153.

36. Friml J, Benkova E, Blilou I, Wisniewska J, Hamann T, et al. (2002) AtPIN4 mediates sink-driven auxin gradients and root patterning in Arabidopsis. Cell 108: 661–673.

37. Shirakawa M, Ueda H, Shimada T, Nishiyama C, Hara-Nishimura I (2009) Vacuolar SNAREs function in the formation of the leaf vascular network by regulating auxin distribution. Plant Cell Physiol 50: 1319–1328.

38. Sassi M, Lu Y, Zhang Y, Wang J, Dhonukshe P, et al. (2012) COP1 mediates the coordination of root and shoot growth by light through modulation of PIN1- and PIN2-dependent auxin transport in Arabidopsis Development 139: 3402–3412.

39. Jaillais Y, Fobis-Loisy I, Miege C, Rollin C, Gaude T (2006) AtSNX1 defines an endosome for auxin-carrier trafficking in Arabidopsis. Nature 443: 106–109.

40. Kleine-Vehn J, Leitner J, Zwiewka M, Sauer M, Abas L, et al. (2008) Differential degradation of PIN2 auxin efflux carrier by retromer-dependent vacuolar targeting. Proc Natl Acad Sci U S A 105: 17812–17817.

41. Cheng Y, Qin G, Dai X, Zhao Y (2008) NPY genes and AGC kinases define two key steps in auxin-mediated organogenesis in Arabidopsis. Proc Natl Acad Sci U S A 105: 21017–21022.

42. Dhonukshe P, Huang F, Galvan-Ampudia CS, Mahonen AP, Kleine-Vehn J, et al. (2010) Plasma membrane-bound AGC3 kinases phosphorylate PIN auxin carriers at TPRXS(N/S) motifs to direct apical PIN recycling. Development 137: 3245–3255.

43. Laxmi A, Pan J, Morsy M, Chen R (2008) Light plays an essential role in intracellular distribution of auxin efflux carrier PIN2 in Arabidopsis thaliana. PLoS One 3: e1510.

44. Wisniewska J, Xu J, Seifertova D, Brewer PB, Ruzicka K, et al. (2006) Polar PIN localization directs auxin flow in plants. Science 312: 883.

45. Sorefan K, Girin T, Liljegren SJ, Ljung K, Robles P, et al. (2009) A regulated auxin minimum is required for seed dispersal in Arabidopsis. Nature 459: 583–586.

46. Harper RM, Stowe-Evans EL, Luesse DR, Muto H, Tatematsu K, et al. (2000) The NPH4 locus encodes the auxin response factor ARF7, a conditional regulator of differential growth in aerial Arabidopsis tissue. Plant Cell 12: 757–770.

47. Okushima Y, Overvoorde PJ, Arima K, Alonso JM, Chan A, et al. (2005) Functional genomic analysis of the AUXIN RESPONSE FACTOR gene family members in Arabidopsis thaliana: unique and overlapping functions of ARF7 and ARF19. Plant Cell 17: 444–463.

48. Garbers C, DeLong A, Deruere J, Bernasconi P, Soll D (1996) A mutation in protein phosphatase 2A regulatory subunit A affects auxin transport in Arabidopsis. EMBO J 15: 2115–2124.

9

Estimating Gene Flow between Refuges and Crops: A Case Study of the Biological Control of *Eriosoma lanigerum* by *Aphelinus mali* in Apple Orchards

Blas Lavandero[1]*, **Christian C. Figueroa**[3], **Pierre Franck**[2], **Angela Mendez**[1]

1 Laboratorio de Interacciones Insecto-Planta, Universidad de Talca, Talca, Chile, 2 Plantes et Systèmes de culture Horticoles, INRA, Avignon, France, 3 Facultad de Ciencias, Instituto de Ecología y Evolución, Universidad Austral de Chile, Valdivia, Chile

Abstract

Parasitoid disturbance populations in agroecosystems can be maintained through the provision of habitat refuges with host resources. However, specialized herbivores that feed on different host plants have been shown to form host-specialized races. Parasitoids may subsequently specialize on these herbivore host races and therefore prefer parasitizing insects from the refuge, avoiding foraging on the crop. Evidence is therefore required that parasitoids are able to move between the refuge and the crop and that the refuge is a source of parasitoids, without being an important source of herbivore pests. A North-South transect trough the Chilean Central Valley was sampled, including apple orchards and surrounding *Pyracantha coccinea* (M. Roem) (Rosales: Rosacea) hedges that were host of *Eriosoma lanigerum* (Hemiptera: Aphididae), a globally important aphid pest of cultivated apples. At each orchard, aphid colonies were collected and taken back to the laboratory to sample the emerging hymenopteran parasitoid *Aphelinus mali* (Hymenoptera: Aphelinidae). Aphid and parasitoid individuals were genotyped using species-specific microsatellite loci and genetic variability was assessed. By studying genetic variation, natural geographic barriers of the aphid pest became evident and some evidence for incipient host-plant specialization was found. However, this had no effect on the population-genetic features of its most important parasitoid. In conclusion, the lack of genetic differentiation among the parasitoids suggests the existence of a single large and panmictic population, which could parasite aphids on apple orchards and on *P. coccinea* hedges. The latter could thus comprise a suitable and putative refuge for parasitoids, which could be used to increase the effectiveness of biological control. Moreover, the strong geographical differentiation of the aphid suggests local reinfestations occur mainly from other apple orchards with only low reinfestation from *P. coccinea* hedges. Finally, we propose that the putative refuge could act as a source of parasitoids without being a major source of aphids.

Editor: Jeffrey A. Harvey, Netherlands Institute of Ecology, Netherlands

Funding: This study was funded by the International Foundation for Science Grant No C/4023-2 and FONDECYT (Project Number 11080013). The funders had no role in study design, data collection and analysis, decision to publish, or preparation of the manuscript.

Competing Interests: The authors have declared that no competing interests exist.

* E-mail: blavandero@utalca.cl

Introduction

Natural enemies of insect pests are constantly disturbed in agroecological systems, and classical management practices can severely reduce parasitoid populations. The use of habitat refuges, offering shelter and alternative hosts for these organisms, has been proposed for maintaining high density of parasitoids close to cultivated plants, acting as a constant source to control agricultural pests [1]. At larger scales, landscape heterogeneity has been proposed to have a positive effect on natural enemy populations and parasitism rates in general [2]. Nevertheless, one must have enough evidence that parasitoids do disperse between the refuges and the crop, and that they exert an effect on the herbivore populations.

Ecological specialization of herbivore insects could affect their relationship with the third trophic level. Specialist herbivores that feed on different host plants have been shown to form host-specialized races, evidenced through reduced migration and gene flow [3]. The effect on the next trophic level (the natural enemies) can follow the specialization of their herbivore host, resulting in the formation of specialized parasitoid races, in a process termed sequential radiation [4]. In fact, as herbivorous insects and their parasitoids interact with their environment on a fine spatial and temporal scale, sequential radiation may be quite common [5]. Thus, parasitoids coming from a refuge may not readily forage on the crop or they may be totally isolated if gene flow between the refuge and the crop is absent, in which case the refuge would not constitute a real source of parasitoids for improving biocontrol.

Genetic markers, particularly highly polymorphic ones such as microsatellites, have been widely used to study several aspects of insect ecology. These DNA markers provide the raw data to estimate genetic diversity and gene flow between insect populations or to reconstruct migration routes and colonization history. Using appropriate bioinformatic tools to analyze DNA marker data, gene flow and genetic diversity within insect species can be quantified, which is critical for explaining population structure and dynamics in time and space (for a review see [6]). For instance, microsatellites in combination with powerful analytical tools [7] have proven to be useful for describing movement of insect pests between continents (for the western corn rootworm see [8]; for the tobacco aphid see [9]), between different production areas (for the codling moth see Fuentes–Contreras et al. [10]; for the woolly apple aphid see [11]), and between native and introduced ranges of parasitoids [12]. To our knowledge, however, there are no studies using neutral genetic variation to estimate natural enemy migration (movement and reproduction) between a putative refuge and the crop.

Here, using neutral genetic variation, we show the existence of geographical natural barriers to aphids in a main apple production area. The level of host specialization of this aphid pest is shown to have no influence on the population differentiation of its most important parasitoid wasp, due to the high gene flow observed among plant species and locations. We argue that the proposed refuge could act as a source of parasitoids without being a major source of the aphid pest.

Results

The aphids

Aphids were found in apple orchards and at four *P. coccinea* hedge sites, irrespective of pest management practices (organic vs. conventional orchards) (Table 1). A total of 581 aphid colonies were sampled and 471 different multilocus genotypes characterized (for a list of multilocus genotypes see Table S1). Twenty six genotypes were found more than once. Frequency of these multicopy genotypes was low in most sites (less than 10%), with the exception of site *Cato* where 44.8% of the colonies belonged to the same genotype. The genotypic diversity was high and similar among all sites as evidenced by the indices of Shannon, Simpson and their evenness (Table 1). Mean standardized allelic richness per site varied from 2.7 to 4.1. Heterozygosity ranged between 0.68 and 0.95 and gene diversity between 0.53 and 0.71 (see Table 1). Significant and frequent departures from Hardy-Weinberg Equilibrium were found in most of the sampled sites due to heterozygote excess. No evidence for null alleles was found (data not shown).

The genetic differentiation of populations (*Phi-pt*) between sites ranged from 2 to 23%. Analyses of Molecular Variance (AMOVA) of the aphid populations revealed different genetic structures that can be explained both by differences among the sites (22%) and differences between the host plants (2%) (p = 0.01). Pairwise comparisons between pairs of neighbouring *Pyracantha* hedges and their corresponding apple orchards showed a significantly high differentiation, ranging between 12.3% for *Colin* (site 9 and C, Figure 1) and 39% for *Cañadilla* (site 3 and A, Figure 1). Further analyses using TESS suggested that the aphid colonies were grouped into seven geographically related clusters, where sites close to each other shared more ancestry than those further apart (represented in Figure 2 and 3 (top) by different colours). The Bayesian clustering method showed different genetic clusters between neighbouring collection sites including samples from different host plants. This was confirmed after analyzing a smaller comparable scale (*P. coccinea* sites A, B, C and D; Apple sites 3, 8 and 9, in Figure 1), revealing a high differentiation between host-plants (5%; *p* = 0.01), although the greater differences among populations were independent of the host (21%; *p* = 0.01). Further analyses using TESS confirmed the AMOVA results by showing almost no admixis between host plants or sites (Figure 4). Analyses using shared allelic distance between individuals at the site *Cañadilla* suggest that aphids from the same host plant are more closely related (Figure 5).

The parasitoids

A total of 1018 parasitoid specimens were obtained (one to three parasitoids emerged from each aphid colony sampled) and 902

individuals were successfully genotyped and considered for analyses. Mean standardized allelic richness per site varied from 3.1 to 4.0. Allelic richness of the parasitoid was independent of the geographical distance between sites (Partial Mantel test; r = −0.1, *p* = 0.46). The proportion of heterozygotes ranged between 0.26 and 0.50, while gene diversity ranged between 0.39 and 0.54 (see Table 2). Slight heterozygote deficiencies were detected in most sites, probably due to null alleles (frequency of null alleles was under 19% for all loci). AMOVA evidenced significant but very low variation between sites (1%) and within host plants (1%), suggesting great gene flow between sites and host plants at the landscape level (see further details of pairwise *Fst* in Table 3). Further analyses using the Bayesian structuring algorithm implemented in TESS and considering all individuals independent of their collection sites, suggested no host or geographically-associated differentiation for the parasitoids (see Figure 3).

Kinship analysis also detected numerous full-sib pairs between parasitoids collected from different aphid colonies sampled from either the same or different trees. Furthermore, parasitoid females emerging from the same aphid colony were usually not full-sibs (Table 4). Parasitism levels ranged from 67.3% to 100%, with no significant differences between organic or conventional orchards (*p* = 0.897). In contrast, parasitism levels were significantly higher on aphids collected from *P. coccinea* than those collected from apples (see Table 2).

Aphid-parasitoid complex

Mean standardized allelic richness for the parasitoids per site were inversely correlated with the parasitism rates per orchard (Spearman r = −0.5, p = 0.038). Parasitism rates were independent of geographical distance when controlling for allelic richness (r = −0.11, p = 0.14). When estimating parasitism rates for the *Malus* sites per genetic cluster according to TESS (Mean ± SE: *Blue* 81.5±4.2; *Dark Yellow* 100±0; *Green* 91.8±3.03; *Pink* 87.7±4.02; *Red* 97.4±1.67 and *Yellow* 81.6±8.59), clusters *Blue* and *Yellow* (Figure 2) had significantly lower parasitism rates (Z values and correspondent p-values for paired comparisons with the *Blue* cluster for the *Dark Yellow* z = 6.266 *Green* z = 5.239 *Pink* z = 2.909; *Red* z = 6.303 and *Yellow* z = 0.001; *p* = 3.70e-10; *p* = 1.61e-07; *p* = 0.00363; *p* = 2.92e-10 and *p* = 0.99951).

Analyses using shared allelic distance between individuals at the site level for the populations from *Cañadilla* suggested that aphids from the same host plant were more closely related; however, the comparable tree for the parasitoids (constructed with individuals emerged from those same aphids), showed no significant grouping of parasitoids per tree or host plant (Figure 5).

Discussion

The very low genetic differentiation among *A. mali* populations suggests that individuals do disperse between sites and host plants, although there is still no clear evidence that this can exert a difference in the herbivore abundances on the crop. The partitioning of molecular variance of the parasitoids revealed very low levels of variation between sites (i.e. orchards), especially considering that parasitoids reproduce sexually. Since no host or geographically-associated structuring was evident for the parasitoid, the natural barriers affecting aphids [11] seem not to be affecting the parasitoids. Moreover, the kinship analysis of parasitoids suggests that oviposition does not occur in a patchy or aggregated fashion. Thus, female parasitoids would lay eggs far away from each other, reducing the endogamy between points by increasing gene flow, at least at the orchard level, thus supporting the idea of a higher dispersal and gene flow between sites. Bayesian

Table 1. Site Number, Location, Host plant, Management conditions (O = Organic, C = Conventional), sample size, Number of Genotypes, Unique vs. Multicopy genotypes (U/M), Shannon diversity (H) and its evenness (VH), Simpson diversity (D) and its evenness (ED), Gene Diversity (1-Q), Inbreeding coefficient (Fis) and significance (p-value), Loci under disequilibrium and allelic richness (A) of *Eriosoma lanigerum* females per site.

Site N°	Location	Host plant	Manag.	n	Genotypes	U/M	H	VH	D	ED	(1-Q)	*Fis*	p-value	LD	A
1	Villa Alemana	*Malus*	O	30	28	26/2	3,309	0,993	0,995	0,519	0,867	−0,349	>0,01	2/22	3,6
2	Graneros	*Malus*	C	19	13	11/2	2,347	0,915	0,924	0,481	0,895	−0,642	>0,01	3/22	3
3	Cañadilla	*Malus*	O	13	13	13/0	2,565	1,000	1,000	−1,000	0,890	−0,492	>0,01	1/22	3,1
4	San Fernando	*Malus*	O	29	29	29/0	3,367	0,999	1,000	−1,000	0,823	−0,267	>0,01	4/22	3,4
5	Los Niches	*Malus*	O	51	50	49/1	3,905	0,998	0,999	0,000	0,790	−0,189	>0,01	7/22	3,7
6	Panguilemo	*Malus*	C	28	24	23/1	3,045	0,958	0,974	0,000	0,888	−0,460	>0,01	2/22	3,3
7	Maiten Huapi	*Malus*	O	58	55	53/2	3,980	0,993	0,998	0,453	0,867	−0,316	>0,01	5/22	3,7
8	Las Rastras	*Malus*	C	30	27	25/2	3,245	0,985	0,991	0,462	0,895	−0,351	>0,01	3/22	3,7
9	Colin	*Malus*	C	30	27	25/2	3,245	0,985	0,991	0,462	0,805	−0,256	>0,01	4/22	3,4
10	Las Lomas	*Malus*	C	27	27	27/0	3,296	1,000	1,000	−1,000	0,783	−0,185	>0,01	1/22	3,8
11	Pataguas	*Malus*	C	30	18	13/5	2,691	0,931	0,949	0,824	0,867	−0,554	>0,01	2/22	2,9
12	Miraflores	*Malus*	C	30	28	27/1	3,291	0,988	0,993	0,000	0,810	−0,219	>0,01	8/22	3,8
13	Ancoa	*Malus*	C	37	36	35/1	3,573	0,997	0,998	0,000	0,865	−0,329	>0,01	2/22	3,7
14	Huaquivilo	*Malus*	O	36	35	34/1	3,545	0,997	0,998	0,000	0,679	−0,070	NS	5/22	3,8
15	Miraríos	*Malus*	O	26	26	26/0	3,258	1,000	1,000	−1,000	0,769	−0,113	NS	10/22	4,1
16	Cato	*Malus*	C	29	12	9/3	1,913	0,770	0,786	0,498	0,828	−0,592	>0,01	13/22	2,8
17	Mulchén	*Malus*	O	28	26	25/1	3,214	0,987	0,992	0,000	0,745	−0,245	>0,01	3/22	3,4
			SUBTOTAL	531	474	451/26	6,072	0,984	0,999	0,930	0,830		>0,001	19/22	
A	Cañadilla	*Pyracantha*	n/a	12	7	5/2	1,748	0,898	0,864	0,560	0,917	−0,682	>0,001	0/22	2,7
B	Las Rastras	*Pyracantha*	n/a	5	5	5/0	1,609	1,000	1,000	−1,000	0,971	−0,432	>0,001	0/22	3,9
C	Colin	*Pyracantha*	n/a	19	19	19/0	2,944	1,000	1,000	−1,000	0,895	−0,397	>0,001	1/22	3,5
D	Manzano	*Pyracantha*	n/a	8	6	5/1	1,667	0,931	0,893	0,000	0,929	−0,526	>0,001	3/22	3,3
			SUBTOTAL	44	37	34/3	3,508	0,972	0,987	0,671	0,928		>0,001	1/22	
			Whole sample	575	511	485/29	6,146	0,985	0,999	0,941	0,879				

grouping algorithms revealed no geographic or host-driven structuring for the parasitoid, although the aphid host showed seven geographically related groups, where sites close to each other shared more ancestry than those further apart.

As reported before, aphids show low levels of gene flow at the landscape scale, with significant barriers between geographical areas [11]. The high levels of Heterozygosity, and few linked loci, suggest the occurrence of sexual reproduction in *E. lanigerum* in Chile, although this aphid species has not been found on its primary host where sexual reproduction is reported to occur (*Ulmus americana*) [13]. As suggested by Sandanayaka and Bus [14], sexual reproduction could indeed occur on apple, but further studies are necessary to determine the environmental conditions needed to trigger sexual reproduction, and to screen for the presence of sexual morphs in Chile. Interestingly, environmental conditions such as short days and below-zero temperatures (the factors that trigger sexuality in many aphid species [13,15]), could affect parasitism rates through an increased genetic diversity in the aphid host. In any case, this seems not be enough to affect the parasitoids genetic structure.

The genetic diversity of the woolly apple aphid is clearly geographically structured; however, some of the genetic variation can be also be explained by the different host plants used by the aphids. Analyses comprising only those sites where neighbouring *Pyracantha* hedges are found, suggest a higher differentiation between host plants. Interestingly, the genetic clusters at each *Malus* site were different compared to their corresponding *Pyracantha* hedge. Evidence obtained from TESS, AMOVA and the neighbour-joining tree analyses, clearly separate individuals coming from different host plants. When the survival and preference of females were compared in reciprocal-transference experiments, *E. lanigerum* from *M. domestica* showed a stronger preference for its own natal branch as compared with other *M. domestica* or *P. coccinea* trees (Lavandero, unpublished data). In contrast, aphids born on *P. coccinea* had no significant preference for its natal host, showing a lower rejection for the *M. domestica* host. This could be the case for *E. lanigerum* aphids coming from *Malus*, which are not able to disperse into neighbouring *P. coccinea* hedges, although some individuals from *P. coccinea* may successfully colonize apple trees. This suggests that although *P. coccinea* could potentially become a source of some recolonizing aphids, it should not act as a significant source, as there seems to be a restricted and biased migration between both host plants. Hence, our results are indicative of no sequential radiation in this aphid-parasitoid system; however, aphids still exhibit geographical and some host-driven genetic structure.

Parasitism rates varied greatly among the studied sites; however, the management of the orchards (organic or conventional) did not explain these differences as expected. The literature suggests that the main explanation for parasitism decrease and aphid popula-

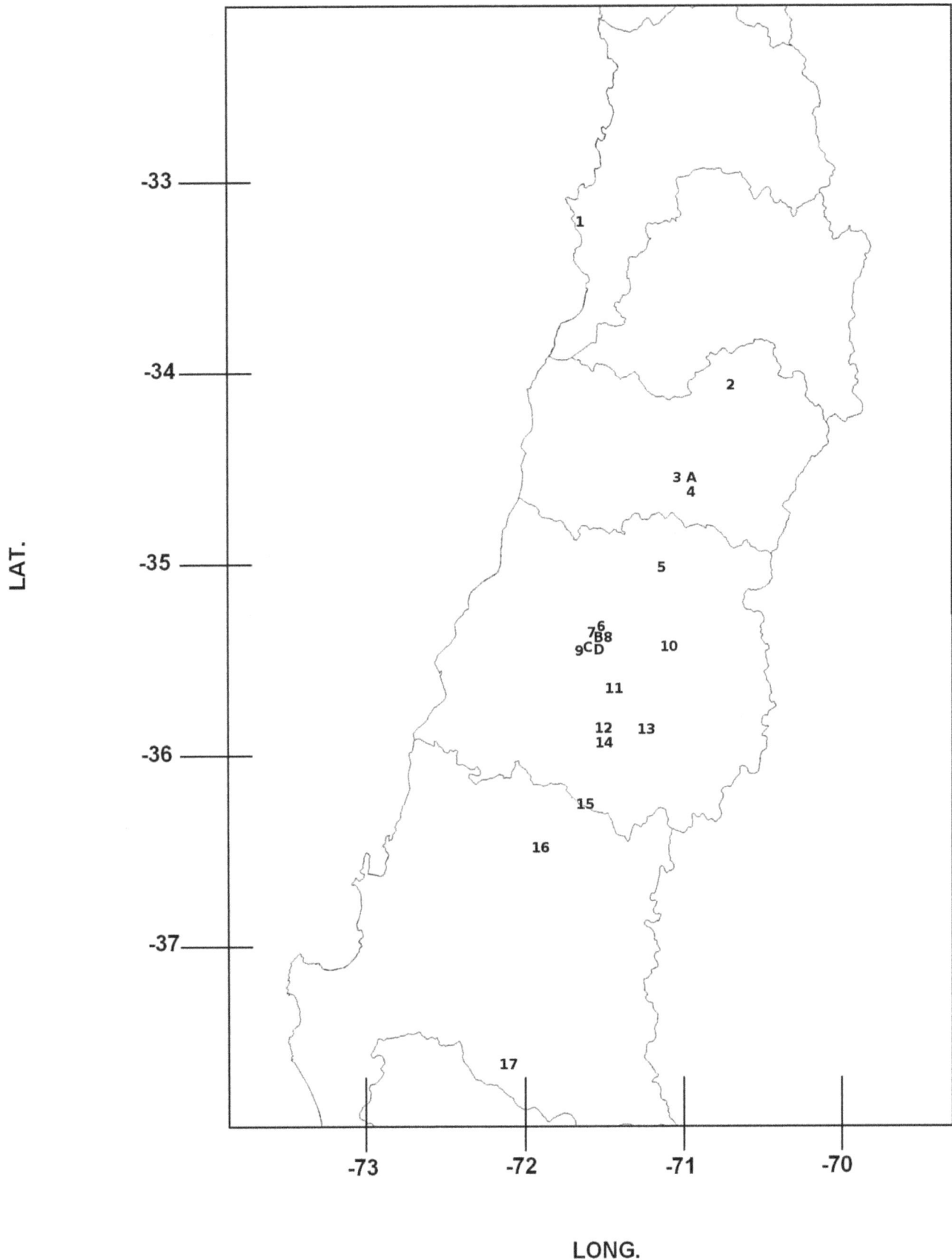

Figure 1. Collection sites of apple orchards (*Malus domestica*) (numbers) and *Pyracantha coccinea* sampling sites (letters). 1 Villa Alemana, 2 Graneros, 3 Cañadilla, 4 San Fernando, 5 Los Niches, 6 Panguilemo, 7 Maiten Huapi, 8 Las Rastras, 9 Colin, 10 Las Lomas, 11 Pataguas 12 Miraflores, 13 Ancoa, 14 Huaquivilo, 15 Miraríos, 16 Cato, 17 Mulchén, A Cañadilla, B Las Rastras, C Colin, D Manzano.

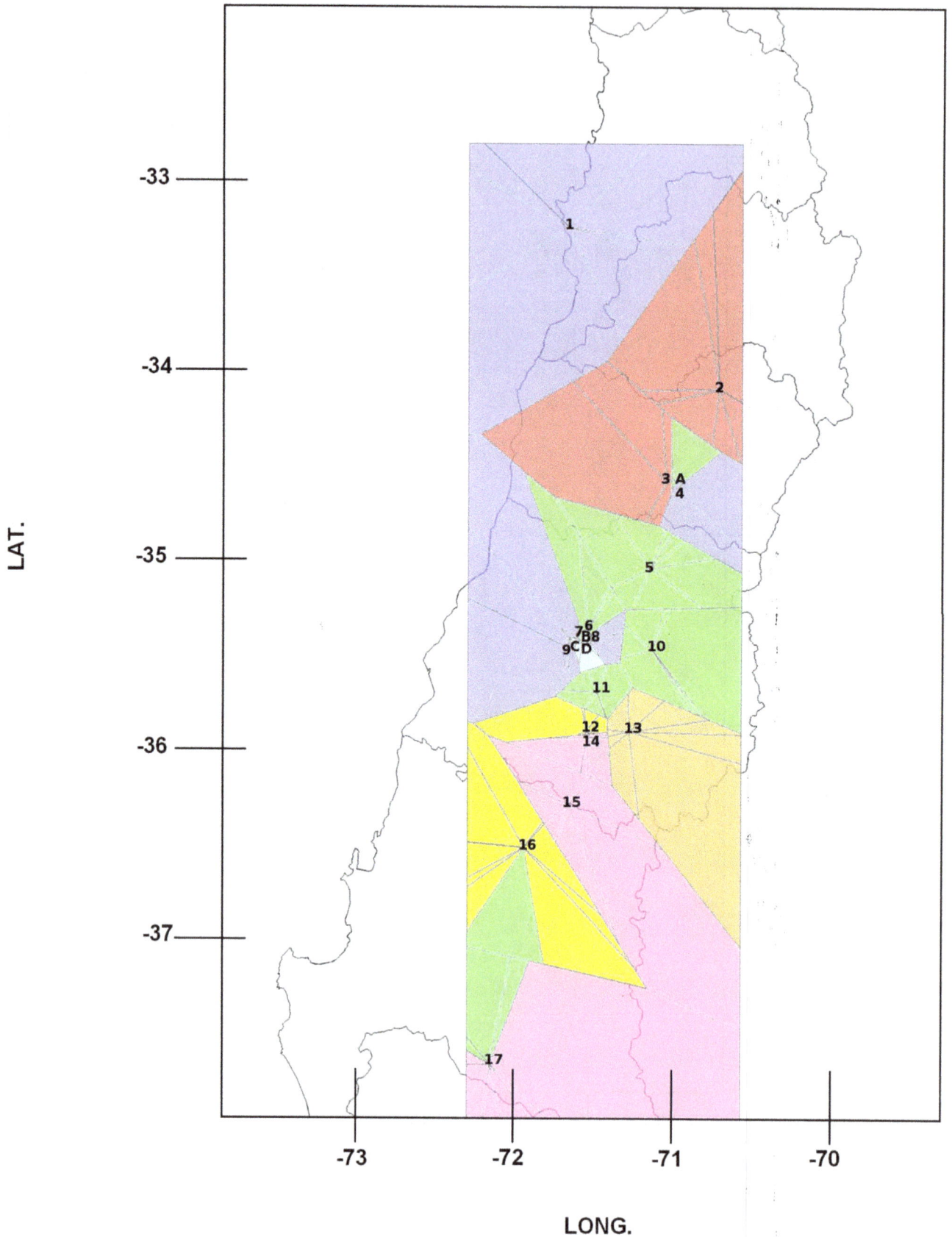

Figure 2. Membership of individuals of *Eriosoma lanigerum* **based on 50,000 sweeps using TESS, assuming no admixis, between sites.** Tessellation is ordered from North to South.

Figure 3. Average assignment probability of individuals of *Eriosoma lanigerum* **(aphid host), independent of sampling origin.** Assignment is based on 100 repetitions of 50,000 sweeps using TESS showing K = 7 genetic clusters and the correspondent structure for its parasitoid *Aphelinus mali*. Individuals (bars) are from North to South.

tion outbreaks are due to the susceptibility of the parasitoids to pesticides (organophosphates and pyrethroids), sulphur and kaoline [16–18]. In both management systems, however, management practices alone cannot account for the differences found (67.3% to 100% rates of parasitism). Indeed, parasitism rates were not related with geographical distance between sites, even considering allelic richness, which could be used as an estimator of effective population sizes [19]. In our study, the allelic richness of the parasitoids was negatively correlated with the parasitism rates per site, which suggest inverse density dependence, meaning that parasitoids are effectively controlling the aphid populations up to a threshold where the rate of increase of aphid populations is greater than the parasitoid ability to exert control. The thermal biology of these organisms could explain this pattern, as the parasitoid has a greater thermal developmental threshold than its aphid host, translating into a lower growth rate (GR) compared to its host (GR = 0.1 parasitoid, 0.14–0.27 for the aphid at 20°C) [20]. On the other hand, aphid populations showed different genetic structures, some genetic clusters showing more susceptibility to *A. mali* parasitism than others, with no significant effect of management practices (i.e. genetic cluster grouped aphids coming from both conventional and organic orchards). Other factors such as land use and nectar availability for parasitoids, among others, need to be further analyzed, as well as the possible interaction between aphid and defense endosymbiont bacteria as found for other aphid species [21].

In conclusion, the lack of genetic differentiation of the parasitoids suggest the existence of a single large and panmictic population, which could parasitise aphids on apple orchards and on *P. coccinea* hedges, the latter being a suitable and putative refuge for parasitoids to increase their effectiveness in biological control. Moreover, the strong geographical differentiation of the aphid suggests that local reinfestations occur mainly from other apple orchards, with little reinfestation occurring from *P. coccinea* hedges. Further mark-recapture studies should be conducted to quantify dispersal, frequency and intensity of aphid infestations in apple orchards coming from both host plants. Quantification of the actual effect of this putative refuge on the population dynamics of the pest across several seasons will be critical if any effort for improving biocontrol is attempted using *P. coccinea*. Overall, we have shown that neutral genetic variation is a useful tool for

addressing population dynamics between host plant species of pests and their parasitoids, determining potential refuges for natural enemies.

Materials and Methods

Study system

Aphids are important pests and disease vectors for a variety of crops, and parasitoids are often introduced for aphid biological control. The woolly apple aphid (*Eriosoma lanigerum* (Haussman)) (Hemiptera: Aphididae) native to North America, is a globally-important pest of apple orchards (*Malus domestica* Borkh). This aphid forms colonies on roots, trunks, branches and shoots, with greatest damage occurring at the shoot level [22]. Other associated damage is cosmetic, as fruits become covered with honeydew leading to subsequent fungus colonization, which reduces their commercial value. Although *M. domestica* is its most common host, this aphid also attacks other Rosacea species, notably *Pyracantha coccinea* (M. Roem) (Rosales: Rosacea), which is a very common plant distributed along farm hedges.

The wooly apple aphid (*E. lanigerum*) was first introduced into Chile during the 19th century, most probably as root colonies from plant material. As the damage to apple orchards in Chile reached dramatic levels, in 1920 the chalcidoid parasitoid *Aphelinus mali* (Hymenoptera: Aphelinidae) was introduced. Although this parasitoid is the main species controlling *E. lanigerum* in Chile, it has been determined that under the current management conditions (conventional agriculture), aphid population outbreaks still occur [23]. There are several reasons for aphid population outbreaks, the most important probably being organophosphates, pyrethroids, sulfur and even kaolin treatments that affect its main parasitoid, *A. mali* [16–18]. In order to improve the effectiveness of the parasitoid, the use of host-plant refuges such as *Pyracantha coccinea* is proposed to attract and maintain parasitoid populations. Indeed, *E. lanigerum* is frequently observed at high densities on *P. coccinea*, with high parasitism rates by *A. mali*. This proposed refuge could be a source of parasitoids when the pest is not present in the orchard or as protection after pesticide use. However, evidence is required that the parasitoids are able to move between the refuge (*P. coccinea*) and the crop (apple), thereby determining its suitability as a source or sink for both aphids and parasitoids.

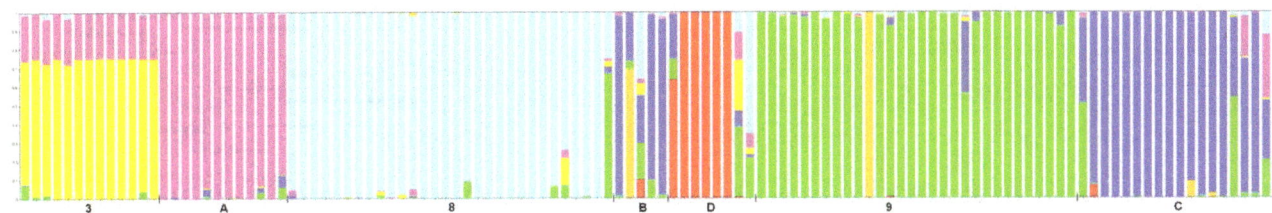

Figure 4. Average assignment probability of a subsample of individuals of *Eriosoma lanigerum* **(aphid host), independent of sampling origin, based on 100 repetitions of 50,000 sweeps using TESS showing K = 7 genetic clusters, ordered showing neighbouring sites between both host plants (***Malus domestica*** and ***Pyracantha coccinea***).** 3 = Cañadilla-*Malus*, A = Cañadilla-*Pyracantha*, 8 = Las Rastras-*Malus*, B = Las Rastras-*Pyracantha*, D = Los Manzanos–*Pyracantha*, 9 = Colin-*Malus*, C = Colin-*Pyracantha*.

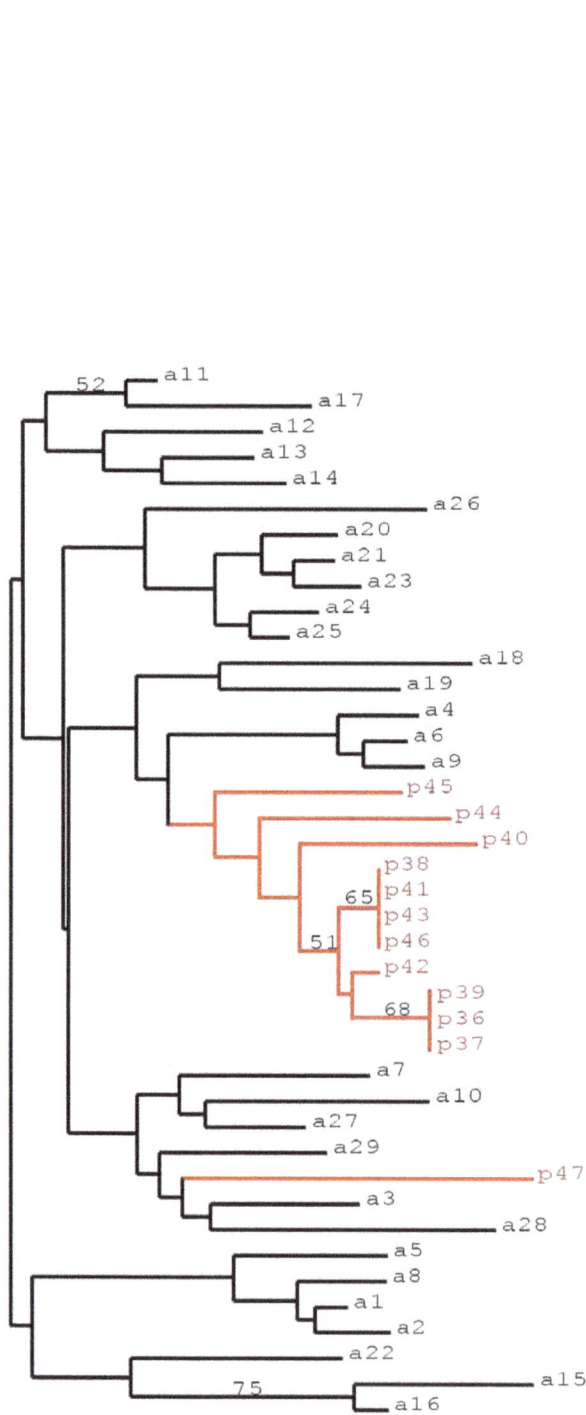

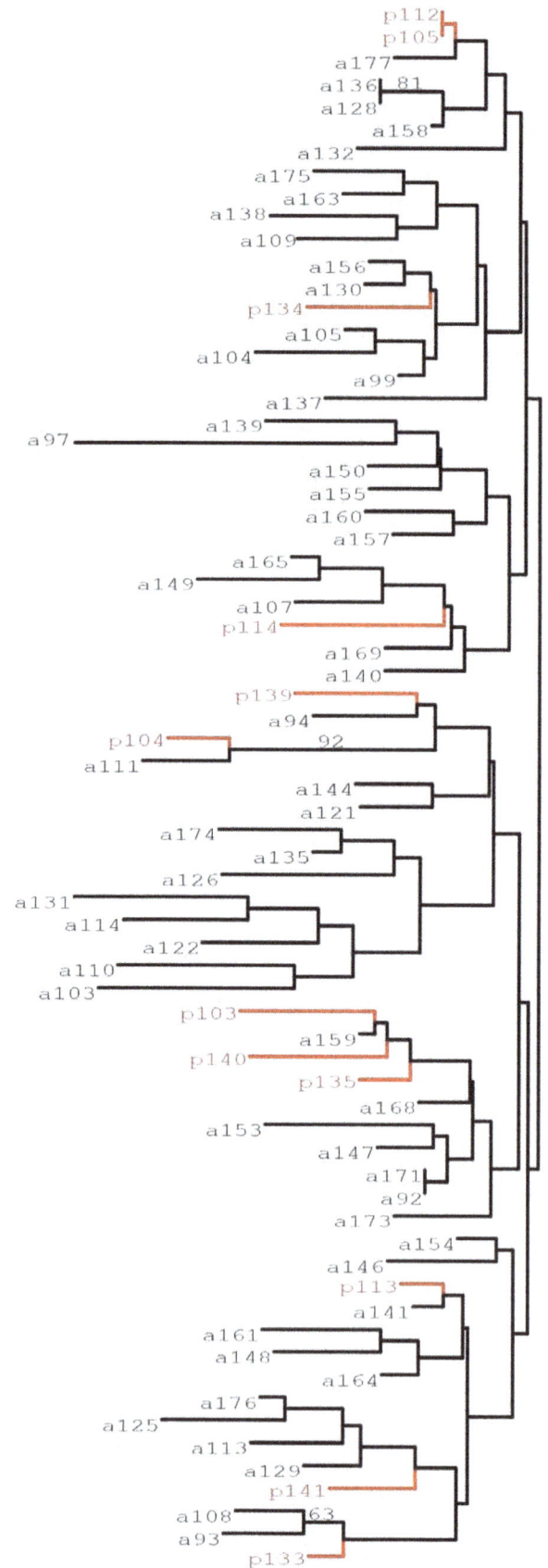

Eriosoma

Aphelinus

Figure 5. A neighbor-joining tree constructed using the shared allele distance between individuals of a single site (*Cañadilla*, site 3 Figure 1) among individuals collected on different host plants *Pyracantha* (in red) and *Malus* (black). Bootstrap values were computed over 2000 replications resampling the microsatellite loci. On the left side the tree for the aphid *Eriosoma lanigerum* and on the right the tree for the emerged parasitoids (*Aphelinus mali*).

A 700 Km. North-South transect was sampled, including 17 apple orchards and surrounding *Pyracantha coccinea* hedges at four of the 17 chosen sites (33.19 S 71.733 W to 37.721 S 72.244 W). Orchards were all over 30 ha in size, planted with the Granny Smith apple cultivar. Permissions for entering and taking samples at conventional orchards were issued as part of an ongoing center within Universidad de Talca, Centro de Pomaceas (Stone fruit center), which gives the university the faculty of sampling in their farms (more details at http://pomaceas.utalca.cl/html/index.html). All orchards sampled are members of this center. Permission for entering and using materials of organic orchard were issued as part of an ongoing agreement between Comercial Greenvic Ltda and Universidad de Talca, through their branch Huertos Organicos de Chile S:A: (more details at http://www.huertosorganicosdechile.cl/). All organic orchards sampled are members of this industry-university research agreement. At each orchard, up to 40 colonies of *E. lanigerum* were collected on different apple trees, while all available colonies on the *P. coccinea* hedges were sampled. Each aphid colony was georeferenced and taken back to the laboratory to determine parasitism rates under controlled conditions (20±1°C, 65±10% RH y 16:8 hrs. day/

night cycle). Parasitism rates per orchard were assessed for 10 trees (one colony per tree) per orchard. Colonies taken from the field were individually caged and reared under controlled conditions for two weeks. The number of aphids per colony and emerged parasitoids were registered from each cage. A single wingless adult aphid female per colony was preserved in 95% alcohol for subsequent DNA extraction. Parasitism rates were assessed by rearing aphids on 9 cm long shoots placed on a damp tissue paper inside plastic boxes with top ventilation. At emergence, parasitoids were identified to the species level, and up to three *A. mali* females per colony were preserved in 95% alcohol for DNA extraction. Genomic DNA was obtained following the 'salting out' protocol from [24]. Aphid and parasitoid individuals were genotyped using seven (aphids) and six (parasitoids) microsatellite (SSR) markers described in [25] and [26], respectively. The reverse primer for each pair of primers was fluorescently labeled, and PCR products analyzed on a MegaBASE 1000 automatic DNA Sequencer.

The microsatellite data were checked for null alleles and technical artifacts like stuttering bands and large allele dropout using the MICRO CHEKER v.2.2.3 software [27]. Deviations from the Hardy-Weinberg equilibrium (HWE) and linkage

Table 2. Management conditions, Host plant, Parasitism rates, allelic richness, Observed Heterozygocity (Ho) and Gene Diversity of *Aphelinus mali* females per site.

Site N°	Location	Management	Host plant	Parasitism rates mean	SD	Samples N	Allelic Richness Mean	SD	Ho Mean	SD	Gene diversity Mean	SD
1	Villa Alemana	O	*Malus*	70%	48%	16	4,1	0,9	0,28	0,07	0,52	0,11
2	Graneros	C	*Malus*	95%	31%	15	3,4	1,2	0,31	0,13	0,42	0,21
3	Cañadilla	O	*Malus*	79%	47%	27	3,4	0,8	0,33	0,14	0,52	0,21
4	San Fernando	O	*Malus*	73%	26%	61	3,8	1,0	0,35	0,15	0,49	0,16
5	Los Niches	O	*Malus*	82%	42%	60	3,7	1,1	0,38	0,15	0,53	0,23
6	Panguilemo	C	*Malus*	96%	9%	70	3,7	1,1	0,39	0,14	0,49	0,19
7	Maiten Huapi	O	*Malus*	95%	9%	66	3,7	1,2	0,40	0,16	0,53	0,19
8	Las Rastras	C	*Malus*	88%	18%	21	3,1	1,1	0,38	0,19	0,39	0,15
9	Colin	C	*Malus*	94%	14%	48	4,0	1,0	0,38	0,18	0,49	0,19
10	Las Lomas	C	*Malus*	91%	31%	65	4,0	1,0	0,37	0,15	0,51	0,20
11	Fundo Pataguas	C	*Malus*	96%	13%	57	3,7	1,1	0,36	0,13	0,50	0,16
12	Miraflores	C	*Malus*	67%	50%	29	3,9	1,4	0,34	0,11	0,52	0,17
13	Ancoa	C	*Malus*	100%	/	72	3,6	1,2	0,31	0,11	0,46	0,15
14	Huaquivilo	O	*Malus*	86%	40%	49	3,7	1,2	0,37	0,12	0,48	0,18
15	Miraríos	O	*Malus*	84%	32%	24	4,1	0,8	0,50	0,18	0,51	0,17
16	Cato	C	*Malus*	94%	15%	56	4,0	1,3	0,42	0,20	0,51	0,22
17	Mulchén	O	*Malus*	93%	23%	73	3,9	1,3	0,43	0,17	0,54	0,16
A	Cañadilla	-	*Pyracantha*	100%	/	12	4,0	1,0	0,26	0,09	0,48	0,12
B	Las Rastras	-	*Pyracantha*	100%	/	18	3,5	1,3	0,33	0,12	0,54	0,18
C	Colin	-	*Pyracantha*	100%	/	29	3,7	1,2	0,39	0,16	0,45	0,22
D	Manzanos	-	*Pyracantha*	100%	/	22	3,8	0,9	0,33	0,15	0,46	0,20
			Pyracantha									
	Mean			90%			3,7	1,1	0,36	0,14	0,49	0,18

Table 3. Pairwise Population *Fst* Values, *Aphelinus mali.*

	Pop1	Pop2	Pop3	Pop4	Pop5	Pop6	Pop7	Pop8	Pop9	Pop10	Pop11	Pop12	Pop13	Pop14	Pop15	Pop16	Pop17	Pop18	Pop19	Pop20	Pop21
Pop1		0.460	0.110	0.010	0.010	0.430	0.270	0.450	0.300	0.140	0.050	0.290	0.030	0.430	0.290	0.290	0.170	0.070	0.510	0.260	0.380
Pop2	0.000		0.270	0.010	0.160	0.210	0.030	0.010	0.030	0.050	0.030	0.030	0.020	0.300	0.040	0.020	0.390	0.010	0.280	0.150	0.390
Pop3	0.005	0.002		0.010	0.160	0.010	0.020	0.010	0.140	0.040	0.050	0.060	0.010	0.060	0.120	0.110	0.460	0.010	0.060	0.010	0.370
Pop4	0.034	0.034	0.031		0.020	0.010	0.020	0.010	0.010	0.040	0.040	0.010	0.020	0.020	0.010	0.030	0.010	0.010	0.020	0.010	0.010
Pop5	0.022	0.008	0.006	0.042		0.010	0.010	0.010	0.030	0.010	0.010	0.030	0.010	0.010	0.010	0.010	0.430	0.010	0.170	0.010	0.160
Pop6	0.000	0.002	0.014	0.043	0.038		0.310	0.420	0.030	0.040	0.200	0.180	0.010	0.390	0.180	0.020	0.040	0.010	0.200	0.510	0.060
Pop7	0.004	0.014	0.009	0.029	0.047	0.000		0.170	0.030	0.040	0.470	0.470	0.520	0.260	0.450	0.040	0.010	0.100	0.020	0.020	0.040
Pop8	0.000	0.012	0.017	0.035	0.043	0.000	0.004		0.010	0.140	0.470	0.470	0.520	0.190	0.450	0.040	0.010	0.030	0.020	0.020	0.040
Pop9	0.002	0.006	0.003	0.054	0.022	0.012	0.015	0.015		0.140	0.010	0.180	0.010	0.020	0.410	0.040	0.010	0.030	0.160	0.090	0.450
Pop10	0.006	0.007	0.007	0.022	0.024	0.009	0.007	0.014	0.024		0.120	0.040	0.090	0.310	0.010	0.210	0.010	0.010	0.110	0.050	0.070
Pop11	0.010	0.015	0.015	0.027	0.060	0.004	0.000	0.013	0.019	0.008		0.170	0.470	0.050	0.500	0.070	0.020	0.010	0.020	0.050	0.060
Pop12	0.002	0.010	0.009	0.044	0.027	0.002	0.000	0.005	0.009	0.011	0.007		0.470	0.160	0.070	0.010	0.050	0.030	0.080	0.010	0.100
Pop13	0.010	0.016	0.015	0.034	0.036	0.010	0.000	0.018	0.021	0.005	0.000	0.000		0.040	0.030	0.020	0.010	0.020	0.030	0.020	0.020
Pop14	0.000	0.001	0.007	0.038	0.026	0.000	0.003	0.003	0.010	0.001	0.012	0.004	0.008		0.200	0.380	0.040	0.440	0.240	0.230	0.160
Pop15	0.000	0.012	0.006	0.025	0.045	0.005	0.000	0.000	0.003	0.012	0.000	0.009	0.012	0.004		0.320	0.490	0.010	0.020	0.010	0.450
Pop16	0.001	0.011	0.005	0.034	0.026	0.011	0.009	0.012	0.009	0.003	0.012	0.012	0.010	0.001	0.002		0.020	0.020	0.030	0.020	0.240
Pop17	0.008	0.000	0.000	0.065	0.001	0.019	0.031	0.029	0.000	0.035	0.042	0.021	0.032	0.019	0.025	0.017		0.010	0.260	0.010	0.400
Pop18	0.012	0.025	0.029	0.075	0.062	0.015	0.010	0.018	0.028	0.025	0.031	0.012	0.018	0.000	0.025	0.020	0.047		0.080	0.050	0.030
Pop19	0.000	0.002	0.010	0.050	0.012	0.007	0.026	0.008	0.015	0.015	0.036	0.009	0.022	0.004	0.026	0.012	0.006	0.020		0.150	0.320
Pop20	0.005	0.007	0.033	0.050	0.062	0.000	0.018	0.009	0.032	0.019	0.022	0.023	0.020	0.003	0.026	0.021	0.051	0.017	0.010		0.060
Pop21	0.000	0.000	0.002	0.058	0.010	0.015	0.034	0.026	0.000	0.027	0.023	0.019	0.025	0.010	0.009	0.014	0.000	0.035	0.002	0.024	

Fst Values below diagonal.
Probability values based on 9999 permutations are shown above diagonal.

Table 4. Percentage of parasitoid full sibs from the same aphid colony, from aphid colonies collected on the same host plant (for *Pyracantha* and *Malus*) and on different host plants at the Colin and Cañadilla sites.

	% full sibs same Colony	% full sibs *Pyracantha*	% full sibs *Malus*	% full sibs other Host
Colin	7.25	18.84	33.33	47.83
Cañadilla	18.32	2.29	68.70	29.01

disequilibrium (LD) were tested using GENEPOP v.3.2a software [28]. To analyze genotypic data and test for clonality in the aphid populations, the number of genotypes, the rate of unique vs./ multicopy genotypes, Shannon diversity and its evenness, Simpson diversity and its evenness, gene diversity, inbreeding coefficient (Fis) and significance (p-value), loci under disequilibrium and allelic richness per site were estimated using the GenClone 2.0 software [29]. Observed heterozygocity, gene diversity and allelic richness of *A. mali* per site were estimated using HP-RARE 1.0 [30]. Population structure of both species (parasitoids and aphids) was examined first using a hierarchical analysis of molecular variance (AMOVA) assuming asexuality for the aphids (*Phi-pt*; significant deviations from HWE) and sexuality for the parasitoids (*Fst*) as implemented in Genalex v 6.41 [31], with two levels (host plants and total effect). In addition, the population-genetic structure was assessed for aphids and parasitoids using the aggregation Bayesian algorithm implemented in TESS 2.3 [32]. The admixture model was compared with a non-admixture model as suggested by [33], because admixture models are robust to an absence of admixture in the sample, but non-admixture models are robust when admixture is present between some individuals. The TESS algorithm was run with 10,000 sweeps, discarding the first 5,000 with 20 independent iterations for each model for maximum clusters (Kmax) varying from 2 to 12 for both aphids and parasitoids. The highest likelihood runs were selected based on the Deviance Information Criterion (DIC) and graphed against Kmax (as suggested by [32]), allowing selection of the number of hypothetical clusters (K). Then the program was run 100 times for the selected Kmax with 50,000 sweeps discarding the first 10,000. The 10 highest likelihood runs were then averaged. Population genetic structure was assessed again on a subsample consisting of sampling sites with neighboring *P. coccinea* (sites 3, 8 and 9 for *Malus* and A, B, C and D for *P. coccinea* in Figure 1) using the aggregation Bayesian algorithm implemented in TESS, as described before. At the *Cañadilla* site (site 3 on Figure 1) a neighbor-joining tree [34] was constructed using the shared allele distance [35] between individuals, in order to visualize the genetic similarity among individuals collected on different host plants (*P. coccinea* and apple), as site 3 was the only site where aphids were found on hedges of *P. coccinea* inside an apple orchard. Bootstrap values were computed over 2000 resamplings of the microsatellite loci. In order to assess the ability of a parasitoid female to lay eggs grouped or dispersed among the aphid colonies, parasitoids that emerged from the same aphid colony were tested for being

daughters from a single or many females. This was done using a kinship analysis on parasitoids that emerged from aphids sampled at sites where neighboring *P. coccinea* hedges are found (sites *Cañadilla* and *Colin*, 3 and 9 in Figure 1, respectively). Analyses were carried out using the full likelihood method [36,37], as implemented in the software COLONY v 2.0, with data from six SSR loci.

In order to test the hypothesis that parasitoids respond to aphid population structure independently from geographical or sampling effects, a series of partial and simple Mantel tests were carried out. The significance of these correlations were assessed using *zt* version 1.0 [38], with 10.000 permutations [39]. The tested variables were parasitoid allelic richness as an estimate of population sizes, geographical distance between sites, sample size between sites, and parasitism rates per site. Spearman correlation was also carried out between parasitism rates per site and allelic richness of the parasitoids, using R version 2.10.1. Once the number of genetic clusters was estimated for the aphids, parasitism rates per cluster were estimated to asses the influence of the aphid's genetic background on the efficiency of the parasitoid. A generalized linear model (GLM) assuming a Poisson distribution was carried out [40] with the *glm* function in the base package of R version 2.10.1 written by Simon Davies. Mean values per cluster were then compared to the lowest mean value in a series of paired comparisons, and significances were estimated.

Supporting Information

Table S1 List of multilocus genotypes of sampled *Eriosoma lanigerum*.

Acknowledgments

The authors would like to thank Jason Tylianakis for useful comments and help with English and Marcos Dominguez for assistance during laboratory and field work. P.F. would like to thank Jérôme Olivares for his help in the laboratory.

Author Contributions

Conceived and designed the experiments: BL. Performed the experiments: BL PF. Analyzed the data: BL PF CCF AM. Contributed reagents/ materials/analysis tools: BL PF. Wrote the paper: BL PF CCF.

References

1. Thies C, Tscharntke T (1999) Landscape structure and biological control in agroecosystems. Science 285: 893–895.
2. Bianchi F, Booij CJH, Tscharntke T (2006) Sustainable pest regulation in agricultural landscapes: a review on landscape composition, biodiversity and natural pest control. Proceedings of the Royal Society B-Biological Sciences 273: 1715–1727.
3. Forbes AA, Powell THQ, Stelinski LL, Smith JJ, Feder JL (2009) Sequential Sympatric Speciation Across Trophic Levels. Science 323: 776–779.
4. Abrahamson W, Blair C (2008) Sequential radiation through host-race formation: herbivore diversity leads to diversity in natural enemies. In: Tilmon K, ed.

Specialization, Speciation, and Radiation: The Evolutionary Biology of Herbivorous Insects. BerkeleyCalifornia: University of California Press. pp 188–202.
5. Feder JL, Forbes AA (2010) Sequential speciation and the diversity of parasitic insects. Ecological Entomology 35: 67–76.
6. Behura SK (2006) Molecular marker systems in insects: current trends and future avenues. Molecular Ecology 15: 3087–3113.
7. Guillemaud T, Beaumont MA, Ciosi M, Cornuet JM, Estoup A (2010) Inferring introduction routes of invasive species using approximate Bayesian computation on microsatellite data. Heredity 104: 88–99.

8. Ciosi M, Miller NJ, Kim KS, Giordano R, Estoup A, et al. (2008) Invasion of Europe by the western corn rootworm, Diabrotica virgifera virgifera: multiple transatlantic introductions with various reductions of genetic diversity. Molecular Ecology 17: 3614–3627.

9. Zepeda-Paulo FA, Simon JC, Ramirez CC, Fuentes-Contreras E, Margaritopoulos JT, et al. (2010) The invasion route for an insect pest species: the tobacco aphid in the New World. Molecular Ecology 19: 4738–4752.

10. Fuentes-Contreras E, Espinoza JL, Lavandero B, Ramirez CC (2008) Population genetic structure of codling moth (Lepidoptera : Tortricidae) from apple orchards in central Chile. Journal of Economic Entomology 101: 190–198.

11. Lavandero B, Miranda M, Ramirez CC, Fuentes-Contreras E (2009) Landscape composition modulates population genetic structure of Eriosoma lanigerum (Hausmann) on Malus domestica Borkh in central Chile. Bulletin of Entomological Research 99: 97–105.

12. Hufbauer RA, Bogdanowicz SM, Harrison RG (2004) The population genetics of a biological control introduction: mitochondrial DNA and microsatellie variation in native and introduced populations ofAphidus ervi, a parisitoid wasp. Molecular Ecology 13: 337–348.

13. Blackman RL, Eastop VF (2000) Aphids on the world's crops: an identification and information guide. Aphids on the world's crops: an identification and information guide. pp x+466.

14. Sandanayaka WRM, Bus VGM (2005) Evidence of sexual reproduction of woolly apple aphid, Eriosoma lanigerum, in New Zealand. Journal of Insect Science 5.

15. James BD, Luff ML (1982) Cold-Hardiness and Development of Eggs of Rhopalosiphum-Insertum. Ecological Entomology 7: 277–282.

16. Cohen H, Horowitz AR, Nestel D, Rosen D (1996) Susceptibility of the woolly apple aphid parasitoid, Aphelinus mali (Hym: Aphelinidae), to common pesticides used in apple orchards in Israel. Entomophaga 41: 225–233.

17. Penman DR, Chapman RB (1980) Woolly Apple Aphid Homoptera, Aphididae Outbreak Following Use of Fenvalerate in Apples in Canterbury, New-Zealand. Journal of Economic Entomology 73: 49–51.

18. Marko V, Blommers LHM, Bogya S, Helsen H (2008) Kaolin particle films suppress many apple pests, disrupt natural enemies and promote woolly apple aphid. Journal of Applied Entomology 132: 26–35.

19. Wang J (2005) Estimation of effective population sizes from data on genetic markers. Philosophical Transactions of the Royal Society of London, Biological Sciences Series B 360: 1395–1409.

20. Asante SK (1997) Natural enemies of the woolly apple aphid, Eriosoma lanigerum (Hausmann) (Hemiptera: Aphididae): A review of the world literature. Plant Protection Quarterly 12: 166–172.

21. Castaneda L, Sandrock C, Vorburger C (2010) Variation and covariation of life history traits in aphids are related to infection with the facultative bacterial endosymbiont Hamiltonella defensa. Biological Journal of the Linnean Society 100: 237–247.

22. Weber DC, Brown MW (1988) Impact of Woolly Apple Aphid (Homoptera, Aphididae) on the Growth of Potted Apple-Trees. Journal of Economic Entomology 81: 1170–1177.

23. Moreno V (2003) Dinámica poblacional y enemigos naturales del Pulgón Lanígero (Eriosomalanigerum) en manzano Braeburn/Franco sometidos a manejo tradicional y confusiónsexual para la polilla de la manzana. Talca: Universidad de Talca.

24. Sunnucks P, Hales D (1996) Numerous transposed sequences of mitochondrial cytochrome oxidase I-II in aphids of the genus Sitobion (Hemiptera: Aphididae). Molecular Biology and Evolution 13: 510–524.

25. Lavandero B, Dominguez M (2010) Isolation and characterization of nine microsatellite loci from Aphelinus mali (Hymenoptera: Aphelinidae), a parasitoid of Eriosoma lanigerum (Hemiptera: Aphididae). Insect Science 17: 549–552.

26. Lavandero B, Figueroa CC, Ramirez CC, Caligari PDS, Fuentes-Contreras E (2009) Isolation and characterization of microsatellite loci from the woolly apple aphid Eriosoma lanigerum (Hemiptera: Aphididae: Eriosomatinae). Molecular Ecology Resources 9: 302–304.

27. Van Oosterhout C, Hutchinson WF, Wills DPM, Shipley P (2004) MICRO-CHECKER: software for identifying and correcting genotyping errors in microsatellite data. Molecular Ecology Notes 4: 535–538.

28. Raymond M, Rousset F (1995) Genepop (Version-1.2) - Population-Genetics Software for Exact Tests and Ecumenicism. Journal of Heredity 86: 248–249.

29. Arnaud-Haond S, Belkhir K (2007) Genclone: a computer program to analyse genotypic data, test for clonality and describe spatial clonal organization. Molecular Ecology Notes 7: 1471–8286.

30. Kalinowski S (2005) HP-Rare: a computer program for performing rarefaction on measures of allelic diversity. Molecular Ecology Notes 5: 187–189.

31. Peakall R, Smouse PE (2006) GENALEX 6: genetic analysis in Excel. Population genetic software for teaching and research. Molecular Ecology Notes 6: 288–295.

32. Chen C, Durand E, Forbes F, Francois O (2007) Bayesian clustering algorithms ascertaining spatial population structure: a new computer program and a comparison study. Molecular Ecology Notes 7: 747–756.

33. Francois O, Durand E (2010) Spatially explicit Bayesian clustering models in population genetics. Molecular Ecology Resources 10: 773–784.

34. Saitou N, Nei M (1987) The Neighbor-Joining Method - a New Method for Reconstructing Phylogenetic Trees. Molecular Biology and Evolution 4: 406–425.

35. Chakraborty R, Jin L (1993) Determination of Relatedness between Individuals Using DNA-Fingerprinting. Human Biology 65: 875–895.

36. Wang JL (2004) Sibship reconstruction from genetic data with typing errors. Genetics 166: 1963–1979.

37. Wang J, Santure AW (2009) Parentage and Sibship Inference From Multilocus Genotype Data Under Polygamy. Genetics 181: 1579–1594.

38. Bonnet E, Van de Peer Y (2002) zt: a software tool for simple and partial Mantel tests. Journal of Statistical Software 7: 1–12.

39. Frantz AC, Pourtois JT, Heuertz M, Schley L, Flamand MC, et al. (2006) Genetic structure and assignment tests demonstrate illegal translocation of red deer (Cervus elaphus) into a continuous population. Molecular Ecology 15: 3191–3203.

40. Dobson AJ (2002) An Introduction to Generalized Linear Models CRC Press.

β-Amylase from Starchless Seeds of *Trigonella Foenum-Graecum* and Its Localization in Germinating Seeds

Garima Srivastava, Arvind M. Kayastha*

School of Biotechnology, Faculty of Science, Banaras Hindu University, Varanasi, India

Abstract

Fenugreek (*Trigonella foenum-graecum*) seeds do not contain starch as carbohydrate reserve. Synthesis of starch is initiated after germination. A *β*-amylase from ungerminated fenugreek seeds was purified to apparent electrophoretic homogeneity. The enzyme was purified 210 fold with specific activity of 732.59 units/mg. M_r of the denatured enzyme as determined from SDS-PAGE was 58 kD while that of native enzyme calculated from size exclusion chromatography was 56 kD. Furthermore, its identity was confirmed to be β-amylase from MALDI-TOF analysis. The optimum pH and temperature was found to be 5.0 and 50°C, respectively. Starch was hydrolyzed at highest rate and enzyme showed a K_m of 1.58 mg/mL with it. Antibodies against purified Fenugreek *β*-amylase were generated in rabbits. These antibodies were used for localization of enzyme in the cotyledon during different stages of germination using fluorescence and confocal microscopy. Fenugreek *β*-amylase was found to be the major starch degrading enzyme depending on the high amount of enzyme present as compared to α-amylase and also its localization at the periphery of amyloplasts. A new finding in terms of its association with protophloem was observed. Thus, this enzyme appears to be important for germination of seeds.

Editor: Alberto G. Passi, University of Insubria, Italy

Funding: This work was supported by the Council of Scientific and Industrial Research (CSIR), New Delhi, India, in the form of Junior and Senior research fellowships to Garima Srivastava and University Grants Commission-University with Potential for Excellence Grant (UGC-UPE) for innovative research from Banaras Hindu University, Varanasi, India, to Arvind M. Kayastha. The funding agency (CSIR and UGC-UPE) had no role in study design, data collection and analysis, decision to publish, or preparation of the manuscript.

Competing Interests: The authors have declared that no competing interests exist.

* E-mail: kayasthabhu@gmail.com

Introduction

β-Amylase (E.C. 3.2.1.2), member of family 14 of glycosyl hydrolases, catalyses the release of successive *β*-maltose from the non-reducing ends of α-1,4 linked oligo and poly glucans. The enzyme is distributed in higher plants and some micro-organism. *β*-Amylase is major contributor of diastatic power (i.e. the combined α-amylase, *β*-amylase, debranching enzyme and α-glucosidase activities) of malt. Exclusive maltose production is utilised in pharmaceutical industry for dispensing, production of maltose rich syrups, and non-digestible sweetner, maltitol [1].

The enzyme is synthesised and accumulated during grain development [2]. In germination of cereal seeds, *β*-amylase is known to play a vital role, where it is present in free and bound forms [1]. *De-novo* synthesis of the enzyme is reported during early germination of rice [3]. *β*-Amylase has also been shown to play a more important role as compared to α-amylase during early hours of germination in wheat scutella [4].

In plants, *β*-amylase is known to play primary role in hydrolysis of starch. The identification of *β*-maltose as a predominant sugar exported from the chloroplast at night, points towards role of *β*-amylase in degradation of transient starch [5]. Furthermore, study of transgenic and mutant plants lacking enzyme also showed its importance in starch hydrolysis [6]. Its association with starch granules was shown in endosperm of rice [7]; also the enzyme was found in the starchy endosperm of barley [8]. Abiotic stresses are also known to affect the activity and expression of *β*-amylases [9,10,11]. During stress, stimulation of starch degradation takes place in leaves for supporting respiration under condition of low photosynthesis and to protect cell structure by raising the levels of maltose as an osmoprotectant [11].

But, this role cannot be extended to all plant *β*-amylase, where it is known to perform various physiological roles. Its role as a storage protein was suggested in case of taproots of alfalfa (*Medicago sativa*) [12], cereal endosperm [1] and hedge bindweed rhizome (*Calystegia sepium*) [13]. In barley seeds *β*-amylase may have a nitrogen-storage function, since the profile of accumulation of *β*-amylase in developing seed resembles to that of the storage protein, hordein, and the synthesis of *β*-amylase responds to increased supply of nitrogen [14].

The extra chloroplastic localization of enzyme in spinach cells [15] and *Arabidopsis* leaves [16], provides evidence for its non-amylolytic function. Nine genes of *Arabidopsis* are known to encode for *β*-amylase isozymes. Six members of the family are predicted to be extrachloroplastic isozymes and three contain plastid transit peptides [17]. In leaves of pea (*Pisum sativum*) and wheat (*Triticum aestivum*) [18], the enzyme was found to be present in vacuole, where the assessment of its role is difficult, as the identification of vacuolar substrate remains to be determined. Furthermore, studies of proper germination of *β*-amylase deficient lines of soybean (*Glycine max*) [19] and rye (*Secale cereal*) [20] demonstrated its non-essential role in carbohydrate metabolism. *In vitro* studies have shown that native starch granules are not degraded by β-amylase without prior digested by other enzymes, making its role unclear [21]. The localization of enzyme was also found in phloem, by raising monoclonal antibody against sieve elements of tissue

Table 1. Purification of β-amylase from *Trigonella foenum-graecum* (fenugreek) (50 g).

Steps	Total Activity (units)	Total Protein (mg)	Specific Activity (units/mg)	Purification (fold)	Recovery (%)
Crude	2845.34	769.8	3.49	–	–
Acetone fractionation (50–65%)	2273.87	75.8	29.59	8.28	79.8
Ion Exchange (DEAE-Cellulose)	1365.76	6.8	200.85	57.54	48
Glycogen precipitation	586.07	0.8	732.59	209.91	20.6

Fold purification calculated with respect to the specific activity of the crude extract.

culture of *Streptanthus tortuosus* into BALB/c mice. The antibody identified enzyme to be present in phloem forming tissue cultures but absent in cultures lacking phloem. This antibody also cross-reacted with the major form of *A. thaliana* β-amylase and showed binding with its sieve elements [22].

Fenugreek (*Trigonella foenum-graecum* L. Leguminosae) is one of the oldest medicinal plant originating in India and North Africa. It has a long history of medicinal uses in Ayurvedic and Chinese medicine. It has been used for labour induction, aiding digestion and for improving metabolism and health. Apart from its use as a medicinal plant it is commonly used in cuisine. The seed contains many nutrients including protein, carbohydrates, fat in the form of volatile and fixed oil, vitamin and minerals as well as enzymes, fiber, saponins, choline and trigonelline [23]. During fenugreek seed germination and seedling development three different types of carbohydrates (galactomannans, soluble sugars including galacto-syl sucrose sugars and starch) are utilized [24]. Fenugreek seeds are starchless; galactomannans being the main carbohydrate reserve. Synthesis of starch starts only after germination [25]. Cotyledon was found to be the major site of starch accumulation [24]. Thus, making fenugreek seed an interesting material to study the potential role of β-amylase in carbohydrate mobilization, following germination.

Here, an attempt was made to study the physiological role of β-amylase in fenugreek seed and during seedling development, by purifying it from ungerminated seeds and characterizing it. Further, its immunolocalization studies were carried out in cotyledons till 62 h after germination.

Materials and Methods

Chemicals and Plant Materials

Fenugreek (*Trigonella foenum-graecum*) seeds were purchased from local market.

DEAE-cellulose, glycogen, amylopectin, maltose, pullulan, α-cyclodextrin, β-cyclodextrin, trypsin profile 1GD kit were purchased from Sigma Chemical Co., St. Loius; Mo, U.S.A.

Molecular markers for FPLC were from Pharmacia, Sweden.

All solutions were prepared in Milli Q (Millipore, Bedford, MA, U.S.A.) water.

All the chemicals for buffer were of analytical or electrophoresis grade from Merck Eurolab GmbH Darmstadt, Germany.

Enzyme and Protein Assays

Enzyme activity was measured using Bernfeld's method [26]. Reaction mixture was prepared by taking 0.5 mL of suitably diluted enzyme and 1% starch prepared in 50 mM sodium acetate buffer, pH 5.0. This was incubated at 30°C for 3 min, reaction was stopped by addition of 1 mL of 3,5-dinitrosalicylic acid. Test tubes were then placed in boiling water bath for 5 min and were

allowed to cool down to room temperature, followed by addition of 10 mL of distilled water. Absorbance was recorded at 540 nm. One unit of β-amylase is defined as the amount required for release of 1 μM of β-maltose per min at 30°C and pH 5.0 under the specified condition.

For measurement of α-amylase activity, the extract was heated at 70°C for 15 min and assayed as described above. The activity was also checked using starch azure as substrate.

Amount of protein present in sample was determined by Folin's method using crystalline BSA as standard protein. The protein profiles in column chromatography were followed by measuring the absorbance of the eluates at 280 nm.

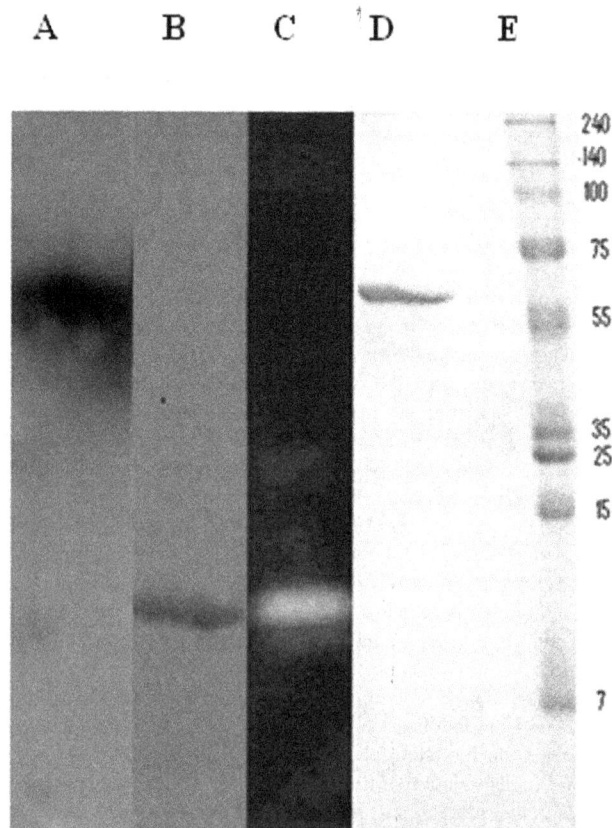

Figure 1. Electrophoresis pattern of Fenugreek β-amylase. Lane A : Western blotting of β-amylase, B: Native PAGE, C:Activity staining, D: SDS-PAGE of purified β-amylase, E: Molecular weight markers.

Table 2. Sorted peptide according to their residue number and masses along with peptide match.

Start-End	Observed	Mr(expt)	Mr(calc)	Delta	Miss	Sequence
150–160	1333.5499	1332.5426	1332.6271	−0.0845	0	R.TAIEIYSDYMK.S
161–171	1339.5571	1338.5498	1338.6601	−0.1103	1	K.SFRENMSDLLK.S
324–331	919.4445	918.4372	918.4923	−0.0551	0	R.DGYRPIAK.I
324–331	919.4445	918.4372	918.4923	−0.0551	0	R.DGYRPIAK.I
336–348	1641.6852	1640.6779	1640.7915	−0.1136	0	R.HHAILNFTCLEMR.D
336–348	1657.6980	1656.6907	1656.7864	−0.0957	0	R.HHAILNFTCLEMR.D
374–386	1401.6140	1400.6067	1400.6895	−0.0828	0	R.ENIEVAGENALSR.Y
387–406	2249.0198	2248.0125	2248.1600	−0.1474	0	R.YDATAYNQIILNARPQGVNK
414–421	1002.4468	1001.4395	1001.5004	−0.0609	0	R.MYGVTYLR.L
414–421	1002.4468	1001.4395	1001.5004	−0.0609	0	R.MYGVTYLR.L
414–421	1018.4418	1017.4345	1017.4953	−0.0608	0	R.MYGVTYLR.L
414–421	1018.4418	1017.4345	1017.4953	−0.0608	0	R.MYGVTYLR.L

Enzyme Purification

All the steps were performed at 4°C and centrifugation was carried out at 8,720 g, unless stated otherwise.

50 g of fenugreek seeds were soaked in 50 mM Tris-HCl buffer, pH 7.2, for 12 h. Seeds were coarsely crushed using waring blender in extraction buffer (50 mM Tris-HCl buffer, pH 7.2 containing 1 mM EDTA and 1 mM PMSF), and then filtered through two layers of pre-washed muslin cloth. The extract thus prepared was centrifuged for 20 min.

Crude extract was subjected to 50–65% acetone fractionation at −15°C. The precipitated proteins obtained after centrifuging it for 20 min were made free of acetone and then dissolved in extraction buffer. Solubilized proteins were loaded onto DEAE-cellulose column equilibrated with 25 mM Tris-HCl buffer, pH7.0. Enzyme was eluted and 2 mL fractions were collected with an increasing gradient of 0–0.2 M NaCl prepared in the same buffer, with a flow rate of 0.3 mL/min. The enzyme active fractions were pooled and dialysed against extraction buffer.

The dialyzed sample was used for affinity precipitation by following the method as described by Silvanovich and Hill [27] with some modifications. The sample was saturated with ethanol at −15°C. Precipitated and centrifuged for 20 min, the proteins precipitated were discarded. Glycogen solution (2%) was added dropwise to the supernatant with stirring at 4°C for 20 min followed by centrifugation for 15 min. The pellet thus obtained was washed twice with extraction buffer containing 40% ethanol. The washed pellet was suspended in 1 mL extraction buffer and kept at 37°C for 1 h for hydrolysis of glycogen and dialyzed it against extraction buffer. After dialysis it was centrifuged for 15 min. The supernatant was collected and small molecular oligosaccharides were removed by dialyzing against extraction buffer, overnight.

Electrophoresis

SDS-PAGE was used to check the homogeneity of each fraction. Analytical SDS-PAGE of 12% polyacrylamide gel was performed according to Laemmli's method [28], using vertical gel electrophoresis apparatus (Monokin, India). Proteins were visualized by staining with Coomassie Brilliant Blue R-250 (CBB R-250). The apparent molecular mass of β-amylase was determined by comparing with relative mobility of protein standards (molecular mass ranging from 72 kD to 290 kD). For, detection of enzyme activity, Native-PAGE (8%) was carried at 4°C in the same manner as SDS-PAGE but in the absence of SDS. Activity staining was performed by incubating gel at 30°C in 1% starch prepared in 0.1 M sodium-acetate buffer, pH 5.4, for 15 min followed by staining with 0.01 N I_2-KI solution.

To determine the pI of β-amylase, the glycogen precipitated sample was subjected to Isoelectric Focussing (IEF) with an Ettan IPG phor 3 instrument (GE Healthcare Lifesciences Ltd., Uppsala, Sweden) using an 11 cm IPG strip (pH range 3–10) by following protocol described elsewhere [29]. Protein was visualized using CBB R-250.

Native Molecular Mass Determination

Size-exclusion chromatography using Sephacryl S-200 (Hiprep 16/60 High resolution column, GE Healthcare Lifesciences Ltd., Uppsala, Sweden) on AKTA FPLC was performed for determination of native molecular weight of purified protein. The column was pre-equilibrated with 50 mM sodium acetate pH 5.0 containing 0.1 M NaCl. The column was calibrated with five proteins in molecular range of (12.5–240 kD). The enzyme was eluted with the same buffer at a flow rate of 0.5 mL/min. Absorbance was recorded at 280 nm and fractions corresponding to peak were checked for enzyme activity as described above. The native molecular mass was determined from a plot of Ve/Vo against log of molecular mass of standard proteins taken.

Table 3. The effect of various substrates on the enzyme activity.

S. No.	Substrate	Relative Activity (%)
1	Starch (solubilized)	100.00
2	Amylopectin	99.12
3	Amylose	50.30
4	Glycogen	43.00
5	Pullulan	0.04
6	Starch (insolubilized)	0.00

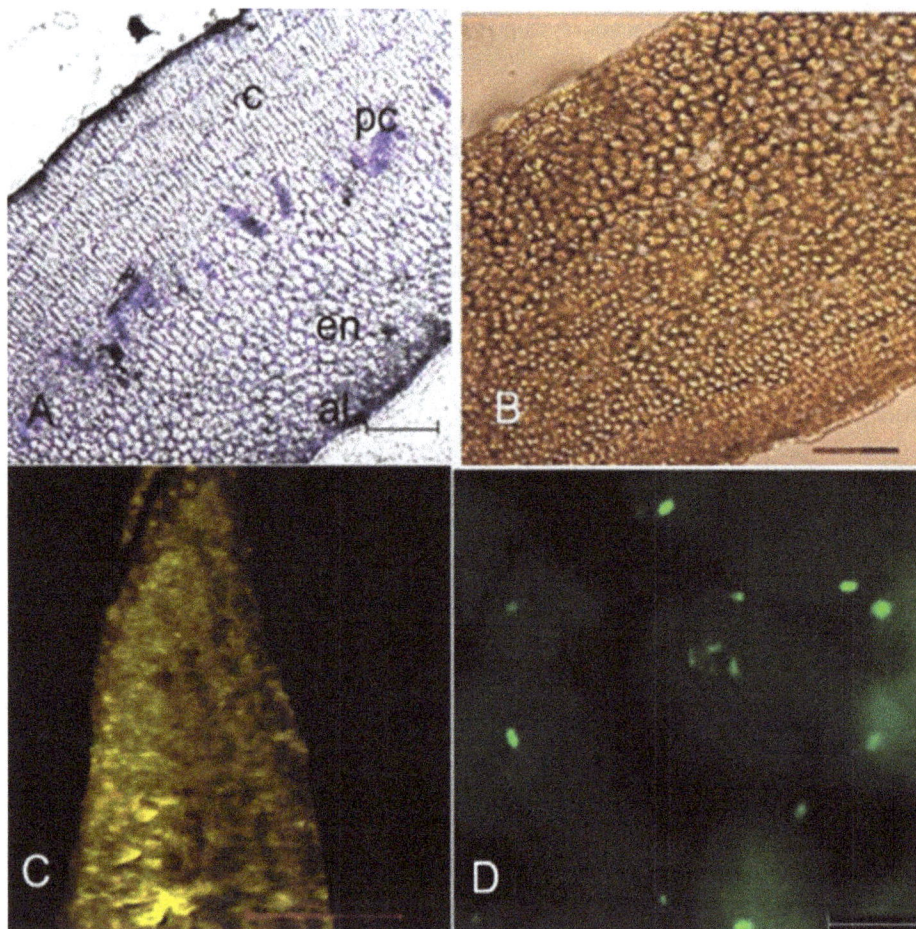

Figure 2. General histology of fenugreek seed and immunohistochemical localization of Fenugreek β-amylase. Thin sections of seeds of *Trigonella foenum-graecum* (A) stained with toluidine blue for observing general histology (B) stained with I₂-KI solution for visualization of starch (C) section labelled with anti-(β-amylase) IgG showed localization of enzyme in the endosperm (D) enlargement of (C) showing part of the endosperm. Bars represent 100 μm in A and B, 200 μm in C and 20 μm in D. cotyledon cells (*c*), procambial cells (*pc*), endosperm cells (*en*), aleurone cells (*al*).

Glycoprotein Properties

In order to ascertain whether enzyme is glycoprotein, 0.2 mL of purified β-amylase was taken in a test tube and to this 0.1 mL of 10 mM sodium metaperiodate was added and incubated at 30°C for 10 min. Exposure to air was avoided and 0.3 mL of 0.5% Schiff's reagent was added to the test tube and incubated at 30°C for 1 h, absorbance was recorded at 550 nm.

MALDI-TOF Mass Spectrometry

The identification and molecular weight determination of the protein band obtained after SDS-PAGE was performed using Matrix Assisted Laser Desorption Mass Spectrometry (MALDI; Voyager-DE-STR instrument; Applied Biosystems, USA). The protein band was excised from SDS-PAGE gel, destained, washed and digested with MALDI grade trypsin (Sigma, St. Louis, USA) overnight, following the protocol described by Kishore et al. [30]. The peptides obtained were analysed by matching with the sequence available in database. The amino acid sequence of β-amylase from *Trigonella foenum-graecum* is not available in protein database. Therefore, the peptide mass fingerprint obtained after MALDI-TOF was for matched with β-amylase from other plant sources using MASCOT (www.matrixscience.com) and applying following parameters: (a) fixed modifications carbamidomethyl (C);

(b) variable modifications oxidation (M); (c) cleavage by trypsin: cuts C-term side of KR unless next residue is P.

Substrate Specificity

The ability of amylase to degrade various polysaccharides was determined by performing assay using 1% solution of starch (insolubilized and solubilized), amylopectin, glycogen, pullulan, and starch azure. For performing the assay using chromogenic substrate, starch azure, following procedure was followed: 4.5 mL of 2% starch azure suspension was prepared in 0.02 M sodium phosphate buffer, pH 7.0, containing 0.05 M NaCl and equilibrated at 37°C, followed by addition of 0.5 mL of suitably diluted enzyme. It was mixed by swirling and incubated at 37°C for 15 min in a water bath. Reaction was stopped by addition of 20 mL of 2.75 M acetic acid. The suspension was filtered through Whatman filter paper no. 2 and its absorbance was recorded at 595 nm [31].

Effect of EDTA

The amylase was dialyzed overnight against buffer containing 5 mM EDTA, followed by determination of the residual activity under standard condition.

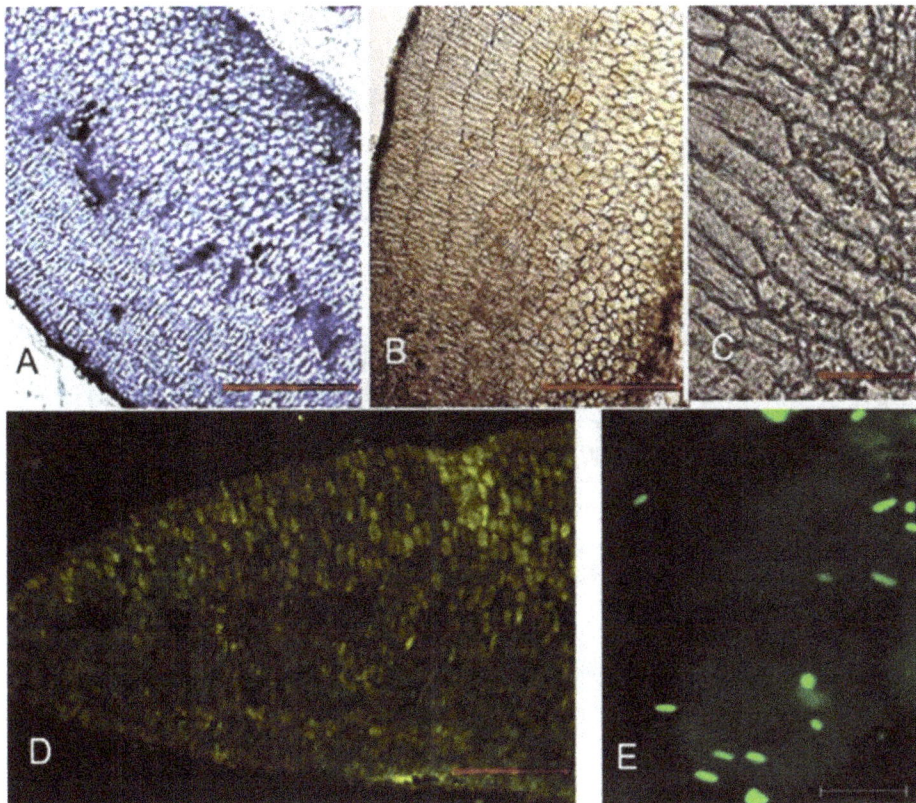

Figure 3. Thin longitudinal sections of cotyledons of *Trigonella foenum-graecum* after 14 h germination. (A) Toluidine blue stained section (B) Section stained with I$_2$-KI solution showed presence of starch in few cotyledon cells (C) Detail of (B) showing part of cotyledon cells (D) Immunolocalization shows presence of enzyme in endosperm and few cotyledon cells (E) Enlargement of (D) showing endosperm cells. Bars represent 120 μm in A, B and D, 45 μm in C and 20 μm in E.

Paper Chromatography

Identification of low molecular weight products of starch hydrolysis formed due to enzyme action, was performed by paper chromatography using butanol:pyridine:water (6:4: 3; v/v) solvent system. 5% starch solution was incubated with enzyme overnight, which was then kept in boiling water bath for 5 min to denature the enzyme, and was removed by centrifugation for 20 min. The supernatant was loaded onto the Whatman paper and chromatograms were developed for 4 h. Reducing sugars were visualized by treating it with a solution of 0.5% 3,5-dinitrosalicylic acid reagent and 4% NaOH [32,33].

Kinetic Studies

Enzyme obtained after glycogen precipitation was used for the kinetic studies. Effect of pH was observed by using the following buffers 0.05 M sodium-acetate buffer (pH range 3.6 to 5.6), 0.05 M phosphate buffer (pH range 6.0 to 7.0), 0.05 M Tris-HCl buffer (pH range 7.2 to 9.0). Enzyme activity was assayed by preparing starch solution in the appropriate buffers and following the procedure described earlier. Optimum temperature was determined by carrying out assay procedure in temperature range 20 to 70 ($\pm$1)°C, and activation energy (E_a) for β-amylase was calculated from the slope of the curve using an Arrhenius plot in the range of 20 to 50 ($\pm$1)°C. Thermal inactivation study for the enzyme was performed by maintaining enzyme at 52°C in a water bath (Multitemp, Pharmacia, Sweden). Small aliquots withdrawn at different time intervals were assayed for enzyme activity. Effect of substrate concentration was observed by using starch and amylopectin in the range of 0–10 mg/mL (pH 5.0), data thus obtained was used for determination of Michaelis constant (K_m) and maximum velocity (V_{max}) using Lineweaver-Burk plot. Enzyme was incubated for 10 min at 30°C with Schardinger dextrins, sucrose and maltose, so as to observe their effect on enzyme activity (for determining effect of maltose on enzyme, assay was carried out according to Fuwa's method [34]). Inhibition constant (K_i) for α-cyclodextrin and sucrose was determined using Dixon plot. All parameters were the mean of triplicate determinations from three independent preparations.

Preparation of Specific Antibodies against β-amylase

Antibody against Fenugreek β-amylase was raised by IMGE-NEX, Bhubneshwar, India; by injecting purified protein into rabbits, consequently the polyclonal antiserum was developed and antibody generated was purified using Protein-A affinity chromatography. These antibodies were used for blotting analysis.

Seed Germination

50 g of dry seeds were surface sterilised with 0.5% hydrogen peroxide, washed four times with water and left in water to imbibe for 12 h at 30°C. The seeds which had swollen were set out to germinate over moist filter paper on moist sand bed.

Western Blot Analysis

Protein (30 μg) from crude extract was denatured and resolved by 10% SDS-PAGE and transferred onto PVDF membrane (Sigma, St. Louis,USA). Transfer efficiency was checked by

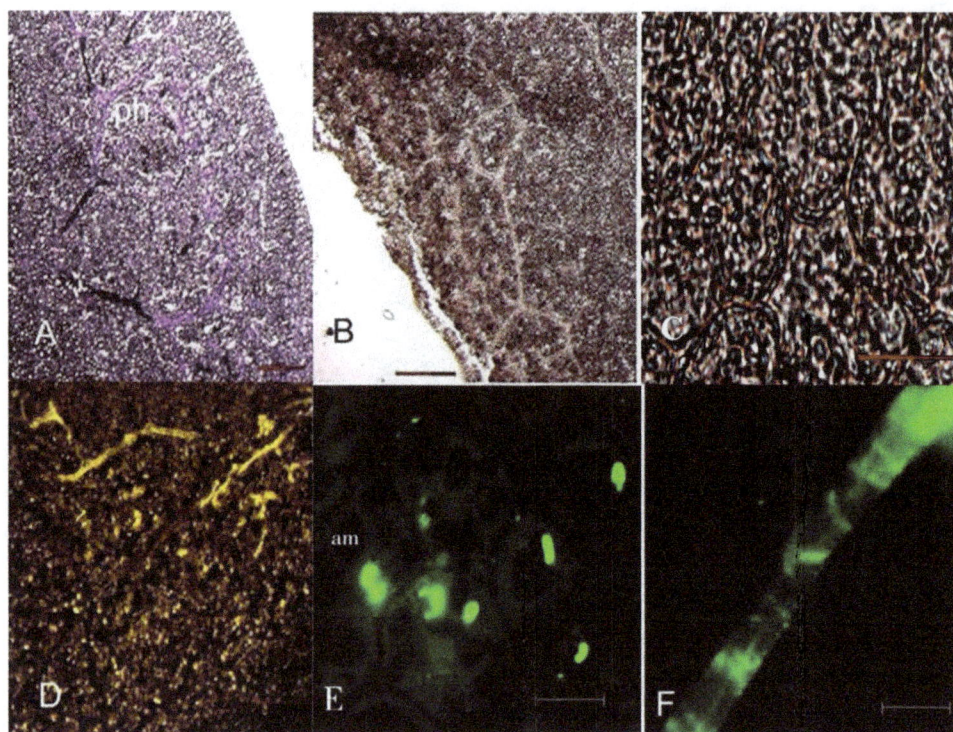

Figure 4. Longitudinal sections of cotyledon of *Trigonella foenum-graecum* after 31 h germination. (A) Toluidine blue stained section (B) I$_2$-KI solution stained section (C) Detail of (B) showing amyloplasts (D) Immunolabelling of β-amylase by using anti-(β-amylase) (E) Enlargement of (D) shows presence of enzyme associated with amyloplasts and in (F) with protophloem. Bars represent 100 μm in A, B and D, 45 μm in C and 20 μm in E and F. phloem (*ph*), amyloplasts (*am*). No fluorescence was seen in the sections concomitant to the sections stained when incubated with FITC-conjugated secondary antibody without prior treatment with primary antibody.

Ponceau-S staining, which was later completely removed from the membrane by repeated washing in water. The membrane was blocked with 5% non-fat milk in Phosphate Buffer Saline (PBS) overnight at room temperature, followed by incubation with anti-β-amylase polyclonal antibody raised in rabbit (1:1000) for 6 h. Subsequently, the blot was washed two times (5 min each) in PBS-0.1% Tween 20 and incubated with goat anti-rabbit IgG-horse-radish peroxidase conjugate (1:2000) (Banglore Genei, India) for 3 h at room temperature. The blot was washed thrice (5 min each) in PBS-0.1% Tween 20 and detected by enhanced chemilumi-nescence (ECL).

Immunohistochemistry

Slices (2 mm) of cotyledon tissue were fixed in 3% glutaralde-hyde in 0.2 M phosphate buffer (pH 7.2), dehydrated and embedded in paraffin wax. For light and fluorescence microscopy, sections were prepared using Leica RM2245 semi-motorized rotary microtome. Approximately 6 μm thick sections were cut and collected on polylysine coated slides.

Prior to staining sections were deparaffinised using xylene and rehydrated using 100%, 90%, 70%, 50% and 30% alcohol, respectively in sequential order, for 5 min each.

Sections were pre-incubated in 5% (w/v) BSA in PBS at pH 7.4 for 30 min, then incubated for 4 h in primary antibody diluted in PBS containing 1% BSA and 0.32% Tween 20 at 1:500 dilution. Slides were rinsed four times for 15 min with PBS-Tween, then incubated for 2 h, in the dark with secondary antibody (Goat Anti-rabbit IgG (H+L)-FITC conjugate) diluted 1:300 in PBS, 1% BSA, 0.32% Tween 20. Slides were then rinsed 4 times with PBS-Tween for 15 min each [35]. Longitudinal sections were analysed

using flurescence microscope (Nikon DS-Fi1) or Laser Scanning Confocal Microscope (Nikon Eclipse Ti microscope) at 488 nm excitation wavelength for fluorescein isothiocyanate (FITC). Images were processed using the software MetaMorph and CorelDRAW Graphics Suite X6.

Photomicroscopy

Sections approximately 6 μm thick were cut and collected on polylysine coated slides. Staining of section with 0.1% (w/v) toluidine blue in 1% (w/v) borax, pH 11 was done for general structure observations [35].

To determine the location of starch deposits in cotyledons of ungerminated and germinated seeds, semithin sections of seeds were stained with I$_2$-KI solution [25]. Slides were observed under light microscope (Nikon DS-Fi1).

Results

Purification

Crude extract was prepared from ungerminated seeds, as described under experimental section. The fractionation of proteins by acetone followed by anion-exchange chromatography using DEAE-cellulose and glycogen precipitation, as outlined in table 1, resulted in purification of Fenugreek β-amylase 210 fold with a yield of 21%. Homogeneity of the sample obtained after glycogen precipitation was checked on SDS-PAGE (Fig. 1 D), which revealed a single band after CBB staining. M_r of enzyme was determined to be 58 kD, when compared to standard molecular weight markers. Activity staining showed development of clear band (site of activity of β-amylase) over a dark background

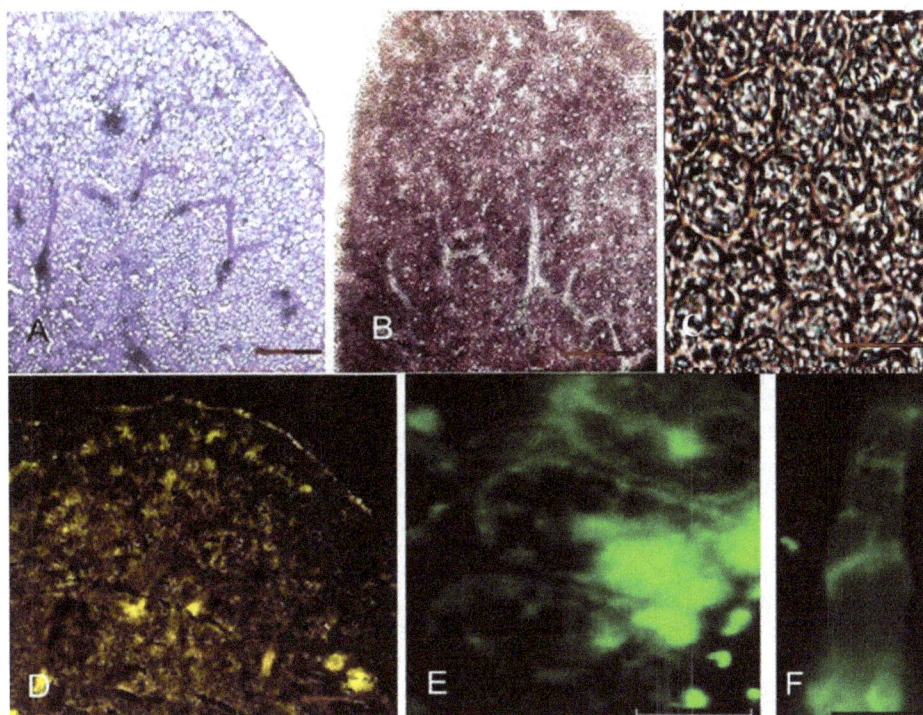

Figure 5. Longitudinal sections of cotyledons of *Trigonella foenum-graecum* after 62 h germination. (A) Sections stained with toluidine blue for general histology visualization (B) I_2-KI stained sections show distribution of starch (C) Detail of (B) showing amyloplasts (D) Immunolocalization of β-amylase (E) Enlargement of (D) shows association with amyloplasts and in (F) with proto-phloem. Bars represent 100 μm in A, B and D, 45 μm in C and 20 μm in E and F. Fluorescence was not observed in the sections treated with Goat Anti-rabbit IgG (H+L)-FITC conjugate antibodies without prior treatment with primary antibody.

formed due to complex of starch and I_2-KI solution (Fig. 1 C), corresponding protein band on Native-PAGE is shown in Fig. 1 B. The purified protein gives a single peak from Sephacryl S-200 gel filtration column and the molecular mass of native Fenugreek β-amylase determined from the mean of two experiments with it was 56 kD, which is in accordance with the mass obtained on SDS-PAGE, hence suggesting monomeric nature of enzyme. pI of Fenugreek β-amylase was determined to be 5.2 from IEF. The peaks obtained after MALDI-TOF analysis and the peptide mass fingerprints were used for database searching using MASCOT.

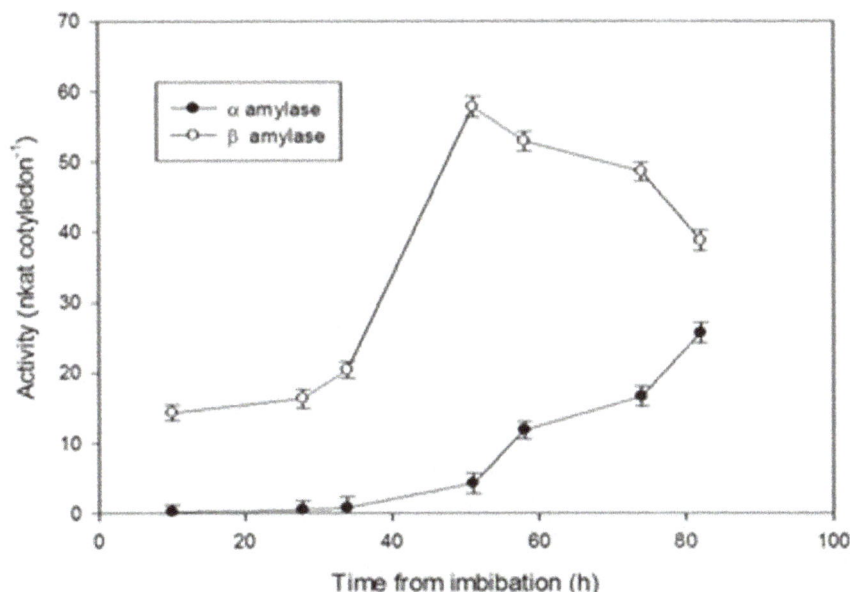

Figure 6. α- and β-amylase activity of crude extract of fenugreek from the seeds imbibed for upto 80 h.

The most significant match was found with β-amylase of *Medicago sativa* (alfalfa; Accession no. T09300) with a score of 82 and E-value of 0.0015 (Table 2).

Characterization of Fenugreek β-amylase

Enzyme hydrolyzed starch at highest rate followed by amylopectin, amylose and glycogen. Pullulan was not hydrolyzed indicating that enzyme failed to cleave α-1,6 bond. Also, the inability to give colour with assay performed using starch azure as substrate suggested that it was endoamylase. Fenugreek β-amylase was not able to degrade native starch (Table 3). There was no loss in enzyme activity after dialysis against buffer containing 5 mM EDTA for overnight, which indicates that Fenugreek β-amylase is not a metallo-enzyme. The characteristic property of β-amylase to release exclusively maltose as end product of starch hydrolysis was identified by paper chromatography. Periodic acid-Schiff's reagent method showed absence of any carbohydrate in the purified enzyme, indicating its non-glycosylated nature. The effect of pH on purified β-amylase was examined in the pH range 3.6 to 8.0. Maximum activity was observed at pH 5.0 in 50 mM sodium acetate buffer and enzyme was found to be fairly stable for a week in pH range of 3.0–7.5. Optimum temperature was found to be 50°C. The enzyme activity showed decline on increasing temperature beyond 50°C; complete loss of enzyme activity was observed when it was kept at 70°C for 15 min. The value of activation energy (E_a) was calculated to be 6.21 kcal/mole from Arrhenius plot. The K_m value for starch and amylopectin as a substrate was found to be 1.58 mg/mL and 2.86 mg/mL, respectively. The highest specific activity of β-amylase gave a turnover number of 137.93 min^{-1}. Thermal inactivation studies were carried out at 52°C, and it was found to follow first-order kinetics with rate constant of 0.0198 min^{-1} and $t_{1/2}$ equal to 35 min. When enzyme was incubated with 1 mM of Schardinger dextrins; α-cyclodextrin showed inhibitory effect and was found to be a competitive inhibitor from Dixon plot method with K_i 3.97 mM. Sucrose was also found to be a competitive inhibitor of Fenugreek β-amylase with K_i 2.32 mM, while maltose did not show any inhibitory effect on enzyme activity even at a concentration of 25 mM.

Localization of the β-amylase in the Cotyledons of Fenugreek

Fenugreek seeds were soaked in water for 12 h and then germinated on sand bed for different time intervals. Radicles were cleaved from the cotyledons and were fixed in 3% glutaraldehyde, further they were embedded in paraffin wax followed by sectioning. Sections were collected on polylysine coated slides. Toluidine blue staining was done for observing general histology of cotyledons after 14, 31 and 62 h of germination. Comparable samples of seed were stained with I$_2$-KI solution for detection of starch. To investigate the cellular localization of β-amylase during ungerminated and germination stages, immunohistochemistry was performed. The sections obtained from paraffinized blocks were immune-reacted with the purified antibodies raised against Fenugreek β-amylase. Comparable samples were also treated in the same manner with non-immunized rabbit serum as control. The specificity of the affinity purified antibodies was checked by western blot analysis of crude extract of seed. Antibodies specifically reacted with the 56 kD polypeptide corresponding to the Fenugreek β-amylase (Fig. 1 A).

The toluidine stained longitudinal section of fenugreek seeds showed one cell layer of aleurone cells surrounding the endosperm. The cotyledon cells were present below the endosperm (Fig. 2 A). The enzyme was found to be localised in the

endosperm of ungerminated seeds (Fig. 2 C, D). Fluorescence was not observed in the cotyledonary cells and procambial cells. Comparable sections stained with I$_2$-KI showed absence of starch (Fig. 2 B). Germination being defined as penetration of the seed coat by the radicle, occurred after about 10 h in the seeds set out for germination. No change in general histological structure of the seed was observed after 14 h of germination (Fig. 3 A). Starch could now be seen in cotyledonary cells (Fig. 3 B, C). β-Amylase was found to be present mainly in endosperm cells at this stage also, although some of the cotyledon cells also show fluorescence. Both amount of starch and β-amylase showed an increase at the stage of 31 h after germination. Toluidine blue stained sections showed dissolution of endosperm cell and the reserve stored (Fig. 4 A). Starch was predominantly present (Fig. 4 B, C). Fluorescence staining showed association of β-amylase with amyloplasts (Fig. 4 D, E). Enzyme also showed its association with protophloem at this stage of germination (Fig. 4 D, F). After 62 h β-amylase could still be seen in phloem and at the periphery of amyloplasts (Fig. 5 D, E, F). Although, at this stage decline in the amount of enzyme (Fig. 5 D) and starch was observed (Fig. 5 B, C).

The activity profile of α-amylase and β-amylase during germination of seed was observed by performing assay procedure as described above. α-Amylase activity was negligible up till 31 h and showed an increase after that, while β-amylase activity was comparatively quite higher and showed an increasing trend upto 41 h, thereafter decline was observed (Fig. 6).

Discussion

The plant material used for purification of β-amylase consisted of ungerminated fenugreek seeds. Enzyme was purified by using steps as summarized in table. Affinity precipitation using glycogen proved to be a useful step for purification. β-Amylase from pea [36], maize [37] and radish [38] have also been purified using this method. Monomeric β-amylase are more commonly found in plants, having molecular mass in range of 42 to 65 kD [8,33,37,38], β-amylase from sweet potato [39], leaves of *Vicia faba* [40] and hedge bindweed [13] are known to be tetramer having subunits of 64.7 kD, 26 kD and 56.07 kD, respectively. Fenugreek β-amylase was also found to be monomeric in nature. Fenugreek β-amylase resembled other plant β-amylase in terms of highest affinity for starch as compared to other substrates, pI, kinetic parameters and α-cyclodextrin being competitive inhibitor [41,36,13]. However, end product inhibition by maltose was not observed.

Fenugreek is a leguminous seed, having major carbohydrate reserve in the form of galactomannans stored in endosperm cells. During course of germination the hydrolysis of galactomannans take place and the break down products are absorbed by the cotyledons in which sugar increases and starch is formed [25,42]. ADPGppase (adenosine diphosphate glucose pyrophosphorylase) is known to be a key regulatory enzyme for the formation of glycosyl donor for starch synthesis. Partial regulation of starch accumulation within cotyledons was suggested to occur by the transcription of ADPGppase genes, which shows positive correlation with the synthesis and decay of starch [24]. Temporary storage of starch takes place which gets remobilized during later time of seedling development [42]. β-Amylase was found to be present in the ungerminated seed, which did not contain starch. To study the probable role of β-amylase in fenugreek seeds, they were allowed to germinate for different time periods. In ungerminated seeds, enzyme was present in the endosperm cells. β-Amylase could be seen in the endosperm of seeds after 14 h of germination. Some of the cotyledon cells at this stage showed the presence of starch which were near endosperm cells. Thus, the substrate was inaccessible to the enzyme. The dissolution of endosperm cells was completed by 31 h and the cells showed presence of high amount of starch. β-Amylase content also showed an increase along with α-amylase (Fig. 6). β-Amylase could be seen at the periphery of amyloplasts. Similar to other plant β-amylases, Fenugreek β-amylase also showed inability towards degradation of the native starch granule. Therefore, presence of not high but significant amount of α-amylase may be required to initiate the reaction, followed by hydrolysis by β-amylase. At this stage of germination, starch acts as carbohydrate reserve and thus its hydrolysis is needed for germination and growth of seedling. The high amount of β-amylase present as compared to α-amylase and its presence at the periphery of amyloplasts defines the key role of β-amylase in hydrolysis of starch. Similar observations were made during germination of rice seeds [7] and barley seeds [43], where β-amylase was shown to be associated with starch granule. In sweet potato, cell walls and starch granules were labelled by the polyclonal antibodies raised against an inhibitor of starch phosphorylase, which was later identified as β-amylase [44]. After 62 h of germination amyloplasts were still surrounded by the enzyme. A decline in the content of starch and β-amylase was observed at this stage. The correlation between starch and β-amylase further confirms its role in starch hydrolysis.

The fluorescence could be seen in protophloem in the cotyledons after 31 h and 62 h of germination. Phloem is the primary transport for organic compounds in plants and cotyledons. The transported organic compounds serve as major substrate for plant growth. During the process of vascular differentiation, some of the provascular cells divide longitudinally to give rise to procambial cells, out of which some are destined to become phloem precursor cells [45]. Protophloem cells specified during embryogenesis differentiate with in the first three days of germination [46]. In *Arabidopsis* it has been shown that during embryogenesis phloem differentiation in the cotyledons occur earlier than in the axis [47]. Association of β-amylase with the protophloem after 31 h of germination can be related to its important role in transportation. There is an earlier report on presence of phloem specific β-amylase in *A. thaliana* and *Streptanthus tortuous* stems. It was suggested that the enzyme may have a role in prevention of starch build up during translocation of sugars in phloem sieve elements [22]. Probably, similar role can be attributed to Fenugreek β-amylase.

Conclusion

The Fenugreek β-amylase was found to be major starch degrading enzyme of fenugreek based on localization studies. This study also lead to the new finding of association of β-amylase with the protophloem, which can further be explored for providing an insight into the phloem physiology.

Acknowledgments

We thank Dr. Madhu Yashpal for technical assistance in immunohistochemistry. Imaging data were collected at National Facility for Laser Scanning Confocal Microscope, Department of Zoology, Banaras Hindu University, Varanasi, U.P., India.

Author Contributions

Conceived and designed the experiments: GS AMK. Performed the experiments: GS. Analyzed the data: GS AMK. Contributed reagents/materials/analysis tools: AMK. Wrote the paper: GS.

References

1. Ziegler P (1999) Cereal β-amylases. J Cereal Sci 29: 195–204.
2. Giese H, Hejgaard J (1984) Synthesis of salt-soluble proteins in barley. Pulse-labeling study of grain filling in liquid-cultured detached spikes. Planta 161: 172–177.
3. Okamoto K, Akazawa T (1980) Enzymic mechanism of starch breakdown in germinating rice seeds 9. de novo synthesis of β-amylase. Plant Physiol 65: 81–84.
4. Nandi S, Das G, Sen-mandi S (1995) β-Amylase activity as an index for germination potential in rice. Ann Bot-London 75: 463–467.
5. Weise SE, Weber APM, Sharkey TD (2004) Maltose is the major form of carbon exported from the chloroplast at night. Planta 218: 474–482.
6. Monroe JD, Preiss J (1990) Purification of a β-amylase that accumulates in *Arabidopsis thaliana* mutants defective in starch metabolism. Plant Physiol 94: 1033–1039.
7. Okamoto K, Akazawa T (1979) Enzymic mechanism of starch breakdown in germinating rice seeds 8. Immunohistochemical localization of β-amylase. Plant Cell Physiol 64: 337–340.
8. Bilderback DE (1974) Amylases from aleurone layers and starchy endosperm of barley seeds. Plant Physiol 53: 480–484.
9. Nielsen TH, Deiting U, Stitt M (1997) A β-amylase in potato tubers is induced by storage at low temperature. Plant Physiol 113: 503–510.
10. Todaka D, Matsushima H, Morohashi Y (2000) Water stress enhances β-amylase activity in cucumber cotyledons. J Exp Bot 51: 739–745.
11. Kaplan F, Guy CL (2004) β-Amylase induction and the protective role of maltose during temperature shock. Plant Physiol 135: 1674–1684.
12. Gana JA, Kalengamaliro NE, Cunningham SM, Volenec JJ (1998) Expression of β-amylase from alfalfa taproots. Plant Physiol 118: 495–506.
13. Van Damme EJM, Hu J, Barre A, Hause B, Baggerman G, et al. (2001) Purification, characterization, immunolocalization and structural analysis of the abundant cytoplasmic β-amylase from *Calystegia sepium* (hedge bindweed) rhizomes. Eur J Biochem 268: 6263–6273.
14. Qi JC, Zhang GP, Zhou MX (2006) Protein and hordein content in barley seeds as affected by nitrogen level and their relationship to β-amylase activity. J Cereal Sci 43: 102–107.
15. Okita TW, Greenberg E, Kuhn DN, Preiss J (1979) Subcellular localization of the starch degradative and biosynthetic enzymes of spinach leaves. Plant Physiol 64: 187–192.
16. Lin T, Spilatro SR, Preiss J (1988) Subcellular localization and characterization of amylases in Arabidopsis leaf. Plant Physiol 86: 251–259.

17. Sparla F, Costa A, Schiavo FL, Pupillo P, Trost P (2006) Redox regulation of a novel plastid-targeted β-Amylase. Plant Physiol 141: 840–850.
18. Ziegler P, Beck E (1986) Exoamylase activity in vacuoles isolated from pea and wheat leaf protoplasts. Plant Physiol 82: 1119–1121.
19. Hildebrand DF, Hymowitz T (1981) Role of β-amylase in starch metabolism during soybean seed development and germination. Physiol Plantarum 53: 429–434.
20. Daussant J, Zbaszyniak B, Sadowski J, Wiatroszak I (1981) Cereal β-amylase: immunochemical study on two enzyme-deficient inbred lines of rye. Planta 151: 176–179.
21. Beck E, Ziegler P (1989) Biosynthesis and degradation of starch in higher plants. Annul Rev Plant Physiol 40: 95–117.
22. Wang Q, Monroe J, Sjolund RD (1995) Identification and characterization of a phloem-specific β-amylase. Plant Physiol 109: 743–750.
23. Basch E, Ulbricht C, Kuo G, Szapary P, Smith M (2003) Therapeutic applications of fenugreek. Altern Med Rev 8: 20–27.
24. Dirk LMA, van der Krol AR, Vreugdenhil D, Hilhors HWM, Bewley JD (1999) Galactomannan, soluble sugar and starch mobilization following germination of *Trigonella foenum-graecum* seeds. Plant Physiol and Biochem 37: 41–50.
25. Reid JSG (1971) Reserve carbohydrate metabolism in germinating seeds of *Trigonella foenum-graecum* L. (Leguminosae). Planta 100: 131–142.
26. Bernfeld P (1955) Amylases α and β In: Methods in Enzymology Vol. 1. Academic Press Inc Publishers New York: 149–158.
27. Silvanovich MP, Hill RD (1974) α-Amylases from Triticale 6A190: Purification and Characterization. Cereal Chem 54(6): 1270–1281.
28. Laemmli UK (1970) Cleavage of structural proteins during the assembly of the head of bacteriophage T4. Nature 227: 680–685.
29. Tiwari N, Woods L, Haley R, Kight A, Goforth R, et al. (2010) Identification and characterization of native proteins of *Escherichia coli* BL-21 that display affinity towards immobilized metal affinity chromatography and hydrophobic interaction chromatography matrices. Protein Expres and Purif 70: 191–195.
30. Kishore D, Kayastha AM (2012) A β-galactosidase from chick pea (*Cicer arietinum*) seeds: Its purification, biochemical properties and industrial applications. Food Chem 134: 1113–1122.
31. Kumari A, Singh VK, Fitter J, Polen T, Kayastha AM (2010) α-Amylase from germinating soybean (*Glycine max*) seeds-purification, characterization and sequential similarity of conserved and catalytic amino acid residues. Phytochemistry 71: 1657–1666.

32. Jeanes A, Wise CS, Dimler RJ (1951) Improved techniques in paper chromatography of carbohydrates. Anal Chem 23: 415–420.

33. Doehlert DC, Duke SH, Anderson L (1982) β-Amylases from alfalfa (*Medicago sativa* L.) roots. Plant Physiol 69: 96–102.

34. Fuwa H (1954) A new method for microdetermination of amylase activity by the use of amylose as the substrate. J Biochem (Tokyo) 41, 583–603.

35. Tosi P, Parker M, Gritsch CS, Carzaniga R, Martin B, et al. (2009). Trafficking of storage proteins in developing grain of wheat. J Exp Bot 60: 979–991.

36. Lizotte PA, Henson CA, Duke SH (1990) Purification and characterization of pea epicotyl β-amylase. Plant Physiol 92: 615–621.

37. Subbarao KV, Datta R, Sharma R (1998) Amylases synthesis in scutellum and aleurone layer of maize seeds. Phytochemistry 49: 657–666.

38. Hara M, Sawada T, Ito A, Ito F, Kuboi T (2009) A major β-amylase expressed in radish taproots. Food Chem 114: 523–528.

39. Thoma JA, Spradlin JE, Dygert S (1971) Plant and animal amylases. In PS Boyer, ed, The Enzymes, Ed 3, Vol 5. Academic Press, New York: 115–189.

40. Chapman GW, Pallas JE, Mendicino J (1972) The hydrolysis of maltodextrins by a β-amylase isolated from leaves of *Vicia faba*. Biochim Biophys Acta 276: 491–507.

41. Lundgard R, Svensson B (1987) The four major forms of barley β-amylase. Purification, characterization and structural relationship. Carlsberg Res Commun 52: 313–326.

42. Bewley JD, Leung DWM, Macisaac S, Reid JSG, Xu N (1993) Transient starch accumulation in the cotyledons of fenugreek seeds during galactomannan mobilization from the endosperm. Plant Physiol and Biochem 31(4): 483–490.

43. Shen-Miller J, Kreis M, Shewry PR, Springall DR (1991) Spatial distribution of β-amylase in germinating barley seeds based on the avidin-biotin-peroxidase complex method. Protoplasma 163: 162–173.

44. Chang TC, Su JC (1986) Starch phosphorylase inhibitor from sweet potato. Plant Physiol 80: 534–538.

45. Esau K (1969) The phloem. Stuttgart: Gebrüder Bornträger, Berlin.

46. Busse JS, Evert RF (1999) Pattern of differentiation of the first vascular elements in the embryo and seedling of *Arabidopsis thaliana*. Int J Plant Sci 160: 1–13.

47. Bauby H, Divol F, Truernit E, Grandjean O, Palauqui JC (2007) Protophloem differentiation in early *Arabidopsis thaliana* development. Plant Cell Physiol 48: 97–109.

11

Ectopic Expression of the RING Domain of the Arabidopsis PEROXIN2 Protein Partially Suppresses the Phenotype of the Photomorphogenic Mutant De-Etiolated1

Mintu Desai[1¤], Navneet Kaur[1], Jianping Hu[1,2]*

1 Michigan State University-Department of Energy Plant Research Laboratory, Michigan State University, East Lansing, Michigan, United States of America, **2** Plant Biology Department, Michigan State University, East Lansing, Michigan, United States of America

Abstract

The Arabidopsis CONSTITUTIVE PHOTOMORPHOGENIC/DE-ETIOLATED 1/FUSCA (COP/DET1/FUS) proteins repress photomorphogenesis by degrading positive regulators of photomorphogenesis, such as the transcription factor LONG HYPOCOTYL5 (HY5). The gain-of-function mutant *ted3*, which partially suppresses the *det1* mutant, contains a missense mutation of a Val-to-Met substitution before the C-terminal RING finger domain of the peroxisomal membrane protein PEROXIN2 (PEX2). We hypothesized that a truncated PEX2 protein, which only contains the C-terminal RING domain, is initiated by the *ted3* mutation and by-passes the function of DET1 in the nucleus. Although we have not been able to detect this hypothetic peptide *in vivo*, we show in this study that, when fused with a fluorescent protein and overexpressed, the PEX2 RING domain can localize to the nucleus, where it is able to interact with HY5, and PEX2 RING domain overexpression in *det1* also partially suppresses the *det1* phenotype. Compared with *det1*, *ted3 det1* plants have significantly decreased levels of the HY5 protein and the expression of most of the analyzed HY5 target genes is altered to levels comparable to those in *hy5*. We conclude that compromised activity of HY5 may have been mainly responsible for the partial reversal of the *det1* phenotype in *ted3 det1*. Our data support the notion that, when appropriately localized, some RING finger domains may be able to achieve neomorphic effects in the cell.

Editor: Vladimir N. Uversky, University of South Florida College of Medicine, United States of America

Funding: This work was supported by the National Science Foundation (http://www.nsf.gov), MCB 0618335 and MCB 1330441, to JH and the Chemical Sciences, Geosciences and Biosciences Division, Office of Basic Energy Sciences, Office of Science, U.S. Department of Energy (http://science.energy.gov/) (DE-FG02-91ER20021) to JH. The funders had no role in study design, data collection and analysis, decision to publish, or preparation of the manuscript.

Competing Interests: The authors have declared that no competing interests exist.

* Email: huji@msu.edu

¤ Current address: Tyton BioEnergy Systems, LLC, Danville, VA, United States of America

Introduction

In response to the changing light regime, seedlings of higher plants undergo two drastically different programs. Skotomorphogenesis (etiolation) takes place in the dark, during which seedlings develop a long hypocotyl and hooked/undeveloped cotyledons. When exposed to light, seedlings go through photomorphogenesis (de-etiolation), where hypocotyl growth is inhibited, cotyledons open, chloroplasts develop, and genes involved in photosynthesis and light-regulated development are expressed. Light signals are transduced from photoreceptors, early signaling factors, central integrators, to downstream effectors, resulting in changed expression of hundreds of genes [1,2,3].

As central regulators of photomorphogenesis, CONSTITUTIVE PHOTOMORPHOGENIC/DE-ETIOLATED 1/ FUSCA (COP/DET1/FUS) proteins comprise three distinct protein complexes in a ubiquitin (Ub)-proteasome system. This system targets key positive regulators of light response, such as the photoreceptor phytochrome A (phyA) and transcription factors

Long Hypocotyl5 (HY5)/HY5 Homolog (HYH), Long Hypocotyl in Far-Red1 (HFR1), and Long After Far-Red Light1 (LAF1), for degradation. COP1 is a RING finger-containing E3 ligase that acts as a central component of a CULLIN4 (CUL4)-based E3 complex. Besides being a chromatin regulator to repress gene expression, DET1 is part of another CUL4-based E3 complex that functions to enhance the activity of the COP1 complex with the help of the COP9 signalosome (CSN) [4,5]. The bZIP transcription factor HY5 is a master regulator of photomorphogenesis that controls the expression of a repertoire of light-response genes [6]. In the dark, COP1 transits from cytoplasm to the nucleus, where it interacts with HY5 and mediates its ubiquitination by the concerted activity of COP/DET1/FUS protein complexes, resulting in significant reduction of HY5 abundance due to protein degradation by the 26S proteasome [7]. As such, dark-grown loss-of-function mutants of most *COP/DET1/FUS* genes show developmental patterns akin to that in light-grown wild-type seedlings (i.e. de-etiolated), whereas seedlings of photomorphogenesis-promoting factors such as HY5 often have long hypocotyls in the light [4].

Light regulates the development and function of subcellular organelles as well. In addition to its well-known impact on chloroplasts, light has also been linked to peroxisomes, essential eukaryotic organelles that mediate a variety of metabolic processes, such as photorespiration, fatty acid β–oxidation, and biosynthesis and metabolism of hormones in plants [8,9]. Light up-regulates the expression of genes encoding enzymes involved in photorespiration – a process that accompanies photosynthesis, while it represses genes involved in fatty acid β–oxidation and the glyoxylate cycle – processes that provide energy to seedling establishment before photosynthesis begins [10]. Light also promotes the proliferation of peroxisomes in Arabidopsis seedlings through phyA and the bZIP transcription factor HYH, the latter of which directly binds to the promoter and presumably activates the expression of the peroxisome proliferation factor gene *PEX11b* [11,12]. This is consistent with the idea that during photomor-phogenesis, an increase in peroxisomal population takes place besides the activation of the expression of photorespiratory genes and the import of their products into the peroxisome.

Before the discovery of DET1 as part of the protein complexes that degrade positive regulators of photomorphogenesis, the *det1-1* allele was used as the background to isolate extragenic suppressors to investigate the function of the DET1 protein [13]. One partial suppressor, *ted3* (for reversal of *det*), turned out to carry a gain-of-function mutation in the peroxisome biogenesis factor *PEROXIN2* (*PEX2*) [14]. PEX2 is a conserved RING finger domain-containing peroxisomal membrane protein in-volved in peroxisomal protein import in diverse species. PEX2 or its RING domain possesses E3 ubiquitin ligase activity in yeast *Saccharomyces cerevisiae* [15], mammals [16], and Arabidopsis [17].

Multiple models have been proposed to explain the partial suppression of *det1* by *ted3* [10]. One model postulated that DET1 is a key positive regulator of peroxisomal functions and that *ted3* possesses enhanced peroxisomal activities to suppress *det1-1*. Some of the phenotypes in *det1-1* are similar to those in peroxisomal β-oxidation mutants, such as sugar-dependent seedling establishment and partial resistance to indole-3-butyric acid (IBA), a protoauxin that is converted to the bioactive auxin indole-3-acetic acid (IAA) by β–oxidation [14]. However, viable loss-of-function peroxisomal mutants do not have opened cotyle-dons like *det1* despite having shorter hypocotyls on media without sucrose, arguing that peroxisomes do not play a major role in photomorphogenic development but rather represent one of the many downstream branches in DET1's regulatory network in growth and development. In addition, DET1 represses photo-morphogenesis yet light activates photorespiration and peroxi-somal proliferation, suggesting that DET1 is not a primary regulator of general peroxisomal function.

A second hypothesis favored the scenario that *ted3* encodes a gain-of-function product, which bypasses the function of DET1 in photomorphogenesis. The *ted3* mutation contains a G-to-A transition that leads to a Val-to-Met substitution one amino acid upstream from the first Cys of the C-terminal RING finger domain [14]. It is conceivable that in *ted3 det1*, this new Met may initiate the translation of a cryptic peptide that comprises the RING finger domain. Alternatively, changing from Val to Met may increase the accessibility of the protein to cytoplasmic proteases, which cleave off the cytosolically exposed RING domain of PEX2. This RING domain from PEX2, which has been shown to contain E3 ubiquitin ligase activity *in vitro* [17], may be mobilized to the nucleus because of its small size (~6 kDa)

and substitute for the function of the COP1-DET1 E3 ligase complexes in degrading some of the positive regulators of photomorphogenesis. We have not been able to detect this small peptide *in vivo*. However, in this study we have provided evidence that the RING domain of PEX2 when overexpressed is able to partially rescue the *det1* phenotype. PEX2's RING domain can enter the nucleus, where it interacts with the transcription factor HY5 and presumably reduces its function. We postulate that this alteration of HY5 activity may largely account for the partial reversal of the *det1* phenotypes in the *ted3 det1* dominant mutant during photomorphogenesis.

Materials and Methods

Plant growth, light conditions and genetic crosses

The wild-type Arabidopsis plants used in this study were from the Columbia-0 (Col-0) ecotype. *hy5-1*, *cop1-4*, *det1-1* and *ted3 det1* were in the Col-0 background. These mutants were confirmed by their respective dark-grown phenotypes, and genotyped by PCR analysis to ensure their homozygosity. Seeds were surface sterilized with 20% Clorox and 0.025% Triton X-100, washed 5 times with sterile water. To measure hypocotyl length, sterilized seeds were plated on 0.5X MS medium supplemented with 0.5% sucrose and solidified with 0.6% phytagar, stratified at 4°C for 3d, exposed to white light (100 μm m^{-2}s^{-1}) for 1 h to induce synchronous germination, and returned to the darkness for 4d at 22°C. Hypocotyl lengths of >30 seedlings from each genotype were measured using ImageJ software (http://imagej.nih.gov/ij/). Three biological replicates were undertaken. After having acquired their first true leaves, the seedlings were transferred to soil and grown in growth chambers with 100 μm m^{-2}s^{-1} white light, 16/8 h photoperiod, and at 22°C.

Confocal laser scanning microscopy (CLSM) and epifluorecence microscopy

Plant tissues (as indicated in the text) were incubated with DAPI (Invitrogen, Carlsbad, CA) at 300 nM concentration in 1X PBS at room temperature, covered with aluminum foil for 15 min followed by 3–4 washes to remove excess stain, and directly mounted in distilled water to be analyzed by CLSM (Zeiss LSM 510 META). A 488-nm, 514-nm argon ion laser and 401-nm diode were used for excitation; emission filters of 505–530 nm, 520–555 nm band-pass and 433-nm long-pass were used for GFP, YFP and DAPI respectively. Images were acquired at 63X with oil. Epifluorescence microscopy was performed with an Axio Imager M1 microscope (Carl Zeiss) for visualization of the BiFC between HY5-YFPct and YFPnt-PEX2RF proteins (excitation 500±12 nm; emission 542±13.5 nm).

RT-PCR analyses

Total RNA was isolated from 4d dark-grown seedlings using SV total RNA isolation system kit (Promega, Madison, WI). For RT-PCR analysis, 2 μg total RNA was reverse-transcribed with the Omniscript RT kit (Qiagen, Valencia, CA). PEX2 RF-specific primers FW (5′-GTGACTTGCCCTATTTGC-3′) and RE (5′-TCATTTGCCACTTGAAAC-3′) were used to amplify a 0.1-kb product that covered the entire C-terminal end containing the RF domain from *PEX2* cDNA. *UBQ10*-FW (5′-TCAATTCTCTC-TACCGTGATCAAGATGCA-3′) and *UBQ10*-RE (5′-GGTGTCAGAACTCTCCACCTCAAGAGTA-3′) from the *UBQ10* gene (At4g05320) were used to amplify a product of ~320 bp that served as an internal control. For *PEX2* RF domain and *UBQ10* amplification, PCR was performed with the following

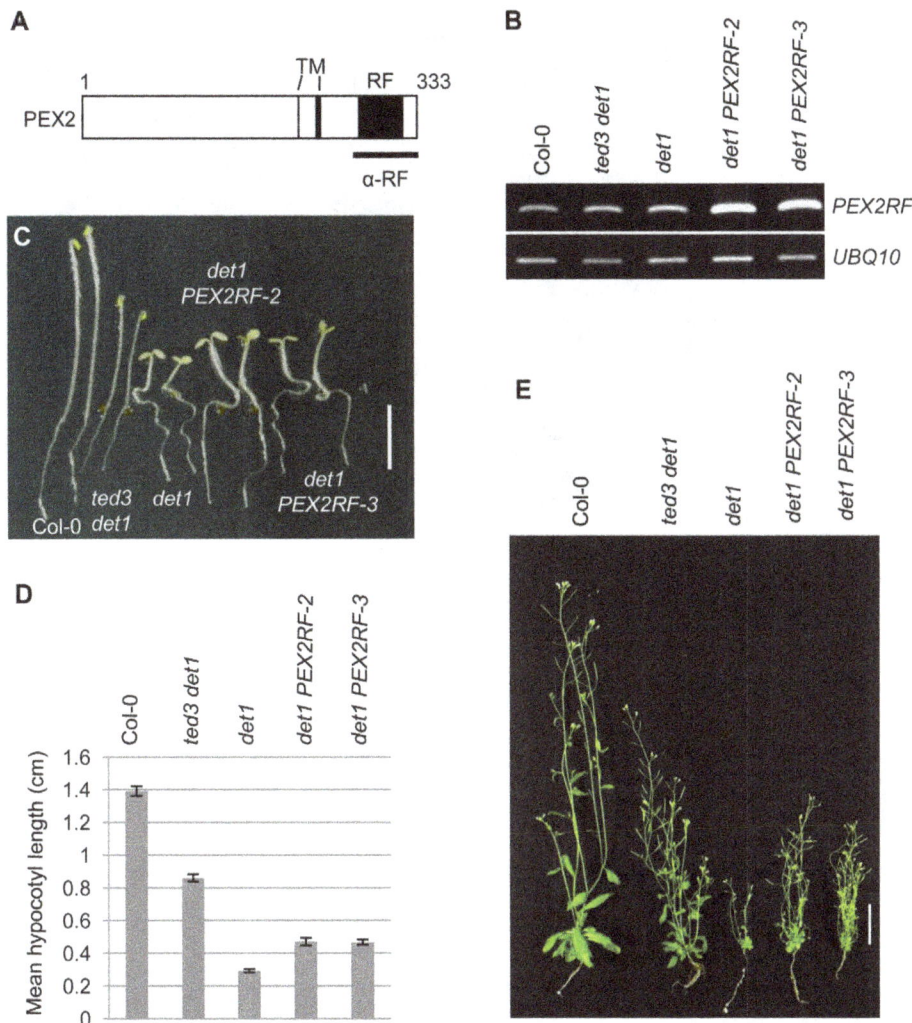

Figure 1. Overexpression of a peptide that contains the PEX2 RING domain suppresses *det1*. (A) Schematic of the Arabidopsis PEX2 protein, showing positions of the transmembrane (TM) and RING finger (RF) domains, and the region (indicated by horizontal bar) used as PEX2RF and as antigen for antibody generation. (B) RT-PCR analyses of the *PEX2RF* transcript in two transgenic lines (lines2 & 3) overexpressing PEX2RF in the *det1* background. *UBQ10* is the internal control. (C) Phenotype of 4d dark-grown seedlings grown on 0.5X MS supplemented with 0.5% sucrose. Scale bar = 0.5 cm. Two seedlings are shown for each genotype. (D) Hypocotyl length measurements of 4d dark-grown seedlings shown in (C). n>30 for each genotype. Student *t*-test, P<0.0001 for all lines vs. *det1*. Error bars indicate s.e.m. (E) Four-week plants. Scale bar = 3 cm.

conditions: 94°C for 2 min, 30 cycles of 94°C for 30 s, 57°C for 30 s, 72°C for 30 s, and a final extension at 72°C for 4 min.

Quantitative real-time PCR

For transcript analysis, whole Arabidopsis seedlings grown under constant light at 22°C on 0.5X MS media plates were used. Harvested seedling samples were frozen in liquid N_2 and total RNA was extracted using RNeasy plant mini kits (Qiagen, Valencia, CA) followed by treatment with DNase I (Qiagen, Valencia, CA) according to manufacturer's instructions. Synthesis of cDNA was performed with the Omniscript Reverse Transcription system (Qiagen, Valencia, CA) using random primers with 0.1 µg of total RNA in a 20 µl volume RT reaction, and incubated for 1 hr at 42°C. The RT reaction mixture was diluted 10-fold and 1 µl was used as a template in 10-µl PCR reaction, using the Applied Biosystems FAST7500 Real-Time PCR systems in fast mode and FAST SYBR GREEN PCR Master Mix (Applied Biosystems, Foster City, CA), following the manufacturer's protocol. Cycling conditions were as follows: 8 min at 95°C, 40

cycles of 10 s at 95°C, 30 s at 58°C, and 30 s at 72°C, followed by a 60 to 95°C dissociation protocol. The primers for transcript analysis were designed by the primer express software (Applied Biosystems, Foster City, CA) and are listed in Table S1. All reactions were performed in triplicate and the products were checked by melting curve analysis. Sequence of the PCR products had been confirmed. The transcript level was measured by normalizing the level with that of the *UBQ10 as* reference transcript. Each experiment was repeated at least 2 times. The values are average of three biological replicates which yielded consistent results.

Plasmid construction

For all the plasmid construction, PFU turbo (Invitrogen, Carlsbad, CA) was used. PEX2RF was amplified from pCHF3-PEX2 [14] by PCR with primers introducing a *Kpn*-I site at the 5′ of FW and *Sac*-I at the 3′ end of RE. The amplified PCR product was confirmed by sequencing and cloned into pCHF3:GFP [14] to generate pHU006 and into a pCAMBIA vector (Cambia,

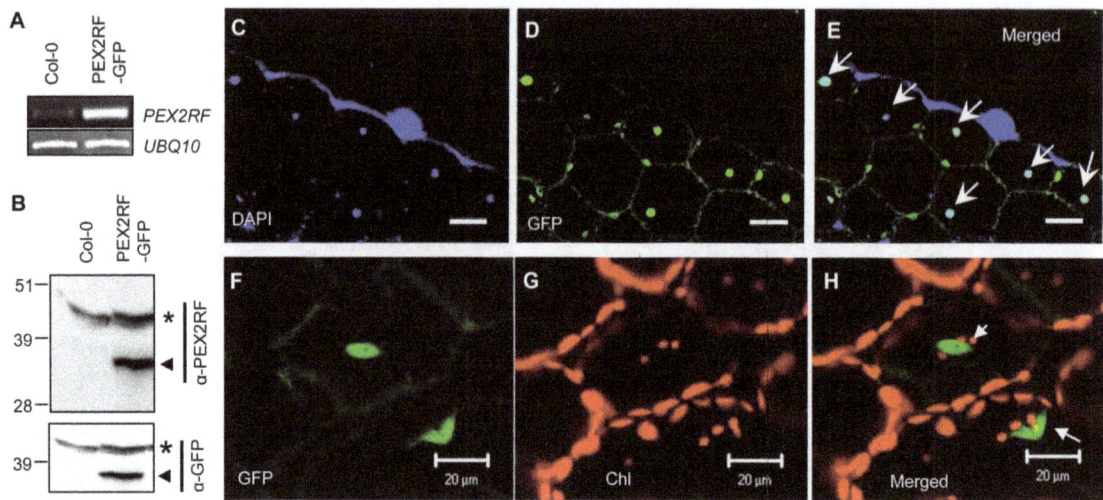

Figure 2. Nuclear localization of PEX2RF-GFP in transgenic plants. (A) RT-PCR analysis of *PEX2RF* mRNA and the *UBQ10* control in Col-0 and 35S::PEX2RF-GFP lines. (B) Immunoblot analyses of proteins from Col-0 and PEX2RF-GFP-expressing plants, using α–PEX2RF and α–GFP antibodies respectively. Asterisks indicate cross-reacting bands, and arrowheads point to the PEX2RF-GFP fusion protein. Numbers on the left indicate molecular weight markers in kDa. (C–H) Confocal images of transgenic plants from hypocotyl cells of 10d seedlings (C–E) and leaf mesophyll cells of two-week plants (F–H). DAPI stains the nucleus, green signals are from PEX2RF-GFP, and red signals are from chlorophyll autofluorescence. Arrows in the merged images indicate the nucleus. Scale bars = 10 μm in (C–E) and = 20 μm in (F–H).

Canberra) to generate pHU007. By floral dipping [18], pHU006 was transformed into Col-0 for PEX2RF-GFP the localization study, using Hygromycin for selection, and pHU007 was transformed into *det1-1* for the complementation study, using Kanamycin for selection. T2 transgenic plants were used for further analyses.

To express the PEX2RF protein for antibody generation, specific oligonucleotides were synthesized and cloned at *Nco* I and *Xho* I sites. For amplification of the PEX2 RING finger domain, 5′-CATGCCATGGGGCATGACTTGCCCTATTTGC-3′ and 5′-CCGCTCGAGTCATTTGCCACTTGAAAC-3′ were used to PCR-amplify PEX2RF with the *pfu* turbo enzyme (Stratagene,

La Jolla, CA) from Arabidopsis total cDNA from light-grown seedlings. The product was cloned into *Nco*I and *Xho*I sites of the bacterial pET28a+ expression vector (Novagen, Madison, WI) to generate pHU010. Insert was confirmed by sequencing. Recombinant PEX2RF fused to 6xHis in the pET28a+ vector was expressed in bacteria and purified with nickel nitrilotriacetic acid agarose (Qiagen, Valencia, CA) according to the manufacturer's protocol.

Constructs pHU011 and pHU012 were made by cloning the coding region of PEX2 RING finger and HY5, which had been amplified using the following PCR primer sets: 5′-GCGCAG-GAGCTCATGACGCCGTCTACGCCTGC-3′ and 5′-GAC-

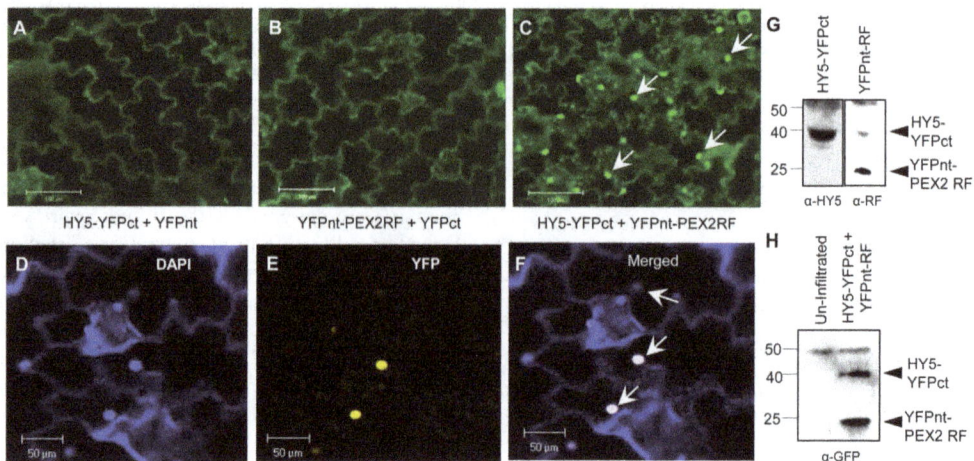

Figure 3. PEX2RF and HY5 interact in the nucleus. (A–C) Epifluorescence micrographs of tobacco leaf epidermal cells infiltrated with the indicated gene constructs. Strong YFP signals (BiFC) in the nucleus, as indicated by arrows in (C), were observed only when HY5-YFPct and YFPnt-PEX2RF were co-expressed. Scale bars = 100 μm. (D–F) Confocal micrographs of tobacco leaf epidermal cells co-infiltrated with HY5-YFPct and YFPnt-PEX2RF constructs. DAPI stains the nucleus (D), and BiFC signals are indicated by YFP fluorescence (E). Arrows in the merged image (F) indicate the overlaps of DAPI and BiFC. Scale bars = 50 μm. (G–H) Immunoblot analyses showing expression of HY5-YFPct and YFPnt-PEX2RF proteins in tobacco tissue. In (G), tissues were from plants shown in (A) and (B) respectively and α-HY5 (left) and α-RF (right) antibodies were used. In (H), tissue was from plant shown in (C), and α-GFP was used. Molecular weight markers in kDa are shown to the left of the blots.

A

B

α-c-Myc

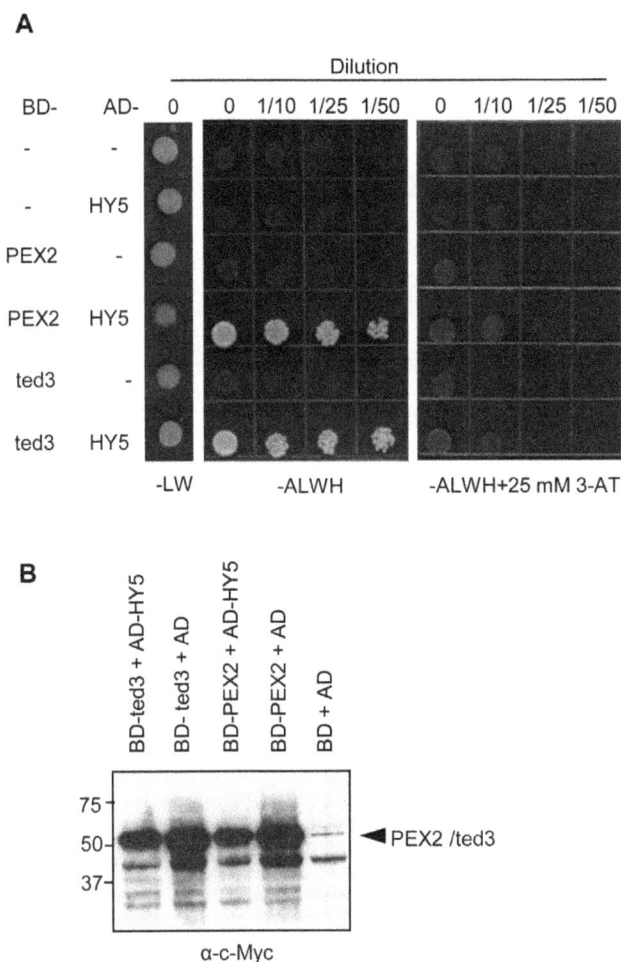

Figure 4. PEX2 and HY5 interact in yeast two-hybrid assays. (A) Yeast two-hybrid assays to show interaction between PEX2 and HY5. Yeast transformants containing the indicated GAL4 DNA binding domain (BD) and GAL4 activation domain (AD) fusion constructs were grown overnight in liquid culture and spotted on selection media plates (lacking leucine and tryptophan; –LW) and interaction media plates (lacking adenine, leucine, tryptophan and histidine; –ALWH, or –ALWH supplemented with 25 mM 3-amino-1,2,4-triazole; –ALWH+25 mM 3-AT). Growth on –ALWH and –ALWH+25 mM 3-AT media indicates protein interaction. (B) Immunoblot analysis of BD fusion constructs. Proteins extracted from transformed yeast cells shown in (A) were subjected to immunoblotting using α-c-Myc antibody. Numbers on the left of the blot indicate protein molecular weight markers in kDa.

Figure 5. Immunoblot analysis of the HY5 protein in various genetic backgrounds. Proteins were extracted from 4d dark-grown seedlings exposed to 1 hr white light and detected with the α–HY5 antibody. Purified HY5–6xHis from our previous study [11] was used as a control. Asterisks indicate non-specific bands. A cross-reacting band indicated by a double asterisk served as the loading control.

Antibody production

Polyclonal antibody was raised in rabbit against the PEX2 RING finger domain (aa 275–333) that had been purified to homogeneity from *E. coli* cells expressing the PEX2RF. ImmunoPure (Protein A) IgG Purification Kit (Thermo Fisher Scientific, Rockford, IL) was used to isolate IgG from the rabbit sera according to the manufacturer's instructions. Purified IgG was desalted using Zeba desalt column using phosphate buffer (pH 7.2). A 1:500 dilution of the desalted IgG fraction was used for all subsequent immunoblot assays.

Yeast two-hybrid analysis

Full-length HY5, PEX2, ted3, and PEX2RF were restriction cloned into pGBKT7 (PEX2/ted3/PEX2RF) and pGADT7 (HY5) plasmids of the GAL4 Y2H system (Clontech, Mountainview, CA), using the method as previously described [17]. The yeast strain Y190 was transformed with the respective constructs and transformants selected on minimal media lacking leucine and tryptophan (–LW). Interactions were assessed by growing transformants in liquid culture at 30°C and spotting serial dilutions on –LW, –ALWH and –ALWH+25 mM 3-AT media. Plates were imaged after 2d of growth at 30°C. Immunoblotting of yeast extracts was carried out as previously described [17].

Immunoblot analysis

Plant tissues (as indicated in the text) were ground to fine powder with liquid N_2 and resuspended in 200 µl of buffer (400 mM sucrose, 50 mM Tris-HCl pH 7.5, 2.5 mM EDTA, 10 mM PMSF). Total extract was cleared by centrifugation and supernatant was mixed with 5x Lamelli buffer and resolved in a PAGE (polyacrylamide gel electrophoresis). Resolved protein was then transferred to PVDF membrane and blocked with 5% milk and 0.5% Tween-20 for 2 hr at room temperature and subsequently incubated with 1:500 dilution of α-PEX2RF (Covance, Princeton, NJ), 1:200 dilution of α-HY5 (Xing Wang Deng lab), or 1:20,000 dilution of α-GFP (Abcam) overnight at 4°C. 1:20,000 goat anti-rabbit IgG (Thermo Fisher Scientific, Rockford, IL) was used as the secondary antibody. The PVDF membrane was washed four times with 1X TBST for 10 min each time before the signals were visualized with SuperSignal West Dura Extended duration substrate (Thermo Fisher Scientific, Rockford, IL).

TAGTTCATTTGCCACTTGAAACACCTTC-3' for PEX2RF with *Sac* I and *Spe* I sites (restriction sites are underlined); and 5'-GCGCAGGAGCTCATGCAGGAACAAGCGAC-TAGCTCTTTAGC-3' and 5'-CATGACCGTCGA-CAAAAGGCTTGCATCAGCATTAGAAC-3' for HY5 (At5g11260) with *Sac* I and *Sal* I sites (restriction sites underlined). Restriction enzyme-digested PCR product was cloned at the *Sal* I and *Sac* I sites of pSY735 to generate pHU011 and *Sac* I and *Spe* I sites to generate pHU012. Both these constructs were verified by sequencing and subsequently digested with *Hind* III and subcloned into binary vector pZP221 for generating BiFC constructs pHU014 and pHU015. All constructs were confirmed by sequencing.

Vectors used in this study are described in Table S2.

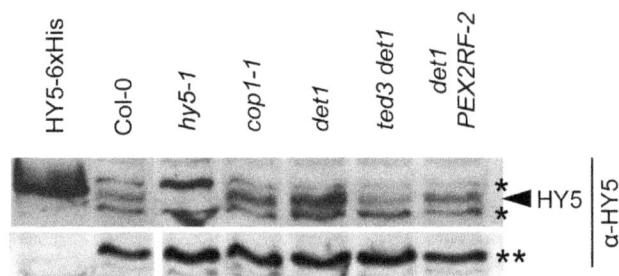

At5g13930 (CHS)

At5g08640 (FLS)

At5g44110 (POP1)

At5g54510 (DLF1)

At2g35300 (LEA1)

At1g53170 (ERF8)

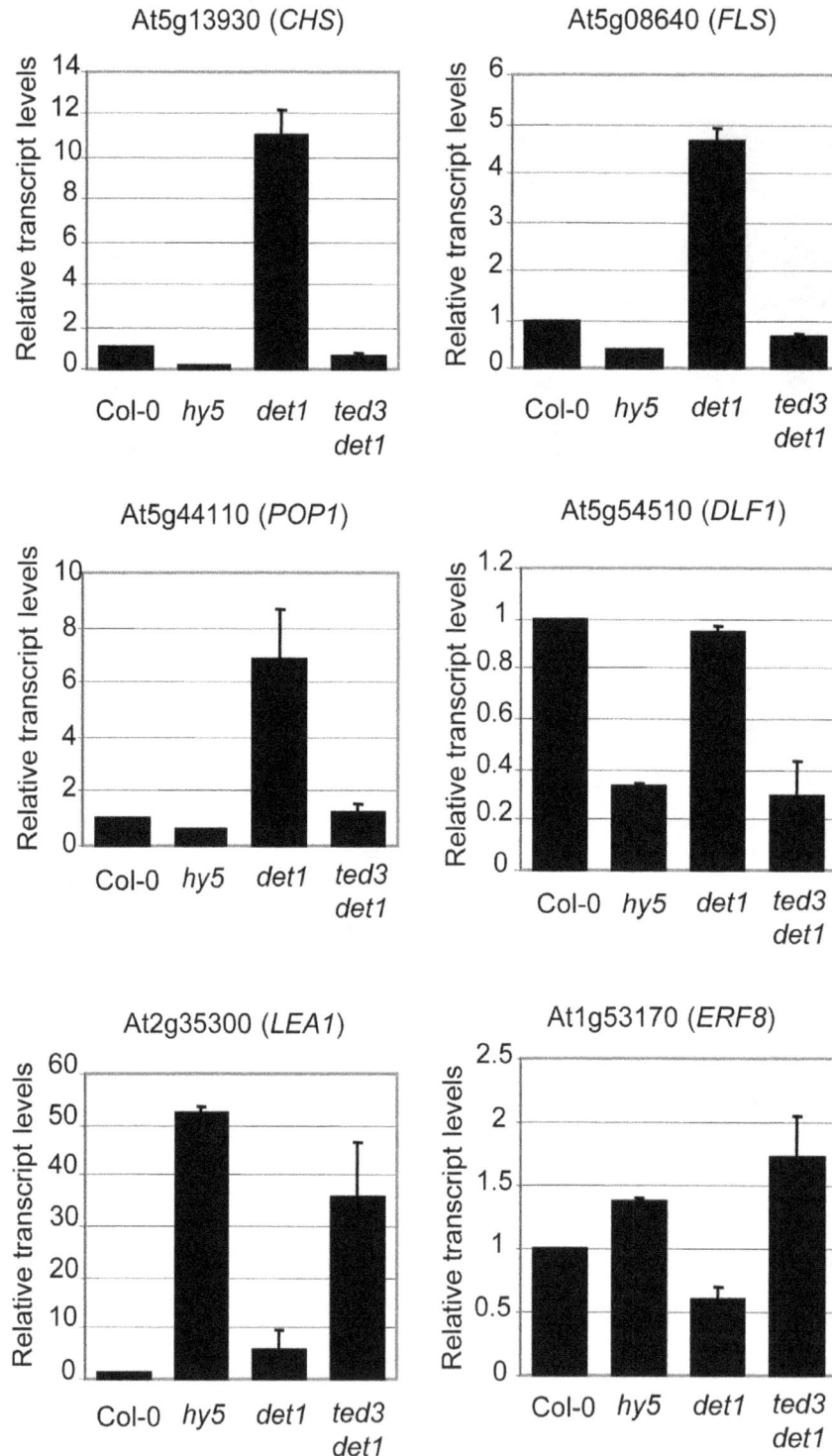

Figure 6. qRT-PCR analysis of the expression of some of HY5's target genes. RNA was extracted from 4d dark-grown seedlings in different genetic backgrounds. Three biological replicates of qRT reaction were performed for individual primer sets. The transcript level of each gene in the mutant is represented as arbitrary unit relative to the transcript level of the same gene in the wild-type plant, which was set to 1.0. The transcript level (relative expression) is the ratio between the transcript abundance of the studied gene and the transcript abundance of *UBQ10*. Values correspond to the mean and s.d. of three biological replicates. The experiments were repeated twice with consistent results.

Transient protein expression assays

Agrobacterium tumefaciens (strain GV3101) were transformed with the BiFC constructs and transformants were selected with 50 µg/ml kanamycin and 30 µg/ml gentamycin. Overnight

bacterial cultures (28°C) of GV3101 containing the plasmid of interest was harvested by centrifugation, washed in water and resuspended in induction medium. Leaf infiltration was done as previously described [19]. Infiltrated plants were grown for 2 to

3 d in growth chambers before the leaf epidermal cells were examined for BiFC with epifluorescence or confocal microscopy.

Results

Overexpression of PEX2's RING finger domain partially rescues det1

To test the hypothesis that *ted3* creates a small peptide containing PEX2's RING finger domain, which can translocate to the nucleus to partially compensate for the loss of a functional DET1, we first tested whether this RING domain is able to rescue the mutant phenotypes of *det1*. To this end, the *det1-1* mutants were transformed with a construct containing the RING finger (RF) domain of PEX2 (aa 275 to 333, Figure 1A) under the control of the 35S constitutive promoter. After RT-PCR analysis, two transgenic lines showing increases in the expression of *PEX2RF* mRNA compared with the *det1* control were selected for further analysis (Figure 1B). Dark-grown *det1* seedlings had short hypocotyls and opened cotyledons, whereas transgenic seedlings overexpressing PEX2RF had longer hypocotyls (Figure 1C). Quantification of the hypocotyl lengths of the transgenic seedlings proved this longer-hypocotyl phenotype to be significant (P<0.0001; Figure 1D). Further, adult transgenic plants were on average two times taller than *det1-1* although smaller than *ted3 det1-1* (Figure 1E). These results suggested that the seedling and adult phenotypes of *det1* can be partially suppressed by overexpression of PEX2's RING finger domain.

PEX2 RING-GFP localizes to the nucleus

To determine whether the RING domain of PEX2 is capable of entering the nucleus, we generated a construct that expressed the PEX2RF-containing peptide (aa 275$^{Val->Met}$ to 333) and fused it in-fame with a C-terminal green fluorescent protein (GFP). After generating transgenic lines expressing 35S::PEX2RF-GFP, semi quantitative RT-PCR analysis was performed to check for gene overexpression (Figure 2A). We also checked the presence of the PEX2RF-GFP protein with immunoblots, using a polyclonal antibody generated against PEX2's RING domain (see Methods). This antibody detected the presence of overexpressed PEX2RF-GFP protein in plants (Figure 2B) and the overexpressed MBP-PEX2RF protein in yeast cells (Figure S1), but it failed to detect the hypothetical endogenous small peptide that contains PEX2RF in *ted3 det1*.

Transgenic plants expressing PEX2RF-GFP were subjected to confocal laser-scanning microscopy. Besides some localization in the cytosol, PEX2RF-GFP was primarily found in the nucleus in seedling hypocotyl (Figure 2C–2E) and leaf mesophyll cells (Figure 2F–2H). The presence of PEX2 RING domain in the nucleus and PEX2RF's ability to partially suppress the *det1* phenotypes together suggested that this small peptide may be able to function in the nucleus to play a positive role in skotomorphogenesis, i.e. etiolation in the dark.

PEX2RF interacts with HY5 in the nucleus

Since HY5 is a key nuclear regulator of photomorphogenesis, we hypothesized that the nuclear localized PEX2RF may have an effect on HY5's function. For example, it may physically interact with HY5 and allosterically modify its activity or stability. To determine whether PEX2's RING finger domain and HY5 physically interact, we performed a Bimolecular Fluorescence Complementation (BiFC) assay [20] using tobacco (*Nicotiana tabacum*) plants. HY5 and PEX2RF were fused to the C- and N-terminal halves of YFP respectively to generate HY5-YFPct and YFPnt-PEX2RF. Epifluorescence and confocal microscopy anal-

yses of infiltrated tobacco leaves revealed strong YFP complementation signals (BiFC) only when both proteins were expressed, and these YFP signals were enriched in the nucleus labeled by DAPI (Figure 3). These results confirmed that HY5 and PEX2RF were able to interact in the nucleus, where HY5 normally performs its function.

We also employed yeast two-hybrid assays to test the interaction between PEX2RF and HY5 by fusing HY5 into the prey vector and PEX2/ted3/PEX2RF into the bait vector (see Methods). However, constructs containing PEX2RF autoactivated (Figure S2), so we focused on PEX2 and ted3 (i.e. PEX2 containing the 275$^{Val->Met}$ substitution) instead. Both PEX2 and ted3 proteins were able to interact with HY5 (Figure 4), supporting the conclusion that PEX2 can physically interact with HY5 and that this interaction is likely mediated by the RING domain of PEX2.

HY5's function in photomorphogenesis is compromised in ted3 det1

To explore the possible physiological relevance of this protein-protein interaction between HY5 and PEX2RF, we checked the abundance of HY5 in dark-grown seedlings in various genetic backgrounds. HY5 is the target for degradation by the COP-DET1 complexes in the dark; lack of or significant reduction of the level of this protein leads to long hypocotyls in light-grown seedlings [6]. Conversely, in mutants of the COP-DET1 complexes such as *det1* and *cop1*, HY5 is stabilized and thus dark-grown seedlings display a de-etiolated phenotype by having short hypocotyls [7]. Similar to what had been shown previously, HY5 showed higher accumulation in *cop1-1* and *det1-1* mutants when compared with wild-type Col-0, whereas this higher accumulation was reduced in *ted3 det1-1* and to lesser degrees, in *det1* mutant overexpressing PEX2RF (Figure 5). These results led us to speculate that PEX2's RING finger domain in the nucleus may be involved in inactivation and/or turnover of HY5 directly or indirectly.

Given the significant reduction of the level of HY5 in *ted3 det1*, we reasoned that the downstream events regulated by HY5 may also be reversed in this suppressor to revert *det1*'s phenotype. To test this, we selected six light-regulated genes that are known to be direct targets of HY5, and performed expression profiling by quantitative real-time PCR of these genes in Col-0, *hy5-1*, *det1-1*, and *ted3 det1-1*. Those positively regulated by HY5 included genes that encode chalcone synthase (CHS) and flavonol synthase (FLS), which are involved in anthocyanin/flavonoid biosynthesis [21], the ABC transporter POP1 (P-loop containing nucleoside triphosphate hydrolases superfamily protein) [22], and the auxin signal transduction component Dwarf in Light1 (*DFL1*) [23]. The two genes negatively regulated by HY5 encoded the late embryogenesis abundant protein LEA1 and ethylene response factor ERF8 [22]. As expected, transcript levels of *CHS, FLS, POP1, and DLF1* decreased in *hy5-1* but increased in *det1*., For the genes negatively regulated by HY5, ERF8 was up-regulated in *hy5-1* and down-regulated in *det1*, whereas *LEA1* was up-regulated in both *hy5-1* and *det1* (Figure 6). In *ted3 det1*, the altered expression pattern shown in *det1* was reversed for five of the six genes to levels similar to those in the *hy5* mutant (Figure 6), supporting the notion that HY5 activity in *ted3 det1* is compromised, which may be a major cause for the partial reversal of the *det1* phenotype.

Discussion

COP/DET1/FUS are global repressors of light-regulated development, functioning in the proteolysis of positive regulators of photomorphogenesis in the nucleus [4,5]. The identification of a dominant peroxisomal mutation that suppressed the de-etiolated phenotype of *det1* was intriguing [14]. One plausible explanation was that the Met created at position 275 in *ted3* initiated the translation of a small peptide that contains the C-terminal RING finger domain of PEX2. Alternatively, this Val-to-Met change renders the protein more susceptible to proteases, which cleave off the RING domain of PEX2 in the cytosol. The RING domain-containing peptide then translocates to the nucleus to substitute the function of DET1 in photomorphogenesis. We have not been able to unequivocally prove the above hypothesis in this study, as we could not detect the hypothetical small peptide derived from the Val-to-Met substitution in *ted3 det1* or *ted3* overexpressors. This is possibly due to insufficient avidity of the PEX2 antibody we generated and/or the low abundance/instability of this peptide. However, we have shown in this study that overexpression of a small peptide containing PEX2 RING domain in *det1* can indeed partially suppress the *det1* phenotype. Majority of the PEX2RF-GFP protein was seen in the nucleus, and only a small portion of the fusion protein was visible in the cytoplasm. We speculate that PEX2RF passively enters the nucleus due to its small size (~6 kDa), although we do not rule out the possibility that it goes to the nucleus through active targeting or other mechanisms.

The RF domains of PEX2 and COP1 both belong to the C_3HC_4 type. When overexpressed in wild-type Arabidopsis plants, an N-terminal fragment of COP1 that contained both the RING finger and coiled-coil domains was found in the nucleus and conferred a dominant negative effect that mimicked the phenotype of *cop1*. The phenotype was believed to be caused by the interaction between this peptide and the endogenous COP1, which resulted in the interference with COP1's normal function [24]. In our study, PEX2 RF alone was overexpressed in the mutant *det1* background and conferred phenotype opposite to that of the COP1 study. Our BiFC and yeast two-hybrid assays demonstrated the interaction between PEX2 and HY5. In addition, the accumulation of the HY5 protein in *det1* was reduced in *det1 PEX2RF* and *ted3 det1*. Furthermore, the altered expression of five out of the six analyzed HY5 target genes in *det1* was reversed in *ted3 det1*, prompting us to speculate that this reduced activity of HY5 was at least in part responsible for the suppression of *det1*. This is also consistent with a previous report, which showed that HY5 inactivation in *cop1* and *det1* mutants resulted in reversal of their dark-grown phenotypes [25]. How does ted3 reduce the level of HY5? Given that PEX2RF contains E3 ubiquitin ligase activity [17], ted3 may be directly involved in the degradation of HY5. *ted3* also partially suppressed *cop1* but not *det2*, a de-etiolated mutant deficient in an enzyme in brassinosteroid biosynthesis [14]. Therefore, *ted3* seems to have some specificity toward the COP/DET1-associated photomorphogenic pathway, which makes sense given that HY5 is a major target of the COP/DET1 proteolytic complexes. Finally, since HY5 is not the only target of DET1's function, reducing HY5 activity may not be sufficient to completely rescue the *det1* mutant phenotypes.

Although we have not been able to prove this hypothesis, we predict that the partial suppression of *det1* by *ted3* is primarily due to the creation of a RING finger-containing peptide that replaces the function of DET1 in the nucleus, and not due to changes in

peroxisomal function. Replacement of a Val, which is nonpolar, by the partially charged Met was shown to affect the function of proteins related in human diseases [26,27,28]. Similarly, substitution of Val by Met may distort the overall configuration of the cytoplasmic end of the PEX2 protein thus affecting the activity of the RING finger domain, resulting in a PEX2 protein with mildly reduced activity in peroxisome biogenesis.

RING-type E3 ligases mediate ubiquitination and are implicated in diverse developmental processes across kingdoms [29]. Our work supports the possibility that a gain-of-function mutation in a peroxisomal gene can have a marked effect on the function of a nuclear protein. Given the conservation of the RING finger domain among proteins in various genomes, it is interesting to speculate that some other RING domains when appropriately localized may also cause neomorphic phenotypes.

Supporting Information

Figure S1   Specificity of the PEX2RF antibody. (A) SDS-PAGE gel showing induction of the expression of the fusion of maltose binding protein (MBP) and PEX2RF in bacterial protein lysates. U, Is and Ip stand for uninduced, soluble and pellet fractions, respectively. Protein expression constructs have been previously described in Kaur *et al.*, 2013 [17]. Arrow and arrowhead point to MBP alone and MBP-PEX2RF respectively. (B) Immunoblot analysis of MBP-RF expression in induced bacterial protein lysates, as detected by the PEX2RF antibody. Numbers on the left of the blots indicate molecular weight markers in kDa.

Figure S2   PEX2RF auto-activates in yeast two-hybrid assays. (A) Yeast cells transformed with BD and AD constructs were spotted on selection (−LW) and interaction media (−ALWH and −ALWH+25 mM 3-AT). Strains containing BD-PEX2RF grow on interaction media even in the absence of HY5, indicating that the RF autoactivates. (B) Immunoblot analysis to detect the expression of BD-PEX2RF fusion proteins in yeast cells, using anti-c-Myc and anti-PEX2RF antibodies respectively.

Table S1   Primers used in qRT-PCR.

Table S2   Vectors used in this study.

Acknowledgments

We would like to express our thanks to Dr. Vandana Yadav for help with the qRT-PCR analysis and comments on the manuscript, Jilian Fan for assistance with the construction of pHU006, the Arabidopsis Biological Resources Center for providing all the mutant seeds, and Dr. Xing-Wang Deng for sharing the HY5 antibody.

Author Contributions

Conceived and designed the experiments: MD NK JH. Performed the experiments: MD NK. Analyzed the data: MD NK JH. Contributed reagents/materials/analysis tools: MD NK JH. Contributed to the writing of the manuscript: MD NK JH.

References

1. Kami C, Lorrain S, Hornitschek P, Fankhauser C (2010) Light-regulated plant growth and development. Curr Top Dev Biol 91: 29–66.
2. Chen M, Chory J, Fankhauser C (2004) Light signal transduction in higher plants. Annu Rev Genet 38: 87–117.
3. Jiao Y, Lau OS, Deng XW (2007) Light-regulated transcriptional networks in higher plants. Nat Rev Genet 8: 217–230.
4. Lau OS, Deng XW (2012) The photomorphogenic repressors COP1 and DET1: 20 years later. Trends Plant Sci 17: 584–593.

5. Nezames CD, Deng XW (2012) The COP9 signalosome: its regulation of cullin-based E3 ubiquitin ligases and role in photomorphogenesis. Plant Physiol 160: 38–46.

6. Oyama T, Shimura Y, Okada K (1997) The Arabidopsis HY5 gene encodes a bZIP protein that regulates stimulus-induced development of root and hypocotyl. Genes Dev 11: 2983–2995.

7. Osterlund MT, Hardtke CS, Wei N, Deng XW (2000) Targeted destabilization of HY5 during light-regulated development of Arabidopsis. Nature 405: 462–466.

8. Hu J, Baker A, Bartel B, Linka N, Mullen RT, et al. (2012) Plant peroxisomes: biogenesis and function. Plant Cell 24: 2279–2303.

9. Beevers H (1979) Microbodies in higher plants. Ann Rev Plant Physiol 30: 159–193.

10. Kaur N, Li J, Hu J (2013) Peroxisomes and photomorphogenesis. Subcell Biochem 69: 195–211.

11. Desai M, Hu J (2008) Light induces peroxisome proliferation in Arabidopsis seedlings through the photoreceptor phytochrome A, the transcription factor HY5 HOMOLOG, and the peroxisomal protein PEROXIN11b. Plant Physiol 146: 1117–1127.

12. Hu J, Desai M (2008) Light control of peroxisome proliferation during Arabidopsis photomorphogenesis. Plant Signal Behav 3: 801–803.

13. Pepper AE, Chory J (1997) Extragenic suppressors of the Arabidopsis det1 mutant identify elements of flowering-time and light-response regulatory pathways. Genetics 145: 1125–1137.

14. Hu J, Aguirre M, Peto C, Alonso J, Ecker J, et al. (2002) A role for peroxisomes in photomorphogenesis and development of Arabidopsis. Science 297: 405–409.

15. Platta HW, El Magraoui F, Baumer BE, Schlee D, Girzalsky W, et al. (2009) Pex2 and Pex12 function as protein-ubiquitin ligases in peroxisomal protein import. Mol Cell Biol 29: 5505–5516.

16. Okumoto K, Noda H, Fujiki Y (2014) Distinct modes of ubiquitination of peroxisome-targeting signal type 1 (PTS1)-receptor Pex5p regulate PTS1 protein import. J Biol Chem 289: 14089–14108.

17. Kaur N, Zhao Q, Xie Q, Hu J (2013) Arabidopsis RING Peroxins are E3 Ubiquitin Ligases that Interact with Two Homologous Ubiquitin Receptor Proteins(F). J Integr Plant Biol 55: 108–120.

18. Clough SJ, Bent AF (1998) Floral dip: a simplified method for Agrobacterium-mediated transformation of Arabidopsis thaliana. Plant J 16: 735–743.

19. Sparkes IA, Runions J, Kearns A, Hawes C (2006) Rapid, transient expression of fluorescent fusion proteins in tobacco plants and generation of stably transformed plants. Nat Protoc 1: 2019–2025.

20. Bracha-Drori K, Shichrur K, Katz A, Oliva M, Angelovici R, et al. (2004) Detection of protein-protein interactions in plants using bimolecular fluorescence complementation. Plant J 40: 419–427.

21. Song YH, Yoo CM, Hong AP, Kim SH, Jeong HJ, et al. (2008) DNA-binding study identifies C-box and hybrid C/G-box or C/A-box motifs as high-affinity binding sites for STF1 and LONG HYPOCOTYL5 proteins. Plant Physiol 146: 1862–1877.

22. Lee J, He K, Stolc V, Lee H, Figueroa P, et al. (2007) Analysis of transcription factor HY5 genomic binding sites revealed its hierarchical role in light regulation of development. Plant Cell 19: 731–749.

23. Nakazawa M, Yabe N, Ichikawa T, Yamamoto YY, Yoshizumi T, et al. (2001) DFL1, an auxin-responsive GH3 gene homologue, negatively regulates shoot cell elongation and lateral root formation, and positively regulates the light response of hypocotyl length. Plant J 25: 213–221.

24. McNellis TW, Torii KU, Deng XW (1996) Expression of an N-terminal fragment of COP1 confers a dominant-negative effect on light-regulated seedling development in Arabidopsis. Plant Cell 8: 1491–1503.

25. Ang LH, Deng XW (1994) Regulatory hierarchy of photomorphogenic loci: allele-specific and light-dependent interaction between the HY5 and COP1 loci. Plant Cell 6: 613–628.

26. Kazemi-Esfarjani P, Beitel LK, Trifiro M, Kaufman M, Rennie P, et al. (1993) Substitution of valine-865 by methionine or leucine in the human androgen receptor causes complete or partial androgen insensitivity, respectively with distinct androgen receptor phenotypes. Mol Endocrinol 7: 37–46.

27. Murray EW, Giles AR, Lillicrap D (1992) Germ-line mosaicism for a valine-to-methionine substitution at residue 553 in the glycoprotein Ib-binding domain of von Willebrand factor, causing type IIB von Willebrand disease. Am J Hum Genet 50: 199–207.

28. Orth U, Fairweather N, Exler MC, Schwinger E, Gal A (1994) X-linked dominant Charcot-Marie-Tooth neuropathy: valine-38-methionine substitution of connexin32. Hum Mol Genet 3: 1699–1700.

29. Metzger MB, Pruneda JN, Klevit RE, Weissman AM (2014) RING-type E3 ligases: master manipulators of E2 ubiquitin-conjugating enzymes and ubiquitination. Biochim Biophys Acta 1843: 47–60.

12

The Establishment of Genetically Engineered Canola Populations in the U.S.

Meredith G. Schafer[1], **Andrew A. Ross**[2], **Jason P. Londo**[1], **Connie A. Burdick**[3], **E. Henry Lee**[3], **Steven E. Travers**[2], **Peter K. Van de Water**[4], **Cynthia L. Sagers**[1]*

1 Department of Biological Sciences, University of Arkansas, Fayetteville, Arkansas, United States of America, 2 Department of Biological Sciences, North Dakota State University, Fargo, North Dakota, United States of America, 3 Western Ecology Division, National Health and Environmental Effects Research Laboratory, U.S. Environmental Protection Agency, Corvallis, Oregon, United States of America, 4 Earth and Environmental Sciences Department, California State University, Fresno, California, United States of America

Abstract

Concerns regarding the commercial release of genetically engineered (GE) crops include naturalization, introgression to sexually compatible relatives and the transfer of beneficial traits to native and weedy species through hybridization. To date there have been few documented reports of escape leading some researchers to question the environmental risks of biotech products. In this study we conducted a systematic roadside survey of canola (*Brassica napus*) populations growing outside of cultivation in North Dakota, USA, the dominant canola growing region in the U.S. We document the presence of two escaped, transgenic genotypes, as well as non-GE canola, and provide evidence of novel combinations of transgenic forms in the wild. Our results demonstrate that feral populations are large and widespread. Moreover, flowering times of escaped populations, as well as the fertile condition of the majority of collections suggest that these populations are established and persistent outside of cultivation.

Editor: Alfredo Herrera-Estrella, Cinvestav, Mexico

Funding: Provided by: USDA CREES NRI 35615-19216, "Ecological Impacts from the Interactions of Climate Change, Land-Use Change and Invasive Species." The funders had no role in study design, data collection and analysis, decision to publish, or preparation of the manuscript.

Competing Interests: The authors have declared that no competing interests exist.

* E-mail: csagers@uark.edu

Introduction

Crop and forage species now cover more than one quarter of the Earth's land surface [1], but the ecological and evolutionary influences of agricultural species on native and weedy plants have been difficult to measure. The commercial release of GE crops has provided novel genetic markers to track crop-to-weed gene flow [2,3] raising both awareness of the difficulties of transgene confinement and concerns about the ecological consequences of transgenes in the environment [4,5]. Genetically engineered varieties could influence the population ecology of wild species by introducing novel, beneficial traits, or lead to detrimental effects such as extirpation of native alleles or declines of natural populations [6]. The escape of crops or crop alleles is no longer in doubt [7], but reports of transgene escape are few and are limited in the U.S. to the case of creeping bentgrass, *Agrostis stolonifera* (Poaceae), from a field trial in central Oregon, USA [8,9]. Given that biotech crops cover more than 130Mha globally [10], the rarity of reported escapes has led some to question the environmental risks of genetically engineered crops [11,12].

Canola (*Brassica napus* L. (Brassicaceae)) is an oilseed crop grown on approximately 31Mha globally [13]. *Brassica napus*, an allotetraploid formed by the hybridization of *B. rapa* L. and *B. oleraceae* L., is sexually compatible with more than 15 other mustard species [14], a number of which are considered noxious weeds [15]. Canola cultivars engineered for glyphosate and glufosinate herbicide resistance escaped cultivation shortly after their unconditional commercial release in Canada in 1995 [16] and more recent research has documented widespread escape and persistence of transgenic canola in Canadian roadside populations [17,18]. Since these discoveries, feral canola populations or non-engineered populations expressing biotech traits have been reported from Great Britain, France, Australia and Japan [2,3,19–21]. In the U.S., GE canola was first approved for commercial release in 1998 and now most (>90%) of the acreage planted in the U.S. is genetically engineered for herbicide resistance [10].

The objective of this study was to document the extent of feral canola populations in North Dakota, the dominant canola growing region of the United States. We used roadside surveys and commercially available test strips evaluate the distribution of transgenic canola growing outside of cultivation in the U.S.

Materials and Methods

We conducted systematic roadside surveys to quantify the presence and abundance of feral GE and non-GE canola populations in North Dakota, USA, beginning 4 June and continuing through 23 July 2010. Field crews established east-west transects on major roads throughout the state. A 1×50 m quadrat was established every 8.05 km (5 miles) of roadway on one or both sides of the road, where traffic permitted, in which all identifiable *B. napus* plants were counted. We drove a total of 5600 km and sampled 63.1 km of roadside habitats (1.1% of the distance driven) Sampling was conducted early in the summer prior to the onset of flowering of cultivated canola. When canola was present at a sampling site, one randomly selected plant was collected, photographed and archived as a voucher specimen. Leaf fragments from voucher specimens were tested for the presence of CP4 EPSPS protein (confers tolerance to glyphosate herbicide) and PAT protein (confers tolerance to glufosinate herbicide) with TraitChek[TM] immunological lateral flow test strips (Strategic Diagnostics, Inc., Newark, DE). Previous studies have demonstrated the utility of the lateral flow strips in detecting the expression of transgenes from field samples [8,22]. Test strips are not available for a third, non-GE resistance trait, resistance to Clearfield[TM] herbicide, which comprises approximately 10% of the canola grown in the region (R Beneda, pers comm). At random intervals, single plants were tested with multiple test strips to assure that test results were repeatable and reliable. No failures were detected during the course of the study. To determine if populations of escaped canola are composed of multiple genotypes, multiple plants were sampled and tested for the presence of CP4 EPSPS or PAT proteins at 9 randomly selected, large canola populations Test strips and plant voucher specimens are archived at the University of Arkansas. GPS locations and transgene state values for each collected plant are available in Table S1.

Results

The escape of GE *B. napus* in North Dakota is extensive (Fig. 1). *Brassica napus* was present at 45% (288/634) of the road survey sampling sites. Of those, 80% (231/288) expressed at least one transgene: 41% (117/288) were positive for only CP4 EPSPS (glyphosate resistance); 39% (112/288) were positive for only PAT (glufosinate resistance); and 0.7% (2/288) expressed both forms of herbicide resistance, a phenotype not produced by seed companies (Table 1). Densities of *B. napus* plants at collection sites ranged

from 0 to 30 plants m^{-2} with an average of 0.3 plants m^{-2}. Among the archived specimens, 86.8% were sexually mature varying in developmental stage from flower bud to mature fruit with seeds. At the time of roadside sampling, in-field canola was non-flowering having matured to the 4-leaf to pre-bolting stage (JPL pers. obs.). This striking difference in flowering phenology suggests that flowering canola in roadside habitats may have originated from the previous generation's seed bank rather than from seed spill during the current growing season.

Populations of transgenic canola were denser along major transport routes, at construction sites and in regions of intense canola cultivation (Fig. 1). At a finer scale, feral populations appeared denser at junctions between major roadways, access points to crop fields and bridges, and intersections of roadways with railway crossings. At these sites, seed spill during transport is a likely mechanism for the escape of transgenic canola. Nonetheless, feral *B. napus* plants were occasionally found at remote locations far from canola production, transportation, or processing facilities. Populations were also observed at roadsides that had recently been mowed or treated with herbicide. Although our sampling protocol stipulated that a single plant be tested at each collection site, multiple sampling of additional plants revealed a mix of both herbicide resistant phenotypes, or a mix of herbicide resistant and vulnerable phenotypes in all randomly-tested large populations (Table S1).

Discussion

To date there have been relatively few reports of the escape from cultivation of genetically engineered varieties leading some researchers to discount the environmental risks of biotech crops. Concurrently, public demonstrations have led to a consumer backlash against genetically engineered foods. A first step toward understanding the environmental impact of biotech crops is to identify the incidence and extent of their escape from cultivation.

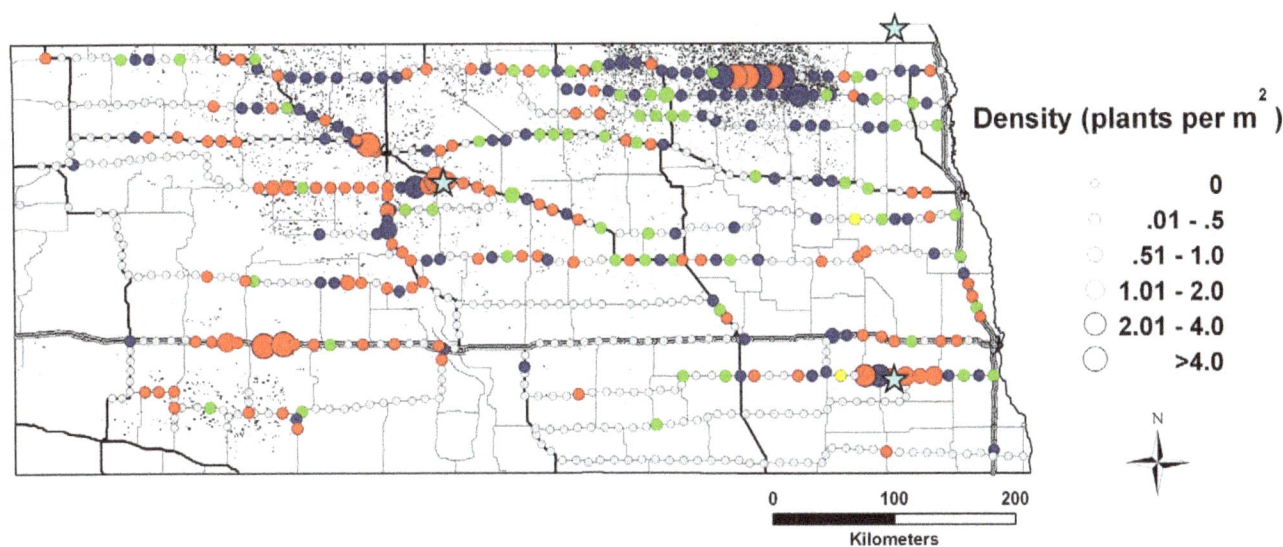

Figure 1. Distribution and density of feral canola populations in North Dakota road surveys (2010). Circles indicate locations of sampling sites; diameter of circle indicates plant density; gray circles indicate no canola present. The presence of genetically engineered protein in the vouchered specimen is shown by color: red – glyphosate resistance; blue – glufosinate resistance; yellow – dual resistance traits; green – non-transgenic. Canola fields are indicated by stippling based on 2009 USDA National Agricultural Statistics Service report (http://www.nass.usda.gov/Statistics_by_Subject/index.php?sector=CROPS). Stars show the locations of oilseed processing plants (3). Solid lines illustrate interstate, state and county highways.

Table 1. Distribution of transgenic and non-transgenic canola in North Dakota transects.

	# of sites	Percent
Total transects	634	
Canola present	288	0.454
Transgenic	231	0.802
Liberty Link+	112	0.389
Roundup Ready+	117	0.406
LL+ and RR+	2	0.007
Non-Transgenic		
Null	57	0.198

We conducted this study to document feral populations of genetically engineered canola and to evaluate potential mechanisms of persistence outside of crop fields.

The escape of canola from cultivation is not particularly surprising. *Brassica napus* is thought to have been domesticated very recently, in the last 300–400 years [23]. As a consequence, "wild" traits, such as seed shattering and partial seed dormancy, are still expressed in commercial canola and may contribute to escape from cultivation. For example, up to 30% of a seed crop may be lost each year by shattering during harvest [24] and canola seeds may remain dormant for up to three years [25]. The combined effects of seed loss on harvest and seed dormancy rapidly stock the soil seed bank, which can lead to frequent re-seeding of marginal soils [17].

Surprising from our study is the widespread distribution of feral canola outside of cultivated areas both near and far from cultivated fields over much of North Dakota and the likely persistence of these populations beyond single years. Additionally, these populations occur both in habitats with selection pressure (e.g., roadsides sprayed with glyphosate) and also in habitats without obvious selection pressure. Although canola cultivation in North Dakota occurs primarily in the northeastern counties, we identified transgenic canola populations in parts of North Dakota with little or no known canola production. Our results suggest a number of routes by which canola plants may be introduced to the wild. Feral canola populations were found in high densities along major trucking routes but not smaller tributaries suggesting that feral canola populations are established by seed spill. Similar results have been reported in studies of feral canola in Canada [17,18]. The mixture of phenotypes that we found in 9 large populations, further suggests that multiple seed spills or dispersal events can occur at a given location. In addition, we identified large, continuous populations of feral transgenic canola (population IDs 215–216) growing on fill dirt at highway construction zones that clearly did not result from seed shatter or seed spill (JPL pers. obs.). We suggest that canola may colonize repositories of fill dirt and rapidly establish a soil seed bank. The movement of contaminated fill dirt to remote construction sites provides an additional mechanism for the dispersal of transgenic canola far beyond field margins.

Movement by transport is likely to explain the current distribution of feral canola populations in North Dakota, but re-seeding by fertile plants further contributes to population persistence. Our evidence that these populations persist outside of cultivation includes the striking difference in flowering phenology between feral and commercial populations. Flowering times differed by approximately four weeks, indicating that field and feral populations originated from different sources. Further evidence for persistence is found in our statewide collections of fertile plants with viable seeds. Metapopulation dynamics by which feral populations are fed by seed transport but supplemented by *in situ* seed production are likely at play here as described by [18] for feral canola populations in Canada.

The occurrence of novel resistance phenotypes may provide additional evidence that these populations can persist outside of cultivation. When transgenic resistance genotypes grow in sympatry, varieties may hybridize to create novel combinations of traits, as we found at two locations. Because resistance to multiple herbicides has not been commercially developed in canola, the discovery of "stacked" traits in feral canola plants is evidence that biotech varieties have hybridized. Hybridization could possibly have occurred by pollen flow between fields of transgenic canola varieties, followed by seed spill along roadsides. Alternatively, hybridization could have occurred by pollen movement among resistant phenotypes within roadside populations, because feral populations were frequently found to include multiple phenotypes, or by flow of transgenic pollen from other feral populations or crop fields. By whatever mechanism, hybridization among genetically engineered varieties is not uncommon. Although we sampled a relatively small number of plants (N = 288) from a small percentage of the total potential habitat along roadways in North Dakota (1.1%), we nonetheless identified two individuals expressing novel stacked traits (0.7%). Furthermore, the incidence of crop-crop hybridization is undersampled in this survey because test strips for a third commercial form of herbicide resistant canola, Clearfield™, are not available.

These results support the hypothesis that roadside populations of canola in the U.S. are likely persistent from year to year, are capable of hybridizing to produce novel genotypes, and that escaped populations can contribute to the spread of transgenes outside of cultivation. Reports in Canada of feral populations of GE canola emerged soon after its commercial release there. Confirmation of GE pollen and crop movement among fields in Australia, U.K., Germany and France and Japan followed shortly thereafter. Ours is the first report of feral canola in the U.S. more than a decade after its commercial release. This delay raises questions of whether adequate oversight and monitoring protocols are in place in the U.S. to track the environmental impact of biotech products. At issue is the need to re-evaluate previous assumptions about crop systems: that crop genotypes outside of agriculture are not competitive; that protocols designed to reduce or prevent escape and proliferation of feral transgenic crops are effective; and that current tracking and monitoring of GE organisms are sufficient. Emerging pressures on agricultural systems by the accelerating growth of human populations argues that we take full advantage of the tools that biotechnology and conventional varietal development make available. It is essential that researchers, regulatory agencies and industry cooperate to ensure the continued security of food systems worldwide. The challenges of feeding a burgeoning global population in the face of limited and eroding natural resources requires substantial investments by all stakeholders. We must safely engage all tools available to us to advance food, fuel and fiber alternatives as modern agriculture rises to the challenges of the next decades.

Supporting Information

Table S1 Supplemental table of all collected *B. napus* populations.

Acknowledgments

We owe thanks to the L.P.L., P.R.L., and R. Beneda for their kindness during our field visits and to R. Finlay and additional reviewers for their comments. The research described in this article has been subjected to peer and administrative review by the U.S. Environmental Protection Agency and has been approved for publication. Mention of trade names or commercial products does not constitute endorsement or recommendation for use.

Author Contributions

Conceived and designed the experiments: MGS AAR JPL EHL CLS. Performed the experiments: MGS AAR JPL SET CLS. Analyzed the data: MGS CAB EHL JPL CLS. Contributed reagents/materials/analysis tools: EHL PKV CLS. Wrote the paper: MGS JPL CAB EHL CLS. Publication costs: PKV CLS.

References

1. Corvalan C, Hales S, McMichael A (2005) Ecosystems and human well-being: health synthesis: a report of the Millennium Ecosystem Assessment. Geneva: World Health Organization. 64 p.
2. Rieger MA, Lamond M, Preston C, Powles SB, Roush RT (2002) Pollen-mediated movement of herbicide resistance between commercial canola fields. Science 296: 2386–2388.
3. Beckie HJ, Warwick SI, Nair H, Seguin-Swartz G (2003) Gene flow in commercial fields of herbicide-resistant canola (Brassica napus). Ecol Appl 13: 1276–1294.
4. Wolfenbarger LL, Phifer PR (2000) The ecological risks and benefits of genetically engineered plants. Science 290: 2088–2093.
5. Lu B-R, Snow AA (2005) Gene flow genetically modified rice and its environmental consequences. BioScience 55: 669–678.
6. Levin D, Francisco-Ortega J, Jansen R (1996) Hybridization and the extinction of rare plant species. Conserv Biol 10: 10–16.
7. Ellstrand NC, Prentice HC, Hancock JF (1999) Gene flow and introgression from domesticated plants into their wild relatives. Annu Rev Ecol Syst 30: 539–563.
8. Reichman JR, Watrud LS, Lee EH, Burdick CA, Bollman, et al. (2006) Establishment of transgenic herbicide-resistant creeping bentgrass (Agrostis stolonoifera L.) in nonagronomic habitats. Mol Ecol 15: 4243–4255.
9. Zapiola ML, Campbell CK, Butler MD, Mallory-Smith CA (2008) Escape and establishment of transgenic, glyphosate-resistant creeping bentgrass Agrostis stolonifera in Oregon, USA: a 4-year study. J Appl Ecol 45: 486–494.
10. Sinemus K (2009) European Commission Sixth Framework Programme. Available: http://www.gmo-compass.org/eng/database/plants/63.rapeseed. html. Accessed 2011 Sep 12.
11. Dale PJ, Clarke B, Fontes EMG (2002) Potential for the environmental impact of transgenic crops. Nat Biotechnol 20: 567–574.
12. Federoff NV, Battisti DS, Beach RN, Cooper PJM, Fischhoff DA, et al. (2010) Radically rethinking agriculture for the 21st century. Science 327: 833–834.
13. Food and Agriculture Organization of the United Nations (2009) Available: http://faostat.fao.org/site/567/DesktopDefault.aspx?PageID = 567#ancor. Accessed 2011 Sep 12.
14. FitzJohn RG, Armstrong TT, Newstrom-Lloyd LE, Wilton AD, Cochrane M (2007) Hybridisation within Brassica and allied genera: evaluation of potential for transgene escape. Euphytica 158: 209–230.
15. United States Department of Agriculture, National Resources Conservation Service (2011) The PLANTS Database. Available: http://plants.usda.gov, Accessed 2011 Sep12.
16. Warwick SI, Simard M-J, Légère A, Beckie HJ, Braun L, et al. (2003) Hybridization between transgenic Brassica napus L. and its wild relatives: B. rapa L., Raphanus raphanistrum L., Sinapis arvensis L., and Erucastrum gallicum (Willd.) O. E. Schulz. Theor Appl Genet 107: 528–539.
17. Knispel AL, McLachlan SM, Van Acker RC, Friesen LF (2008) Gene flow and multiple herbicide resistance in escaped canola populations. Weed Sci 56: 72–80.
18. Knispel AL, McLachlan SM (2010) Landscape-scale distribution and persistence of genetically modified oilseed rape (Brassica napus) in Manitoba, Canada. Environ Sci Pollut R 17: 13–25.
19. Crawley MJ, Brown SL (1995) Seed limitation and the dynamics of feral oilseed rape on the M25 motorway. P R Soc B 256: 49–54.
20. Pessel FD, Lecomte J, Emeriau V, Krouti M, Messean A, Gouyon PH (2001) Persistence of oilseed rape (Brassica napus L.) outside of cultivated fields. Theor Appl Genet 102: 841–846.
21. Aono M, Wakiyama S, Nagatsu M, Nakajima N, Tamaoki M, et al. (2006) Detection of feral transgenic oilseed rape with multiple-herbicide resistance in Japan. Environ Biosafety R 5: 77–87.
22. Watrud LS, Lee EH, Fairbrother A, Burdick C, Reichman JR, et al. (2004) Evidence for landscape-level, pollen-mediated gene flow from genetically modified creeping bentgrass with CP4 EPSPS as a marker. Proc Natl Acad Sci–Biol 101: 14533–14538.
23. Gómez-Campo C, Prakash S (1999) Origin and domestication. In: Gómez-Campo C, ed. Biology of Brassica coenospecies. Netherlands: Elsevier. pp 33–58.
24. Gulden RH, Shirtliffe SJ, Thomas AG (2003) Harvest losses of canola (Brassica napus) cause large seedbank inputs. Weed Sci 51: 83–86.
25. Gulden RH, Shirtliffe SJ, Thomas AG (2003) Secondary seed dormancy prolongs persistence of volunteer canola in western Canada. Weed Sci 51: 904–913.

13

The B-Box Family Gene *STO* (*BBX24*) in *Arabidopsis thaliana* Regulates Flowering Time in Different Pathways

Feng Li[1,2], Jinjing Sun[1], Donghui Wang[1], Shunong Bai[1], Adrian K. Clarke[2]*, Magnus Holm[†2]

1 Peking University-Yale Joint Research Center of Agricultural and Plant Molecular Biology, National Key Laboratory of Protein Engineering and Plant Gene Engineering, College of Life Sciences, Peking University, Beijing, China, **2** Department of Biological and Environmental Sciences, Gothenburg University, Gothenburg, Sweden

Abstract

Flowering at the appropriate time is crucial for reproductive success and is strongly influenced by various pathways such as photoperiod, circadian clock, *FRIGIDA* and vernalization. Although each separate pathway has been extensively studied, much less is known about the interactions between them. In this study we have investigated the relationship between the photoperiod/circadian clock gene and *FRIGIDA/FLC* by characterizing the function of the B-box *STO* gene family. STO has two B-box Zn-finger domains but lacks the CCT domain. Its expression is controlled by circadian rhythm and is affected by environmental factors and phytohormones. Loss and gain of function mutants show diversiform phenotypes from seed germination to flowering. The *sto-1* mutant flowers later than the wild type (WT) under short day growth conditions, while over-expression of *STO* causes early flowering both in long and short days. *STO* over-expression not only reduces *FLC* expression level but it also activates *FT* and *SOC1* expression. It also does not rely on the other B-box gene *CO* or change the circadian clock system to activate *FT* and *SOC1*. Furthermore, the *STO* activation of *FT* and *SOC1* expression is independent of the repression of *FLC*; rather *STO* and *FLC* compete with each other to regulate downstream genes. Our results indicate that photoperiod and the circadian clock pathway gene *STO* can affect the key flowering time genes *FLC* and *FT/SOC1* separately, and reveals a novel perspective to the mechanism of flowering regulation.

Editor: Yuehui He, Temasek Life Sciences Laboratory & National University of Singapore, Singapore

Funding: The authors have no support or funding to report.

Competing Interests: The authors have declared that no competing interests exist.

* E-mail: adrian.clarke@bioenv.gu.se

† Deceased.

Introduction

There are several key developmental changes during the plant lifecycle. One of these is flowering, the correct timing of which is critical for reproductive success [1]. Physiological and genetic studies have shown that multiple pathways can promote or repress flowering [1] [2] [3]. The floral pathway integrators *FLOWERING LOCUS C* (*FLC*), *FLOWERING LOCUS T* (*FT*), *SUPPRESSOR OF CONSTANS1* (*SOC1*) and *LEAFY* (*LFY*) are all involved in this transition. The MADS box gene *FLC* is a central repressor of flowering in *Arabidopsis* [4] [5], and its expression is regulated by vernalization and an autonomous pathway involved in chromatin regulation, transcription level and co-transcriptional RNA metabolism [6]. Vernalization pathway genes (*VRN1*, *VRN2*, *VIN3*, *VIL1/VRN5*, *atPRMT5*) repress *FLC* expression by histone modification of *FLC* during and after the cold treatment [7] [8] [9] [10] [11] [12].

Arabidopsis has two antagonistic pathways that regulate *FLC* expression. The *FRIGIDA* (*FRI*) pathway is a positive regulator, while a group of genes that belong to the autonomous floral-promotion pathway are negative regulators (i.e., *LD*, *FLD*, *FCA*, *FY*, *FVE*). *FRI* is a unique plant gene that encodes a nuclear-localized protein with a coiled-coil domain [13]. The functional allele of *FRI* is only found in the winter-annual *Arabidopsis*, which requires vernalization to flower rapidly in the spring through repression of *FLC*. In rapid-cycling *Arabidopsis* there is no

functional *FRI*; *FLC* expression is kept at low levels and the photoperiod pathway accelerates flowering. *FLC* directly binds and represses the two important flowering genes *FT* and *SOC1* [14]. *FT* is the "florigen" acting as a long distance signal that is transported from leaves to the shoot meristem [15] [16] [17] [18]. The MADS-box gene *SOC1* was initially cloned as a suppressor of the *CONSTANS1* (*CO*) overexpressor [19] [20]. *SOC1* regulation integrates inputs from multiple flowering pathways including photoperiod, vernalization, aging and GA [12] [21]. Besides GA, other phytohormones (BR, ethylene and ABA) also play important roles in flowering, but the underlying mechanisms are less well understood [22] [23] [24]. *SOC1* can also be considered as a meristem-identity gene because it maintains the meristem in a floral state [25]. Both *FT* and *SOC1* are regulated by different flowering pathways [20] [26]. The effect of day length within the photoperiod pathway has been extensively studied [21] [27].

Arabidopsis senses changes in day length by the circadian clock, which in turn regulates the transcription factor *CO*. CO belongs to a subfamily of the zinc finger protein family that is now known as the B-Box Zinc Finger Family (BBX) [28]. This family consists of 32 genes divided into five structural groups from I to V. Proteins in group I (including CO) all contain a B-box B1, a B-box B2, and a C-terminal CCT (*CO/COL/TOC1*) domain. Group II members are similar to group I, and contain both B1, B2 and CCT domains, but have minor differences in their B2 domains. Proteins from group III only contain a B1 and CCT domain, while group

IV members have B-box B1 and B2 domains but lack the CCT domain (including STO, STH, STH2 and STH3). Proteins in group V have only a single B1 domain [28].The temporal and spatial regulation of *CO* on a transcriptional and protein stability level is the most important element of the photoperiod pathway. Under long day conditions (LD), *CO* transcription is regulated by the circadian clock and accumulates late in the day. *FLAVIN-BINDING, KELCH REPEAT, F-BOX 1 (FKF1)* combined with *GIGANTEA (GI)* and *CYCLING DOF FACTOR1 (CDF1)* not only promote *CO* expression but also stabilize the CO protein in the afternoon in LD [29] [30] [31]. During the night, the CO protein is degraded by COP1 [32]. Based on the complex regulation of *CO*, CO protein accumulates at the end of the day and activates the downstream genes such as *FT* and *SOC1* to promote flowering in LD. In contrast, the CO protein is not stably produced under short day (SD) conditions [21].

The flowering-time integrators *FT* and *SOC1* are common targets of distinct pathways, but the relationship between the different pathways is still unclear. There is little evidence for a direct connection between *FRI/FLC* and the photoperiod/circadian clock pathway in flowering time except that they antagonistically regulate common gene targets (*FT* and *SOC1*). In this article, we have studied the function of another B-box family gene *Salt Tolerance* (*STO* or *BBX24*), and show that *STO* (*BBX24*) links the *FRI/FLC* and photoperiod/circadian clock pathways.

Materials and Methods

Growth conditions and plant material

Plants were grown in soil under controlled conditions of LD (16 h light/8 h dark) or SD (8 h light/16 h dark) at 22°C. The level of photosynthetic active radiation was 60 µmol photons $m^{-2} s^{-1}$ under both LD and SD conditions. Plants were grown on MS plates, 4 d in dark at 4°C before moved to LD or SD at 22°C. The Columbia (Col-0) ecotype was used. In the *STO* over-expression studies, phenotypic analysis of all transgenic lines and controls were conducted on plates with MS with addition of 50 µg ml^{-1} kanamycin for transgenic plant selection. The mutant seeds are listed in Table S1.

Measurement of hypocotyl length

Seeds were sterilized in 70% ethanol and 0.01% Triton X-100 for 15 min, followed by 95% ethanol for 10 min. After steriliza-tion, seeds were suspended in 0.1% low-melting-point agarose and spotted on plates containing MS medium (Gibco/BRL) and 0.8% phytagar (Gibco/BRL). Seeds on plates were then stratified in the dark at 4°C for 4 d. Plants were transferred to 22°C. Hypocotyl lengths from 20 seedlings were measured on day 4 with NIH Image 1.62.

Flowering time determination

Flowering time was determined by counting the number of rosette leaves after bolting. Data were reported as mean leaf number (± S.D.) and were measured from homozygous lines.

Brassinolide (BR) treatment

Plants were grown on the 1/2 MS plates for 5 d with 0–1000 nM brassinolide (BL), which is the most biologically active BR before measuring hypocotyl length. For the RNA extractions, all seedlings were grown on 1/2 MS plates and treated with 100 nM BL for 3 h or 4 d prior to harvest, with all samples harvested at the same time.

Ethylene triple response

Dark-grown WT and *sto-1* seedlings were treated with 0–10 µM ACC (the precursor of ethylene) for 3.5 d. SE values were determined from 20 to 30 seedlings.

Vernalization treatment

Seeds were germinated on agar plates for 4 d at 22°C and vernalized for 4 weeks at 4°C under SD (8 h light/16 h dark). Post vernalization samples continued to grow on agar plates at 22°C under SD.

RNA extraction and quantitative real-time PCR

Total RNA was isolated from seedlings (4, 8 and 10 d old) or mature rosette leaves using the RNAeasy extraction kit (Qiagen). First-strand cDNA synthesis was performed on 1–2 µg of RNA using the M-MLV System for RT-PCR (Fermentas) followed by PCR amplification with dream Taq DNA Polymerase (Fermentas) or by quantitative PCR (Bio-Rad). Reactions were performed using the primers described in Table S2. For cDNA synthesis, the poly-dT primer was used. The quantitative real-time PCRs were performed with at least three independent RNA samples. For *STO* expression analysis by real-time PCR, all the samples were harvested 6 h into the photoperiod when the peak level of *STO* expression is reached.

DNA extraction

Plant tissue (200 mg) was ground to a fine paste in 500 µl of CTAB buffer. The CTAB/plant extract mixture was transferred to a microfuge tube and incubated for 15 min at 55°C in a recirculating water bath. After incubation, the CTAB/plant extract mixture was centrifuged at 12000 *g* for 5 min to pellet cell debris. The supernatant was transferred to clean microfuge tubes and 250 µl of chloroform/isoamylalcohol (24:1) was added to each, and then mixed by inversion. After mixing, tubes were centrifuged at 13000 *g* for 10 min. The upper aqueous phase containing the DNA was transferred to a clean microfuge tube and placed at −20°C for 1 h after the addition of ethanol to precipitate DNA. The precipitated DNA was then pelleted, washed twice in 70% ethanol and then resuspended in sterile DNase-free water.

Results

Developmental phenotypes of the STO loss- and gain-of-function mutants

STO was originally found to increase the tolerance of yeast to both Li and Na ions [33]. STO, along with STH (*BBX25*) interacts with the WD40 domain of COP1 [34], with *COP1* repressing the transcription of *STO* and contributing to STO protein destabil-ization in etiolated seedlings [35]. Overall, *STO* acts as a negative regulator in the early photomorphogenesis response to red, far-red, blue and UV-B light signaling [32] [35] [36]. To further uncover the biological function of *STO*, we examined the phenotypes of the *sto-1* T-DNA insertion knockout mutant (SALK_067473) and *STO* over-expression line *STO-OE* (*35s:: STO*) at different developmental stages. Early in development, *STO* appears to repress the rate of seed germination (24 h after transfer to light), with the *sto-1* mutant exhibiting faster and the *STO-OE* line slower germination rates than the WT (Figure 1A). Despite this variation, 90–100% of seeds for all lines studied germinated after 48 hours. At the seedling stage, *sto-1* and *STO-OE* hypocotyls were shorter and longer than WT, respectively (Figure 1B). The difference in hypocotyl length between *sto-1* and *STO-OE* appears due to elongation post germination rather than to the rate of germination itself, and *STO* plays an important role in this process. Moreover, both the *sto-1* and *STO-OE* lines lost sensitivity to low and moderate concentrations of BL (0–100 nM), but all responded to a higher concentration (1000 nM) (Figure 1C and E). Under dark conditions, *sto-1* had slightly longer hypocotyls than the WT and was less sensitive to low concentrations of ACC (0–0.1 μM) (the precursor of ethylene). However, *sto-1* still showed a triple-response phenotype at higher concentrations (Figure 1D and F; it should be noted that the *STO-OE* germination rate was very poor in darkness and ACC treatments). In the adult rosette leaves, *sto-1* accumulated much more anthocyanin on the abaxial surface than WT, while *STO-OE* had much less anthocyanin than WT. In *STO-OE*, the purple color concentrated in the main vascular tissue, while the remaining leaf blade was much greener than WT. *STO* affects not only mature rosette leaf pigmentation but also leaf morphology. In *sto-1*, adult leaves were narrower and more curled than WT, but less serrated, whereas the STO-OE leaves had much deeper serrations than WT (Figure 1G).

In addition to changes in leaf characteristics, flowering time was also affected by *STO* in both LD and SD conditions (Figure 1H–K). Over-expression of *STO* promoted flowering in both LD (Figure 1H and J) and SD (Figure 1I and K) while loss of *STO* delayed flowering relative to the WT but only in SD; both *sto-1* and WT were almost bolting at the same time in LD. These results confirm that *STO* is an important gene affecting different developmental stages throughout the plant lifecycle.

STO expression has diurnal transcript characteristics and is affected by environmental factors and phytohormones

To gain insights into the biological function of *STO*, its spatial and temporal expression characteristics were analyzed by qPCR. *STO* is expressed in major tissues of *Arabidopsis* and the highest expression level was found in the shoot apex (Figure S4). Previous studies have shown that expression of *STO* is under circadian clock control [36]. The *STO* transcript has diurnal characteristics in both LD and SD conditions (Figure 2A). The highest *STO* expression level was at 6 h into the photoperiod under both LD and SD conditions, after which a rapid decline in LD and SD was observed. During the dark period, however, *STO* expression increased earlier in SD than LD. In the *STO-OE* line, *STO* expression was maintained at high levels in both light and dark conditions (Figure 2B). *STO* expression in WT was also up-regulated by exposure to low temperature (Figure 2C), similar to that previously shown for UV-B treatment [32]. This cold-induced increase in *STO* transcripts, however, was reversed upon returning the plants to the standard growth temperature of 22°C, dropping almost to the levels observed in the control plants (Figure 2C). In contrast, *STO* expression was repressed by treatment with the phytohormone BR (Figure 2D). BR is an essential hormone that regulates a wide range of developmental and physiological processes including cell expansion, vascular differentiation, etiolation, flowering and male fertility [37]. Light and BR antagonistically regulate the developmental process in de-etiolation of plants [38]. When seedlings were grown in 1/2 MS medium and treated with BL (100 nM) for 3 h and 4 d, respectively, *STO* expression was repressed during both exposure times (Figure 2D). Taken together, these phenotypes and expression characteristics indicate that *STO* is an important gene during the development of the plant and its response to certain environmental cues.

STO represses FLC expression

Although the role of *STO* in early photomorphogenesis has been well studied [32] [35] [36], little is known about its function during the late developmental stages. Earlier microarray analysis revealed

that the level of *FLC* expression increases in *sto-1* etiolated seedlings (unpublished data), results which were confirmed in this study by qPCR at different time points in etiolated seedlings (Figure S1). Both *sto-1* and *STO-OE* have flowering phenotypes; *FLC* was selected for further investigation to examine how this B box family gene affects *Arabidopsis* flowering time.

We investigated *FLC* expression in different lines and light conditions (i.e., LD, SD, darkness) (Figure 3A). We found that over-expression of *STO* strongly repressed *FLC* expression in all three light conditions, whereas *FLC* expression in *sto-1* seedlings was higher than in WT Col-0 and *STO-OE* in SD and darkness. In the LD condition, however, the level of *FLC* expression in *sto-1* was similar to that in WT seedlings. Moreover, over-expression of *STO* in *FRIGIDA* and *fld-3*, which have high levels of *FLC* transcripts, did not further repress *FLC* expression (Figure 3B). At same time, in *FRI* and *flc-3/FRI* background, *STO* expression level did not show any statistically significant difference compared with WT (Figure S2). These results suggest *STO* is an upstream regulator that represses *FLC* expression.

STO promotes FT and SOC1 expression

In addition to *FRI* and *FLC*, the circadian clock system, photoperiod pathway (*CO*), and flowering integrators (*FT, SOC1*) also play important roles in the transition from vegetative to reproductive phases. The circadian clock consists of at least three interlocked transcriptional feedback loops. The LATE ELONGATED HYPOCOTYL (LHY) and CIRCADIAN CLOCK–ASSOCIATED1 (CCA1) proteins are important components of the circadian clock system and have partially redundant functions [39] [40]. *LHY* and *CCA1* not only repress the floral transition under LD and SD conditions, they also accelerate flowering in continuous light by promoting *FT* expression [41]. *FT* and *SOC1* are both *CO* target genes. *CO, FT* and *SOC1* are deeply affected by the circadian clock system at different levels [40].

To further investigate how *STO* regulates flowering time, we next examined how it affects expression of the aforementioned genes. Several different time points under the SD condition, where the *sto-1* knock-out mutant and *STO* over-expressor showed significant differences in flowering time were investigated (Figures 1 H–K). *CCA1* expression level did not change either in amount or rhythm in Col, *sto-1* and *STO-OE* (Figure 4A). In *sto-1*, *CO* expression level was equal to that in Col during most time points and slightly lower in the dark (15–18 h). In the *STO-OE* line, *CO* expression level was higher compared to that in Col and *sto-1* at the end of the day period (after 6–8 h) but lower during the dark (Figure 4B). The expression level of *FT* and *SOC1* first peaked in the middle of the photoperiod (6 h) and then again in the dark (15 h) (Figures 4C and D). Since both CO and STO proteins are degraded by COP1 in the dark [32] [34] [42], the second peak of *FT* and *SOC1* expression in *STO-OE* might be a result of the extra STO protein overwhelming the ability of COP1 to degrade it, thereby activating *FT* and *SOC1* expression. Also in the *STO-OE* line, *FT* and *SOC1* expression increased earlier than *CO* and started to decrease at the time when *CO* had only just begun to increase (i.e., after 6 h; Figures 4C and D). These results suggest that *STO* activates *FT* and *SOC1* expression in a *CO* and circadian rhythm-independent manner, although this does not exclude the possibility that *CO* affects *STO* at the protein level. In the *STO-OE* line, moreover, the promotion of *FT* expression mainly occurred in cotyledons, whereas the expression of *FT* in *sto-1* cotyledons was lower than in Col. Furthermore, the level of *FT* expression did not differ in the shoot apex, hypocotyl and root (Figure S5).

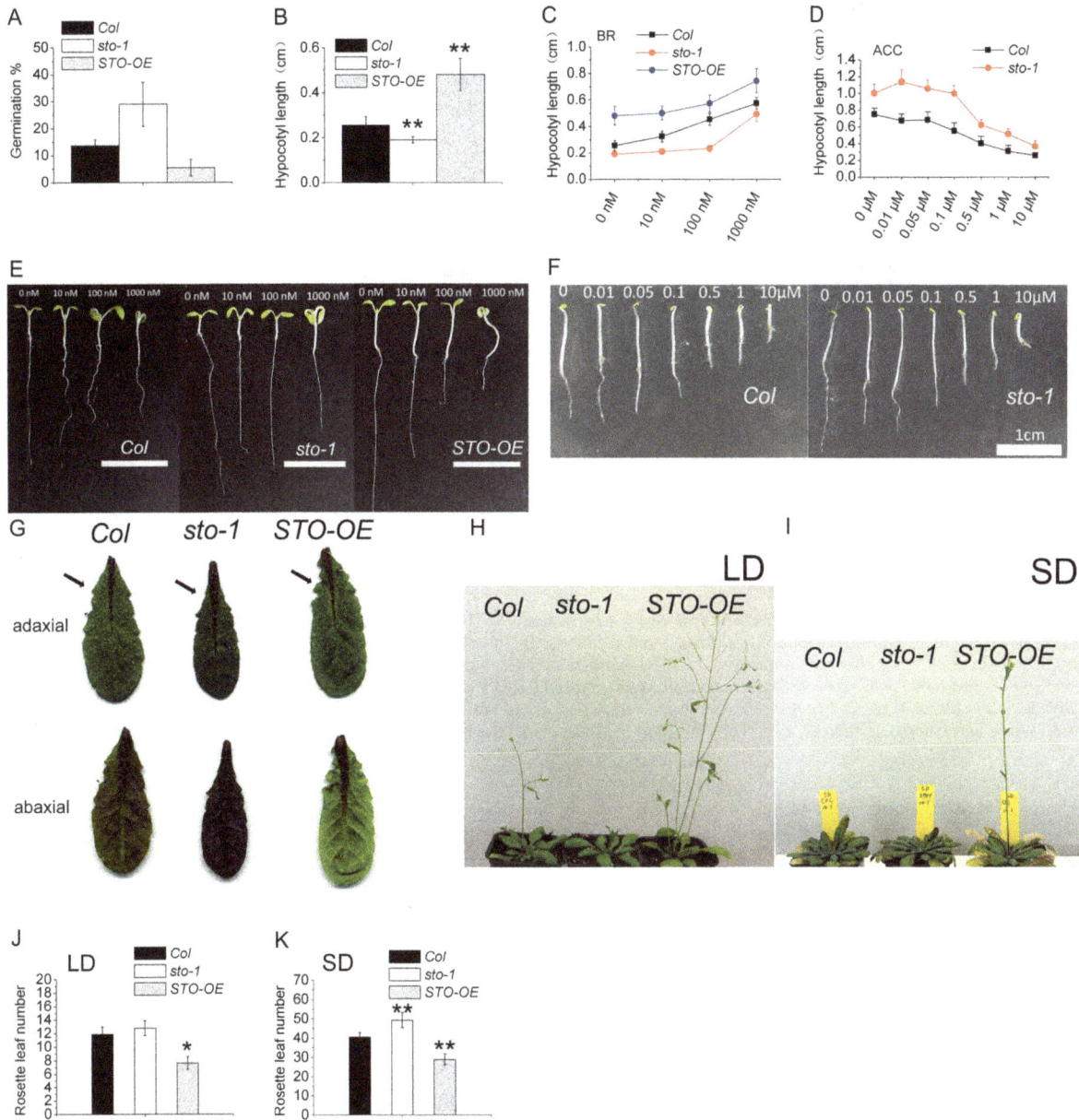

Figure 1. sto-1 and STO-OE show diverse phenotypes. Col, sto-1 and STO-OE were grown under SD conditions and different growth characteristics analyzed. (A) Seed germination as a percentage of total seeds analyzed; (B) hypocotyl length; (C) and (E) Seedling growth as measured by average hypocotyl length (n = 20–30) after 5 d on1/2 MS plates treated with BL (0–1000 nM) in SD; (D) and (F) Seedling growth as measured by average hypocotyl length (n = 20–30) after treatment with ACC (0–10 μM) for 3.5 days in darkness; (G) Appearance of adult rosette leaves on both adaxial and abaxial surfaces, with arrows indicating the position of leaf serration; (H–K) Flowering time of Col, sto-1 and STO-OE in LD and SD conditions. The average number of rosette leaves in LD (J) and SD (K) was also calculated (n = 15–28, * means p<0.05 and ** means p<0.01 in TTEST).

STO competes with FLC to promote FT and SOC1 expression

Since *FLC* is known to block *FT* by directly binding to its chromatin [14], there are two possible explanations why *STO* can promote flowering: that either *STO* directly represses *FLC* and activates *FT* and *SOC1* expression simultaneously, or that *STO* only represses *FLC* that then leads to the activation of both *FT* and *SOC1*. To investigate these two scenarios, we overexpressed *STO* in different genetic backgrounds (Figure S3) to hopefully reveal the regulatory relationship between *STO* and *FLC* (Figure 5A). In the *flc-3/FRI* lines, which lack full length *FLC* but have functional *FRI*,

FT and *SOC1* expression was slightly higher than in *FLC/FRI* under SD conditions. In the *STO-OE/flc-3/FRI* lines (individual lines 4 and 15), *FT* and *SOC1* expression significantly increased (Figures 5B and C), showing that *STO* activates *FT* and *SOC1* expression independently of the repression of *FLC*. If *STO* activates *FT* and *SOC1* expression by only reducing *FLC* expression, there should be similar amounts of *FT* and *SOC1* transcripts in both the *STO-OE/flc-3/FRI* and *flc-3/FRI* lines. Moreover, these two lines (*STO/flc-3/FRI* -4 and 15) also flowered earlier under SD conditions (Figure 5D and Table S 3). In contrast, in the *STO-OE/FLC/FRI* line (line11) that has high levels of *FLC* and over-

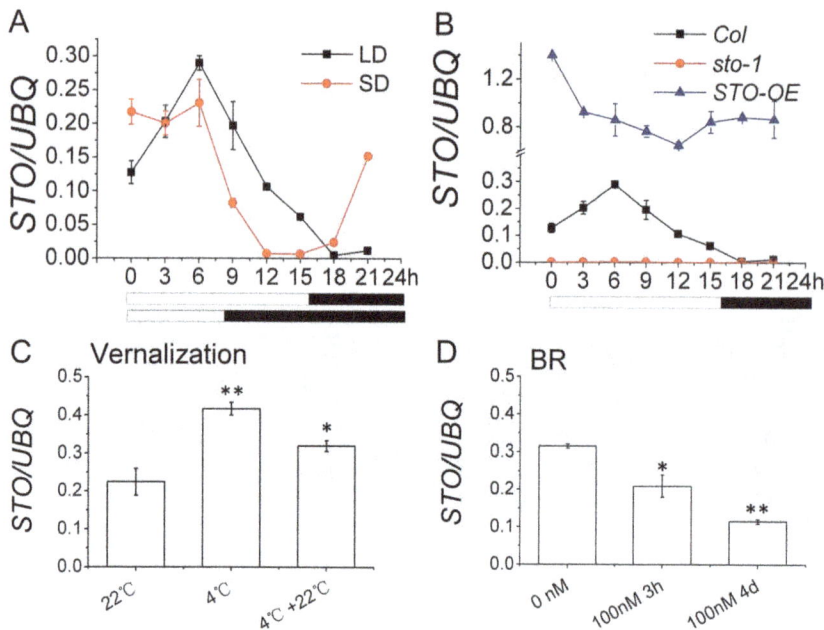

Figure 2. *STO* is regulated by environmental factors and phytohormones. (A) The diurnal expression pattern of *STO* under both LD (16 h light/8 h dark) and SD (8 h light/16 h dark) conditions in eight-day-old seedlings. (B) The diurnal expression pattern of *STO* in eight-day-old seedlings of Col, *sto-1* and *STO-OE* in LD. (C) The effect of vernalization on *STO* expression. Seedlings were either warm-treated (22°C, 2 d, in SD), cold-treated (4°C, 4 weeks, in SD) or cold-treated (4°C, 4 weeks, in SD) and then warm-treated (22°C, 2 d, in SD). (D) The effect of BR on *STO* expression. Seedlings were treated with BL (100 nM) for either 3 h or 4 d. All experiments were performed with two (A–B) or three (C–D) independent replicates, *UBQ10* used as a control. White and black bars represent the light and dark periods, respectively. * means p<0.05 and ** means p<0.01 in TTEST.

expresses *STO*, the level of *FT* expression was only slightly higher than in the *FRI* line, and *SOC1* transcripts were no longer up-regulated (Figures 5B and C).

Discussion

In this work, we have examined in detail the biological function of *STO* from seed germination to flowering by the use of loss- and gain-of-function mutants. *STO* loss- and gain-of-function mutants both lost sensitivity to low concentrations of phytohormones such as BR and ethylene. Both BR and ethylene affect flowering time. We showed that *STO* was involved in those signaling pathways at least in the seedling stage (Figure 1C–F). In *Arabidopsis*, BR biosynthetic and signaling pathway mutants exhibit delayed flowering phenotypes [43] [44]. BR can promote flowering time

by affecting circadian clock and *FLC* at both the transcription and chromatin modification levels [44] [45]. BR and *STO* have similar functions in the repression of photomorphogenesis, increased hypocotyl length and promotion of flowering (Figure 1 and [24]). However, the relationship between *STO* and BR signaling is rather complicated, as *STO* expression was repressed by BL treatment (Figure 2D). The reason for this conflicting observation could be that BL treatment results in accumulation of STO protein and feedback inhibition of *STO* transcription. Another explanation is that they may have common downstream genes, such as *GATA2* and *GATA4*, which are positive regulators of photomorphogenesis and are repressed by BR signaling [38]. *GATA2* and *GATA4* expression levels in *sto-1* and *STO-OE* were higher and lower than WT, respectively (unpublished data). BR signaling could through repressing *STO* achieve the appropriate expression level of

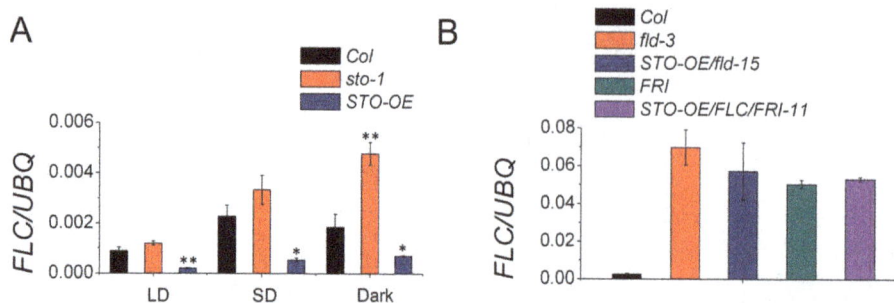

Figure 3. *STO* represses the level of *FLC* expression. (A) Expression of *FLC* in four-day-old seedlings of Col, *sto-1* and *STO-OE* grown in either LD, SD or darkness. (B) Changes in *FLC* expression in four-day-old seedlings in SD when *STO* is over-expressed in the *FRI* or *fld-3* background. The level of *FLC* expression in Col and the *FRI* and *fld-3* lines was also included for comparison. All experiments were performed with at least three independent biological replicates, with *UBQ10* used as a control. * means p<0.05 and ** means p<0.01 in TTEST.

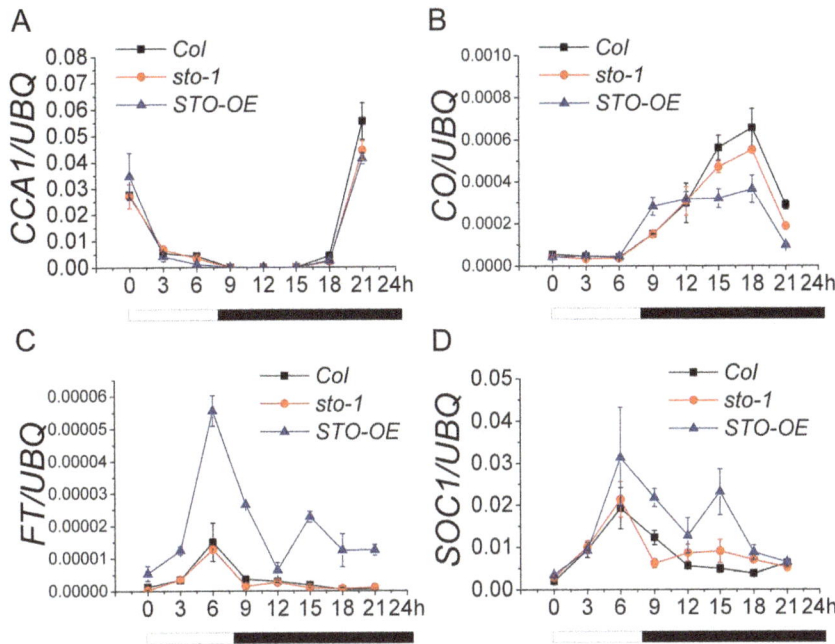

Figure 4. Over-expression of *STO* promotes *FT* and *SOC1* expression independent of *CO* or *CCA1*. The level of *CCA1* (A), *CO* (B), *FT* (C) and *SOC1* (D) expression in eight-day-old seedlings under SD conditions. Data from two independent replicates are shown, with *UBQ10* used as a control. White and black bars represent the light and dark periods, respectively.

downstream genes. In addition, *STO* expression was also regulated by environmental factors, such as the photoperiod and cold temperature (Figure 2B and C). All of those environmental and endogenous factors substantially affect plant flowering time.

Important was the observation that over-expression of *STO* produced an early flowering phenotype under both LD and SD conditions, whereas loss of *STO* caused late flowering under SD. Altogether, *STO* appears to function more than just a negative

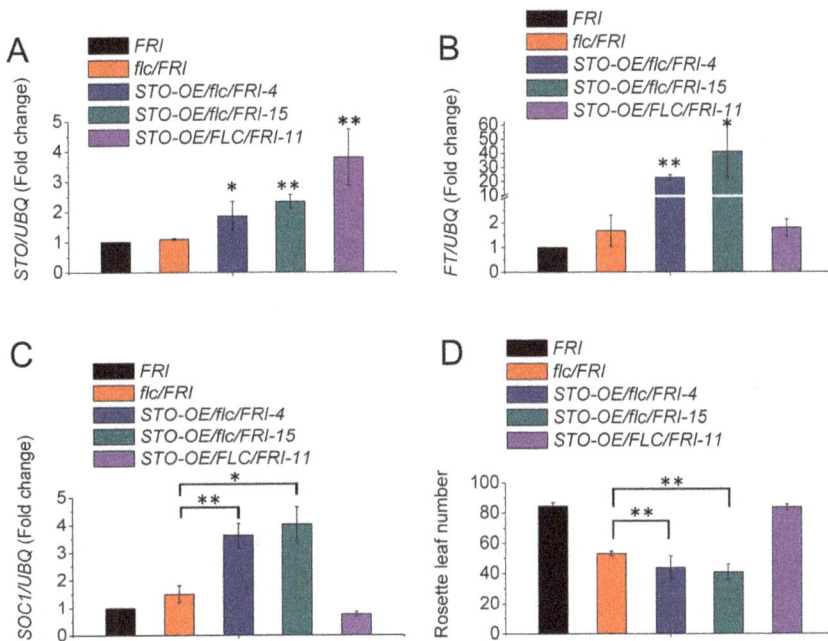

Figure 5. Over-expression of *STO* promotes *FT* and *SOC1* expression independent of *FLC* repression. The level of *STO* (A), *FT* (B) and *SOC1* (C) expression in eight-day-old seedlings of the indicated genotypes. In the analysis, the data from STO-OE/flc/FRI-4 and -15 were compared to that of flc/FRI, while the data for STO-OE/FLC/FRI-11 was compared to that of FRI. Data from three independent replicates are shown, with *UBQ10* used as a control. (D) The number of rosette leaves in each indicated genotype at the time of flowering. All seedlings and mature plants were grown in SD. ** means p<0.01 in TTEST.

regulator during early photomorphogenesis, and instead to play a crucial role connecting different signaling pathways throughout the plant lifecycle.

The observations from this study support the following conclusions: (i) *STO* represses the flowering repressor *FLC*; (ii) *STO* can upregulate *FT* and *SOC1* expression, and; (iii) *STO* competes with *FLC* in their regulation of *FT* and *SOC1*. *STO* can repress *FLC* in LD, SD and darkness. However, in the *FRI* and *fld-3* genetic background, over-expressed *STO* can no longer repress *FLC*. Since *STO* lacks the transcriptional repression/activation domain, it could be that it interacts with another complex to regulate *FLC* expression. *FLC* is not only a repressor of flowering time, but it is also functional throughout the lifecycle of the plant [14], such as promoting seed germination [46], lengthening the circadian period, and vegetative development [47]. Interestingly, the *Ler* background line that has a low-expressing *FLC* allele exhibited high dormancy. A high-expressing *FLC* allele produced a significantly higher germination rate at cool temperatures (10°C), but only slightly higher at warmer temperature (22°C) [46]. We found that the *STO-OE* line which has low *FLC* expression level also displayed slower germination rates, whereas the *sto-1* germination rate was slightly higher than the WT (Figure 1 A, B). As a consequence, we propose that *STO* as a photoperiod/circadian clock controlled gene is involved in the regulation of *FLC*, although the molecular mechanism by which this occurs remains unclear.

We have demonstrated that *STO* activates *FT* and *SOC1* expression, thereby promoting flowering. Given that *STO* expression is controlled by photoperiod/circadian rhythm, it would appear that it is a new component within this regulatory pathway. Moreover, *STO* stimulation of *FT* and *SOC1* expression does not rely on repression of *FLC* but rather *STO* competes with *FLC* to regulate *FT* and *SOC1* expression and thereby promotes flowering. Overall, these characteristics of *STO* reveal new relationships between *FRI/FLC* and the photoperiod/circadian clock pathway.

The combination of phenotype, expression characteristics and genetic results extends our knowledge on the biological function of the B-box zinc finger family in plant development, especially in the transition from vegetative to reproductive phase. Given the central roles of *FLC*, *FT* and *SOC1* in flowering-time regulation in *Arabidopsis*, these findings suggest that the B-box family gene *STO*, which is without the CCT domain, plays an important role in the control of flowering in *Arabidopsis*. The behavior of *STO* in promoting flowering gives us new clues in understanding of how plants integrate environmental and developmental signals. We can therefore expect that the relationship between the different pathways to be more complex than first suspected.

The homologous B-box zinc-finger family genes are widely conserved in higher plants [36] [48] [49]. Despite this, the study of flowering time only focuses on the first group (*CO* and *CO*-like *COL*'S [48]) because previous research has indicated that both B-box motifs and CCT domains are important for promoting flowering under LD conditions [50] [51]. *CO* and *COL* genes have been well studied. While resembling *STO* and *STH1-3*, they also show the opposite function. Flowering in the *CO* mutant is delayed in LD while overexpression of *CO* results in the acceleration of flowering in both LD and SD. Over-expression of *COL5* can induce flowering in SD but *col5* mutants do not show altered flowering [52]. Over-expression of *COL9* resulted in delayed flowering, whereas the *col9* mutant flowered earlier under LD conditions. *COL9* negatively regulates *CO* expression, but it does not appear to directly affect flowering time [53]. Moreover, *CO*, *COL5* and *COL9* influence flowering time mainly through *FT* and *SOC1* and no reports to date have shown that they can affect *FLC* expression. Interestingly, a recent study has shown that *COL1* and

-2, which are the most closely-related genes to *CO*, do not affect flowering time due to the amino acid differences coded for within the first exon [54]. The first B-box domain may have a more important role than others in affecting flowering time. *COL* genes (including *STO*) have evolved rapidly in the *Brassicaceae* family [48] [55]. It is possible that the effect of *STO* on flowering time independent of the CCT domain is not a unique case. It will be interesting to investigate if other members that lack the CCT domain within the B-box family are also involved in regulating flowering.

Supporting Information

Figure S1 Confirmation of microarray data. Col, *sto-1* and *STO-OE* were grown in darkness, with *FLC* expression level checked at 2, 4 and 6 d. *FLC* expression levels of the indicated genotypes were checked, and all lines were grown under SD.

Figure S2 *STO* expression levels in different lines. The level of *STO* expression in the indicated genotypes was checked in four-day-old seedlings grown under SD. Data from three independent replicates are shown, with *UBQ10* used as a control.

Figure S3 Double mutant genomic PCR and RT-PCR confirmation. (A) Genomic PCR test was performed in F2 *FRI/flc-3 x STO-OE*, with the lines 4, 7, 10 and 15 being *flc* homozygous. (B) Functional *FRI* allele's have a BsmFI restriction site. Restriction endonuclease BsmFI was used to test *FRI* homozygous in F2 *FRI/flc-3 x STO-OE*, with the lines 4,7,10 and 15 having a functional *FRI*. (C) RT-PCR test of full-length *FLC* expression, with the lines 4, 10 and 15 having no *FLC* mRNA. (D) RT-PCR test of *FLD* expression level in F2 *fld-3 x STO-OE*, with *UBQ10* as the control. The lines 11, 13, 15, 19 and 21 were *fld* homozygous. (E) Restriction endonuclease BsmFI was used to test *FRI* homozygous in F2 *FRI x STO-OE*. The lines 9, 11, 12, 15 and 17 were *FRI* homozygous. All lines were grown for two weeks under SD prior to genomic PCR or RT-PCR.

Figure S4 Determination of *STO* expression levels in different plant tissues. The level of *STO* expression in different tissues of *Arabidopsis* was analyzed in ten-day-old seedlings (cotyledon, shoot apex, hypocotyl and root) and adult plants (rosetta leaf and flower meristem). Data from three or four independent replicates are shown, with *UBQ10* used as a control. * means $p<0.05$ and ** means $p<0.01$ in TTEST. Plants were grown under LD.

Figure S5 Increased expression of *FT* in cotyledons and hypocotyls. The level of *FT* expression in different tissues (cotyledon, shoot apex [including young leaf primordial], hypocotyl and root) of ten-day-old seedlings of Col, *sto-1* and *STO-OE*. . Data from three independent replicates are shown, with *UBQ10* used as a control. * means $p<0.05$ and ** means $p<0.01$ in TTEST. Plants were grown under LD.

Table S1 Seed list.

Table S2 Primer list.

Table S3 Rosette leaf number at flowering time for the indicated genotypes.

Acknowledgments

We thank Cornelia Spetea Wiklund, Anders Nilsson, Lisa Adolfsson and Jenny Carlsson for critical reading and comments on the manuscript. We also thank Caroline Dean for kindly providing the *flc-3* and *FRI* seeds. This article is dedicated to the memory of our colleague, fellow author and dear friend Magnus Holm, who sadly passed away in September 2012.

Author Contributions

Conceived and designed the experiments: FL MH. Performed the experiments: FL JS DW SB. Analyzed the data: FL JS DW SB AKC MH. Contributed reagents/materials/analysis tools: MH. Wrote the paper: FL AKC.

References

1. Baurle I, Dean C (2006) The timing of developmental transitions in plants. Cell 125: 655–664.
2. Bernier G, Havelange A, Houssa C, Petitjean A, Lejeune P (1993) Physiological Signals That Induce Flowering. Plant Cell 5: 1147–1155.
3. Simpson GG, Dean C (2002) Arabidopsis, the Rosetta stone of flowering time? Science 296: 285–289.
4. Michaels SD, Amasino RM (1999) FLOWERING LOCUS C encodes a novel MADS domain protein that acts as a repressor of flowering. Plant Cell 11: 949–956.
5. Sheldon CC, Burn JE, Perez PP, Metzger J, Edwards JA, et al. (1999) The FLF MADS box gene: a repressor of flowering in Arabidopsis regulated by vernalization and methylation. Plant Cell 11: 445–458.
6. Crevillen P, Dean C (2011) Regulation of the floral repressor gene FLC: the complexity of transcription in a chromatin context. Curr Opin Plant Biol 14: 38–44.
7. Henderson IR, Dean C (2004) Control of Arabidopsis flowering: the chill before the bloom. Development 131: 3829–3838.
8. Bastow R, Mylne JS, Lister C, Lippman Z, Martienssen RA, et al. (2004) Vernalization requires epigenetic silencing of FLC by histone methylation. Nature 427: 164–167.
9. Sung S, Schmitz RJ, Amasino RM (2006) A PHD finger protein involved in both the vernalization and photoperiod pathways in Arabidopsis. Genes Dev 20: 3244–3248.
10. Schmitz RJ, Sung S, Amasino RM (2008) Histone arginine methylation is required for vernalization-induced epigenetic silencing of FLC in winter-annual Arabidopsis thaliana. Proc Natl Acad Sci U S A 105: 411–416.
11. Sung S, Amasino RM (2004) Vernalization in Arabidopsis thaliana is mediated by the PHD finger protein VIN3. Nature 427: 159–164.
12. Kim DH, Doyle MR, Sung S, Amasino RM (2009) Vernalization: winter and the timing of flowering in plants. Annu Rev Cell Dev Biol 25: 277–299.
13. Johanson U, West J, Lister C, Michaels S, Amasino R, et al. (2000) Molecular analysis of FRIGIDA, a major determinant of natural variation in Arabidopsis flowering time. Science 290: 344–347.
14. Deng WW, Ying H, Helliwell CA, Taylor JM, Peacock WJ, et al. (2011) FLOWERING LOCUS C (FLC) regulates development pathways throughout the life cycle of Arabidopsis. Proceedings of the National Academy of Sciences of the United States of America 108: 6680–6685.
15. Kardailsky I, Shukla VK, Ahn JH, Dagenais N, Christensen SK, et al. (1999) Activation tagging of the floral inducer FT. Science 286: 1962–1965.
16. Kobayashi Y, Kaya H, Goto K, Iwabuchi M, Araki T (1999) A pair of related genes with antagonistic roles in mediating flowering signals. Science 286: 1960–1962.
17. Wigge PA, Kim MC, Jaeger KE, Busch W, Schmid M, et al. (2005) Integration of spatial and temporal information during floral induction in Arabidopsis. Science 309: 1056–1059.
18. Mathieu J, Warthmann N, Kuttner F, Schmid M (2007) Export of FT protein from phloem companion cells is sufficient for floral induction in Arabidopsis. Curr Biol 17: 1055–1060.
19. Lee H, Suh SS, Park E, Cho E, Ahn JH, et al. (2000) The AGAMOUS-LIKE 20 MADS domain protein integrates floral inductive pathways in Arabidopsis. Genes Dev 14: 2366–2376.
20. Samach A, Onouchi H, Gold SE, Ditta GS, Schwarz-Sommer Z, et al. (2000) Distinct roles of CONSTANS target genes in reproductive development of Arabidopsis. Science 288: 1613–1616.
21. Srikanth A, Schmid M (2011) Regulation of flowering time: all roads lead to Rome. Cellular and Molecular Life Sciences 68: 2013–2037.
22. Achard P, Baghour M, Chapple A, Hedden P, Van Der Straeten D, et al. (2007) The plant stress hormone ethylene controls floral transition via DELLA-dependent regulation of floral meristem-identity genes. Proc Natl Acad Sci U S A 104: 6484–6489.
23. Kotchoni SO, Larrimore KE, Mukherjee M, Kempinski CF, Barth C (2009) Alterations in the endogenous ascorbic acid content affect flowering time in Arabidopsis. Plant Physiol 149: 803–815.
24. Li J, Nagpal P, Vitart V, McMorris TC, Chory J (1996) A role for brassinosteroids in light-dependent development of Arabidopsis. Science 272: 398–401.
25. Melzer S, Lens F, Gennen J, Vanneste S, Rohde A, et al. (2008) Flowering-time genes modulate meristem determinacy and growth form in Arabidopsis thaliana. Nat Genet 40: 1489–1492.
26. Amasino RM, Michaels SD (2010) The timing of flowering. Plant Physiol 154: 516–520.
27. Kobayashi Y, Weigel D (2007) Move on up, it's time for change–mobile signals controlling photoperiod-dependent flowering. Genes Dev 21: 2371–2384.
28. Khanna R, Kronmiller B, Maszle DR, Coupland G, Holm M, et al. (2009) The Arabidopsis B-box zinc finger family. Plant Cell 21: 3416–3420.
29. Imaizumi T, Schultz TF, Harmon FG, Ho LA, Kay SA (2005) FKF1 F-box protein mediates cyclic degradation of a repressor of CONSTANS in Arabidopsis. Science 309: 293–297.
30. Sawa M, Nusinow DA, Kay SA, Imaizumi T (2007) FKF1 and GIGANTEA complex formation is required for day-length measurement in Arabidopsis. Science 318: 261–265.
31. Song YH, Smith RW, To BJ, Millar AJ, Imaizumi T (2012) FKF1 conveys timing information for CONSTANS stabilization in photoperiodic flowering. Science 336: 1045–1049.
32. Jiang L, Wang Y, Li QF, Bjorn LO, He JX, et al. (2012) Arabidopsis STO/BBX24 negatively regulates UV-B signaling by interacting with COP1 and repressing HY5 transcriptional activity. Cell Res 22: 1046–1057.
33. Lippuner V, Cyert MS, Gasser CS (1996) Two classes of plant cDNA clones differentially complement yeast calcineurin mutants and increase salt tolerance of wild-type yeast. J Biol Chem 271: 12859–12866.
34. Holm M, Hardtke CS, Gaudet R, Deng XW (2001) Identification of a structural motif that confers specific interaction with the WD40 repeat domain of Arabidopsis COP1. EMBO J 20: 118–127.
35. Indorf M, Cordero J, Neuhaus G, Rodriguez-Franco M (2007) Salt tolerance (STO), a stress-related protein, has a major role in light signalling. Plant J 51: 563–574.
36. Kumagai T, Ito S, Nakamichi N, Niwa Y, Murakami M, et al. (2008) The common function of a novel subfamily of B-Box zinc finger proteins with reference to circadian-associated events in Arabidopsis thaliana. Biosci Biotechnol Biochem 72: 1539–1549.
37. Kim TW, Wang ZY (2010) Brassinosteroid signal transduction from receptor kinases to transcription factors. Annu Rev Plant Biol 61: 681–704.
38. Luo XM, Lin WH, Zhu S, Zhu JY, Sun Y, et al. (2010) Integration of light- and brassinosteroid-signaling pathways by a GATA transcription factor in Arabidopsis. Dev Cell 19: 872–883.
39. Harmer SL (2009) The circadian system in higher plants. Annu Rev Plant Biol 60: 357–377.
40. Imaizumi T (2010) Arabidopsis circadian clock and photoperiodism: time to think about location. Current Opinion in Plant Biology 13: 83–89.
41. Fujiwara S, Oda A, Yoshida R, Niinuma K, Miyata K, et al. (2008) Circadian clock proteins LHY and CCA1 regulate SVP protein accumulation to control flowering in Arabidopsis. Plant Cell 20: 2960–2971.
42. Jang S, Marchal V, Panigrahi KC, Wenkel S, Soppe W, et al. (2008) Arabidopsis COP1 shapes the temporal pattern of CO accumulation conferring a photoperiodic flowering response. EMBO J 27: 1277–1288.
43. Clouse S (1997) Molecular genetic analysis of brassinosteroid action Plant Physiology 702–709.
44. Li J, Li Y, Chen S, An L (2010) Involvement of brassinosteroid signals in the floral-induction network of Arabidopsis. J Exp Bot 61: 4221–4230.
45. Domagalska MA, Schomburg FM, Amasino RM, Vierstra RD, Nagy F, et al. (2007) Attenuation of brassinosteroid signaling enhances FLC expression and delays flowering. Development 134: 2841–2850.
46. Chiang GC, Barua D, Kramer EM, Amasino RM, Donohue K (2009) Major flowering time gene, flowering locus C, regulates seed germination in Arabidopsis thaliana. Proc Natl Acad Sci U S A 106: 11661–11666.
47. Willmann MR, Poethig RS (2011) The effect of the floral repressor FLC on the timing and progression of vegetative phase change in Arabidopsis. Development 138: 677–685.
48. Griffiths S, Dunford RP, Coupland G, Laurie DA (2003) The evolution of CONSTANS-like gene families in barley, rice, and Arabidopsis. Plant Physiol 131: 1855–1867.
49. Almada R, Cabrera N, Casaretto JA, Ruiz-Lara S, Gonzalez Villanueva E (2009) VvCO and VvCOL1, two CONSTANS homologous genes, are regulated during flower induction and dormancy in grapevine buds. Plant Cell Rep 28: 1193–1203.
50. Putterill J, Robson F, Lee K, Simon R, Coupland G (1995) The CONSTANS gene of Arabidopsis promotes flowering and encodes a protein showing similarities to zinc finger transcription factors. Cell 80: 847–857.
51. Robson F, Costa MM, Hepworth SR, Vizir I, Pineiro M, et al. (2001) Functional importance of conserved domains in the flowering-time gene CONSTANS demonstrated by analysis of mutant alleles and transgenic plants. Plant J 28: 619–631.
52. Hassidim M, Harir Y, Yakir E, Kron I, Green RM (2009) Over-expression of CONSTANS-LIKE 5 can induce flowering in short-day grown Arabidopsis. Planta 230: 481–491.

53. Cheng XF, Wang ZY (2005) Overexpression of COL9, a CONSTANS-LIKE gene, delays flowering by reducing expression of CO and FT in Arabidopsis thaliana. Plant J 43: 758–768.

54. Kim SK, Park HY, Jang YH, Lee JH, Kim JK (2013) The sequence variation responsible for the functional difference between the CONSTANS protein, and the CONSTANS-like (COL) 1 and COL2 proteins, resides mostly in the region encoded by their first exons. Plant Sci 199–200: 71–78.

55. Lagercrantz U, Axelsson T (2000) Rapid evolution of the family of CONSTANS LIKE genes in plants. Mol Biol Evol 17: 1499–1507.

14

Gibberellin Overproduction Promotes Sucrose Synthase Expression and Secondary Cell Wall Deposition in Cotton Fibers

Wen-Qin Bai[9], Yue-Hua Xiao[9], Juan Zhao, Shui-Qing Song, Lin Hu, Jian-Yan Zeng, Xian-Bi Li, Lei Hou, Ming Luo, De-Mou Li, Yan Pei*

Biotechnology Research Center, Southwest University, Beibei, Chongqing, China

Abstract

Bioactive gibberellins (GAs) comprise an important class of natural plant growth regulators and play essential roles in cotton fiber development. To date, the molecular base of GAs' functions in fiber development is largely unclear. To address this question, the endogenous bioactive GA levels in cotton developing fibers were elevated by specifically up-regulating GA 20-oxidase and suppressing GA 2-oxidase via transgenic methods. Higher GA levels in transgenic cotton fibers significantly increased micronaire values, 1000-fiber weight, cell wall thickness and cellulose contents of mature fibers. Quantitative RT-PCR and biochemical analysis revealed that the transcription of sucrose synthase gene GhSusA1 and sucrose synthase activities were significantly enhanced in GA overproducing transgenic fibers, compared to the wild-type cotton. In addition, exogenous application of bioactive GA could promote GhSusA1 expression in cultured fibers, as well as in cotton hypocotyls. Our results suggested that bioactive GAs promoted secondary cell wall deposition in cotton fibers by enhancing sucrose synthase expression.

Editor: Baohong Zhang, East Carolina University, United States of America

Funding: This work was supported by the National Natural Science Foundation of China (31130039 to Y. P. and 31271769 to Y. H. X.) and Ph.D. Programs Foundation of Southwest University, China (Kb2009007 to W. Q. B.). The funders had no role in study design, data collection and analysis, decision to publish, or preparation of the manuscript.

Competing Interests: The authors have declared that no competing interests exist.

* E-mail: peiyan3@swu.edu.cn

[9] These authors contributed equally to this work.

Introduction

Cotton is the leading natural fiber for textile industry worldwide. Biologically, cotton fibers are extremely elongated single-celled trichomes originating from outermost layer of ovule epidermis [1–4]. The development of cotton fiber may be divided into 4 stages, i.e. initiation, elongation, secondary cell wall deposition and maturation. Secondary cell wall deposition starts at around 14–17 days post anthesis (dpa) and lasts for over 30d [1,3,5]. In this stage, cellulose is intensely deposited to form a thick secondary cell wall. At maturation, cotton fiber consists primarily of secondary cell wall and over 90% dry weight of fiber may exist as cellulose. Therefore, carbon partitioning to cellulose biosynthesis is a key determinant of fiber weight and qualities, such as fiber strength and fineness [1,3,5–8]. Many efforts have been taken to reveal the role of genes involved in the regulation of secondary cell wall deposition and to manipulate them by genetically modification for improvement of cotton yield and quality [1,6,7,9,10]. Recently, Jiang and coworkers showed that over-expressing a cotton sucrose synthase gene, GhsusA1, enhanced thickening of secondary cell wall and fiber qualities, suggesting an important role of sucrose synthase in controlling carbon partitioning to cellulose biosynthesis in cotton fibers [6].

Gibberellins (GA) are a class of important plant hormones involved in many physiological and developmental processes,

including seed germination, cell elongation, photomorphogenesis, flowering and seed development [11]. In the last two decades, the molecular base of GA biosynthesis pathway and its regulation have been largely clarified in model plants [12–14]. Endogenous bioactive GA contents are regulated mainly through three 2-oxoglutarate-dependent dioxygenases, i.e. GA 20-oxidase (GA20ox), GA 3-oxidase (GA3ox) and GA 2-oxidase (GA2ox) [13,14]. GA20ox and GA3ox catalyze the last two steps to synthesize bioactive GAs, while GA2ox convert bioactive GAs and their precursors to inactive 2-hydroxylated forms. A wealth of evidence demonstrated that both up-regulating GA20ox and suppressing GA2ox could significantly increase endogenous bioactive GA levels and lead to GA overproduction phenotypes in plants [15–20].

Physiological and molecular studies have revealed that GAs played important roles in fiber development. Exogenous application of GAs in vitro and in planta promoted fiber initiation and elongation [21,22]. Recently, we showed that over-expression of GhGA20ox1 in cotton significantly increased bioactive GA level and promote fiber initiation and elongation at early stage [17]. However, global up-regulation of GAs leaded to overgrowth of plant and somewhat negatively affected fiber development, especially at the late developmental stage. Instead, it is reasonable to elucidate GA roles in fiber development by tissue specific regulation of GA levels in developing fibers. To this end, we elevated the endogenous active GA levels in cotton fibers by tissue-specific up-regulation of GA20ox gene and down-regulation of GA2ox gene. We found that enhancement of GA production in fibers promoted sucrose synthase expression and secondary cell wall deposition. Our results implied that GAs might enhance carbon partitioning to cellulose and secondary cell wall synthesis via up-regulating sucrose synthase expression in cotton fibers.

Materials and Methods

Plant material and growth condition

Upland cotton (*Gossypium hirsutum* L. *cv.* Jimian No. 14) was used for cotton transformation and GA treatment. Cotton seedlings were grown in a greenhouse with a 16h/8h (light/dark) schedule and temperature kept at 26–30°C. Fibers and ovules were collected from field-grown cotton plants at growing season in Chongqing, China.

For GA_3 treatment of hypocotyls, 4-day-old seedlings were immersed in distilled water (pH6.0) or GA_3 solutions of various concentrations (0.05 mM, 0.1 mM and 0.5 mM, pH6.0) for 48 h. Then the hypocotyls were measured and collected for expression analyses and cellulose determination.

For GA treatment of *in vitro* fibers, cotton ovules were cultured as described by Beasley [23]. Cotton bolls were harvested at 0-dpa and surface-sterilized in 75% (v/v) ethanol for 1 min, rinsed in sterile water, then soaked in 0.1% w/v $HgCl_2$ solution for 12 min for sterilization, followed by rinsing with sterile water for six times. Ovules were separated, floated on BT media containing 5 μM IAA and GA_3 of various concentrations (0.5, 2.5, 10 and 25 μM), and then incubated in darkness at 32°C for 20d. Fibers were striped from ovules and used for RNA extraction.

Vector construction and plant transformation

To construct specific expression vector of cotton GA20ox (*SCFP::GhGA20ox1*), the CaMV 35S promoter in an over-expression vector (*35S::GhGA20ox1*) [17] was replaced with SCFP promoter[24]. The BAN promoter was amplified from Arabidopsis with a forward primer (5′-TCTAGATAACAGAACCTTAC TGTAACACTATT-3′) and a reverse primer (5′-ACTAGT-GATTGTACTTTTGAAATTACAGAG AT-3′) and cloned into TA cloning vector pMD19-T (TaKaRa, Dalian, China). After sequencing, the BAN promoter was digested from the cloning vector by *Hind*III and *Bam*HI and inserted into a basic expression vector p5 vector [25] digested with the same enzymes to generate the vector p5-BAN. An intron-containing hairpin RNA construct of cotton GA20ox gene (*GhGA2ox2RNAi*) was amplified from cotton genomic DNA as previously described [26]. The 25-μl PCR mixture included 100 ng cotton genomic DNA, 10×Ex Taq buffer (TaKaRa), 200 μM each dNTPs, 2 mM $MgCl_2$, 400 nM flanking primer (5′-GTATTGGTCTGGTGGGACTG-3′), 40 nM bridge primer (5′-CAAGTATCTCACATGCC AAGACCC-GAATTCTCCTTG-3′), 1.5U Ex Taq DNA polymerase. The PCR thermo cycling parameters were as follows: 94°C for 5 min, followed by 35 cycles of 94°C for 30 s, 56°C for 30 s and 72°C for 30 s, and a final extension of 10 min at 72°C. The *GhGA2ox2RNAi* fragment was cloned and sequenced, then inserted into p5-BAN using *Bam*HI and *Kpn*I. Transgenic plants were generated using *Agrobacterium*-mediated transformation as described [25]. Based on expression analysis of target genes in transgenic cottons, two homologous transgenic lines were obtained by self-crossing, and their performances were documented at T3 and T4 generations in comparison with untransformed acceptor line (Jimian No. 14) grown in parallel in the field.

RNA extraction and qRT-PCR analyses

Total RNA was extracted from roots, hypocotyls, leaves, petals, anthers, ovules and fibers using a rapid plant RNA extraction kit (Aidlab, Beijing, China). The single-stranded cDNAs were synthesized from total RNA using a cDNA synthesis kit (TaKaRa, Dalian, China). The gene-specific primers used for real-time PCR amplification were list in table S1. Cotton *histon3* gene (AF024716) was amplified as internal standard [27]. Real-time PCRs were performed on a CFX96 real-time PCR detection system with SYBR Green supermix (Bio-Rad, CA, USA). The thermocycling parameters were as follows: 95°C for 2 min, followed by 40 cycles of 95°C for 30 s, 56°C for 30 s and 72°C for 30 s, followed by a standard melting curve to monitor the specificity of PCR products. The reactions were duplicated for 3 times and data were analyzed using the software Bio-Rad CFX Manager 2.0 provided by the manufacturer.

Determination of endogenous GA contents

Cotton fibers (200 mg FW) were ground to fine powder in liquid N_2, extracted overnight in 5 ml 80% methanol at −20°C and deuterium-labeled [17, 17-$^2H_2$] GA_1 and [17, 17-$^2H_2$] GA_4 (each 10 ng) from Prof. L. Mander (Australian National University) were added as internal standards. After centrifugation, supernatants were collected, dried in a rotavapor (BUCHI, Switzerland) at 40°C, and re-suspended in 3 ml 10% methanol. The extracts were applied on Oasis HLB extraction cartridges (60 mg, Waters) pretreated with 3 ml methanol and 3 ml water. After washing with 1 ml 10% methanol, GAs were eluted with 1 ml 90% methanol. The eluates were evaporated to dry, dissolved in 100 μL 10% methanol, and subjected to LC-MS assay. The procedures for LC-MS quantification of GAs were described previously [17].

Measurement of fiber quality

Mature fibers were harvested from the field-grown cotton in the same period (Aug. 20 to Sep. 10). After ginning, fibers were mixed well and 6 repeats of 10 g fibers were randomly sampled for each material. Fiber sample were tested independently for fiber quality traits (fiber length, fiber strength, micronarie value) using at a HVI system (HFT 9000, Uster Technologies, Swiss) in Cotton Fiber Quality Inspection and Testing Center, Ministry of Agriculture of China (Anyang, Henan, China).

Microscopic measurement of fiber cell wall thickness

Statistical analysis of cell wall thickness was performed according to Wang *et al.* [7]. After fixing in FAA (37% formaldehyde: acetic acid: ethanol: water, 10:5:50:35) at 25°C for 12 h, mature cotton fibers were dehydrated gradually in alcohol and tert-butyl alcohol series,and then infiltrated in tert-butyl alcohol/paraffin at 65°C and embedded in paraffin. The samples were sliced into 7-μm sections. The slices were mounted, stained with the Fast Green dye and photographed by a BX41TF light microscope (Olympus, Japan). Image-pro Plus program (Olympus) was employed to measure the thickness of cell wall and 1000 sections were measured for each sample.

Determination of fiber weight

Fibers on seeds were combed straight and striped manually. Approximately 1.5 mg fibers were randomly bundled and weighed precisely (W1). The fiber number of each fiber bundle (N) was counted as described[28]. The weight of 1000 fibers (W2) was calculated from the following equation: W2 = 1000W1/N. For each material, the average 1000-fiber weight was calculated on the basis of 60 fiber bundles from different seeds.

Sucrose synthase activity assays

The sucrose synthase was extracted according to Jiang et al. [6]. Fresh fibers (around 0.5 g) were ground to fine powder in liquid N_2. The grinding continued for 5 min in cold extraction buffer (25 mM Hepes–KOH (pH 7.3), 5 mM EDTA, 1 mM DTT, 0.1% soluble PVP, 20 mM β-mercaptoethanol, 1 mM PMSF and 0.01 mM leupeptin). The homogenate was separated by centrifugation (10000 g, 5 min, 4°C) and the supernatant was used as the crude extracts for assays. Protein concentrations were determined via Bradford method [29]and sucrose sythase activities were assayed as previously described [30,31].

Analyses of soluble sugar contents

Fresh fibers (around 50 mg) were separated from developing bolls and ground to fine powder in liquid N2. The powder was extracted in 2 ml 80% (v/v) ethanol at 80°C?for 15 min. After centrifugation (3000 g, 10 min), supernatants were collected. The pellets were further extracted twice, and supernatants were combined and used for soluble sugar assays. The contents of glucose, fructose and sucrose were measured at 340 nm with a Synergy HT microplate reader (BioTek, Vermont, WS) as described [32].

Determination of cellulose content

Cellulose contents were determined according to Wang *et al.* [7]. Around 0.1 g fiber samples were extracted in 10 ml boiling acetic/nitric reagent (80% acetic/nitric, 10:1) for 1 h, then rinsed three times with distilled water and once with ethanol. Residuals were dried at 105°C for 2 h. The weight ratio of residual to initial samples was regarded as cellulose content. Hypocotyls were excised from seeding (10 hypocotyls per sample) and ground to fine powder in liquid N_2. Samples were dried at 105°C for 2 h and extracted in 10 ml boiling acetic/nitric reagent (80% acetic/nitric, 10:1) for 1 h.Determination of cellulose content of per hypocotyls was carried out as described [33].

Statistical analyses

Performances of transgenic materials were compared to wild-type control and statistical significance of divergence between averages was determined by t test. All statistical calculations were performed using Microsoft Excel.

Results

Enhancement of GA production in cotton fiber

We used two strategies to tissue-specially enhance GA production in fibers, i.e. to promote GA biosynthesis by up-regulation GA 20-oxidase (GA20ox) and suppressing GA deactivation by down-regulation of GA2-oxidase (GA2ox). To this end, we used a fiber-specific promoter (SCFP) [24] and a seed coat- and fiber-specific promoter BAN [28,34] to direct the expression of GA20ox (GhGA20ox1) [17], and GA2ox, respectively. Among SCFP::GhGA20ox1 transgenic cottons (SG20), SG20-1 showed dramatically increase in GhGA20ox1 expression level in developing fibers (Figure 1A and 1B).

We compared the expression pattern of six cotton GA 2-oxidase genes (GhGA2ox1-6, Figure S1~3). Among them, GhGA2ox2 showed predominant expression in fibers. Thus we selected GhGA2ox2 as RNAi target to suppress GA deactivation in fibers, and generated GhGA2oxRNAi transgenic cottons. Real-time RT-PCR revealed that the expression level of GhGA2ox2 was reduced in BAN::GhGA2oxRNAi (BG2i) transgenic cottons (Figure 1C), in which transformant BG2i-2 showed most significant suppression of the target gene in fibers (Figure 1C and 1D).

To detect the effect of GhGA20ox1 up-regulation and GhGA2ox2 down-regulation on GA homeostasis in fibers, we determined the contents of endogenous bioactive GAs (GA_1 and GA_4) in 8- and-20 dpa fibers of SG20-1 and BG2i-2 by LC-MS (Figure 1E and 1F). Compared to the wild-type control, GA_4 contents in the 8- and 20-dpa fibers of SG20-1 fibers increased 83.1% and 178.6%, respectively, while GA_1 content was moderately increased (24.0%) in the 20-dpa fibers(Figure 1E and 1F). In BG2i-2 fibers, GA_1 level was 21.6% and 65.9% higher than the control at 8- and 20-dpa respectively, whereas GA_4 contents remained almost unchanged compared to the control (Figure 1E and 1F).

Effects of elevated GA levels on secondary cell wall thickening of cotton fiber

To clarify the effect of elevated GA levels on fiber development and fiber quality, we compared the agronomy performances of transgenic lines SG20-1 and BG2i-2 with the wild-type control in consecutive two-year field trails. No significant change in plant growth, yield traits and fiber length and strength (Figure 2A and S4; Table S2 and S3) was found between the transgenic lines and the wild type, except micronaire value. The micronaire values of SG20-1 and BG2i-2 fibers were significantly higher than that of the wild type (Figure 2C). Micronaire value is a composite measure of fiber maturity and fineness. To clarify whether the fiber fineness was increased in transgenic cotton, we measured the weight per 1000 fibers. The weights of SG20-1 and BG2i-2 mature fibers significantly enhanced in comparison with the control (2.0% and 5.7%, respectively; Figure 2E). Microscopic observation further confirmed that the cell walls of SG20-1 and BG2i-2 mature fibers were thicker (5.4% and 6.6%, respectively) than the wild-type control (Figure 2B and D). Considered that most of cell walls of mature cotton fibers consisted of secondary cell wall [2], it was reasonable that the fineness increase in SG20-1 and BG2i-2 fibers might be mainly attributed to promotion of secondary cell wall deposition. To prove this hypothesis, we determined the cellulose contents in fibers, and found that the contents of SG20-1 and BG2i-2 fibers were significantly higher than the control (Figure 2F). Taken together, these results suggested that elevating bioactive GA levels in cotton fibers promoted secondary cell wall deposition.

Sucrose synthase expression in response to elevated GA levels in fibers and hypocotyls

To reveal the possible mechanism for GAs to control secondary cell wall deposition, we investigated transcript levels of six genes related to secondary cell wall biosynthesis, including GhCesA1, GhCesA2, GhRac13, GhSusA1, GhADF1 and GhCTL1 [6,7,35], in 20-dpa fibers. Only sucrose synthase gene (GhSusA1) showed significant increase in transgenic cottons (Figure 3A). The relative transcript levels of GhSusA1 in SG20-1 and BG2i-2 fibers were 53% and 50% higher than the control, respectively. Biochemical analysis demonstrated that the sucrose synthase activities in 20-dpa fibers of SG20-1 and BG2i-2 increased 8.3% and 10.7%, respectively, compared to the control (Figure 3B). Meanwhile, the concentration of fructose, a direct product of sucrose synthase, was significantly higher in SG20-1 and BG2i-2 fibers (Figure 3C). Furthermore, we found GhsusA1 transcript in cultured fibers was increased with GA_3 concentrations in ovule culture media (Figure 3D). The result of fiber culture, along with the observations on the mature fibers, implied that GA may promote cellulose biosynthesis and secondary cell wall deposition through up-regulation of the expression of sucrose synthase.

Like cotton fibers, hypocotyls that undergo rapid cell elongation require high-speed formation of cellulose. To investigate if same

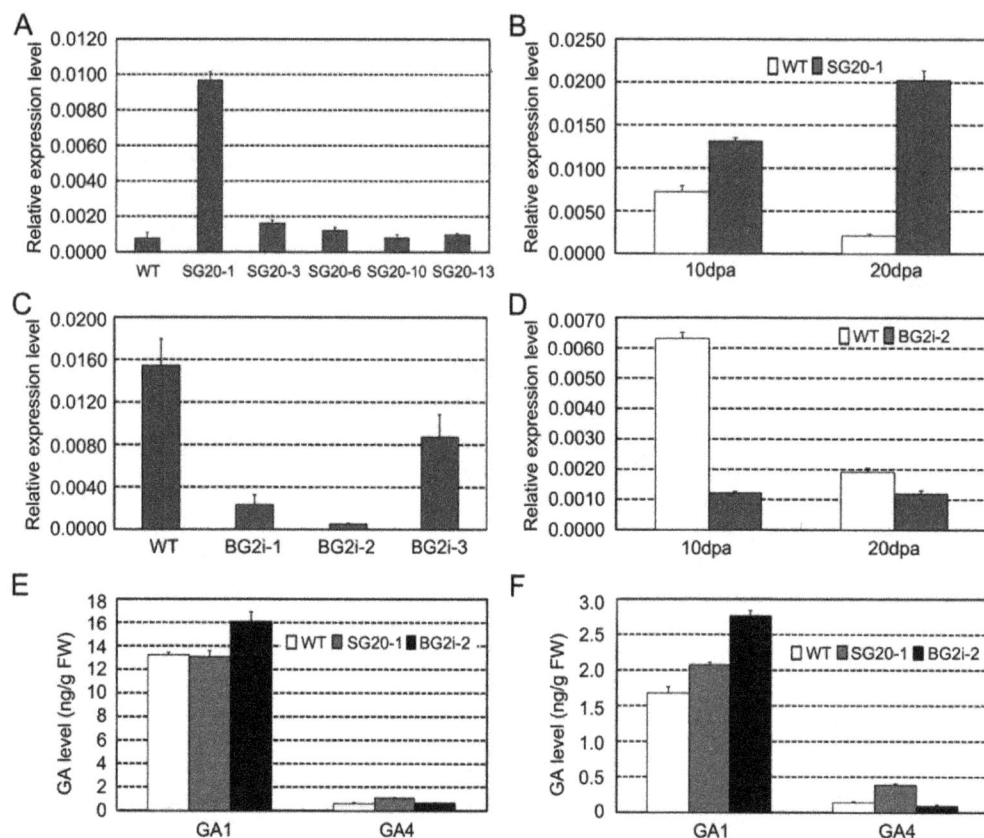

Figure 1. Up-regulation of *GhGA20ox1* or suppression of *GhGA2ox* in cotton fibers. (A) Quantitative RT-PCR analysis of *GhGA20ox1* in 5-dpa fibers of T$_0$*SCPF::GhGA20ox1* lines (SG20) and the wild-type cotton (WT). (B) Quantitative RT-PCR analysis of *GhGA20ox1* in 10-dpa and 20-dpa fibers of SG20-1 (T$_3$generation) and WT. (C) Quantitative RT-PCR analysis of *GhGA2ox2* in 5-dpa fibers of T$_0$*BAN::GhGA2oxRNAi* lines (BG2i) and WT. (D) Quantitative RT-PCR analysis of *GhGA2ox2* in 10-dpa and 20-dpa fibers of BG2i-2 (T$_3$generation) and WT. (E) Endogenous GA$_1$ and GA$_4$ content in 8-dpa fibers of SG20-1, BG2i-2 and WT. (F) Endogenous GA$_1$ and GA$_4$ contents in 20-dpa fibers of SG20-1, BG2i-2 and WT.

response takes place in hypocotyls, we detected the expression of sucrose synthase in the GA-treated hypocotyls. *GhSusA1* transcript levels in hypocotyls were significantly enhanced along with increase of GA$_3$ (Figure 4A). Meanwhile, elongation and cellulose deposition were accelerated (Figure 4B–D). These results further supported that GA promoted cellulose biosynthesis and secondary cell wall deposition through up-regulation of sucrose expression.

Discussion

Genetically manipulating enzymes involved in GA biosynthesis and catabolism provided an effective strategy to regulate GA homeostasis in plants [15–20]. Our work demonstrated that in addition to up-regulation of *GA20ox*, down-regulation of *GA2ox* is another effective way to increase the active GA levels in plant tissues (Figure 1E and F). However, there may be a subtle difference between the two strategies. Up-regulation of GA20ox not only increased the GA level, but also changed the composition of some active GAs by promoting the non-13-hydroxylation pathway for producing 13-H GA$_4$ rather than 13-OH GA$_1$ (Figure 1E and F) [17,19,20]. On the contrary, in our study, down-regulation of GA2ox elevated the GA$_1$ instead of GA$_4$ (Figure 1E and F). The exact physiological effects of different bioactive GAs on cotton fiber development are still to be elucidated.

Cotton fibers that undergo rapid elongation and intense cellulose synthesis represent a strong sink that competes with the developing embryos and endosperms in a single ovule

[1,3,5,6,8,30,36,37]. It was revealed that sucrose synthase played important roles in carbon partition during fiber development. Suppression of sucrose synthase gene (*SS3*) inhibited fiber initiation and elongation [30], while over-expression of a potato sucrose synthase gene in cotton enhanced leaf expansion, early seed development and fiber elongation [5]. Recently, Jiang and coworkers cloned a novel cotton sucrose synthase gene (*GhSusA1*), which might be a key regulator of sink strength in cotton. Over-expression of *GhSusA1* significantly enhanced cell wall thickening during secondary wall formation stage, and improved fiber length and strength [6]. In this study, we revealed that enhancement of GA level in cotton fibers led to an increase of sucrose expression (*GhSusA1*) gene expression, and promoted cellulose biosynthesis and secondary cell wall deposition (Figure 2 and 3). Moreover, GA-induced *GhSusA1* up-regulation also found in cultured fibers and hypocotyls (Figure 4 and S5). GA has long been considered as an important regulator of sink strength in plants, but the molecular basis of how GA enhances the partitioning of carbon assimilates to sink tissues is still unknown [38,39]. Our data offer an experimental evidence for the relationship between GAs and sucrose synthase.

Previous studies showed that exogenous application of GAs or constitutively increased endogenous GA levels promoted fiber initiation and elongation [17,21]. However, in this study we did not find significant improvement in fiber length of the transgenic GA-elevated fibers (TableS2). A possible explanation for this phenomenon is that the enhanced secondary cell wall deposition may fix the morphology of the fiber and, in turn, limit further

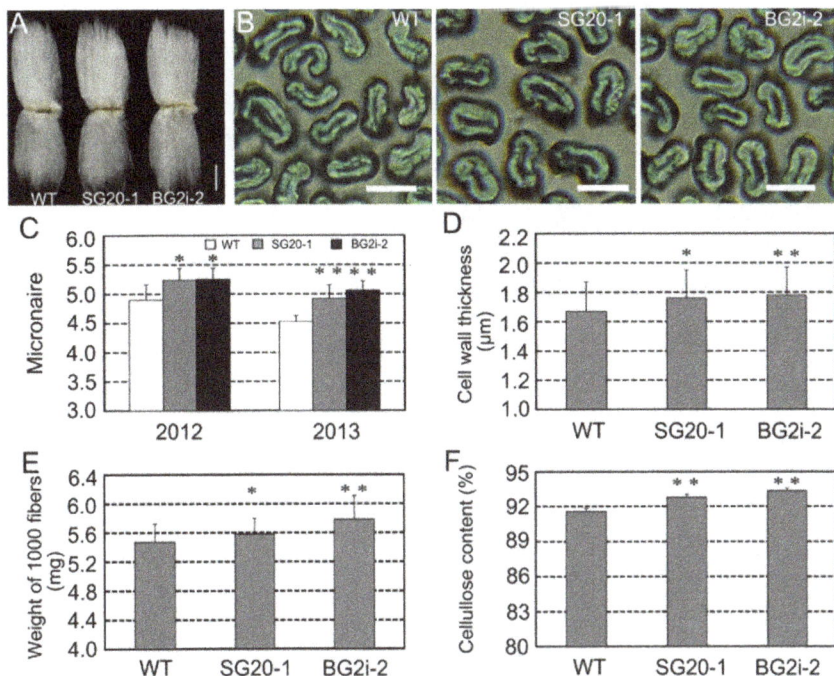

Figure 2. Effects of GA overproduction on secondary cell wall deposition in cotton fibers. (A) Cotton fibers on seeds; bar = 1 cm. (B) Cross-section of mature cotton fibers; Bar = 25 μm. (C) Micronaire values of cotton fibers in 2012 and 2013, n = 6. (D) Cell wall thickness of mature cotton fibers. (E) Weight of 1000 cotton fibers, n = 60. (F)Cellulose contents of mature cotton fibers, n = 6Asterisks (∗) and double asterisks (∗∗) represent significant differences(*t* test) at p = 0.05 and p = 0.01 compared with the wild type, respectively.

elongation of the GA-enhanced fibers at the late stage of fiber elongation. Nevertheless, the finding that GA related regulation of sucrose synthase gives useful information to reveal the mechanism of cotton fiber development and to improve fiber yield and quality for cotton breeders.

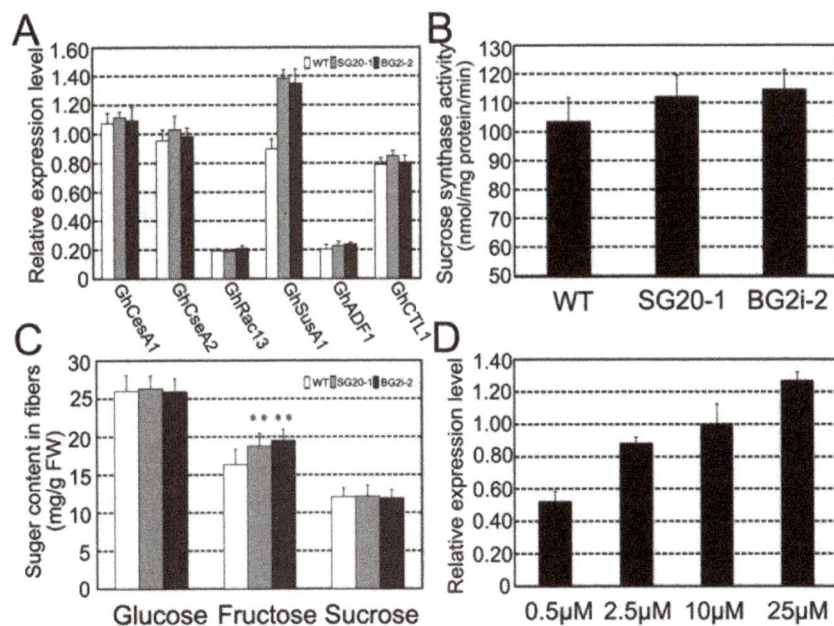

Figure 3. Sucrose synthase expression in GA overproduction cotton fibers. (A) Quantitative RT-PCR expression analyses of secondary cell wall related genes in 20-dpa fibers. *GhCesA1*, U58283; *GhCesA2*, U58284; *GhRac13*, S79308; *GhSusA1*, HQ702817; *GhADF1*, DQ088156 and*GhCTL1*, AY291285. (B) Sucrose synthase activity in 20-dpa cotton fibers. (C) Sugar contents in 20-dpa cotton fibers, n = 3. Double asterisks (∗∗) represent significant differences (*t* test) at p = 0.01 compared with the wild type cotton. (D) *GhSusA1* expression levels of *in vitro* cultured fibers treated with GA₃ of various concentrations.

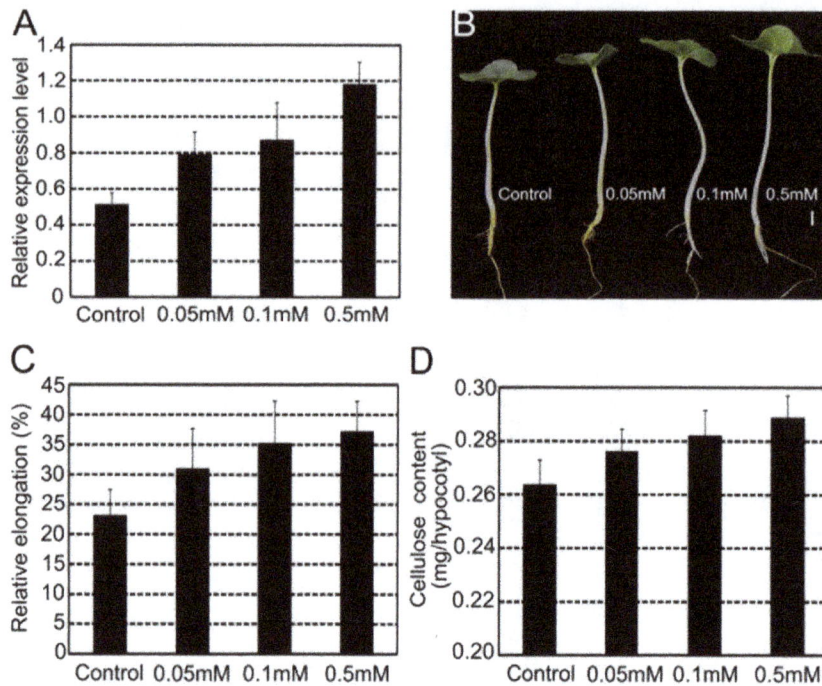

Figure 4. Sucrose synthase expression and cellulose biosynthesis in GA-treated hypocotyls. Seedlings wereimmersed in distilled water (pH6.0) or GA_3 solutions (0.05 mM, 0.1 mM and 0.5 mM, pH6.0).(A) Quantitative RT-PCR analysis of *GhSusA1* transcript in hypocotyls. (B) Effects of GA_3 treatment on cotton seedling growth. (C) Relative elongation of hypocotyls. (D) Cellulose amount in GA-treated hypocotyls.

Supporting Information

Figure S1 Alignment of cotton GA2ox proteins with homologous proteins. The conserved amino acids are highlighted on black background, and similar amino acids are shown on gray background. The symbols (#) indicate Fe-binding sites, and the asterisks(*) indicateputative 2-oxoglutarate-interaction sites. GhGA2ox1-6, Cotton GA 2-oxidases (HQ891930-HQ891935, respectively); SoGA2ox1-3, Spinach GA 2-oxidases (AAN87571, AAN87572 and AAX14674, respectively); At-GA2ox1, 4 and 7, Arabidopsis GA 2-oxidases (CAB41007, AAG51528 and AAG50945, respectively).So, *Spinaciaolracea*; At, *Arabidopsis thaliana*.

Figure S2 Phylogenetic relationship of GhGA2ox with other GA2ox, GA20ox and GA3ox. GenBank accession nos. are as follows: GhGA2ox1-6, HQ891930-HQ891935, respectively; SoGA2ox1-3, AAN87571, AAN87572 and AAX14674, respectively; AtGA2ox1-4 and 6-8, CAB41007, CAB41008, CAB41009, AAG51528, AAG00891, AAG50945 and CAB79120, respectively; CmGA2ox, CAC83090; OsGA2ox3 and 5-9, AK101713, AK106859, AK107142, AK108802, AK101758 and AK059045, respectively; GhGA20ox1-3, AY603789, FJ623273 and FJ623274, respectively. So, *Spinaciaolracea*; At, *Arabidopsis thaliana*; Cm, *Cucurbita maxima*; Os, *Oryza sativa*.

Figure S3 Expression pattern of *GhGA2ox* genes in cotton tissues. The total RNA were prepared from different organs and tissues, including roots(Ro), hypocotyls (Hy), leaves (Le), petals (Pe), anthers (An), 0-dpa ovules (Ov), 6-dpa fibers (Fi).

FigureS4 Field-grown plants of SG20-1, BG2i-2, *GhGA20ox1*-overexpressing (OG20) and wild-type (WT) cottons. All the materials were transplanted to the field in parallel. The representative plants were photographed at 90d post germination. Bar = 30 cm.

Figure S5 Quantitative RT-PCR analysis of fiber elongation related genes in 8-dpa fibers. *GhACT1*, AF305723; *GhEXP1*, AF512539; *GhFLA1*, EF672627; *GhPEL*, DQ073046; *GhVIN1*, FL915120.

Table S1 Primers used in Real-time PCR analyses.

Table S2 Fiber length and strength of mature fibers in two-year successive field trials. Asterisk (*) and double asterisks (**) represent significant differences (*t* test, n = 6) at p = 0.05 and p = 0.01 compared with the wild type, respectively.

Table S3 Plant height, Boll weight, seed index and lint index of transgenic and wild-type cotton in 2013. Seed index, weight of 100 delinted seeds. Fiber index, weight of lints from 100 seeds. Data are shown as average ± SD (n = 10).

Author Contributions

Conceived and designed the experiments: YP YHX WQB. Performed the experiments: WQB YHX JZ SQS LH JYZ. Analyzed the data: YP WQB YHX XBL LH ML DML. Wrote the paper: YP YHX WQB.

References

1. Haigler CH, Zhang D, Wilkerson CG (2005) Biotechnological improvement of cotton fibre maturity. Physiol Plantarum 124: 285–294.
2. Kim HJ, Triplett BA (2001) Cotton fiber growth in planta and in vitro. Models for plant cell elongation and eell wall biogenesis. Plant Physiol 127: 1361–1366.
3. Ruan Y (2007) Rapid cell expansion and cellulose synthesis regulated by plasmodesmata and sugar: insights from the single-celled cotton fibre. Funct Plant Biol 34: 1–10.
4. Qin YM, Zhu YX (2011) How cotton fibers elongate: a tale of linear cell-growth mode. Curr Opin Plant Biol 14: 106–111.
5. Xu SM, Brill E, Llewellyn DJ, Furbank RT, Ruan YL (2012) Overexpression of a potato sucrose synthase gene in cotton accelerates leaf expansion, reduces seed abortion, and enhances fiber production. Mol Plant 5: 430–441.
6. Jiang Y, Guo W, Zhu H, Ruan YL, Zhang T (2012) Overexpression of GhSusA1 increases plant biomass and improves cotton fiber yield and quality. Plant Biotechnol J 10: 301–312.
7. Wang H-Y, Wang J, Gao P, Jiao G-L, Zhao P-M, et al. (2009) Down-regulation of *GhADF1* gene expression affects cotton fibre properties. Plant Biotechnol J 7: 13–23.
8. Haigler CH, Ivanova-Datcheva M, Hogan PS, Salnikov VV, Hwang S, et al. (2001) Carbon partitioning to cellulose synthesis. Plant Mol Biol 47: 29–51.
9. Kurek I, Kawagoe Y, Jacob-Wilk D, Doblin M, Delmer D (2002) Dimerization of cotton fiber cellulose synthase catalytic subunits occurs via oxidation of the zinc-binding domains. P NATL ACAD SCI USA 99: 11109–11114.
10. Potikha TS, Collins CC, Johnson DI, Delmer DP, Levine A (1999) The involvement of hydrogen peroxide in the differentiation of secondary walls in cotton fibers. Plant Physiol 119: 849–858.
11. Olszewski N, Sun T-p, Gubler F (2002) Gibberellin signaling: Biosynthesis, catabolism, and response pathways. Plant Cell 14: S61–S80.
12. Magome H, Nomura T, Hanada A, Takeda-Kamiya N, Ohnishi T, et al. (2013) CYP714B1 and CYP714B2 encode gibberellin 13-oxidases that reduce gibberellin activity in rice. P NATL ACAD SCI USA 110: 1947–1952.
13. Hedden P, Thomas SG (2012) Gibberellin biosynthesis and its regulation. Biochem J 444: 11–25.
14. Yamaguchi S (2008) Gibberellin metabolism and its regulation. Annu Rev Plant Biol 59: 225–251.
15. Bhattacharya A, Ward DA, Hedden P, Phillips AL, Power JB, et al. (2012) Engineering gibberellin metabolism in Solanum nigrum L. by ectopic expression of gibberellin oxidase genes. Plant Cell Rep 31: 945–953.
16. Gou J, Ma C, Kadmiel M, Gai Y, Strauss S, et al. (2011) Tissue-specific expression of Populus C19 GA 2-oxidases differentially regulate above- and below-ground biomass growth through control of bioactive GA concentrations. New Phytol 192: 626–639.
17. Xiao Y-H, Li D-M, Yin M-H, Li X-B, Zhang M, et al. (2010) Gibberellin 20-oxidase promotes initiation and elongation of cotton fibers by regulating gibberellin synthesis. J Plant Physiol 167: 829–837.
18. Dayan J, Schwarzkopf M, Avni A, Aloni R (2010) Enhancing plant growth and fiber production by silencing GA 2-oxidase. Plant Biotechnol J 8: 425–435.
19. Vidal AM, Gisbert C, Talón M, Primo-Millo E, López-Díaz I, et al. (2001) The ectopic overexpression of a citrus gibberellin 20-oxidase enhances the non-13-hydroxylation pathway of gibberellin biosynthesis and induces an extremely elongated phenotype in tobacco. Physiol Plantarum 112:251–260. 112: 251–260.
20. Eriksson ME, Israelsson M, Olsson O, Moritz T (2000) Increased gibberellin biosynthesis in transgenic trees promotes growth, biomass production and xylem fiber length. Nat Biotechnol 18: 784–788.
21. Seagull RW, Giavalis S (2004) Pre- and post-anthesis application of exogenous hormones alters fiber production in Gossypium hirsutum L. cultivar Maxxa GTO. Journal of Cotton Science 8: 105–111.
22. Basra A, Saha S (1999) Growth regulation of cotton fibers. In: ASB, editor. Cotton fibers: developmental diology, quality improvement, and textile processing, New York: Food Products Press (an imprinting of Haworth Press). pp. 47–66.
23. Beasley CA (1973) Hormonal regulation of growth in unfertilized cotton ovules. Science 179: 1003–1005.
24. Hou L, Liu H, Li J, Yang X, Xiao Y, et al. (2008) SCFP, a novel fiber-specific promoter in cotton. Chinese Sci Bull 53: 2639–2645.
25. Luo M, Xiao Y, Li X, Lu X, Deng W, et al. (2007) GhDET2, a steroid 5α-reductase, plays an important role in cotton fiber cell initiation and elongation. Plant J 51: 419–430.
26. Xiao YH YM, Hou L, Pei Y (2006) Direct amplification of intron-containing hairpin RNA construct from genomic DNA. Biotechniques 41: 548–552.
27. Zhu Y-Q, Xu K-X, Luo B, Wang J-W, Chen X-Y (2003) An ATP-binding cassette transporter GhWBC1 from elongating cotton fibers. Plant Physiol 133: 580–588.
28. Zhang M, Zheng X, Song S, Zeng Q, Hou L, et al. (2011) Spatiotemporal manipulation of auxin biosynthesis in cotton ovule epidermal cells enhances fiber yield and quality. Nat Biotechnol 29: 453–458.
29. Bradford MM (1976) A rapid and sensitive method for the quantitation of microgram quantities of protein utilizing the principle of protein-dye binding. Anal Biochem 72: 248–254.
30. Ruan Y-L, Llewellyn DJ, Furbank RT (2003) Suppression of sucrose synthase gene expression represses cotton fiber cell initiation, elongation, and seed development. Plant Cell 15: 952–964.
31. Chourey P (1981) Genetic control of sucrose synthetase in maize endosperm. Mol Gen Genet 184: 372–376.
32. Zhao D, MacKown CT, Starks PJ, Kindiger BK (2010) Rapid analysis of nonstructural carbohydrate components in grass forage using microplate enzymatic assays. Crop Sci 50: 1537–1545.
33. Updegraff DM (1969) Semimicro determination of cellulose in biological materials. Anal Biochem 32: 420–424.
34. Debeaujon I, Nesi N, Perez P, Devic M, Grandjean O, et al. (2003) Proanthocyanidin-accumulating cells in Arabidopsis testa: Regulation of differentiation and role in seed development. Plant Cell 15: 2514–2531.
35. Singh B, Cheek HD, Haigler CH (2009) A synthetic auxin (NAA) suppresses secondary wall cellulose synthesis and enhances elongation in cultured cotton fiber. Plant Cell Rep 28: 1023–1032.
36. Ruan Y-L, Chourey PS (1998) A fiberless seed mutation in cotton is associated with lack of fiber cell initiation in ovule epidermis and alterations in sucrose synthase expression and carbon partitioning in developing seeds. Plant Physiol 118: 399–406.
37. Nolte KD, Hendrix DL, Radin JW, Koch KE (1995) Sucrose synthase localization during initiation of seed development and trichome differentiation in cotton ovules. Plant Physiol 109: 1285–1293.
38. Nadeau CD, Ozga JA, Kurepin LV, Jin A, Pharis RP, et al. (2011) Tissue-specific regulation of gibberellin biosynthesis in developing pea seeds. Plant Physiol 156: 897–912.
39. Iqbal N, Nazar R, Khan MIR, Masood A, Khan NA (2011) Role of gibberellins in regulation of source–sink relations under optimal and limiting environmental conditions. Current Science 100: 998–1007.

15

High Genetic and Epigenetic Stability in *Coffea arabica* Plants Derived from Embryogenic Suspensions and Secondary Embryogenesis as Revealed by AFLP, MSAP and the Phenotypic Variation Rate

Roberto Bobadilla Landey[1], Alberto Cenci[2], Frédéric Georget[1], Benoît Bertrand[1], Gloria Camayo[3], Eveline Dechamp[1], Juan Carlos Herrera[3], Sylvain Santoni[4], Philippe Lashermes[2], June Simpson[5], Hervé Etienne[1]*

1 Unité Mixte de Recherche Résistance des Plantes aux Bioagresseurs, Centre de Coopération Internationale en Recherche Agronomique pour le Développement, Montpellier, France, 2 Unité Mixte de Recherche Résistance des Plantes aux Bioagresseurs, Institut de Recherche pour le Développement, Montpellier, France, 3 Centro Nacional de Investigaciones de Café, Manizales, Colombia, 4 Unité Mixte de Recherche Amélioration Génétique et Adaptation des Plantes Tropicales et Méditerranéennes, Institut National de la Recherche Agronomique, Montpellier, France, 5 Department of Plant Genetic Engineering, Centro de Investigación y de Estudios Avanzados del Instituto Politécnico Nacional, Irapuato, Guanajuato, Mexico

Abstract

Embryogenic suspensions that involve extensive cell division are risky in respect to genome and epigenome instability. Elevated frequencies of somaclonal variation in embryogenic suspension-derived plants were reported in many species, including coffee. This problem could be overcome by using culture conditions that allow moderate cell proliferation. In view of true-to-type large-scale propagation of *C. arabica* hybrids, suspension protocols based on low 2,4-D concentrations and short proliferation periods were developed. As mechanisms leading to somaclonal variation are often complex, the phenotypic, genetic and epigenetic changes were jointly assessed so as to accurately evaluate the conformity of suspension-derived plants. The effects of embryogenic suspensions and secondary embryogenesis, used as proliferation systems, on the genetic conformity of somatic embryogenesis-derived plants (emblings) were assessed in two hybrids. When applied over a 6 month period, both systems ensured very low somaclonal variation rates, as observed through massive phenotypic observations in field plots (0.74% from 200 000 plant). Molecular AFLP and MSAP analyses performed on 145 three year-old emblings showed that polymorphism between mother plants and emblings was extremely low, i.e. ranges of 0–0.003% and 0.07–0.18% respectively, with no significant difference between the proliferation systems for the two hybrids. No embling was found to cumulate more than three methylation polymorphisms. No relation was established between the variant phenotype (27 variants studied) and a particular MSAP pattern. Chromosome counting showed that 7 of the 11 variant emblings analyzed were characterized by the loss of 1–3 chromosomes. This work showed that both embryogenic suspensions and secondary embryogenesis are reliable for true-to-type propagation of elite material. Molecular analyses revealed that genetic and epigenetic alterations are particularly limited during coffee somatic embryogenesis. The main change in most of the rare phenotypic variants was aneuploidy, indicating that mitotic aberrations play a major role in somaclonal variation in coffee.

Editor: Tianzhen Zhang, Nanjing Agricultural University, China

Funding: Financial support for this study was provided by the Mexican Government through a grant to Roberto Bobadilla Landey by the Consejo Nacional de Ciencia y Tecnología (CONACyT) (CVU:1623391) program (http://www.conacyt.mx/), by another grant from the PCP France-Mexico (http://www.pcp-mexique.com/) and by the CIRAD funds for doctoral support (http://www.cirad.fr/). The funders had no role in study design, data collection and analysis, decision to publish, or preparation of the manuscript.

Competing Interests: The authors have declared that no competing interests exist.

* E-mail: herve.etienne@cirad.fr

Introduction

Among micropropagation methods, somatic embryogenesis has the best potential for rapid and large-scale multiplication of selected varieties in a wide range of economically important species. Schematically, the initial step of dedifferentiation leading to the acquisition of embryogenic competence is common to most plant species, whereas for the following step of proliferation of embryogenic material, efficient procedures can be classified under

two main strategies. The first is the proliferation through secondary embryogenesis (SCE) which involves first differentiating the somatic embryos before enhancing their proliferation by adventitious budding (Figure 1). The second consists of establishing embryogenic suspensions (ESP) to favor large-scale embryogenic cell proliferation before the subsequent embryo differentiation step. In order to come up with an industrial procedure, the development of ESP represents the best option to ensure synchronous and massive somatic embryo production [1]. In addition, ESP allows the production of large numbers of embryogenic-competent cells and this process can be easily scaled up. Nevertheless, tissue culture systems such as somatic embryogenesis that involve the acquisition of competence for pluripotentiality and extensive cell division are more risky with respect to genome and epigenome instability [2]. The use of ESP has frequently been associated with an increased risk of genetic instability and somaclonal variation in the regenerated plants [3–5]. Although ESP has been developed for some important crops, it has therefore not been widely applied for commercial purposes. Somaclonal variation in ESP-derived plants is probably related to the presence of 2,4-dichlorophenoxyacetic acid (2,4-D), which is often essential for maintaining proliferating cells in an embryogenic, undifferentiated state [6,7]. This auxin could enhance somaclonal variation through the stimulation of rapid disorganized growth that can influence the mitotic process, resulting in chromosomal aberrations [8,9].

The term 'somaclonal variation' (SV) describes the tissue culture-induced stable genetic, epigenetic or phenotypic variation in clonally propagated plant populations [10]. Somaclonal variation is considered to be one of the main bottlenecks in the development of micropropagation procedures, especially in view of large-scale commercial operations, for which the strict maintenance of genetic and agronomic traits from selected individuals is required. An analysis of the progeny of phenotypic variants showed that some of the variations produced by somatic embryogenesis can occur in the form of stable and heritable mutations [8,11]. In maize, Kaeppler and Phillips [12] also reported stable segregation of somaclonal variant phenotypic qualities in several seed generations. It has also been well documented that somaclonal variants commonly present cytological aberrations such as chromosomal rearrangements (deletions, duplications, inversions and translocations), and sometimes more severe alterations like aneuploidy or polyploidy [12–16].

Although most mutants segregate in a Mendelian fashion upon selfing and outcrossing [12], SV is sometimes present in the form of transient mutations, suggesting the involvement of epigenetic events [11]. Epigenetic traits are heritable changes associated with chemical modification of DNA without alteration of the primary DNA sequence [17]. Cytosine methylation has been proposed as a possible cause of SV [11,12]. Epigenetic modifications (methylation) can mediate the transmission of an active or silent gene in the short-term (mitotic cell division) or long-term (meiotic divisions leading to transmission across generations) [18]. DNA methylation in plants commonly occurs at cytosine (5-methylcytosine, m^5C) bases in all sequence contexts: the symmetric CG and CHG (in which H could be A, T or C) and the asymmetric CHH contexts [17,18]. Molecular marker approaches like methylation-sensitive amplified polymorphism (MSAP) and Met-AFLP have proved efficient in the analysis of methylation patterns [19,20]. The existence of zones susceptible to methylation variations was recently shown in somatic embryogenesis-derived plants (emblings) in grapevine [21] and barley [20]. SV was also associated with the activity of mobile DNA elements or retroelements [22,23]. Novel mechanisms such as RNAi directed demethylation have recently been proposed to explain retrotransposon activation [2,24].

Coffea arabica is an allotetraploid tree species ($2n = 4X = 44$) characterized by low molecular polymorphism [25]. Somatic embryogenesis is currently applied industrially for large-scale and rapid dissemination of selected F1 hybrids that provide a highly significant increase in the yield of high quality coffee [26,27]. Regarding industrial-scale micropropagation, upgrading production to several million vitroplants per production unit would undoubtedly boost economic profitability. This would require switching from an SCE- to an ESP-based protocol. However, former field observations revealed that SV occurs at relatively high rates in ESP-derived *C. arabica* plants [28,29]. Apart from different phenotypic variants easily identifiable through morphological characteristics, we did not discover in trees showing a normal phenotype any variations involving agronomically important quantitative and physiological characteristics [30]. In view of true-to-type propagation of selected *C. arabica* hybrid varieties, we previously developed improved ESP protocols based on the use of low exogenous 2,4-D concentrations and short proliferation periods, allowing reliable somatic embryo mass regeneration [27]. For potential commercial applications, here using two *C. arabica* hybrids we verified the conformity of suspension-derived plants with that of plants obtained by secondary embryogenesis, i.e. the industrial process currently in use. The objectives were to assess large-scale phenotype conformity in commercial field plots, to quantify genetic and epigenetic modifications in the regenerated plants through AFLP (Amplified fragment length polymorphism) and MSAP molecular markers, and to cytologically characterize the karyotype of different phenotypic variants detected in the study.

Results

Frequency of phenotypic variants

Embling batches of hybrids H1 and H3 obtained from both SCE and ESP were checked for phenotype variation at both nursery and field levels. The frequency of phenotypic variants assessed among more than 600 000 plants in the nursery was very low (approx. 0.1%) and not significantly affected by the proliferation system nor the hybrid variety (Table 1). Observation of around 200 000 emblings in the field two years after planting revealed roughly an additional 0.74% of abnormal phenotypes, still without any significant difference between the two proliferation systems and hybrids. Apart from these phenotypic variants, all the other studied trees flowered, grew and produced normally. In conclusion, the overall phenotypic variation rate obtained by pooling the data obtained both in the nursery and in the field was less than 1% and no significant differences were noted between the proliferation systems or between the hybrids used.

Both proliferation systems generated the same kind of phenotypic variants (Figures 2A, B), with the Dwarf and Angustifolia types (Figures 3E, B) being the most frequent. Note that the secondary embryogenesis proliferation system specifically enhanced the occurrence of Dwarf variants, whereas the embryogenic suspensions favored the occurrence of the Angustifolia type. This latter phenotype can easily be detected and eliminated at the nursery level along with the Variegata variant (Figure 3C). A comparison of Figures 2A and 2B clearly shows that elimination in the nursery is not efficient for the Dwarf type. This somaclonal variation is more easily observable 2–3 years after planting in the field thanks to the characteristic grouped canopy morphology and low yield. Similarly, the Giant and Bullata (Figure 3F) phenotypic variants could only be detected in the field on well-developed trees.

Locus specific polymorphisms revealed by AFLP

In order to verify the induction of molecular polymorphism by the somatic embryogenesis process, AFLP analysis (four primer combinations, Table 2) was carried out on mother plants and their derived emblings. From a total of 204 bands obtained, only one polymorphic fragment of 173 bp in size (*Eco*-ACT/*Mse*-AGT) shared by two emblings of hybrid H1 and exhibiting a normal phenotype, was found (Table 3). From a total of 198 bands obtained, no polymorphism was found in emblings of hybrid H3.

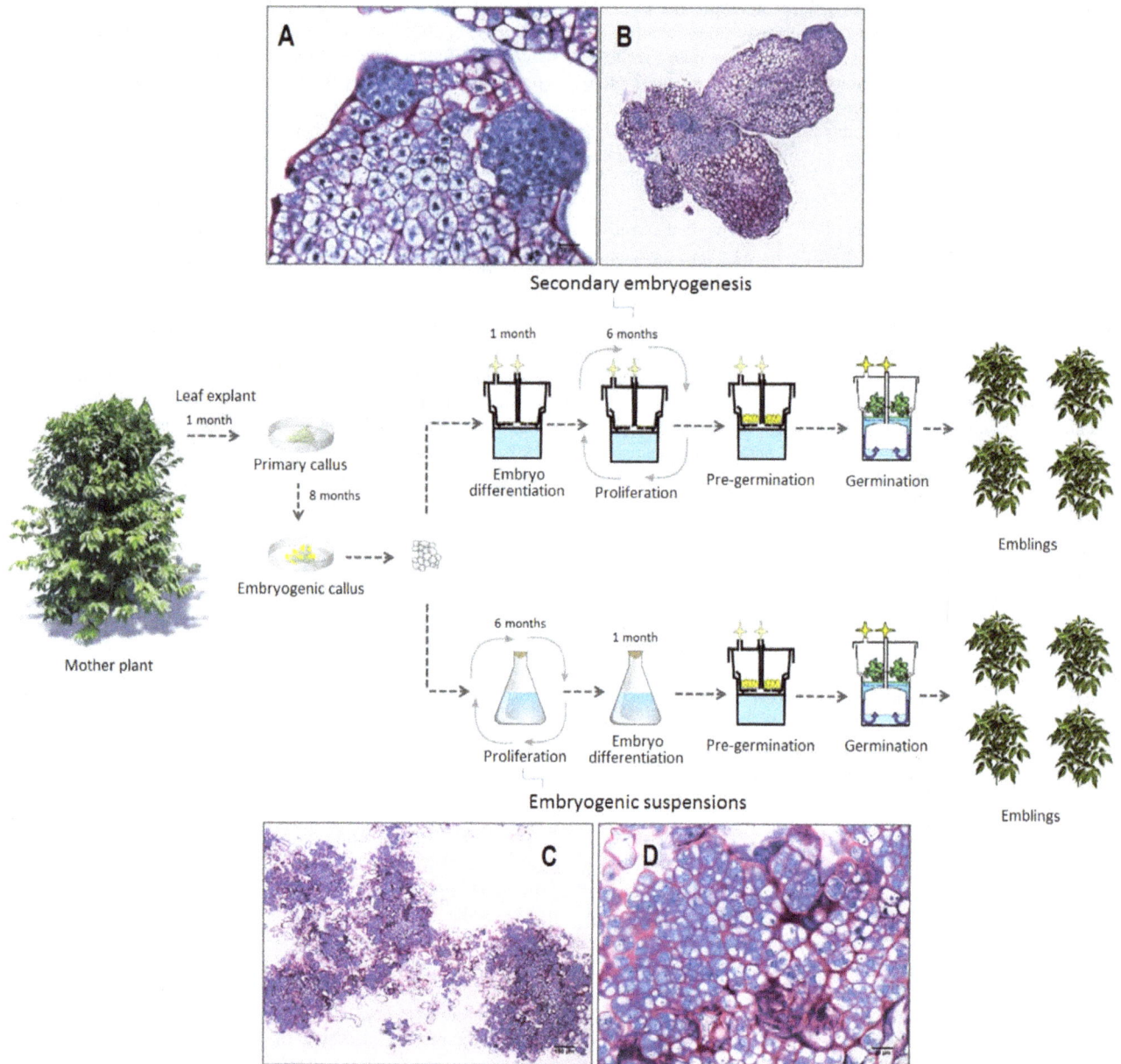

Figure 1. Schematic representation of two somatic embryogenesis processes applied at the industrial level. The first somatic embryogenesis process (upper section of the flow diagram) involved a proliferation step based on secondary embryogenesis in RITA® temporary immersion bioreactors (photos 1A, 1B). The second process (lower section of flow diagram) included a proliferation step based on embryogenic suspensions (photos 1C, 1D). 1A, initial developmental stages of secondary embryos at the root pole of primary somatic embryos; 1B, clusters of primary and secondary embryos; 1C, clusters of embryogenic cells in suspension; 1D, embryogenic cells in suspension.

All variants had the same AFLP pattern as the mother plants. For both hybrids, no significant quantitative effect on AFLP was detected when comparing SCE and ESP.

Methylation changes revealed by MSAP

In order to evaluate the occurrence of possible epigenetic modifications in the micropropagated plants, a study on the alteration of methylation patterns was performed by MSAP analysis using eight primer combinations (Table 2). Only clear and reproducible bands were selected for the analysis. More than 395 fragments were considered. First, MSAP patterns were obtained from DNA digested by the two isoschizomers (*Hpa*II and *Msp*I), as illustrated in Figure 4. They were further compared with those of

mother plants to classify the amplified fragments according to the methylation pattern, as shown in Table 4. The percentages of monomorphic fragments (pattern 1) were elevated and similar for both hybrids at nearly 91%. The remaining 9% of fragments (8.5% for H1 and 9.1% for H3) almost exclusively corresponded to pattern 3.

A comparison of amplification patterns in mother plants and their respective progeny are reported in Table 5. All differences between mother and derived plants were switches between patterns 1 and 3 and vice versa, i.e. likely modifications in the restriction ability of *Hpa*II. Among the polymorphic bands, eight bands corresponded to a change from pattern 3 in mother plants towards the unmethylated pattern 1 in emblings, suggesting

Table 1. Frequency of coffee phenotypic variants detected in the nursery and field on three year-old plants from two *C. arabica* hybrids depending on the type of proliferation system used in the industrial somatic embryogenesis process.

Proliferation system in the somatic embryogenesis process	Hybrid	Observations after 10 months in nursery				Observations after 36 months in field			
		No. of observed emblings	No. of variants	Somaclonal variation frequency (%)	3σ confidence interval*	No. of emblings	No. of variants	Somaclonal variation frequency (%)	3σ confidence interval*
Secondary embryogenesis	H1	117.115	148	0.13	[0.09–0.16]	51.131	373	0.73	[0.62–0.84]
	H3	121.894	159	0.13	[0.10–0.16]	49.126	390	0.79	[0.67–0.91]
	Total	239.009	307	0.13	[0.10–0.15]	100.257	763	0.76	[0.67–0.84]
Embryogenic suspension	H1	204.871	206	0.10	[0.08–0.12]	54.218	402	0.74	[0.63–0.85]
	H3	197.705	183	0.09	[0.07–0.11]	54.566	394	0.72	[0.61–0.83]
	Total	402.576	389	0.09	[0.08–0.11]	108.784	796	0.73	[0.65–0.80]

The variable analyzed was the proportion (p) of variant (p = X/n), where X was the number of variant and n the number of plants observed. A 3σ confidence limit for binomial distribution was calculated using the formula $p\pm3\left(\sqrt{p(1-p/n)}\right)$ with a level of confidence of 99%.

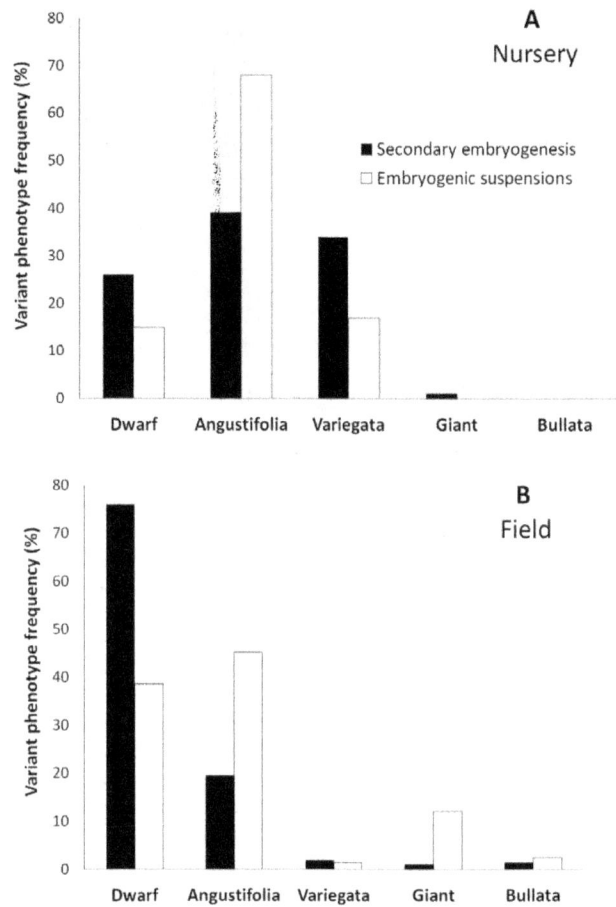

Figure 2. Proportions (%) of the different types of phenotypic variants in comparison to the total number of variants. Variants representing less than 1% of somatic embryogenesis-derived plants were observed in *C. arabica* embling batches at nursery (A) and field (B) levels, depending on the proliferation system used, i.e. secondary embryogenesis (SCE) or embryogenic suspension (ESP). In the nursery, the data were obtained from 239 009 emblings derived from SCE and 402 576 emblings from ESP. In the field, the data were obtained through the observation of 100 257 emblings derived from secondary embryogenesis and 108 784 from embryogenic suspensions.

demethylation. Changes associated with certain polymorphic MSAP bands were more frequent than others (Table 5). Seventy percent of the changes were linked to only five polymorphic bands. The detected polymorphism was very low and similar for both hybrids (Table 6) but slightly higher (0.07–0.18%) when compared to AFLP molecular markers. Similar to the AFLP results, the total MSAP polymorphism was not significantly different between the two proliferation systems nor between the two hybrid varieties.

Most emblings showing changes in MSAP pattern had only one or two methylation polymorphisms (Figure 5). We did not find any emblings with more than three methylation polymorphisms. The same polymorphic bands were shared by plants from both proliferation systems and/or both hybrids in approximately half of the cases (Table 5). No relation was established between the variant phenotype and a particular MSAP pattern (Table 6).

Chromosome counting of somaclonal variants

Chromosome numbers were assessed in 2 phenotypically normal emblings and 11 somaclonal variants (Fig. 6). Table 7 shows that the two normal regenerants exhibited the expected

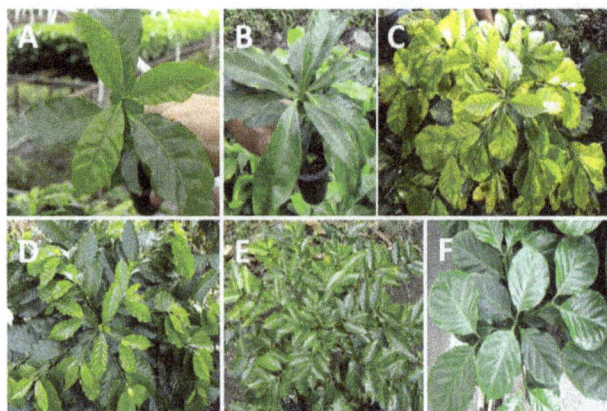

Figure 3. Examples of different *C. arabica* phenotypic variants in plants mass propagated through somatic embryogenesis. A, plant exhibiting a normal phenotype in the nursery; B, Angustifolia variant in the nursery; C, Variegata variant in the field; D, plant showing a normal phenotype in the field; E, Dwarf variant in the field; F, Bullata variant in nursery.

chromosome number for the allotetraploid *C. arabica* species (2n = 4x = 44) whereas 7 of the 11 variant emblings showed a different chromosome number (aneuploids). In almost all cases, the aneuploid karyotypes were characterized by the loss of 1–3 chromosomes. One Angustifolia variant had an extra chromosome. Moreover, the results showed that different chromosome numbers - including the normal number - could be observed for the same variant phenotype and that abnormal chromosome numbers were obtained for most of the variant phenotypes.

Discussion

Until now, relatively high somaclonal variation rates have been reported in *C. arabica* emblings, particularly with embryogenic suspensions [28,29,30]. The variant frequency was found to strongly increase in embryogenic suspensions after 6 months proliferation (25% after 12 months) in the presence of 4.52 µM 2,4-D [29]. The presence of disorganized rapid growth phases in tissue culture, such as callus and cell suspension cultures, has often been considered as one of the factors that cause SV [8,31]. In view of true-to-type propagation, we further established processes with limited disorganized rapid growth phases. The first was based on SCE proliferation in temporary immersion bioreactors enhanced

by the addition of cytokinin, similar to methods described in other woody plants such as rubber [32], oak [33] and tea [34] and is currently used on a commercial scale for coffee. The second involves the proliferation of ESP in the presence of both auxin and cytokinin albeit at a very low level of 2,4-D (1.36 µM) with a short proliferation time (6 months). These conditions allow sufficient amplification of embryogenic material to ensure the cost-effectiveness of the industrial process. A large-scale phenotypic evaluation for each process: 230 000–400 000 emblings in the nursery and 100 000 in the field was performed to obtain valuable and accurate information on genetic stability. The present study showed for the first time that, for both somatic embryogenesis processes, the variant phenotype frequencies were extremely low (less than 1%) and not statistically different. Moreover no statistical difference could be noted between the two studied hybrids. This clearly demonstrates that it is possible to control SV by optimizing the ESP culture conditions. In another cultivated *Coffea* species, i.e. the diploid *C. canephora*, Ducos *et al.* [35] found 2–4% of a low-vigor phenotype off-type among 5 067 emblings derived from 5–7-month-old ESP. In contrast to the *C. arabica* protocol, all somatic embryogenesis steps for *C. canephora* were performed with cytokinins as sole source of growth regulator. Hence, in coffee, both auxin and cytokinin could probably be involved in the generation of SV. The Angustifolia, Variegata and Dwarf variants were the most frequent types, irrespective of the proliferation strategy employed, confirming previous studies conducted on a lesser scale [29,30]. Interestingly, except for the Dwarf variant which seems to be tissue culture specific, all the phenotypic variants observed in the somatic embryogenesis progenies were also observed among *C. arabica* seed-derived progeny. Indeed, phenotypic mutations are frequent in coffee nurseries, and Krug and Carvalho [36] previously characterized the numerous different morphologies in detail.

In several cases, AFLP markers have proved useful in the detection of genetic variation in tissue culture derived plants [20,37–40]. In coffee, the genetic stability of coffee emblings has been poorly evaluated by molecular markers and limited to plants derived from experimental protocols. The present study revealed no or extremely limited mutations at the DNA level in large-scale somatic embryogenesis-propagated plants. From a total sample of 145 plants analyzed belonging to two hybrid varieties, only 1 out of 204 bands was polymorphic in only two SCE-produced plants. ESP-derived plants showed no AFLP polymorphism when compared with mother plants. Our results significantly differed from those previously reported in *C. arabica*. In a first approach on only 27 plants, Rani *et al.* [41] used different DNA markers

Table 2. Primer combinations used for AFLP and MSAP analyses.

AFLP primer combinations (*Eco*+3 [labeled]/*Mse*+3)	MSAP primer combinations (*Eco*−3/*Hpa*+2 [labeled])
Eco-ACT/*Mse*-AGT	C1 *Eco*-AAC/*HPA*-AA
Eco-AGG/*Mse*-AGT	C2 *Eco*-AAC/*HPA*-AT
Eco-CGC/*Mse*-CCA	C3 *Eco*-AGG/*HPA*-AA
Eco-CAC/*Mse*-CCA	C4 *Eco*-AGG/*HPA*-AT
	C5 *Eco*-ACT/*HPA*-CA
	C6 *Eco*-ACT/*HPA*-CT
	C7 *Eco*-AGA/*HPA*-CA
	C8 *Eco*-AGA/*HPA*-CT

Fluorescent dyes for marked primers correspond to 5′-FAM [TM] and 5′-HEX [TM]

Table 3. Summary of AFLP data and observed polymorphisms among mother plants and emblings derived from secondary embryogenesis or embryogenic suspensions.

Proliferation system in the somatic embryogenesis process	Hybrid	No. of analyzed emblings	No. of fragments	Polymorphic fragments		Emblings showing polymorphisms		Total polymorphism **
				No.	(%)	No.	(%)	(%)
Secondary embryogenesis	H1	33	204	1*	0.5	2	6	0.03
	H3	45	198	0	0	0	0	0
Embryogenic suspensions	H1	26	204	0	0	0	0	0
	H3	41	198	0	0	0	0	0

Data were obtained for two C. arabica hybrids (H1 and H3) and compared with the patterns of the mother plants.
*Found in 2 emblings with normal phenotype (N°210 and 232) showing a new AFLP band Eco-ACT/Mse-AGT 173 bp
**Total polymorphism = [No. of polymorphic fragments/(No. of fragments × No. emblings)]×100

(RAPD, random amplified polymorphic DNA and SSR, simple sequence repeat) to assess the genetic integrity of *C. arabica* emblings obtained from embryogenic calli, and they found a higher polymorphism level (4%) in the nuclear genome. By performing RAPD analyses on Norway spruce emblings, Heinze and Schmidt [42] concluded that gross somaclonal variation was absent in their plant regeneration system. In contrast, RAPD and SSR markers allowed the detection of high genetic variation in cotton emblings regenerated in the presence of 2,4-D [39]. AFLP analysis of 24 rye emblings led to the scoring of 887 AFLP markers, among which 8.8% identified the same polymorphism in plants obtained independently, revealing putative mutational hot spots [38].

DNA cytosine methylation plays an important role in plant regulation and development [43,44]. Since the pioneer studies on maize emblings [12], substantial evidence has been obtained which indicates that demethylation can occur at a high frequency during somatic embryogenesis and can be an important cause of tissue culture induced variation [45]. DNA methylation has also been implicated in gene silencing and transposable element reactivation [24,46]. To our knowledge, epigenetic deregulation during coffee micropropagation has not yet been studied. The very low total methylation polymorphism values obtained for both somatic embryogenesis processes and both hybrids (0.07–0.18% range) indicated that the tissue culture procedures employed in coffee weakly affected DNA methylation of the regenerated plants. This finding is in accordance with the 0.87% total polymorphism recently found in *Freesia hybrid* emblings by Gao *et al.* [40]. Moreover, the low number of methylation polymorphisms per embling (range 1–3) confirmed that neither SCE nor ESP induced additional stress at the methylation level during embryogenic material proliferation. In contrast, a significantly higher accumulation of methylation changes in some emblings has been regularly observed in other species [20,21,47,48]. For example, in grapevine, most emblings showed between zero and three changes, similar to our findings, but a few accumulated up to 18 [21]. It has also been well demonstrated that auxin levels strongly alter DNA methylation of embryogenic cell cultures [49]. However, similar to our results, examples of stable MSAP patterns have already been reported using bamboo tissues at different developmental stages of somatic embryogenesis [50]. The timing of plant regeneration from proliferating callus cultures could be crucial for the appearance of variation. In callus-derived hop plants, an increase in the variation was detected by MSAP in prolonged callus cultures [20]. Our results demonstrated that very few changes are possible by limiting both the auxin level and culture duration.

MSAP markers have already been successfully used to demonstrate epigenetic instabilities (methylation alteration) induced by somatic embryogenesis in a great variety of plants, such as the ornamental flower *Freesia*, banana, barley, grapevine and maize [20,21,40,51–53], also indicating that demethylation events were generally the most frequent. Although occurring at low frequency, our results also indicated demethylation events and mainly the loss of methylation in the internal cytosine of the 5'-CCGG-3' sequence to produce a new *Hpa*II band not detected in mother plants but present in the amplification pattern of the isoschizomer *Msp*I. The detection of the same MSAP polymorphic fragments in independent plant samples from different hybrids and somatic embryogenesis processes suggests the existence of hotspots of DNA methylation changes in the genome. The existence of non-randomly behaving methylation polymorphic fragments in micropropagated plants has already been described using Met-AFLP [20,48] and MSAP [21,53–55].

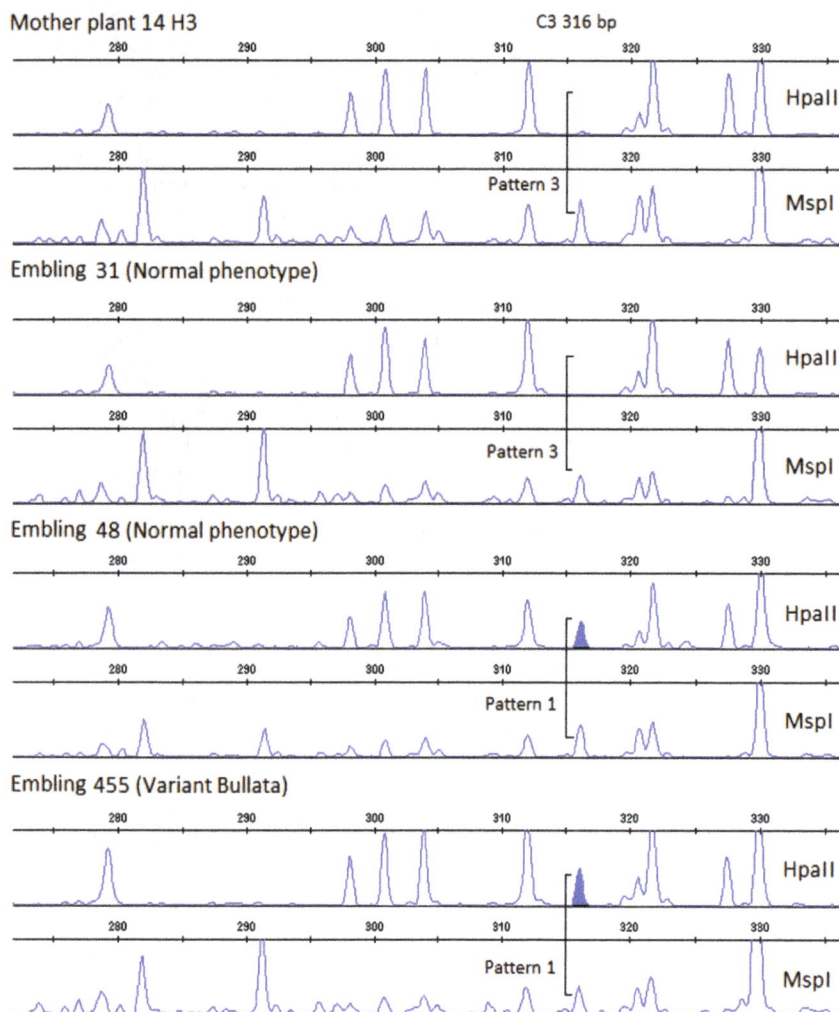

Figure 4. Representation of MSAP electropherograms observed for coffee mother plants and embling progeny using the isoschizomers *Hpa*II **and** *Msp*I. Illustration of the pattern variation obtained for normal and variant phenotypes within the embling progeny.

Gross changes, such as variation in ploidy level, number of chromosomes and structural changes, are mitotic aberrations that represent major genomic alterations of *in vitro* plants often generated during *in vitro* proliferation and differentiation [12,56–58]. Variations in chromosome number and structure have been described among emblings for several species [59–62]. We demonstrated that gross changes occur during coffee somatic embryogenesis and are related to SV, whereas genetic and

Table 4. MSAP patterns corresponding to different methylation states of the symmetric sequence CCGG, as revealed by the specificity of the restriction enzymes *Hpa* II and *Msp* I.

Restriction enzymes	MSAP patterns after enzymatic digestion			
	Pattern 1	**Pattern 2**	**Pattern 3**	**Pattern 4**
Hpa II	1	1	0	0
Msp I	1	0	1	0
Methylation state	Unmethylated	Hemi-methylated	Fully-methylated	Fully-methylated
Methylation position	None	External cytosine	Internal cytosine	External cytosine
Schematic representation	CCGG GGCC	CH3 CCGG GGCC	CH3 CCGG GGCC CH3	CH3 CCGG GGCC CH3

Table 5. MSAP methylation patterns in mother plants and modified patterns in emblings.

Polymorphic MSAP fragments (size in bp)	MSAP methylation patterns		Proliferation system affected by the methylation change		Presence of the methylation change depending on the plant phenotype		No. of methylation changes for each fragment	
	Mother plants	Emblings	Hybrid H1	Hybrid H3	Normal	Variant	No.	(%)
C3- 107 bp	Pattern 3	Pattern 1	SCE, ESP	SCE, ESP	+**	−	8	9.8
C2 -127 bp	Pattern 3	Pattern 1	0	ESP	+	−	1	1.2
C4 -134 bp	Pattern 3	Pattern 1	0	SCE	+	−	4	4.8
C1- 251 bp	Pattern1*	Pattern 3	SCE, ESP	0	+	−	12	14.6
C3- 253 bp	Pattern 3	Pattern 1	0	SCE	+	−	2	2.4
C2- 302 bp	Pattern 3	Pattern 1	0	SCE, ESP	+	+	16	19.5
C3- 316 bp	Pattern 3	Pattern 1	0	SCE	+	+	4	4.8
C6 -370 bp	Pattern 3	Pattern 1	SCE, ESP	ESP	+	+	10	12.2
C8- 370 bp	Pattern 3	Pattern 1	SCE, ESP	SCE, ESP	+	+	12	14.6
C5- 387 bp	Pattern 1	Pattern 3	SCE	SCE	+	−	13	15.8
No. changes							82	

Relation with the type of C. arabica hybrid, type of proliferation system [secondary embryogenesis (SCE) and embryogenic suspension (ESP)] and regenerant phenotype.
*Pattern 1: Fragment present in both HpaII and MspI digests (1:1); Pattern 3: Fragment absent in HpaII digests and present in MspI digests (0:1). ** Relationship with a particular phenotype is indicated with (+) for presence and (−) for absence.

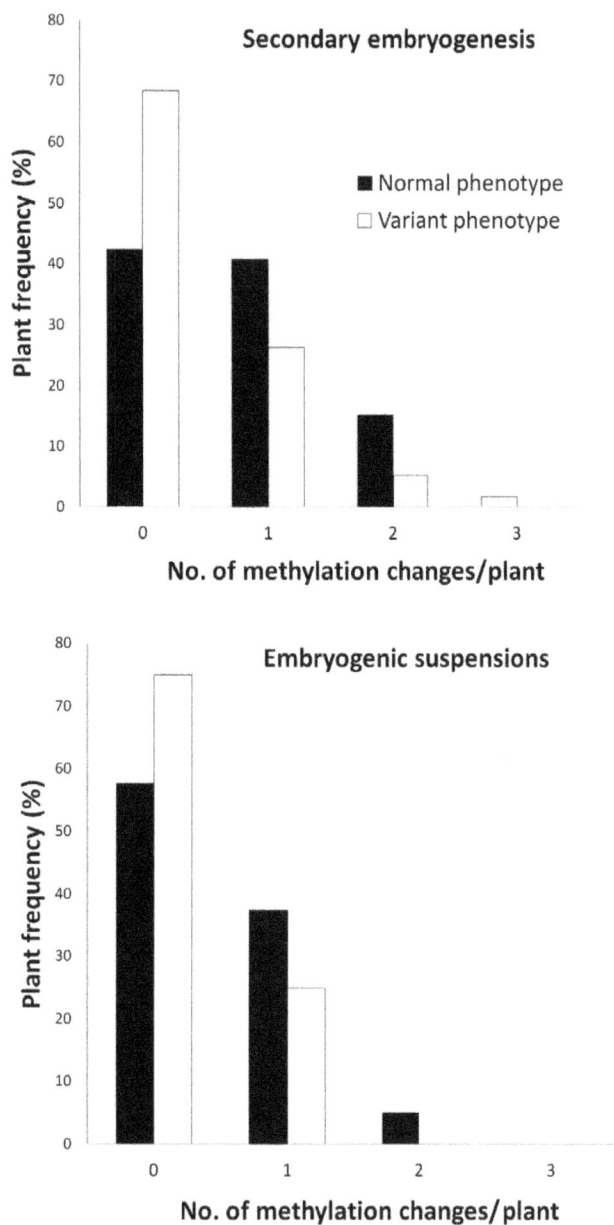

Figure 5. Methylation polymorphism accumulation in coffee emblings showing a normal *vs.* variant phenotype depending on the somatic embryogenesis process used. For the secondary embryogenesis process, data were derived from the analysis of 59 phenotypically normal and 19 variant emblings. For the embryogenic suspension process, 59 phenotypically normal and 8 variant emblings were studied.

epigenetic (methylation) alterations are very weak. Until now, by using flow cytometry analysis, normal ploidy levels were reported in coffee emblings [29,63] but chromosome counting was not performed in these studies. The sensitivity of flow cytometry analysis was probably not sufficient to identify aneuploid plants. Nevertheless, the presence of mitotic aberrations, including double prophase, lagging chromosomes, aneuploids and polyploid cells, has previously been described in leaves [64,65] and embryogenic calli [64] of *C. arabica* but not in the later steps of somatic embryogenesis, and without establishing any relation with SV.

Table 6. Overall MSAP data and methylation polymorphism among mother plants and emblings of *C. arabica* hybrids derived from secondary embryogenesis or embryogenic suspensions.

Proliferation in the somatic embryogenesis process	Hybrid	No. of emblings analyzed	No. of fragments	Methylation polymorphic fragments		Total polymorphism* (%)	3σ confidence intervals**
				No.	Percentage (%)		
Secondary embryogenesis	H1	33	399	5	1.2	0.18	[0.071–0.294]
	H3	45	396	7	1.7	0.16	[0.068–0.246]
		78		12	1.5	0.17	[0.098–0.238]
Embryogenic suspensions	H1	26	399	4	1.0	0.18	[0.057–0.309]
	H3	41	396	5	1.0	0.07	[0.006–0.129]
		67		9	1.1	0.11	[0.051–0.174]

*Total polymorphism = [No. of methylation polymorphic fragments/(No. of fragments × No. emblings)]×100.

**The variable analyzed was the proportion (p) of methylation polymorphisms (p = X/n), where X was the number of methylation polymorphisms and n the total number of fragments. A 3σ confidence limit for binomial distribution was calculated using the formula $p \pm 3\left(\sqrt{p(1-p/n)}\right)$ with a level of confidence of 99%.

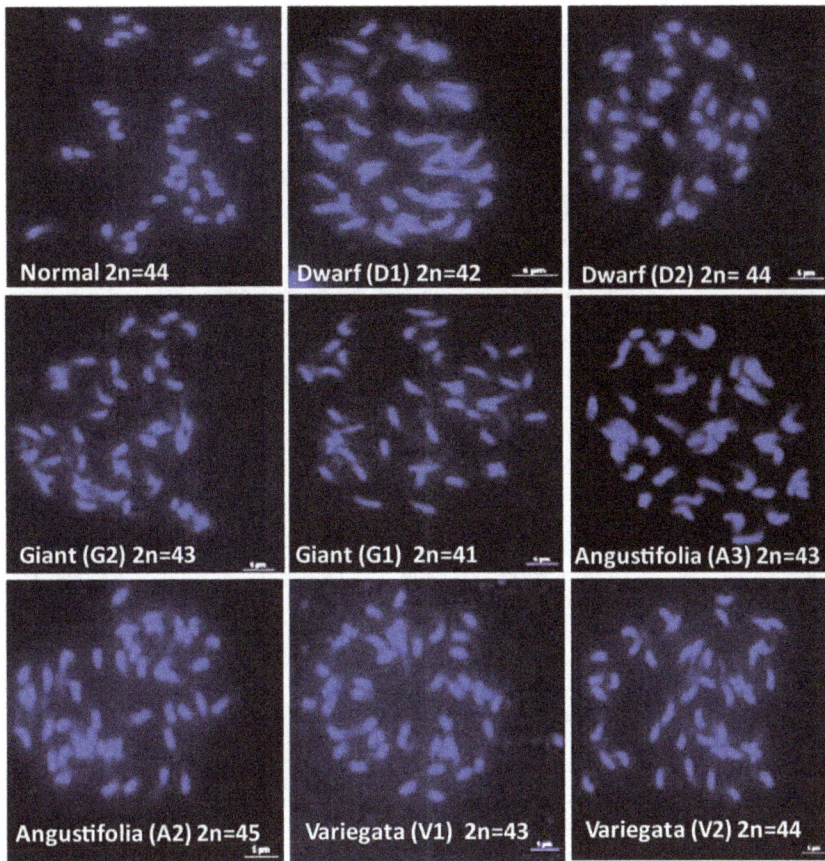

Figure 6. Mitotic cells at metaphase or prometaphase stages and observed ploidy levels of some normal and variant emblings from the allotetraploid *C. arabica* **species (2n = 44).** Karyotype analyses were performed by counting chromosomes on four to eight clear metaphase spreads obtained from root tips of three year-old plants.

Table 7. Summary of chromosome counting in some normal *versus* variant *C. arabica* hybrids derived from somatic embryogenesis.

Embling phenotype (normal or variant)	Somatic embryogenesis process	Code	No. of metaphases analyzed	No. of chromosomes
Normal	ESP	N1	8	44
Normal	SCE	N2	7	44
Angustifolia	ESP	A1	4	44
Angustifolia	SCE	A2	8	45
Angustifolia	ESP	A3	8	43
Bullata	ESP	B1	4	44
Dwarf	SCE	D1	5	42
Dwarf	ESP	D2	4	44
Giant	ESP	G1	6	41
Giant	ESP	G2	6	43
Giant	SCE	G3	6	43
Variegata	ESP	V1	6	43
Variegata	SCE	V2	4	44

The chromosome numbers obtained from root tips are indicated for 3 year-old emblings derived from embryogenic suspensions (ESP) or secondary embryogenesis (SCE) showing normal or abnormal phenotypes.

The presence of aneuploidy has also been well documented in embryogenic calli of *Hordeum vulgare* [66] and sweet orange [60].

The mechanisms underlying SV remain largely theoretical and unclear [16]. Thus it is often difficult to correlate a well-described genetic or epigenetic mechanism to a variant phenotype. For example, although DNA methylation has often been suggested as a possible cause of SV, a number of studies have reported high levels of methylation variation with no effect on the plant phenotype [20,21,52]. Another example is given by oil palm emblings, approximately 5% of which exhibit the 'mantled' phenotype affecting the formation of floral organs in both male and female flowers. Although a decrease in DNA methylation was observed, it was not possible to determine the nature of the epigenetic deregulation [67]. In the present study, it was possible to reveal a large proportion of aneuploid karyotypes in different variant phenotypes, hence showing that chromosomal rearrangements could be directly involved in the occurrence of phenotypic variation. The addition or subtraction of a single chromosome has a greater impact on phenotype than whole genome duplication, i.e. polyploidy [68]. It was clearly demonstrated in *Arabidopsis thaliana* that certain phenotypic traits are strongly associated with the dosage of specific chromosomes and that chromosomal effects can be additive [69]. Similarly, in maize seedlings, trisomic plants showed characteristic features such as reduced stature, tassel morphology changes and the presence of knots on the leaves, suggesting a phenotypic effect caused by the altered copy of specific chromosome related genes [70]. A similar mechanism could explain most of the variant phenotypes observed in *C. arabica*. The observation of a variant phenotype in plants with the expected chromosome number could be explained by the coexistence of monosomic and trisomic chromosomes in the same genome or by other chromosomal-like structural changes associated with undetected deletions, duplications, inversions or translocations of specific chromosomal segments [10,71]. The karyotype analyses performed in the present study were limited to chromosome counting and did not enable observation of such chromosomal alterations.

Conclusions

This report shows that somatic embryogenesis is reliable for true-to-type and large-scale propagation of elite varieties in the *C. arabica* species. Both ESP and SCE ensured high proliferation rates along with very low SV rates, as observed through massive phenotypic observations in a commercial nursery and field plots. Molecular analysis (AFLP and MSAP) performed on 145 emblings derived from two proliferation processes and two different F1 hybrids showed that polymorphism between mother plants and emblings was extremely low. Consequently, it can be concluded that genetic and epigenetic alterations were also particularly limited during somatic embryogenesis in our controlled culture conditions. *C. arabica* is a young allopolyploid still having the most of its genes in duplicated copies [72]. It could be hypothesized that the impact of genetic or epigenetic variations on phenotype was restricted because of the buffer effect due to polyploidy. The main change in most of the rare phenotypic variants was aneuploidy. Although further studies are necessary for an accurate understanding of the chromosome anomalies involved in the acquisition of a particular phenotype, it is now obvious that mitotic aberrations play a major role in SV in coffee. The identification and use of molecular markers at the heterozygous state in mother plants (i.e. polymorphic between the two parental lines) would be required to further investigate this type of chromosome abnormality. Current studies based on the use of long-term embryogenic

cultures [73] are aimed at establishing the full range of cytological, genetic and epigenetic (with a special focus on transposable elements) mechanisms underlying SV.

Materials and Methods

Plant material and somatic embryogenesis

Selected F1 hybrids of *C. arabica* [26] obtained by crossing traditional dwarf American varieties (Caturra, Sarchimor T5296) and wild accessions originating from Ethiopia and Sudan are disseminated in Central America through somatic embryogenesis. In the present study, emblings derived from the two hybrids Sarchimor T5296 x Rume Sudan and Caturra x ET531, named respectively H1 and H3, were analyzed to assess the SV level. Large-scale phenotypic observations were performed in Nicaragua both at the nursery (more than 600 000 young emblings) and field level (more than 200 000 three year-old emblings) on 11 coffee plots belonging to the 'La Cumplida' coffee research experimental sites in the Matagalpa region (Nicaragua). Nursery and field phenotypic observations were done on balanced amounts of plants from hybrids H1 and H3 (Table 1). Field observations were performed for all trees by visual evaluation of growth and morphology, flowering and fruit yield during the first and second production years. No specific permits were required for the described field studies that were performed with the authorization of the Coffee Research Department of 'La Cumplida', owner of the experimental sites. These sites are not protected and the studies did not involve endangered or protected species. Molecular marker analysis was applied on F1 hybrid mother plants propagated by rooted horticultural cuttings (four for each hybrid) and used as source material for *in vitro* propagation, as well as on adult emblings (3 years after planting) for which plants exhibiting an abnormal phenotype (phenotypic variant) were distinguished from plants with a normal phenotype and productivity (Table 8). Molecular marker patterns obtained for emblings were systematically compared with those obtained with mother plants (four plants per hybrid propagated by horticultural cutting) used as reference.

The somatic embryogenesis process (Figure 1) involved the following stages:

1) Production of embryogenic callus: pieces of young leaves were surface-sterilized and used as explants. The explants were cultured for 1 month on a 1/2 strength MS [74] "C" callogenesis medium [75] containing 2.3 μM 2,4-D, 4.9 μM indole-3-butyric acid (IBA) and 9.8 μM iso-pentenyladenine (iP), and then transferred for 6 months to MS/2 "ECP" embryogenic callus production medium [75] containing 4.5 μM 2,4-D and 17.7 μM benzylaminopurine (6-BA). All media were solidified using 2.4 g/l Phytagel (Sigma, Steinheim, Germany). These steps were carried out at 26–27°C in the dark.

2) 6-month multiplication step and embryo regeneration. Fully developed somatic embryos were mass regenerated via two distinct multiplication processes, i.e. either secondary (or repetitive) embryogenesis or embryogenic suspensions (Figure 1).

Secondary embryogenesis (SCE). Two hundred mg of embryogenic aggregates were placed in a 1 liter-RITA® temporary immersion bioreactor (CIRAD, Montpellier, France; [76]) along with 200 ml of "R" MS/2 regeneration medium [75] containing 17.76 μM 6-BA, in darkness for 6 weeks. An immersion frequency of 1 min every 12 h was applied. Proliferating embryo masses were then placed in 1/2 strength MS [74] regeneration medium containing 5.6 μM 6-BA and subcultured once every 6 weeks for three proliferation cycles. Secondary embryos were produced with an immersion frequency of 1 min every 12 h and a high culture

Table 8. Plant material used in molecular marker analyses.

Somatic embryogenesis proliferation step	Type of plant material	No. of plants per *C. arabica* hybrid		Total no. of plants
		H1	H3	
Secondary embryogenesis	Emblings normal phenotype	28	31	59
	Emblings variant phenotype	5	14	19
	Angustifolia (A)	0	4	
	Bullata (B)	2	4	
	Dwarf (D)	0	2	
	Variegata (V)	3	4	
	Total	33	45	78
Embryogenic suspension	Emblings normal phenotype	25	34	59
	Emblings variant phenotype	1	7	8
	Angustifolia (A)	1	2	
	Dwarf (D)	0	1	
	Variegata (V)	0	4	
	Total	26	41	67

Total number of plants analyzed from two F1 *Coffea arabica* hybrid lines (H1 and H3) corresponding to somatic-embryo derived 3 year-old plants (emblings) with normal or variant phenotypes along with their respective mother plants as reference. The numbers of emblings for each variant phenotype are also given.

density (approx. 10 000 embryos). The cultures were kept at 27°C, with a 12 h/12 h photoperiod and 50 µmol m^{-2} s^{-1} photosynthetic photon flux density.

Embryogenic cell suspensions (ESP). Embryogenic calli were transferred to 100 ml Erlenmeyer flasks at a density of 1 g/L in 1/4 MS strength Yasuda liquid proliferation medium [77] with 1.36 µM 2,4-D and 4.4 µM 6-BA. Suspension cultures were maintained by the monthly transfer of 1 g/L of embryogenic aggregates into fresh medium. Six month-old suspensions were used for somatic embryo regeneration. Embryo differentiation was initiated by transferring embryogenic aggregates at a density of 1 g/L in 250 ml Erlenmeyer flasks in a full strength MS medium containing 1.35 µM 6-BA. Fully developed torpedo-shaped embryos were obtained after two 4 week subcultures in such conditions. All suspension cultures were shaken at 110 rpm at 27 °C under 50 µmol m^{-2} s^{-1} photosynthetic photon flux density.

3) Pre-germination in a bioreactor. Germination was triggered by applying a low culture density of around 800–900 embryos per 1 l-RITA® bioreactor. An "EG" embryo germination medium [75] containing 1.33 µM BA was used for 2 months and finally, for 2 weeks, the "EG" culture medium was supplemented with 234 mM sucrose. By the end of the in vitro culture stage, each bioreactor contained around 700 pre-germinated cotyledonary somatic embryos with an elongated embryonic axis and a pair of open, chlorophyllous cotyledons. Pre-germination was conducted in the light (12/12 h, 50 µmol m^{-2} s^{-1}).

4) Plantlet conversion was obtained after direct sowing of mature somatic embryos in the nursery. Mature embryos were sown vertically on top of the substrate (two parts soil, one part sand, one part coffee pulp) sterilized by chemical treatment (Dazomet (DMTT), Union Carbide). The somatic embryo culture density in the plastic boxes (l.w.h = 30/21/10 cm) was approximately 3600 m^{-2}. The cultures were placed under a transparent roof that provided 50% shade, and were watered for 2 min twice daily. Conversion of somatic embryos into plants was generally observed 12 weeks after sowing, and characterized by the emergence of a stem bearing at least two pairs of true leaves.

5) Growth and hardening in the nursery (21 weeks). Plantlets grown from somatic embryos were transferred to 0.3 L plugs on a substrate comprising peat-based growing medium (Pro-mix, Premier Tech Ltd, Canada) and coffee pulp (3/1, v/v) under conventional nursery conditions until they reached the required size for planting in the field (approx. 30 cm). During this stage, the shade (50% light interception) and relative humidity (80%) were gradually reduced over 4 weeks to 0% light interception, with natural RH ranging from 65 to 90%.

Molecular analysis

DNA extraction. Young fully expanded leaves were selected on three year-old plants for molecular analysis. Genomic DNA was isolated from 100 mg of lyophilized leaves using Dellaporta buffer [78] containing sodium dodecyl sulfate (SDS) detergent and sodium bisulfite 1% w/v to avoid leaf oxidation. DNA was purified in spin-column plates as described in the DNeasy plant kit protocol from QIAGEN.

AFLP markers. AFLP analysis was carried out as described by Vos *et al.* [79], with a total of four primer combinations (Table 2), using 5-FAM or 5-HEX fluorescently labeled *Eco*R1 (+3) and unlabeled *Mse*I (+3) primers. A touchdown PCR program for selective amplification was performed in an Eppendorf thermocycler under the following conditions: 3 min at 94°C, 12 cycles of 45 s at 94°C, 12 cycles of 45 s at 65°C and 1 min at 72°C; the annealing temperature was decreased by 0.7°C per cycle from a starting point of 65°C during this stage, with a final round of 25 cycles of 94°C for 45 s, 56°C for 45 s, 72°C for 1 min and a final elongation step of 72°C for 1 min. The same PCR conditions were found to be appropriate for MSAP in preliminary tests.

MSAP markers. MSAP analysis was carried out as described by Reyna-López *et al.* [19] with minor adaptations for capillary electrophoresis. The MSAP protocol is an adaptation of the AFLP method for the evaluation of different states of methylation in the symmetric sequence CCGG. In the MSAP protocol, the frequent cutting endonuclease (*Mse*I) was replaced by the two isoschizomeric restriction enzymes *Hpa*II and *Msp*I with different sensitivity to the methylation state of the symmetric sequence CCGG (Table 4). Specifically, *Hpa*II is able to recognize and cut only when the CCGG sequence is unmethylated or hemi-methylated on the external cytosine. *Msp*I is able to cut when CCGG sequence is unmethylated or if the internal cytosine is fully or hemi-methylated. Both *Hpa*II and *Msp*I are unable to cut if the external cytosine presents full methylation. DNA methylation in plants commonly occurs at cytosine bases in all sequence contexts: the symmetric CG and CHG, in which (H = A, T or C) and the asymmetric CHH contexts [17]. Selective amplification included a total of eight primer combinations per isoschizomer (Table 2). *Hpa*II (+2) primers were fluorescently labeled with 5-FAM or 5-HEX while *Eco*R1 (+3) remained unlabeled. In order to reduce the possibility of technical artifacts, two repetitions using different DNA extractions were performed for each primer combination.

Capillary electrophoresis and data analysis. PCR products were separated by capillary electrophoresis with Pop 7^{TM} polymer in a 16 capillary 3130 XL Genetic Analyzer from Applied Biosystems using an internally manufactured 524 ROX fluorophore as sizing standard. The fragments used for fingerprinting were visualized as electropherograms in applied Biosystems software GeneMapper® version 3.7. Informative fragments were mostly found in the 100–450 bp range. All amplified fragments were classified based on the primer combination used and their size. The sample fingerprint data was converted to binary code, with "1" denoting the presence of the fragment and "0" the absence. Different binary matrices were constructed for comparative analysis depending on the kind of molecular marker.

As shown in Table 4, the MSAP patterns were classified as follows: Pattern 1 when a comigrating amplified fragment was obtained from the DNA template digested by both restriction enzymes *Hpa*II and *Msp*I; Pattern 2 and 3 when an amplification fragment was obtained only from the DNA template digested by *Hpa*II or *Msp*I, respectively.

Before analysis of the embling versus mother plant population, we successfully verified, on a set of plants from the Caturra variety, that the same MSAP patterns were systematically generated whatever the plant age and the leaf developmental stage (data not shown). Hence, a possible developmental variability in the studied plant material does not seem to introduce any additional source of variation in the methylation state. Nevertheless, in all experiments, only leaves from the same developmental stage were chosen.

Slide preparation and karyotyping. Root tips were harvested from individual adult emblings and placed in an aqueous solution of 8-hydroxyquinoline (2.5 mM) used as pre-treatment, for 4 h in darkness (2 h at 4°C plus 2 h at room temperature). A solution of Carnoy (absolute ethanol and glacial acetic acid, 3:1 v/v) was used to fix the tissues for at least 24 h at −20°C. Fixed material was then stored in 70% ethanol at 4°C until use for slide preparation. The stored root tips were used for slide preparations by employing the technique for cell dissociation of enzymatically macerated roots, as described previously by Herrera *et al.* [80]. Preparations were frozen in liquid nitrogen in order to remove the coverslips, stained with 4',6-diamidino-2-phenylindole, DAPI (1µg/mL), and mounted in Vectashield (Vector Laboratories, Peterborough, UK).

In order to determine the occurrence of chromosome modifications, individual plants of *C. arabica* regenerated by somatic embryogenesis showing a normal (2 plants) or variant (11 plants) phenotype were submitted to karyotype analysis. The Angustifolia, Bullata, Dwarf, Giant and Variegata phenotypic variants were analyzed. During slide examination, mitotic cells at metaphase or prometaphase stages were used for chromosome counting. Between 4 and 8 mitotic cells from each individual were analyzed to determine the chromosome number. The best examples were photomicrographed at metaphase to document the chromosome number and morphology. A Nikon Eclipse 90i epi-fluorescence microscope equipped with a digital, cooled B/W CCD camera (VDS 1300B Vosskühler ®) was used with the appropriate filter (UV-2E/C excitation wavelength 340–380).

Acknowledgments

The authors thank Clément Poncon, owner of the Coffee Research Center 'La Cumplida', and José Martin Hidalgo for their interest and invaluable assistance during this work.

Author Contributions

Conceived and designed the experiments: HE. Performed the experiments: RBL FG GC JCH ED SS PL HE. Analyzed the data: RBL AC BB PL JS HE. Contributed reagents/materials/analysis tools: FG ED HE. Wrote the paper: RBL AC BB HE JS.

References

1. Etienne H, Dechamp E, Barry Etienne D, Bertrand B (2006) Bioreactors in coffee micropropagation (Review). Braz J Plant Physiol 18: 45–54.
2. Miguel C, Marum L (2011) An epigenetic view of plant cells cultured *in vitro*: somaclonal variation and beyond. J Exp Bot 62: 3713–3725.
3. Jähne A, Lazzeri PA, Jäger Gussen M, Lörz H (1991) Plant regeneration from embryogenic cell suspensions derived from anther cultures of barley (*Hordeum vulgare* L.). Theor Appl Genet 82: 74–80.
4. Rival A, Beulé T, Barre P, Hamon S, Duval Y, et al. (1997) Comparative flow cytometric estimation of nuclear DNA content in oil palm (*Elaeis guineensis* Jacq.) tissue cultures and seed-derived plants. Plant Cell Rep 16: 884–887.
5. Lu S, Wang Z, Peng X, Guo Z, Zhang G, et al. (2006) An efficient callus suspension culture system for triploid bermudagrass (*Cynodon transvaalensis* x *C. dactylon*) and somaclonal variations. Plant Cell Tiss Org Cult 87: 77–84.
6. Lambé P, Mutambel HSN, Fouche JG, Deltour R, Foidart JM, et al. (1997) DNA methylation as a key process in regulation of organogenic totipotency and plant neoplastic progression. In Vitro Cell Dev Biol Plant 33: 155–162.
7. Von Aderkas P, Bonga J (2000) Influencing micropropagation and somatic embryogenesis in mature trees by manipulation of phase change, stress and culture environment. Tree Physiol 20: 921–928.
8. Karp A (1994) Origins, causes and uses of variation in plant tissue cultures. In: Vasil IK, Thorpe TA, editors. Plant cell and tissue culture. Dordrecht: Kluwer Academic Publishers. pp. 139–152.
9. Bukowska B (2006) Toxicity of 2,4-Dichlorophenoxyacetic acid – Molecular mechanisms. Polish J of Environ Stud 15: 365–374.
10. Larkin PJ, Scowcroft WR (1981) Somaclonal variation: a novel source of variability from cell cultures for plant improvement. Theor Appl Genet 60: 197–214.
11. Kaeppler SM, Kaeppler HF, Rhee Y (2000) Epigenetic aspects of somaclonal variation in plants. Plant Mol Biol 43: 179–188.
12. Kaeppler SM, Phillips RL (1993) DNA methylation and tissue culture-induced DNA methylation variation in plants. In Vitro Cell Dev Bio Plant 29: 125–130.
13. Peschke VM, Phillips RL (1992) Genetic implications of somaclonal variation in plants. Adv Genet 30: 41–75.
14. Duncan RR (1997) Tissue culture-induced variation and crop improvement. Adv Agron 58: 201–240.
15. Sahijram L, Soneji J, Bollamma K (2003) Analyzing somaclonal variation in micropropagated bananas (*Musa* spp.). In Vitro Cell Dev Bio Plant 39: 551–556.
16. Bairu MW, Aremu OA, Van Staden J (2011) Somaclonal variation in plants: causes and detection methods. Plant Growth Regul 63:147–173.
17. Law J, Jacobsen SE (2010) Establishing, maintaining and modifying DNA methylation patterns in plants and animals. Nat Rev Genet 11: 204–220.
18. Saze H (2008) Epigenetic memory transmission through mitosis and meiosis in plants. Sem Cell Dev Biol 19: 527–536.
19. Reyna Lopez GE, Simpson J, Ruiz Herrera J (1997) Differences in DNA methylation pattern are detectable during the dimorphic transition of fungi by amplification of restriction polymorphisms. Mol Gen Genet 253: 703–710.
20. Bednarek PT, Orłowska R, Koebner RMD, Zimny J (2007) Quantification of tissue-culture induced variation in barley (*Hordeum vulgare* L.). BMC Plant Biol 7: 10.

21. Schellenbaum P, Mohler V, Wenzel G, Walter B (2008) Variation in DNA methylation patterns of grapevine somaclones (*Vitis vinifera* L.). BMC Plant Biol 8: 78.

22. Peschke VM, Phillips RL, Gegenbach BG (1987) Discovery of transposable element activity among progeny of tissue culture-derived maize plants. Science 238: 804–807.

23. Mckenzie NL, Wen LY, Dale J (2002) Tissue-culture enhanced transposition of the maize transposable element Dissociation in *Brassica oleracea* var. 'Italica'. Theor Appl Genet 105: 23–33.

24. Slotkin RK, Martienssen R (2007) Transposable elements and the epigenetic regulation of the genome. Nat Rev Genet 8: 272–285.

25. Anthony F, Bertrand B, Quiros O, Lashermes P, Berthaud J, et al. (2001) Genetic diversity of wild coffee (*Coffea arabica* L.) using molecular markers. Euphytica 118: 53–65.

26. Bertrand B, Alpizar E, Llara L, Santacreo R, Hidalgo M, et al. (2011) Performance of Arabica F1 hybrids in agroforestry and full-sun cropping systems in comparison with pure lines varieties. Euphytica 181: 147–158.

27. Etienne H, Bertrand B, Montagnon C, Bobadilla Landey R, Dechamp E, et al. (2012) Un exemple de transfert technologique réussi en micropropagation: la multiplication de *Coffea arabica* par embryogenèse somatique. Cah Agric 21: 115–24.

28. Söndahl MR, Lauritis JA (1992) Coffee. In: Hammerschlag FA, Litz RE, editors. Biotechnology of Perennial Fruit Crops. Wallingford: CAB International. pp. 401–420.

29. Etienne H, Bertrand B (2003) Somaclonal variation in *Coffea arabica*: effects of genotype and embryogenic cell suspension age on frequency and phenotype of variants. Tree Physiol 23: 419–426.

30. Etienne H, Bertrand B (2001) The effect of the embryogenic cell suspension micropropagation technique on the trueness to type, field performance, bean biochemical content and cup quality of *Coffea arabica* trees. Tree Physiol 21: 1031–1038.

31. Rani V, Raina S (2000) Genetic fidelity of organized meristem-derived micropropagated plants: a critical reappraisal. In Vitro Cell Dev Bio Plant 36: 319–330.

32. Hua HW, Huang TD, Huang HS (2010) Micropropagation of self-rooting juvenile clones by secondary somatic embryogenesis in *Hevea brasiliensis*. Plant Breeding 129: 202–207.

33. Mallón R, Covelo P, Vieitez AM (2012) Improving secondary embryogenesis in *Quercus robur*: Application of temporary immersion for mass propagation. Trees Structure and Function 26: 731–741.

34. Akula A, Becker D, Bateson M (2000) High-yielding repetitive somatic embryogenesis and plant recovery in a selected tea clone, 'TRI-2025', by temporary immersion. Plant Cell Rep 19: 1140–1145.

35. Ducos JP, Alenton R, Reano JF, Kanchanomai C, Deshayes A, et al. (2003) Agronomic performance of *Coffea canephora* P. trees derived from large-scale somatic embryo production in liquid medium. Euphytica 131: 215–223.

36. Krug CA, Carvalho A (1951) The genetics of *Coffea*. Advances in Genetics 4: 127–158.

37. Hornero J, Martínez I, Celestino C, Gallego FJ, Torres V, et al. (2001) Early checking of genetic stability of cork oak somatic embryos by AFLP analysis. Int J Plant Sci 162: 827–833.

38. De la Puente R, Gonzalez AI, Ruiz ML, Polanco C (2008) Somaclonal variation in rye (*Secale cereale* L.) analyzed using polymorphic and sequenced AFLP markers. In Vitro Cell Dev Biol Plant 44: 419–426.

39. Jin S, Mushke R, Zhu H, Tu L, Lin Z, et al. (2008) Detection of somaclonal variation of cotton (*Gossypium hirsutum*) using cytogenetics, flow cytometry and molecular markers. Plant Cell Rep 27: 1303–1316.

40. Gao X, Yang D, Cao D, Ao M, Sui X, et al. (2010) *In vitro* micropropagation of *Freesia hybrida* and the assessment of genetic and epigenetic stability in regenerated plantlets. J Plant Growth Regul 29: 257–267.

41. Rani V, Singh KP, Shiran B, Nandy S, Goel S, et al. (2000) Evidence for new nuclear and mitochondrial genome organizations among high-frequency somatic embryogenesis-derived plants of allotetraploid *Coffea arabica* L. (Rubiaceae). Plant Cell Rep 19: 1013–1020.

42. Heinze B, Schmidt J (1995) Monitoring genetic fidelity *vs* somaclonal variation in Norway spruce (*Picea abies*) somatic embryogenesis by RAPD analysis. Euphytica 85: 341–345.

43. Zhang M, Kimatu JN, Xu K, Liu B (2010) DNA cytosine methylation in plant development. Review. J Genet Genomics 37: 1–12.

44. Vanyushin BF, Ashapkin VV (2011) DNA methylation in higher plants: Past, present and future. Biochim Biophys Acta 1809: 360–368.

45. Smulders MJM, de Klerk GJ (2011) Epigenetics in plant tissue culture. Plant Growth Reg 63: 137–146.

46. Feschotte C, Jiang N, Wessler SR (2002) Plant transposable elements: when genetics meets genomics. Nat Rev Genet 3: 329–341.

47. Li X, Yu X, Wang N, Feng Q, Dong Z, et al. (2007) Genetic and epigenetic instabilities induced by tissue culture in wild barley (*Hordeum brevisubulatum* (Trin.) Link). Plant Cell Tiss Org Cult 90: 153–168.

48. Fiuk A, Bednarek PT, Rybczyński JJ (2010) Flow cytometry, HPLC-RP and met AFLP analyses to assess genetic variability in somatic embryo-derived plantlets of *Gentiana pannonica* scop. Plant Mol Biol Rep 28: 413–420.

49. LoSchiavo F, Pitto L, Giuliano G, Torti G, Nuti Ronchi V, et al. (1989) DNA methylation of embryogenic carrot cell cultures and its variations as caused by

50. mutation, differentiation, hormones, and hypomethylating drugs. Theor Appl Genet 77: 325–331.

51. Gillis K, Gielis J, Peters H, Dhooghe E, Orpins J (2007) Somatic embryogenesis from mature *Bambusa balcooa* Roxburgh as basis for mass production of elite forestry bamboos. Plant Cell Tiss Org Cult 91: 115–123.

52. Peraza Echeverria S, Herrera Valencia VA, Kay AJ (2001) Detection of DNA methylation changes in micropropagated banana plants using methylation-sensitive amplification polymorphism (MSAP). Plant Sci 161: 359–367.

53. Yu X, Li X, Zhao X, Jiang L, Miao G, et al. (2011) Tissue culture-induced genomic alteration in maize (*Zea mays*) inbred lines and F1 hybrids. Ann Appl Biol 158: 237–247.

54. Baránek M, Křižan B, Ondružíková E, Pidra M (2010) DNA-methylation changes in grapevine somaclones following *in vitro* culture and thermotherapy. Plant Cell Tiss Org Cult 101: 11–22.

55. Guo WL, Wu R, Zhang YF, Liu XM, Wang HY, et al. (2007) Tissue culture-induced locus-specific alteration in DNA methylation and its correlation with genetic variation in *Codonopsis lanceolata* Benth et Hook. Plant Cell Rep 26: 1297–1307.

56. Díaz Martínez M, Nava Cedillo A, Guzmán López JA, Escobar Guzmán R, Simpson J (2012) Polymorphism and methylation patterns in *Agave tequilana* Weber var. 'Azul' plants propagated asexually by three different methods. Plant Sci 185–186: 321–330.

57. Karp A (1991) On the current understanding of somaclonal variation. In: Miflin BJ, editor. Oxford surveys of plant molecular and cell biology. Vol 17.Oxford : Oxford University Press. pp. 1–58.

58. Larkin PJ (1987) Somaclonal variation: history, method and meaning. Iowa State 1 Res 61: 393–43.

59. Neelakandan AK, Wang K (2012) Recent progress in the understanding of tissue culture-induced genome level changes in plants and potential applications. Plant Cell Rep 31: 597–620.

60. Al Zahim MA, Ford Loyd BV, Newbury HJ (1999) Detection of somaclonal variation in garlic (*Allium sativum* L.) using RAPD and cytological analysis. Plant Cell Rep 18: 473–477.

61. Hao YI, Deng XX (2002) Occurrence of chromosomal variations and plant regeneration from long-term-cultured citrus callus. In Vitro Cell Dev Biol Plant 38: 472–476.

62. Mujib A, Banerjee S, Dev Ghosh P (2007) Callus induction, somatic embryogenesis and chromosomal instability in tissue culture-raised hippeastrum (*Hippeastrum hybridum* cv. United Nations). Propag Ornam Plants 7: 169–174.

63. Leal F, Loureiro J, Rodriguez E, Pais MS, Santos C, et al. (2006) Nuclear DNA content of *Vitis vinifera* cultivars and ploidy level analyses of somatic embryo-derived plants obtained from anther culture. Plant Cell Rep 25: 978–985.

64. Sanchez Teyer LF, Quiroz Figueroa F, Loyola Vargas V, Infante D (2003) Culture-induced variation in plants of *Coffea arabica* cv. Caturra rojo, regenerated by direct and indirect somatic embryogenesis. Mol Biotechnol 23: 107–115.

65. Ménendez Yuffá A, Da Silva R, Rios L, Xena de Enrech N (2000) Mitotic aberrations in coffee (*Coffea arabica* L. 'Catimor') leaf explants and their derived embryogenic calli. Electron J Biotechnol 3: 0–5. ISSN 0717–3458.

66. Zoriniants SE, Nosov AV, Monforte Gonzalez M, Mendes Zeel M, Loyolas Vargas VM (2003) Variation of nuclear DNA content during somatic embryogenesis and plant regeneration of *Coffea arabica* L. using cytophotometry. Plant Sci 164: 141–146.

67. Gözükirmizi N, Ari S, Oraler G, Okatan Y, Ünsal N (1990) Callus induction, plant regeneration and chromosomal variations in barley. Acta Botanica Neerlandica 39: 379–387.

68. Jaligot E, Beulé T, Baurens FC, Billote N, Rival A (2004) Search for methylation-sensitive amplification polymorphism associated with the 'mantled' variant phenotype in oil palm (*Elaeis guineensis* Jacq.). Genome 47: 224–228.

69. Birchler JA, Veitia RA (2007) The gene balance hypothesis: from classical genetics to modern genomics. Plant Cell 19: 395–402.

70. Henry IM, Dilkes BP, Miller ES, Burkart Waco D, Comai L (2010) Phenotypic consequences of aneuploidy in *Arabidopsis thaliana*. Genetics 186: 1231–45.

71. Makarevitch I, Harris C (2010) Aneuploidy causes tissue specific qualitative changes in global gene expression patterns in maize. Plant Physiol 152: 927–938.

72. Fukuoka H, Kawata M, Tkaiwa M (1994) Molecular changes of organelle DNA sequences in rice through dedifferentiation, long-term culture or the morphogenesis process. Plant Mol Biol 26: 899–907.

73. Cenci A, Combes MC, Lashermes P (2012) Genome evolution in diploid and tetraploid *Coffea* species as revealed by comparative analysis of orthologous genome segments. Plant Mol Biol 78: 135–145.

74. Ribas A, Dechamp E, Bertrand B, Champion A, Verdeil JL, et al. (2011) *Agrobacterium tumefaciens*-mediated genetic transformation of *Coffea arabica* (L.) is highly enhanced by using long-term maintained embryogenic callus. BMC Plant Biol 11: 92.

75. Murashige T, Skoog FA (1962) Revised medium for rapid growth and bioassays with tobacco tissue cultures. Physiol Plant 15: 473–497.

76. Etienne H (2005) Protocol of somatic embryogenesis: Coffee (*Coffea arabica* L. and *C. canephora* P.). In: Jain SM, Gupta PK, editors. Protocols for somatic embryogenesis in woody plants. Series: Forestry Sciences. Vol 77. The Netherlands: Springer edn. pp. 167–179.

77. Teisson C, Alvard D (1995) A new concept of plant *in vitro* cultivation liquid medium: temporary immersion. In: Terzi M et al., editors. Current Issues in Plant Molecular and Cellular Biology. Dordrecht: Kluwer Academic Publishers. pp. 105–110.

77. Yasuda T, Fujii Y, Yamaguchi T (1985) Embryogenic callus induction from *Coffea arabica* leaf explants by benzyladenine. Plant Cell Physiol 26: 595–597.

78. Dellaporta SL, Wood J, Hicks JB (1983) A plant DNA miniprepration: version II. Plant Mol Biol Rep 1: 19–21.

79. Vos P, Hogers R, Bleeker M, Rijans M, Van de Lee T, et al. (1995) AFLP: a new technique for DNA fingerprinting. Nucleic Acids Research 23: 4407–4414.

80. Herrera JC, D'hont A, Lashermes P (2007) Use of fluorescence *in situ* hybridization as a tool for introgression analysis and chromosome identification in coffee (*Coffea arabica* L.). Genome 50: 619–626.

miR393 Is Required for Production of Proper Auxin Signalling Outputs

David Windels[1], Dawid Bielewicz[1,2], Miryam Ebneter[1], Artur Jarmolowski[2], Zofia Szweykowska-Kulinska[2], Franck Vazquez[1]*

1 Department of Environmental Sciences, Section of Plant Physiology, University of Basel, Zurich-Basel Plant Science Center, Part of the Swiss Plant Science Web, Basel, Switzerland, **2** Department of Gene Expression, Faculty of Biology, Adam Mickiewicz University, Poznan, Poland

Abstract

Auxins are crucial for plant growth and development. Auxin signalling primarily depends on four partially redundant F-box proteins of the TIR1/AFB2 Auxin Receptor (TAAR) clade to trigger the degradation of AUX/IAA transcriptional repressors. Auxin signalling is a balanced system which involves complex feedback regulations. miR393 regulation of *TAAR* genes is important for different developmental programs and for responses to environment. However, so far, the relevance of the two *MIR393* genes for Arabidopsis leaf development and their significance for auxin signalling homeostasis have not been evaluated. First, our analyses of *mir393a-1* and *mir393b-1* mutants and of *mir393ab* double mutant show that the two genes have only partially redundant functions for leaf development. Expression analyses of typical auxin-induced reporter genes have shown that the loss of miR393 lead to several unanticipated changes in auxin signalling. The expression of *DR5pro:GUS* is decreased, the expression of primary *AUX/IAA* auxin-responsive genes is slightly increased and the degradation of the AXR3-NT:GUS reporter protein is delayed in *mir393ab* mutants. Additional analyses using synthetic auxin and auxin antagonists indicated that miR393 deficient mutants have higher levels of endogenous AUX/IAA proteins, which in turn create a competition for degradation. We propose that the counter-intuitive changes in the expression of *AUX/IAA* genes and in the accumulation of AUX/IAA proteins are explained by the intrinsic nature of *AUX/IAA* genes which are feedback regulated by the AUX/IAA proteins which they produce. Altogether our experiments provide an additional highlight of the complexity of auxin signaling homeostasis and show that miR393 is an important component of this homeostasis.

Editor: Abidur Rahman, Iwate University, Japan

Funding: This work was supported by the Swiss National Science Foundation (Ambizione grant PZ00P3_126329 and PZ00P3_142106 to FV) and by the Rector's Conference of the Swiss Universities (SCIEX-NMS fellowship 11.115 to F.V., Z.S.K. and D.B). The Ph.D. fellowship of DB is part of the International Ph.D. Program "From genome to phenotype: A multidisciplinary approach to functional genomics" funded by the Foundation for Polish Science (FNP). The funders had no role in study design, data collection and analysis, decision to publish, or preparation of the manuscript.

Competing Interests: The authors have declared that no competing interests exist.

* E-mail: franck.vazquez@unibas.ch

Introduction

Auxins are phytohormones important for plant growth, organogenesis and various responses to environmental changes [1,2]. Auxin signalling primarily depends on perception by four partially-redundant auxin receptors of the TRANSPORT INHIBITOR RESPONSE 1 (TIR1)/AUXIN SIGNALLING F-BOX PROTEIN 2 (AFB2) clade [3–5]. Upon binding auxins, these TAAR proteins, which are the specificity-components of SKIP/CULLIN/F-BOX (SCF)-ubiquitin ligase complexes, form a co-receptor complex with AUXIN/INDOLE-3-ACETIC ACID (AUX/IAA) transcriptional repressors [6–8]. AUX/IAA proteins are then ubiquitinated and degraded by the 26S proteasome [7]. This leads to the release of AUXIN RESPONSE FACTOR (ARF) transcription factors to which AUX/IAA were bound and allows ARFs to activate or repress the transcription of primary auxin-responsive genes [9]. The homeostasis of auxin signaling involves feedback-regulation of AUX/IAA genes' expression by the AUX/IAA proteins which they generate. The homeostasis of *TAAR* genes expression was shown to involve the microRNA miR393. This regulation is important for innate immunity [10,11], for root response to nitrate, salinity and drought resistance [12–15] and for

several aspects of rice and Arabidopsis development [15–19]. Our own work has shown that this regulation additionally involves the function of secondary siRNAs, the siTAARs, which are generated from *TAAR* transcripts downstream of the miR393 cleavage site [16,17]. MiR393 is generated from two distinct genes, *MIR393A* and *MIR393B* [20]. *AtMIR393A*, but not *AtMIR393B*, was shown to be induced by stress while *MIR393B*, but not *AtMIR393A*, was shown to be induced by auxin [10,19]. Moreover, our work has suggested that miR393 is primarily produced from *AtMIR393A* in the roots and primarily from *AtMIR393B*, in the aerial parts of plants grown in normal growth conditions [17]. Thus, these observations suggested that *MIR393* genes have major distinct functions: *AtMI393A* being primarily involved in the plant responses to environment, and, *AtMIR393B* being primarily involved in the regulation of auxin homeostasis and of auxin-dependent plant development [10–19]. Intriguingly, *mir393b-1* mutants which accumulate only trace amounts of miR393 in aerial parts of plants exhibit rather mild phenotypes in normal growth conditions essentially characterized by a pronounced leaf epinasty [17]. Thus, this raised the hypothesis that *AtMIR393A* or other pathways compensate for the loss of *AtMIR393B*.

MiR393 is important for the regulation of *TAAR* genes and is part of a complex homeostatic process which involves feedback transcriptional regulations [19]. However, the significance of miR393 for auxin signalling homeostasis has not been evaluated directly in mutants lacking miR393. To gain insights into these important aspects we have obtained single and double mutants of *MIR393* genes and we have analyzed the impact of these mutations on plant development, on physiological response to auxin and at the molecular level of auxin signalling. The data which we obtained show that the two *MIR393* genes have partially redundant functions for leaf polarity with a primarily role for *AtMIR393B*. Moreover, we have observed that the expression level of the artificial reporter gene *DR5pro:GUS* is slightly decreased in these mutants compared to wt plants while the expression of primary auxin-induced genes is increased. Moreover, experiments using synthetic auxin, auxin antagonists and the *HSpro:AXR3-NT:GUS* reporter gene to monitor the degradation of AUX/IAA proteins showed that the degradation rate of the AXR3-NT:GUS protein is longer in the mutants than in wt plants. These unanticipated results demonstrate that the loss of miR393 leads to complex changes and to simultaneously increase the basal expression level of *AUX/IAA* genes and the basal level of AUX/IAA protein accumulation. Together our data show that miR393 is an important component of the auxin signalling homeostasis required for the establishment of proper and timely auxin signalling outputs.

Results

MIR393 Genes Have Distinct, but Partially Overlapping Expression Patterns

The microRNA miR393 is encoded by two genes, *AtMIR393A* (*At2g39885*) and *AtMIR393B* (*At3g55734*) (Fig.1A) [21]. Our earlier studies using *mir393b-1* mutants showed that miR393, which primarily arises from *AtMIR393B* in aerial organs, is important for leaf morphology [17]. However, *mir393b-1* mutants exhibit only mild developmental phenotypes. To test whether the weak amounts of miR393 generated from *AtMIR393A* in *mir393b-1* are responsible for this observation, we identified a mutant in the SALK collection which has a T-DNA insertion located in the proximal region of the *AtMIR393A* promoter. Sequencing of the PCR fragments showed that the insertion is located 74-nt upstream of the transcription start which we have identified by Rapid Amplification of cDNA Ends (RACE) experiments (Fig. 1A and S1). In normal growth conditions, this mutant, which we named *mir393a-1*, accumulated 30% lower levels of miR393 in the roots than the wt plants and accumulated normal levels of miR393 in the leaves (Fig.1B). The double mutant *mir393a-1 mir393b-1* (hereafter noted *mir393ab*) accumulated 50% lower levels of miR393 than the wt plant in the roots and only 1% of the wt level in the leaves (Fig. 1B). These experiments showed that *mir393a-1* is not a null-mutant and that the double mutant accumulates lower levels of miR393 than *mir393a-1* or *mir393b-1* alone. These experiments also showed that the two *MIR393* genes have distinct, but partially overlapping expression profiles.

MIR393 Genes Have Overlapping, Partially Redundant Functions in Leaf Development

A first inspection of the different mutant plants showed that neither *mir393a-1* nor *mir393ab* mutants exhibit a more drastic developmental defects than *mir393b-1*. We analyzed whether the two *MIR393* genes have redundant functions in leaf development by, first, recording the incidence of cotyledon epinasty (ICE) which is a typical auxin hypersensitive response regulated by miR393

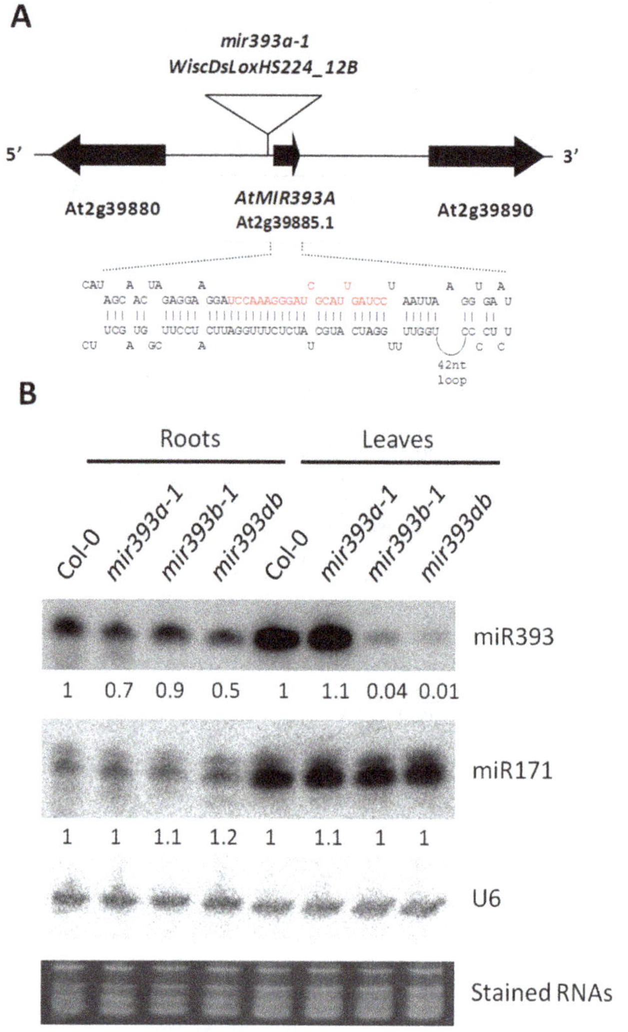

Figure 1. Identification and characterization of *MIR393A* T-DNA insertion mutants. (A) Schematic map showing the 5′-3′ orientation of genes (large arrows and gene names are indicated) flanking *AtMIR393A* on chromosome II. The arrow representing *AtMIR393A* indicates the pri-miRNAs which full-length sequence determined by RACE experiments (given in Fig. S1). The expanded region represents the folded pre-miR393 nucleotide sequence with the miR393 sequence indicated in red. The open triangle represents the T-DNA insertion. The scheme is not drawn to scale. (B) RNA-blot hybridization of RNA prepared from roots and leaves of 34d-old plants. Probed RNAs are indicated on the right. The signal detected for mutants relative to wild-type Col-0 are normalized relative to the *Midori Green* stained RNA signals. Similar results were obtained in three independent experiments.

[17,22]. When grown on standard medium, a high and significantly greater fraction of *mir393a-1* (46%), of *mir393b-1* (39%) and of *mir393ab* mutants (56%) than wild-type plants (16%) exhibited the extreme cotyledon epinasty phenotype (Fig. 2A). In the cases of *mir393a-1* and *miR393b-1*, this ICE was similarly decreased by increasing the concentration of the auxin transport inhibitor NPA (1-N-naphthylphthalamic acid) to 0.1 or 0.5 µM and this ICE was similar to that of wt at a concentration of 1 µM (Fig. 2B) [23]. In *mir393ab* however the incidence of cotyledon epinasty remained significantly higher even at high NPA concentrations. These observations established that the cotyledon

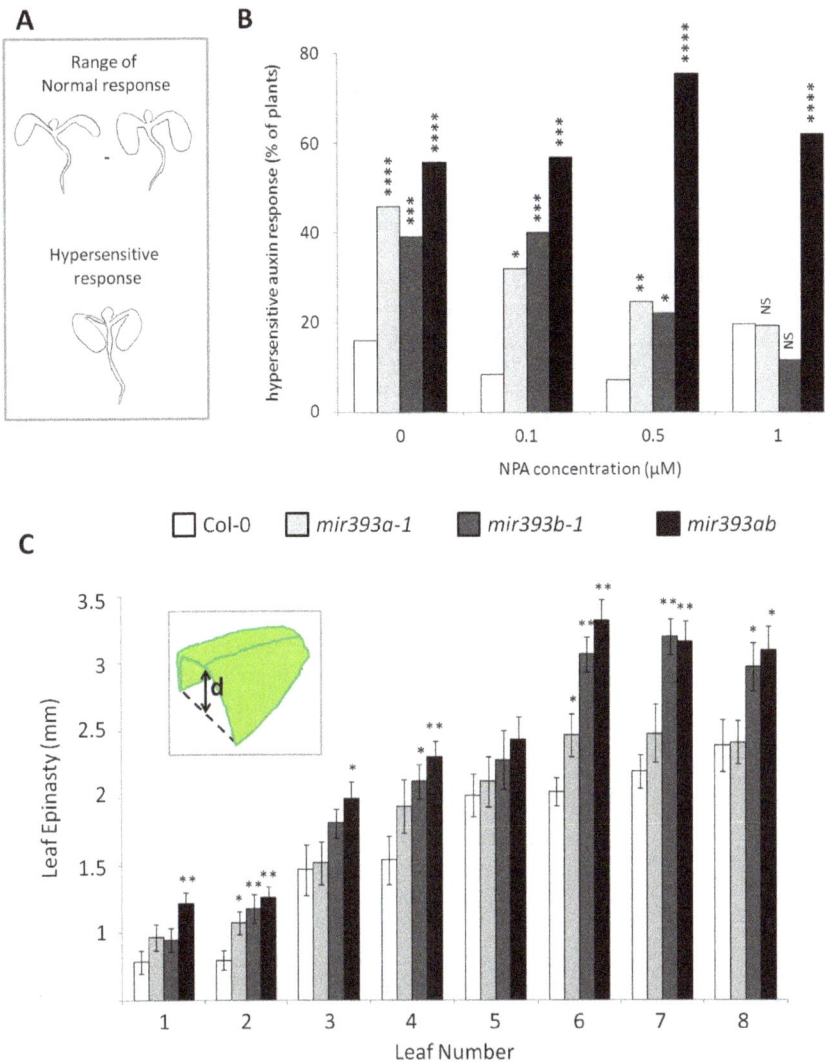

Figure 2. AtMIR393A and AtMIR393B are partially redundant for proper leaf morphogenesis. (A) Schematic representation of the normal range of cotyledon epinasty (top) and the extreme cotyledon epinasty (bottom) typical of the auxin-hypersensitive response. (B) The incidence of cotyledon auxin-hypersensitive response in populations of Col-0 (open bars), mir393a-1 (light grey bars), mir393b-1 (dark grey bars), and mir393ab double mutants (dark bars). Seedlings (n>40 for each condition and genotype) were grown on media containing the concentration of NPA indicated and harvested 4 d after germination. P values (two-tailed Fisher's exact test) for significant differences towards Col-0 are indicated; NS for P>0.05, * for P≤0.05, ** for P≤0.01, *** for P≤0.001, **** for P≤0.0001. P values for significant differences between mutants are given in Fig. S2. (C) Epinasty of leaf number 1 to 8 for Col-0, mir393a-1, mir393b-1 and mir393ab was measured by the vertical distance between the adaxial leaf side and the leaf margin (in mm ± SEM) (see drawing in the insert). Significant difference towards Col-0 is indicated (two-tailed student t-test). * for P≤0.05, ** for P≤ 0.01. N = 10. P values for significant differences between mutants are given in Fig. S2.

epinasty response to auxin depends on the overlapping function of both *MIR393* genes.

Next, we measured the degree of leaf epinasty for the first 8 leaves of the plants. For *mir393a-1* mutants, most of the leaves were more epinastic than wt plants but the difference to wt was significant only for leaves #2 and #6 (Fig. 2C). For *mir393b-1*, the epinasty was greater than that recorded for *mir393a-1* and was significantly different from wt for more than half of the leaves. For *mir393ab* mutants, the leaf epinasty was again greater than that of *mir393b-1* and the difference to wt was highly significant for all leaves except for leaf #5. These observations showed that *AtMIR393B* has a main role in the regulation of leaf epinasty and that *AtMIR393A* contributes in a slightly redundant manner to regulate the underlying developmental process.

MiR393-Deficient Mutants Exhibit Decreased, rather than increased, Expression of DR5pro:GUS

The loss of miR393 and the consequent increase in *TAAR* genes' expression [10,17] is expected to increase the degradation of AUX/IAA proteins and to increase the expression of *AUX/IAA* genes. However, because the expression of *AUX/IAA* genes is feedback-regulated by the AUX/IAA which they generate, it is uncertain at which level the homeostasis of *AUX/IAA* genes will then be maintained in miR393-deficient mutants. To gain insights into the role of miR393 in this homeostasis, we first analyzed the expression of the artificial auxin-induced gene *DR5pro:GUS*, which contains seven soybean auxin response elements (ARE) and serves to report the state of the auxin signalling output [24].

Without treatments, the expression of *DR5pro:GUS* was similar in the distal tips of *mir393a-1*, *mir393b-1* and *mir393ab* and wt leaves while it was slightly lower in the margins of *mir393* mutants' leaves than in those of wt leaves (Fig. 3A). Treatment with 2,4-D, a synthetic diffusible auxin, induced the expression of *DR5pro:GUS* in all plant genotypes. However, its expression was lower in the blades of *mir393* mutants leaves, in the order *mir393a-1>mir393b-1 = mir393ab*, than in those of wt leaves (Fig. 3A). Thus, these results showed that the expression of *DR5pro:GUS* is changed in a counter-intuitive manner by the loss of miR393. *DR5pro:GUS* is more stably repressed in *mir393* mutants than in wt plants. Thus, these results suggest that the loss of miR393 leads to more complex

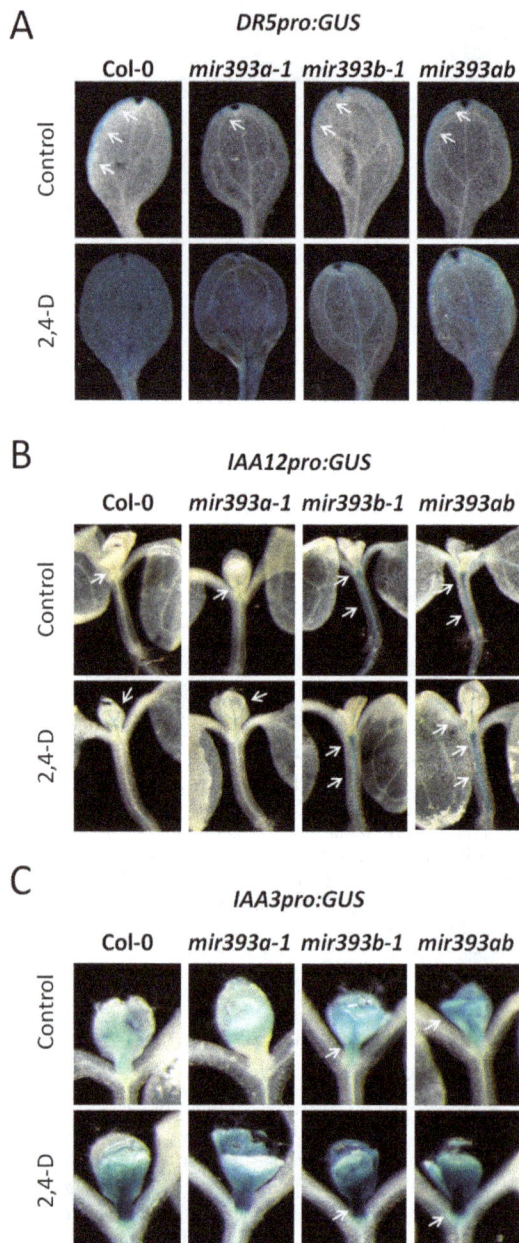

A

DR5pro:GUS

Col-0   mir393a-1   mir393b-1   mir393ab

Control

2,4-D

B

IAA12pro:GUS

Col-0   mir393a-1   mir393b-1   mir393ab

Control

2,4-D

C

IAA3pro:GUS

Col-0   mir393a-1   mir393b-1   mir393ab

Control

2,4-D

Figure 3. Basal expression level of primary auxin-inducible AUX/IAA genes in *mir393* mutants compared to wt plants. Pictures of representative wt and *mir393* mutant plants expressing the *DR5pro:GUS* (A), the *IAA12pro:GUS* (B), or the *IAA3pro:GUS* gene (C) upon treatment or not with 10 μM 2,4-D for 8 h. Arrows highlight the GUS staining detected.

changes than initially anticipated, especially at the level of AUX/IAA proteins.

MiR393-Deficient Mutants Exhibit Increased Basal Expression of AUX/IAA Genes

To better understand the changes triggered by the loss of miR393, we analyzed the expression of artificial genes reporting the activity of the primary auxin-induced promoters of *IAA12/BDL* and *IAA3/SHY2* genes in wt and in miR393-deficient mutants (Fig. 3B–C). Without treatment, the expression level of *IAA12pro:-GUS* was undetectable in wt plants, was weakly detectable in the shoot tips and roots of *mir393a-1* mutants, and was ectopically expressed at high levels in the shoot tips and hypocotyls of *mir393b-1* mutants (Fig. 3B). Interestingly, *IAA12pro:GUS* was also ectopically expressed in the shoot tips, in the hypocotyls and in the leaf veins of *mir393ab* mutants but at even higher levels. For *IAA3pro:GUS* we did not observe ectopic expression in the mutants but it was detected at high and increasing levels in the order *mir393b-1<mir393ab* while it was only weakly expressed in emerging leaves of wt and *mir393a-1* plants (Fig. 3C). Together these observations indicated that the two primary auxin-induced genes *IAA3/SHY2* and *IAA12/BDL* have higher basal expression levels in miR393-deficient mutants than in wt plants. Moreover, the mutation in *AtMIR393B* has a greater effect on *AUX/IAA* genes expression than the mutation in *AtMIR393A*.

Importantly, treatments of plantlets with 2,4-D for 8h induced the expression of *IAA12pro:GUS* and *IAA3pro:GUS* genes in all plant genotypes (Fig. 3B–C). In these conditions, the two reporter genes were also more expressed in *mir393a-1*, *mir393b-1* and *mir393ab* mutants than in wt plants. These data showed that miR393 is important to maintain the proper basal expression level of these two *AUX/IAA* genes and for their proper induction.

MiR393-Deficient Mutants Exhibit Complex Changes in the steady state of AUX/IAA and GH3 mRNAs

To determine whether the observations made with the two reporter genes also apply to endogenous genes, we analyzed the mRNA steady state of few auxin-induced genes (Fig. 4). We selected 5 *AUX/IAA* and 3 *GH3* genes based on their high expression level reported in the *Arabidopsis eFP browser hormone series* (www.bar.utoronto.ca) [25]. Without auxin treatment, *IAA17* and *IAA19* mRNAs accumulated to higher levels in *mir393ab* mutants than in wt plants, and, the expression of the other 6 genes was not detected. The expression of all genes tested was induced by treatment with 2,4-D. In these conditions, *GH3.5* mRNAs accumulated to similar levels in wt and *mir393ab* mutants, *GH3.2*, *IAA19*, *IAA29*, *IAA5* and *IAA2* accumulated to higher levels in *mir393ab* mutants than in wt plants, while *GH3.6* and *IAA17* accumulated to lower levels in *mir393ab* mutants than in wt plants. Thus, although all genes tested were induced by auxin treatments in the *mir393ab* mutants, their accumulation was substantially different from that observed in wt plants. Together with the reporter genes analyses, these data showed that miR393 plays an important role to maintain the proper basal expression and for the accurate induction of *AUX/IAA and GH3* genes.

MIR393 is Necessary for Proper Degradation of AXR3NT:GUS Proteins

Next, we used the heat shock inducible *HSpro:AXR3-NT:GUS* line to assay the level of accumulation of AUX/IAA proteins and the efficiency of their degradation [7]. We anticipated that an increase in AUX/IAA protein steady-state levels should create a competition for ubiquitination and/or for degradation between

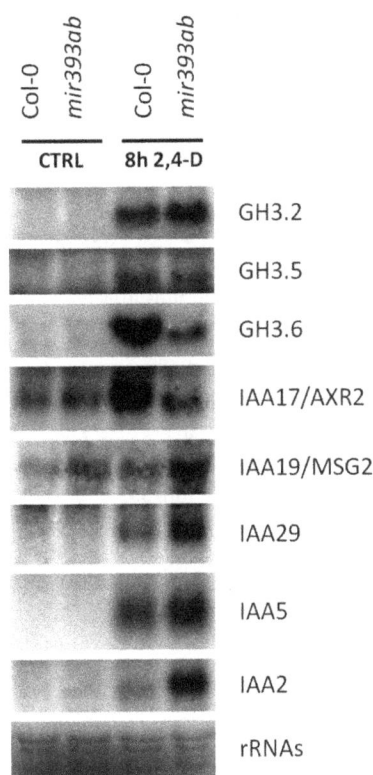

Figure 4. miR393 is required for proper induction of *AUX/IAA* and *GH3* genes. Northern blots of RNAs from wt and *mir393ab* mutant plantlets treated or not with 10 μM 2,4-D for 8 h. Without treatments, *IAA17* and *IAA19* have higher steady state levels in *mir393ab* mutants than in wt plants. Treatment with 2,4-D for 8 h induces the expression of all genes tested. *GH3.5* accumulates to similar levels in wt and mutants, *GH3.2, IAA19, IAA29, IAA5* and *IAA2* accumulate to higher levels in *mir393ab* mutants than in wt plants while *GH3.6* and *IAA17* accumulate to lower levels in *mir393ab* mutants than in wt plants.

AUX/IAAs and might therefore affect the degradation rate of AXR3-NT:GUS. After induction of the *HSpro:AXR3-NT:GUS* gene and without other treatments (Fig. 5; control) the level of AXR3-NT:GUS proteins was consistently higher in the roots of *mir393a-1, mir393b-1* and *mir393ab* mutants than in wt plants (first row). This result showed that in miR393-deficient mutants the AXR3-NT:GUS proteins are not degraded as efficiently as in wt plants. Brief treatment with 2,4-D during induction to enhance AUX/IAA protein degradation lead to a marked decrease in the accumulation of AXR3-NT:GUS proteins in *mir393* mutants compared to the control condition; Although the level was still higher in *mir393* mutants than in wt plants (second row). Similarly, brief treatment with auxinole (α-[2,4-dimethylphenylethyl-2-oxo]-IAA) which prevents the degradation of AUX/IAA proteins by binding to TIR1/AFBs [26] lead to increase the accumulation of AXR3-NT:GUS proteins (Fig. 5; third row). Brief treatment with MG132 which specifically inhibits the function of the proteasome didn't lead to substantial changes in the accumulation of AXR3-NT:GUS proteins (Fig. 5; fourth row). Thus, these experiments showed that the SCF$^{TIR/AFB}$ complexes and the proteasome are functional, although less efficient, in *mir393* mutants than in wt plants. These experiments suggest that the level of auxins is the limiting factor and that the higher steady state level of endogenous AUX/IAA proteins in *mir393* mutants cannot be degraded as fast as in the wt. We further challenged this possibility by treating the plants with 2,4-D four hours before induction of the *HSpro:AXR3-NT:GUS* gene. As we anticipated, chasing the higher levels of AUX/IAA proteins by inducing their degradation with high quantities of auxin lead to restore the proper degradation of the AXR3-NT:GUS protein (Fig. 5; fifth row). Together these experiments showed that miR393 is required to ensure the proper degradation of the AXR3-NT:GUS fusion proteins and of other AUX/IAA proteins. Moreover, miR393-deficient mutants seem to indeed accumulate higher steady state levels of endogenous AUX/IAA proteins.

Discussion

Auxins are major plant hormones with crucial functions in almost every aspect of plant life [1,2]. Identification of mutants affected in synthesis, transport and perception of auxins have demonstrated the importance of auxins in developmental and morphogenetic programs [5,27]. MiR393 is encoded by two distinct genes and regulates the expression of *TIR1/AFB2* auxin receptor genes. According to the crucial role of auxins, we anticipated that miR393-deficient mutants, which we showed in *mir393b-1* mutants, fail to regulate *TAAR* mRNA levels, should exhibit pleiotropic developmental defects. However, no obvious developmental defects other than mild leaf polarity defects were observed in the single mutants *mir393a-1, mir393b-1* or in the double mutants *mir393ab* grown in standard laboratory conditions. We suspect that *mir393a-1* is not a null mutant since this mutation leads to a 30% decrease in the level of mature miR393 in the roots.

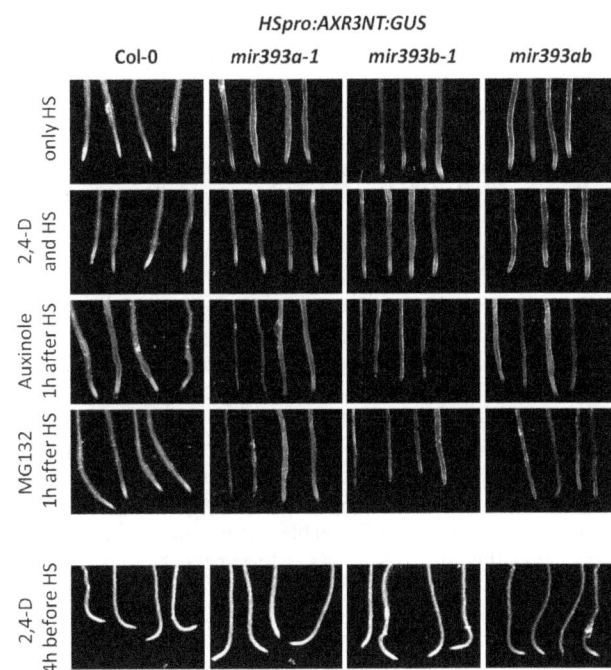

Figure 5. miR393 is required for proper degradation of AXR3-NT:GUS proteins. Representative pictures of wt and *mir393* mutant roots expressing the *HSpro:AXR3-NT:GUS* gene. Without treatments, the AXR3-NT:GUS fusion protein is more stable in *mir393a-1, mir393b-1* and *mir393ab* mutants compared to wt plants. Short treatments with 10 μM auxin induces the degradation of AXR3-NT:GUS, while treatment with 20 μM auxinole or MG132 blocks the degradation of AXR3-NT:GUS. This shows that the SCF complexes and proteasome are functional in the mutants as well. Clearance of presumably high levels of endogenous AUX/IAA proteins by treatment with 2,4-D for 4 h restores the proper degradation of AXR3-NT:GUS.

Another possibility is that a third unidentified *MIR393* gene produces the remaining miR393 in the roots. Chen and colleagues have shown that the expression of *MIR393* and *TAAR* genes are under transcriptional feedback regulations [19]. Thus, it is possible that these feedbacks can compensate for the loss of miR393. However, we do not favor this possibility since we have observed clear effects on expression of primary auxin-induced genes and on degradation of AUX/IAA proteins in miR393-deficient mutants. Thus, we suspect that other unrelated pathways, rather than the above mentioned feedback mechanisms solely, act redundantly with miR393 and somehow compensate for its loss.

Although *mir393a-1* is unlikely a null mutant, our studies with *mir393a-1*, *mir393b-1* and *mir393ab* has shown that both *MIR393* genes contribute in a partially redundant fashion to the establishment of leaf polarity in arabidopsis; with *AtMIR393B* gene playing the predominant role. We found that all miR393-deficient mutants exhibit a hypersensitive auxin response characterized by an extreme epinasty of the cotyledons. For *mir393a-1* and *mir393b-1*, the hypersensitivity phenotype was prevented when the plants were grown on increasing concentrations of the auxin transport inhibitor NPA. For *mir393ab* double mutants the extreme auxin response was not prevented by the highest concentrations of NPA. When we increased the concentration of NPA to 2 μM or higher, the cotyledon epinasty phenotype could not be analyzed appropriately since the plants did not germinate or grow correctly (not shown).

Auxin homeostasis relies on intricate feedback regulations at several levels including synthesis, transport and signaling *via* AUX/IAA proteins. The data which we report here add to the current picture and show that the regulation of *TAAR* genes homeostasis by miR393 plays a significant role for the homeostasis of auxin signaling. We found that the basal steady-state expression level of *IAA12pro:GUS*, of *IAA3pro:GUS* and of several endogenous *AUX/IAA* genes is higher in miR393-deficient mutants than in wt plants. Importantly, this and additional experiments suggest that the increase in *AUX/IAA* gene expression leads to a concomitant increase in the basal steady state level of AUX/IAA proteins. These observations are counter-intuitive on first sight since an increase in the expression level of *AUX/IAA* genes leading to an increase in the level of AUX/IAA proteins should result in a decrease in the expression of *AUX/IAA* genes. Conversely, a decrease in the expression level of *AUX/IAA* genes leading to a decrease in the level of AUX/IAA proteins should result in an increase in the expression of *AUX/IAA* genes. We believe that the counter-intuitive observations are due to the intrinsic homeostatic nature of *AUX/IAA* genes which are feedback regulated by the AUX/IAA proteins which they generate. Indeed, such feedback-regulated systems lead to meta-stable steady state levels. Thus, we believe that the *AUX/IAA* genes, which exhibit a concomitant increase in gene expression and in the level of the corresponding AUX/IAA proteins, have actually reach a higher meta-stable steady-state level in miR393-deficient mutants than in wt plants. We have not been able to challenge this hypothesis by straightforward measurements of endogenous AUX/IAA protein levels because antibodies for AUX/IAA proteins or arabidopsis lines expressing tagged versions of AUX/IAA proteins under their native promoters are not available. We speculate that the expression of a given *AUX/IAA* gene should be even higher in a *mir393ab* mutant expressing non functional AUX/IAA proteins. However, this approach would require the production of mutants with several orders of magnitude. However, we have been able to provide indirect evidences supporting our hypothesis. Indeed, experiments using *HSpro:AXR3-NT:GUS* showed that the AXR3-NT:GUS fusion proteins are not properly degraded in miR393-

deficient mutants. This indicated that the proper degradation of the AXR3-NT:GUS proteins in auxin-limiting conditions is prevented by the high levels of AUX/IAA proteins. This conclusion was also supported by the observations that treatment of plants with 2,4-D for 4 h before inducing the *HSpro:AXR3-NT:GUS* gene, which induces the degradation of endogenous AUX/IAA proteins, could restore the proper degradation of AXR3-NT:GUS. Finally, expression analysis of *DR5pro:GUS*, the universal auxin signalling output marker gene, also support our hypothesis since its expression level was slightly lower in miR393-deficient mutants than in wild type plants, and thus indicate that it is slightly more repressed in mR393-deficient mutants than in wt plants.

The delay of AXR3-NT:GUS degradation is more pronounced in the roots of *mir393b-1* than in those of *mir393a-1*. However, our sRNA blot experiments showed than miR393 accumulates to lower levels in the roots of *mir393a-1* than in those of *mir393b-1*. Thus, this discrepancy suggests that the loss of miR393 observed in the shoots of *mir393b-1* leads to increase the level of competing AUX/IAA proteins in the roots. We speculate that this is achieved either by an increased synthesis or by an increased transport.

Earlier studies had use miR393-resistant target genes and miR393 overexpressers to unravel the function and roles of miR393 [10-12,14,15,19]. However, although miR393 had been shown to regulate the expression of *TAAR* genes and of auxin-responsive genes, and moreover to be involved in important biological processes, its functional significance for auxin signalling and its homeostasis had not really been evaluated directly. Our experiments now clarify the picture and demonstrate that miR393 is necessary to maintain low basal expression levels of *AUX/IAA* genes, low basal levels of AUX/IAA proteins and that these features are important for the proper degradation of AUX/IAA proteins. Thus, miR393 is not only important for *TAAR* genes homeostasis but also for the establishment of proper auxin signalling outputs and for auxin signalling homeostasis *per se*.

Materials and Methods

Plant Material

miR393a-1 mutant was obtained by PCR-based genotyping of plants from the seed batch WiscDsLoxHS224_12B obtained from the NASC stock center. The *mir393ab* double mutant was obtained by PCR-based genotyping of F2 seedlings obtained from the cross of *mir393a-1* and *mir393b-1* [17]. The *DR5pro:GUS*, *IAA12pro:GUS*, *IAA3pro:GUS* and *HSpro:AXR3-NT:GUS* reporters were introgressed in *mir393a-1*, *mir393b-1* and *mir393ab* mutants by crossing and homozygous plants were found by PCR-based genotyping in the F2 population. In all cases, double homozygosis was ascertained on at least 9 plants of the F3 population.

Growth Conditions & Treatments

Arabidopsis thaliana plants used to prepare RNA blots and to measure leaf epinasty were grown on soil in the greenhouse as described previously [28].

Studies of cotyledon epinasty were done as we previously described [17] on Murashige and Skoog (MS) solid medium supplemented as indicated with NPA (1-N-naphthylphthalamic acid, Fluka) in DMSO or DMSO alone as a control.

Measurements of leaf epinasty were made on plants grown for 45 days in short-day (SD) conditions (8 h light/16 h dark). Root measurements were made on plantlets germinated on MS medium and transferred for 8 days in vertically oriented square-plates on solid MS medium supplemented as indicated with 0.1 μM 2,4-D or 20 µM auxinole in SD conditions. For studies of *IAA3pro:GUS*, *IAA12pro:GUS* and *DR5pro:GUS* expression, 7 days old seedlings were incubated in MS medium containing 10 µM 2,4-D in ethanol or ethanol alone as a control for the time indicated before proceeding to staining. For studies of AXR3-NT:GUS fusion protein stability, 7 days old seedlings were placed in water at 37°C for 2 h to induce the expression of the *HSpro:AXR3-NT:GUS* gene and incubated for 30 minutes at room temperature before proceeding to GUS staining. For the treatments, either 10 µM 2,4-D were added to the solution just before the heat shock or 20 µM auxinole were added after 1 h of heat-shock as it was described [7,26].

RNA preparation and RNA Analysis

Extraction of total RNAs, preparation of RNA blots and qRT-PCRs were done as previously described [28].

Histochemical GUS Assays

Arabidopsis seedlings were incubated into staining solution containing 1 mM X-Gluc in 100 mM Na3PO4 (pH 7.2), 0.1% Triton X-100, 5 mM K3Fe(CN)6 and 5 mM K4Fe(CN)6 for 24 h at 37°C in the dark [29]. Seedlings were then cleared in 70% ethanol for 2 days and mounted in 50% v/v glycerol before observations.

Supporting Information

Figure S1 Position of *miR393a-1* T-DNA insertion in *AtMIR393A* (At2g39885). The pri-miRNA sequence (546-nt) which we identified by RACE experiments is indicated in orange color. The pre-miRNA sequence (133-nt) is in typed in capital letters and the miR393 sequence (22-nt) is underlined. The T-

DNA insertion is located between the two nucleotides highlighted in red.

Figure S2 *AtMIR393A* and *AtMIR393B* are partially redundant for proper leaf morphogenesis. (A–B) The incidence of cotyledon auxin-hypersensitive response in populations of Col-0 (open bars), *mir393a-1* (light grey bars), *mir393b-1* (dark grey bars), and *mir393ab* double mutants (dark bars). Seedlings (n>40 for each condition and genotype) were grown on media containing the concentration of NPA indicated and harvested 4 d after germination. *P* values (two-tailed Fisher's exact test) for significant differences towards *mir393a-1* (A) or *mir393b-1* (B) are indicated; NS for *P*>0.05, * for *P*≤0.05, ** for *P*≤0.01, *** for *P*≤0.001, **** for *P*≤0.0001. (C–D) Epinasty of leaf number 1 to 8 for Col-0, *mir393a-1*, *mir393b-1* and *mir393ab* was measured by the vertical distance between the adaxial leaf side and the leaf margin (in mm ± SEM). Significant difference towards *mir393a-1* (C) or *mir393b-1* (D) are indicated (two-tailed student *t*-test). * for *P*≤0.05, ** for *P*≤0.01. N = 10.

Acknowledgments

We thank Thomas Boller for providing laboratory space at the Botanical Institute of the University of Basel. We thank Frederick Meins Jr. and Mark Estelle for helpful comments and discussions, Ken-Ichiro Hayashi for samples of auxinole, Miltos Tsiantis for seeds of *DR5pro:GUS* and Gerd Juergens for seeds of *IAA12pro:GUS* and *IAA3pro:GUS*.

Author Contributions

Conceived and designed the experiments: DW FV. Performed the experiments: DW DB ME FV. Analyzed the data: DW FV. Wrote the paper: DW AJ ZSK FV.

References

1. Mockaitis K, Estelle M (2008) Auxin receptors and plant development: a new signaling paradigm. Annu Rev Cell Dev Biol 24: 55–80.
2. Vanneste S, Friml J (2009) Auxin: a trigger for change in plant development. Cell 136: 1005–1016.
3. Dharmasiri N, Dharmasiri S, Estelle M (2005) The F-box protein TIR1 is an auxin receptor. Nature 435: 441–445.
4. Kepinski S, Leyser O (2005) The Arabidopsis F-box protein TIR1 is an auxin receptor. Nature 435: 446–451.
5. Dharmasiri N, Dharmasiri S, Weijers D, Lechner E, Yamada M, et al. (2005) Plant development is regulated by a family of auxin receptor F box proteins. Dev Cell 9: 109–119.
6. Calderon Villalobos LI, Lee S, De Oliveira C, Ivetac A, Brandt W, et al. (2012) A combinatorial TIR1/AFB-Aux/IAA co-receptor system for differential sensing of auxin. Nat Chem Biol 8: 477–485.
7. Gray WM, Kepinski S, Rouse D, Leyser O, Estelle M (2001) Auxin regulates SCF(TIR1)-dependent degradation of AUX/IAA proteins. Nature 414: 271–276.
8. Lechner E, Achard P, Vansiri A, Potuschak T, Genschik P (2006) F-box proteins everywhere. Curr Opin Plant Biol 9: 631–638.
9. Ulmasov T, Hagen G, Guilfoyle TJ (1999) Activation and repression of transcription by auxin-response factors. Proc Natl Acad Sci U S A 96: 5844–5849.
10. Navarro L, Dunoyer P, Jay F, Arnold B, Dharmasiri N, et al. (2006) A plant miRNA contributes to antibacterial resistance by repressing auxin signaling. Science 312: 436–439.
11. Robert-Seilaniantz A, MacLean D, Jikumaru Y, Hill L, Yamaguchi S, et al. (2011) The microRNA miR393 re-directs secondary metabolite biosynthesis away from camalexin and towards glucosinolates. Plant J 67: 218–231.
12. Vidal EA, Araus V, Lu C, Parry G, Green PJ, et al. (2010) Nitrate-responsive miR393/AFB3 regulatory module controls root system architecture in Arabidopsis thaliana. Proc Natl Acad Sci U S A 107: 4477–4482.
13. Chen H, Li Z, Xiong L (2012) A plant microRNA regulates the adaptation of roots to drought stress. FEBS Lett 586: 1742–1747.
14. Gao P, Bai X, Yang L, Lv D, Pan X, et al. (2011) osa-MIR393: a salinity- and alkaline stress-related microRNA gene. Mol Biol Rep 38: 237–242.

15. Xia K, Wang R, Ou X, Fang Z, Tian C, et al. (2012) OsTIR1 and OsAFB2 downregulation via OsmiR393 overexpression leads to more tillers, early flowering and less tolerance to salt and drought in rice. PLoS One 7: e30039.
16. Windels D, Vazquez F (2011) miR393: Integrator of environmental cues in auxin signaling? Plant Signal Behav 6: 1672–1675.
17. Si-Ammour A, Windels D, Arn-Bouldoires E, Kutter C, Ailhas J, et al. (2011) miR393 and Secondary siRNAs Regulate Expression of the TIR1/AFB2 Auxin Receptor Clade and Auxin-Related Development of Arabidopsis Leaves. Plant Physiol 157: 683–691.
18. Bian H, Xie Y, Guo F, Han N, Ma S, et al. (2012) Distinctive expression patterns and roles of the miRNA393/TIR1 homolog module in regulating flag leaf inclination and primary and crown root growth in rice (Oryza sativa). New Phytol 196: 149–161.
19. Chen ZH, Bao ML, Sun YZ, Yang YJ, Xu XH, et al. (2011) Regulation of auxin response by miR393-targeted transport inhibitor response protein 1 is involved in normal development in Arabidopsis. Plant Mol Biol 77: 619–629.
20. Jones-Rhoades MW, Bartel DP (2004) Computational identification of plant microRNAs and their targets, including a stress-induced miRNA. Mol Cell 14: 787–799.
21. Sunkar R, Zhu JK (2004) Novel and stress-regulated microRNAs and other small RNAs from Arabidopsis. Plant Cell 16: 2001–2019.
22. Hayashi K, Kamio S, Oono Y, Townsend LB, Nozaki H (2009) Toyocamycin specifically inhibits auxin signaling mediated by SCFTIR1 pathway. Phytochemistry 70: 190–197.
23. Scanlon MJ (2003) The polar auxin transport inhibitor N-1-naphthylphthalamic acid disrupts leaf initiation, KNOX protein regulation, and formation of leaf margins in maize. Plant Physiol 133: 597–605.
24. Ulmasov T, Murfett J, Hagen G, Guilfoyle TJ (1997) Aux/IAA proteins repress expression of reporter genes containing natural and highly active synthetic auxin response elements. Plant Cell 9: 1963–1971.
25. Winter D, Vinegar B, Nahal H, Ammar R, Wilson GV, et al. (2007) An "Electronic Fluorescent Pictograph" browser for exploring and analyzing large-scale biological data sets. PLoS One 2: e718.
26. Hayashi KI, Neve J, Hirose M, Kuboki A, Shimada Y, et al. (2012) Rational Design of an Auxin Antagonist of the SCF(TIR1) Auxin Receptor Complex. ACS Chem Biol 7: 590–598.

27. Friml J (2003) Auxin transport - shaping the plant. Curr Opin Plant Biol 6: 7–12.
28. Vazquez F, Blevins T, Ailhas J, Boller T, Meins F Jr (2008) Evolution of Arabidopsis MIR genes generates novel microRNA classes. Nucleic Acids Res 36: 6429–6438.

29. Jefferson RA, Kavanagh TA, Bevan MW (1987) GUS fusions: beta-glucuronidase as a sensitive and versatile gene fusion marker in higher plants. Embo J 6: 3901–3907.

17

Changes in N-Transforming Archaea and Bacteria in Soil during the Establishment of Bioenergy Crops

Yuejian Mao[1,2], Anthony C. Yannarell[1,2,3], Roderick I. Mackie[1,2,4]*

1 Energy Biosciences Institute, University of Illinois, Urbana, Illinois, United States of America, 2 Institute for Genomic Biology, University of Illinois, Urbana, Illinois, United States of America, 3 Department of Natural Resources and Environmental Sciences, University of Illinois, Urbana, Illinois, United States of America, 4 Department of Animal Sciences, University of Illinois, Urbana, Illinois, United States of America

Abstract

Widespread adaptation of biomass production for bioenergy may influence important biogeochemical functions in the landscape, which are mainly carried out by soil microbes. Here we explore the impact of four potential bioenergy feedstock crops (maize, switchgrass, *Miscanthus X giganteus*, and mixed tallgrass prairie) on nitrogen cycling microorganisms in the soil by monitoring the changes in the quantity (real-time PCR) and diversity (barcoded pyrosequencing) of key functional genes (*nifH*, bacterial/archaeal *amoA* and *nosZ*) and 16S rRNA genes over two years after bioenergy crop establishment. The quantities of these N-cycling genes were relatively stable in all four crops, except maize (the only fertilized crop), in which the population size of AOB doubled in less than 3 months. The nitrification rate was significantly correlated with the quantity of ammonia-oxidizing archaea (AOA) not bacteria (AOB), indicating that archaea were the major ammonia oxidizers. Deep sequencing revealed high diversity of *nifH*, archaeal *amoA*, bacterial *amoA*, *nosZ* and 16S rRNA genes, with 229, 309, 330, 331 and 8989 OTUs observed, respectively. Rarefaction analysis revealed the diversity of archaeal *amoA* in maize markedly decreased in the second year. Ordination analysis of T-RFLP and pyrosequencing results showed that the N-transforming microbial community structures in the soil under these crops gradually differentiated. Thus far, our two-year study has shown that specific N-transforming microbial communities develop in the soil in response to planting different bioenergy crops, and each functional group responded in a different way. Our results also suggest that cultivation of maize with N-fertilization increases the abundance of AOB and denitrifiers, reduces the diversity of AOA, and results in significant changes in the structure of denitrification community.

Editor: Jack Anthony Gilbert, Argonne National Laboratory, United States of America

Funding: This work was funded by the Energy Biosciences Institute, Environmental Impact and Sustainability of Feedstock Production Program at the University of Illinois, Urbana. The funders had no role in study design, data collection and analysis, decision to publish, or preparation of the manuscript.

Competing Interests: The authors have declared that no competing interests exist.

* E-mail: r-mackie@illinois.edu

Introduction

Bioenergy derived from cellulosic ethanol is a potential sustainable alternative to fossil fuel-based energy, since the energy from green plants is renewable and largely carbon neutral in comparison to fossil fuel combustion. Perennial grasses, such as switchgrass (*Panicum virgatum*) and *Miscanthus×giganteus*, with large annual biomass production potential, are proposed as biofuel feedstocks that can maximize ethanol production without adversely affecting the market for food crops (e.g. maize). However, our knowledge of the impacts of various bioenergy feedstock production systems on the soil microbial ecosystem is still very limited. The chemistry of perennial crop residues and plant root exudates may stimulate or inhibit the growth and activity of different fractions of the soil microbial community, and thus the planting of different crops can result in distinct microbial communities [1,2,3]. Differences in management techniques between traditional row-crop agriculture and perennial biomass feedstocks represent different soil disturbance regimes, altered water use, differing rates of fertilizer application, etc., and these should have a direct impact on soil microbial dynamics, subsequently influencing the terrestrial biogeochemical cycles. In particular, we predict that the cultivation of high nitrogen-use

efficiency perennial grasses will result in altered nitrogen-transforming microbial communities in comparison to those found under N-fertilized maize.

The biological nitrogen cycle is one of the most important nutrient cycles in the terrestrial ecosystem. It includes four major processes: nitrogen fixation, mineralization (decay), nitrification and denitrification. Because many of the microorganisms responsible for these processes are recalcitrant to laboratory cultivation, previous studies of the distribution and diversity of nitrogen-transforming microorganisms have employed cultivation-independent techniques targeting functional genes: *nifH*, *amoA* and *nosZ* genes, which encode the key enzymes in nitrogen fixation, ammonia oxidization and complete denitrification, respectively [4,5,6,7,8,9].

Biological nitrogen fixation, which converts atmospheric N_2 into ammonium that is available to organisms, is an important natural input of available nitrogen in many terrestrial habitats [10]. Although nitrogen fixation in terrestrial ecosystems is thought to be mainly carried out by the symbiotic bacteria in association with plants, free-living diazotrophs in soils can play important roles in N cycling in a number of ecosystems [11,12]. In average, 2–3 kg N ha^{-1} year^{-1} could be imported by free living N-fixers [13]. Various field experiments have shown that the biomass yield of one candidate biofuel feedstock crop, *Miscanthus×giganteus*, is not significantly increased by the addition of mineral N fertilizer [14]. The lack of response to nitrogen fertilization and the high biomass production suggest that biological nitrogen fixation may play an important role in supplying the nitrogen needs of *Miscanthus* [15]. Plant species have previously been shown to have a significant effect on the composition of diazotrophs in the field; for example, diazotroph diversity is higher in soil under *Acacia tortilis* ssp. *raddiana* (a leguminous tree) than *Balanites aegyptiaca* (a non-leguminous tree) [16]. Plant genotype also has a strong effect on the rhizosphere diazotrophs of rice [17]. Agronomic practices can also influence soil diazotrophs, e.g. application of N-fertilization can reduce the diversity of diazotrophs [17]. Therefore, we hypothesize that the cultivation of maize with inorganic N-fertilizer will reduce the abundance and diversity of diazotrophs in the soil ecosystem, while biofuel feedstocks receiving little or no N-fertilizer (e.g. *Miscanthus*) will encourage the development of active diazotrophic communities.

Nitrification, which converts ammonium to nitrate, includes two steps: ammonia oxidation to nitrite, and nitrite oxidation to nitrate. The production of nitrate in soil not only supplies nutrition for plants, but it can also mobilize nitrogen to groundwater through nitrate leaching. Ammonia oxidation is the first and rate-limiting step of nitrification [18]. It is typically thought to be carried out by a few groups in β- and γ- Proteobacteria, referred to as ammonia-oxidizing bacteria (AOB) [18]. However, recent environmental metagenomic analyses revealed that ammonia monooxygenase α-subunit (*amoA*) genes are also present in archaea (AOA) [19], and archaeal *amoA* has been shown to be widespread in many environments, e.g. soils, hot springs and marine water [6,19,20,21,22]. Recent work has found that AOA can be up to 3000 times more abundant in soil than AOB [22,23,24], meaning that AOA are the most abundant ammonia oxidizing organisms in soil ecosystems [25]. The soil ammonia oxidizing community is known to be influenced by plant types and management, but different segments of this community respond differently [24,26,27]. For example, the abundance and composition of AOB is significantly altered by long-term fertilization, but AOA are rarely affected [24,27]. The nitrification activity in soil ecosystems is known to be correlated with the abundances and structures of ammonia oxidizers [24,28,29]. We therefore hypothesize that different biofuel cropping systems, especially those that rely on N-fertilization, will influence the composition of ammonium oxidizers in soil, with potential consequences for nitrification rates.

Denitrification, which reduces nitrate to N$_2$ gas, is carried out by a diverse group of microorganisms belonging to more than 60 genera of bacteria, archaea, and some eukaryotes [30]. Complete denitrification involves four steps: NO$_3^-$→NO$_2^-$→NO→ N$_2$O→N$_2$. The enzyme nitrous oxide reductase (encoded by *nosZ*) that reduces N$_2$O to N$_2$ is essential for complete conversion of NO$_3^-$ to N$_2$. Approximately 17 Tg N is estimated to be lost from the land surface through denitrification every year [31]. It is known that the structure and activity of denitrifiers in the terrestrial ecosystem could be significantly influenced by the plant species [7,29,32]. In a study of a maize-cropped field, it was found that organic or mineral fertilizer applications could affect both the structure and activity of the denitrifying community in the long term, with changes persisting for at least 14 months [33]. The potential denitrification rate was found to be significantly correlated to the denitrifier density, as estimated by the quantification of *nosZ* gene copy numbers [34]. Denitrification releases mineralized nitrogen in the soil ecosystem to the atmosphere, and thus, the balance between denitrification and N-fixation, can determine the biologically available N for the biosphere (Arp, 2000).

It is known that plant species can change the soil microbial community [1]. However, while much previous work has examined the microbial community differences between the established crops [7,29,34,35,36], less is known about how microbial communities in the agricultural soils develop during the transition from one cropping system to another (e.g. annual row crops to perennial biofuel feedstocks). Thus, to improve our knowledge of the effects of bioenergy feedstock production on the complex N-cycling microbial communities of terrestrial ecosystems, we followed the changes in soil microbial communities during a two-year establishment period of maize, switchgrass, *Miscanthus×giganteus*, and mixed tallgrass prairie. We monitored the abundance of key genes for nitrogen fixation, ammonia oxidation and complete denitrification (*nifH*, bacterial/archaeal *amoA* and *nosZ* as well as the structural changes of these N-cycling genes and bacterial/archaeal 16S rRNA genes using real-time PCR and barcoded pyrosequencing methods respectively.

Materials and Methods

Study site and sampling

The experiment was conducted at the Energy Biosciences Institute's Energy Farm located southwest of Urbana, Illinois, USA. Miscanthus (*Miscanthus×giganteus*, MG), switchgrass (*Panicum virgatum*, PV), maize (*Zea mays*, ZM) and restored tallgrass prairie (used as control, NP) were planted in the spring of 2008. Miscanthus was replanted in the spring of 2009 due to its poor growth in 2008. Each crop was planted in a randomized block design, with a 0.7-ha plot of each crop randomly positioned within four blocks (n = 4 for each crop). Samples were collected in April 2008, before planting of these crops, in order to characterize the background soil microbial communities. Bulk soil samples were collected at monthly intervals during the growing seasons (June–September 2008 and 2009). 10 soil cores (0–10 cm depth, 1.8 cm diameter) were collected from each plot and homogenized in an ethanol-sanitized, plastic bucket. About 60 g of the well-mixed soil was then subsampled into a 50 mL tube for each plot, and kept on ice until brought to the lab and stored in a −80°C freezer. In total, 112 soil samples were collected. Following standard agricultural practices, only maize was fertilized (17 g/m^2 in 2008, 20 g/m^2 in 2009) with a mixture of urea, ammonia and nitrate (28% UAN). Herbicides were applied in these crops except the restored tallgrass prairie: 1.56 l/ha of Roundup (only applied in 2008) and 4.13 l/ha of Lumax for maize; 1.37 l/ha of 2,4-Dichlorophenoxyacetic acid for Miscanthus; 1.37 l/ha of 2,4-Dichlorophenoxyacetic acid for switchgrass in 2008.

DNA extraction and purification

Soil samples were freeze-dried overnight until completely dry and then manually homogenized with a sterile screwdriver. DNA was extracted from 0.3 g soil using the FastDNA SPIN Kit For Soil (MP Biomedicals) according to manufacturer's protocol. Extracted DNA was then purified using CTAB. DNA concentrations were determined using the Qubit quantification platform with Quant-iT^{TM} dsDNA BR Assay Kit (Invitrogen). DNA was diluted to 10 ng/μL and stored in −80°C freezer for the following molecular applications.

Real-time PCR

The abundances of *nifH*, archaeal *amoA*, bacterial *amoA* and *nosZ* genes in all the soil samples were quantified using real-time PCR. Quantitative real-time PCR was performed according to the methods modified from previous studies: *nifH* (as a measure of N-fixing bacteria) used primers PolF (5′- TGC GAY CCS AAR GCB

GAC TC-3′) and PolR (5′-ATS GCC ATC ATY TCR CCG GA-3′) [5]; archaeal *amoA* (as a measure of ammonia-oxidizing archaea) used primers Arch-amoAF (5′-STA ATG GTC TGG CTT AGA CG-3′) and Arch-amoAR (5′-GCG GCC ATC CAT CTG TAT GT-3′) [6]; bacterial *amoA* (as a measure of ammonia-oxidizing bacteria) used primers amoA-1F (5′-GGG GTT TCT ACT GGT GGT-3′) and amoA-2R (5′-CCC CTC KGS AAA GCC TTC TTC-3′) [4]; and *nosZ* (as a measure of denitrification bacteria) used primers nosZ-F (5′-CGY TGT TCM TCG ACA GCC AG-3′) [37] and nosZ 1622R (5′-CGS ACC TTS TTG CCS TYG CG-3′) [7]. Purified PCR products from a common DNA mixture (equal amounts of DNA from all samples collected in August of 2008 and 2009) were used to prepare sample-derived quantification standards as previously described [38]. The copy number of gene in each standard was calculated by DNA concentration (ng/µL, measured by Qubit) divided by the average molecular weight (estimated based on the barcoded-pyrosequencing results) of that gene. In comparison to using a clone (plasmid) as standard, this method avoids the difference of PCR amplification efficiency between standards and samples caused by the different sequence composition in the PCR templates (single sequence in a plasmid for the standard vs. mixture of thousands of sequences in a soil sample). The 10 µL reaction mixture contained 5 µL of 2× Power SYBR Green Master Mix (Applied Biosystems), 0.5 µL of BSA (10 mg/mL, New England Biolabs), 0.4 µL of each primer (10 µM) and 5 ng of DNA template. Real-time amplification was performed in an ABI Prism 7700 Sequence Detector with MicroAmp Optical 384-Well Reaction Plate and Optical Adhesive Film (Applied Biosystems) using the following program: 94°C for 5 min; 40 cycles of 94°C for 45 s, 56°C for 1 min (54°C for *nifH* gene), 72°C for 1 min. A dissociation step was added at the end of the qPCR to assess amplification quality. The specificity of the PCR was further evaluated by running twenty randomly selected samples (for each gene) on a 1% (w/v) agarose gel. The corresponding real-time PCR efficiency for each of these genes was estimated based on a two-fold serial dilution of the common DNA mixture described above. The qPCR efficiency (E) was calculated according to the equation $E = 10^{[-1/slope]}$ [39]. Triplicate qPCR repetitions were performed for each of the gene for all the samples. The real-time PCR amplification efficiency of *nifH*, archaeal *amoA*, bacterial *amoA* and *nosZ* genes was 1.90 ± 0.01, 1.90 ± 0.06, 1.76 ± 0.01 and 1.82 ± 0.01, respectively. The R^2 of all these standard curves was higher than 0.99. The detection limit of this real-time PCR assay was 10 copies/µL.

The copy numbers of these genes per gram of dry soil was calculated by the copy numbers of the gene per ng of DNA multiplied by the amount of DNA contained in each gram of dry soil. The quantities of these genes were corrected assuming a DNA extraction efficiency of 30% [40,41].

T-RFLP

The soil samples collected from four replicated (blocks) plots of the four crops prior to planting, and then August of establishment years 1 and 2 (2008 and 2009; 48 samples in total) were analyzed by terminal restriction fragment length polymorphism (T-RFLP). The *nifH*, archaeal *amoA*, bacterial *amoA*, *nosZ* gene were amplified from these samples with FAM-labeled (on forward primer) primers PolF/R, Arch-amoAF/R, amoA-1F/2R and nosZ-F/1622R (see Table S1). The 16S rRNA gene was amplified with 8F (5′-FAM-AGAGTTTGATCMTGGCTCAG-3′) and 1492R (5′-GGTT-ACCTTGTTACGACTT-3′) [42,43]. The PCR reaction mixture (25 µl) contained 5 µl GoTaq Flexi Buffer (5×), 2 µl MgCl₂ (25 mM), 0.25 µl DNA Polymerase (5 U/µl, Promega, Madison, Wis.), 1.25 µl BSA (10 mg/ml), 1 µl forward primer (10 µM), 1 µl

reverse primer (10 µM), 1.25 µl dNTP Mix (10 mM), 2 µl DNA template (10 ng/µl). The PCR reaction was performed in a thermo cycler (BioRad, Hercules, CA) using a 5-min heating step at 94°C followed by 30 cycles of denaturing at 94°C for 1 min, annealing at 60°C (54°C for *nifH*) for 45 s, and extension at 72°C for 1 min, with a final extension step of 5 min at 72°C. The PCR products were purified by QIAquick PCR Purification Kit (QIAGEN) and digested at 37°C overnight in a 20-µl mixture containing 2 µl NEB Buffer (10×), 0.5 µl AluI/HhaI (20 U/µl) and 5 µl PCR product. 5 µl of the digested product was sent to the Core Sequencing Facility (University of Illinois at Urbana-Champaign) for fragment analysis. ROX1000 was used as inner standard. T-RFLP profiles were analyzed by GeneMarker (v 1.85) according to the manufacturer's instruction. Fragments with sizes between 50 bp and the length of the PCR products and peak area >500 were selected for T-RFLP profile statistical analysis. The profile data were normalized by calculating the relative abundance (percentage) of each fragment (individual peak area divided by the total peak area).

Barcoded pyrosequencing

The same samples used in T-RFLP were also used in barcoded pyrosequencing. The four replicated samples collected from the same crop at the same time were combined in to one composite sample for the construction of sequence libraries. Altogether, nine samples (one sample for background soil [mixed from the soils from all the plots before planting these crops], four samples from each of the different crops at Aug 2008, and four from Aug 2009) were obtained. Furthermore, all of these N-transforming genes of the sample collected from MG at the end of the second year (MG2) were sequenced twice in two different lanes to estimate the variation of the sequencing method. The 16S rRNA gene of ZM2 was also sequenced twice. Details of primers and PCR conditions used in the study are listed in Table S1.

The *nifH*, archaeal *amoA*, bacterial *amoA*, *nosZ* and 16S rRNA genes (V4–V5 region) were amplified using the barcode primers PolF/R, Arch-amoAF/R, amoA-1F/2R, nosZ-F/1622R and U519F/U926R, respectively (primers are shown in Table S1). The primers (HPLC purified, Integrated DNA Technologies) were designed as 5′-Fusion Primer+barcode+gene specific primer-3′ (Fusion Primer A, 5′- CGTATCGCCTCCCTCGCGCCAT-CAG-3′; Fusion Primer B, 5′-CTATGCGCCTTGCCAGC-CCGCTCAG-3′). The PCR conditions were optimized and primers with appropriate barcodes (10 bp) were selected. The barcodes used for each primer are described in NCBI SRA, with accession number SRA023700. The 50 µL PCR mixture contained 10 µL of 5× Phusion HF Buffer (Phusion GC Buffer was used for bacterial *amoA* gene amplification, both buffer contains 7.5 mM MgCl₂), 1 µL of 10 mM dNTPs, 2.5 µL of 10 mg/mL BSA, 0.5 µL of 2 U/µL Phusion Hot Start DNA Polymerase (FINNZYMES), 4 µL of 10 µM forward/reverse primer mixture and 4 µL of 10 ng/µL DNA templates. 1 µL of 100% DMSO was supplemented into the PCR mixture in bacteria *amoA* gene (GC rich) amplification. The PCR amplification was performed in a thermal cycler (BioRad, Hercules, CA) using the program 98°C for 2 min; 30 cycles of 98°C for 10 s, 60°C (54°C for *nifH* gene and 56°C for bacterial *amoA* gene) for 30 s, 72°C for 20 s; 72°C for 5 min. The PCR product was first checked on a 1.2% w/v agarose gel, and then purified by QIAquick PCR Purification Kit (QIAGEN). The DNA concentration of the purified PCR product was measured by Qubit Fluorometer using the Quant-iT™ dsDNA BR Assay Kit (Invitrogen) according to the manual. PCR products of the same gene, to be run together in the same lane (1/16 plate) in 454 sequencing, were mixed in equal mole amounts and run on a 2% w/v agarose gel. The target bands were cut from the gel and purified by QIAquick Gel Extraction Kit (QIAGEN). The DNA concentrations of the purified PCR products were measured by Qubit Fluorometer and adjusted to 50 nM. The *nifH*, archaeal *amoA*, bacterial *amoA* and *nosZ* genes PCR products were then mixed in equal mole amounts and sequenced on a Genome Sequencer FLX Instrument (Roche) using GS FLX Titanium series reagents. The 16S rRNA gene was run in a separate lane.

Sequence analysis

Sequences were first extracted from the raw data according the Genome Sequencer Data Analysis Software Manual (Software Version 2.0.00, October 2008) by the sequencing center (Roy J. Carver Biotechnology Center, University of Illinois at Urbana-Champaign). The sequences with low quality (length <50 bp, which ambiguous base 'N', and average base quality score <20, for detail see manual) were removed. The sequences that fully matched with the barcodes were selected and distributed to separate files for each of the different genes, after removal of the barcode, using RDP Pipeline Initial Process (http://pyro.cme. msu.edu/). For each gene, the sequences that didn't match with the gene specific primers or had a read length shorter than 350 bp were removed. The sequences that matched with the reverse primer were converted to their reverse complement counterparts using BioEdit to make all the sequences forward-oriented.

The 16S rRNA gene sequences were aligned by NAST (Greengenes). The sequences with significant matched minimum length <300 and identity <75% were removed. The aligned 16S rRNA gene sequences were used for chimera check using Bellerophon method in Mothur [44]. Distance matrices were calculated by ARB using the neighbor joining method [45]. A lane mask was used in calculating the 16S rRNA gene sequences to filter out the hyper variable regions. Operational Taxonomic Units (OTUs) were then classified using a 97% nucleotide sequence similarity cutoff and rarefaction curves were constructed based on the distance matrices (both of nucleotide and amino acid sequences) using DOTUR [46]. The phylogenetic affiliation of each 16S rRNA gene sequence was analyzed by RDP CLASSI-FIER (http://rdp.cme.msu.edu/) using confidence level of 80%.

The 16S rRNA gene sequences were also processed by QIIME pipeline and denoised by Denoiser V0.91 according to the manual [47,48]. The results were compared to that obtained by RDP pipeline. In total, 26,431 valid sequences were obtained after denoising using QIIME, which is 12.2% less than that obtained by RDP pipeline (without denoising). Using the 97% similarity cutoff, 8,568 OTUs were obtained, which is 4.7% lower than that observed by RDP pipeline. After random re-sampling to the same sequence depth (1789 sequences per sample) using Daisy_chopper (http://www.genomics.ceh.ac.uk/GeneSwytch/Tools.html), the number of OTUs for each sample obtained by two different processing methods (QIIME, denoised and RDP, non-denoised) was compared (Fig. S1). The estimated number of OTUs after denoising was similar to that obtained by RDP pyrosequencing pipeline (without denoising), showing that the denoising process had a very limited influence on our diversity analysis. The data reported in this paper was analyzed using RPD pipeline described in the previous paragraph.

The *nifH*, archaeal *amoA*, bacterial *amoA* and *nosZ* genes sequences were blasted against a non-redundant protein sequence database (download from NCBI) using BLASTX with an *E*-value cutoff of 0.001. The top 10 closest matches of each sequence were estimated using a custom made Perl script to remove possible chimeras and sequences with sequencing errors causing frameshifts. Sequences

with different regions matching the same sequence in the database but with different frame positions were considered to be frameshifts. Sequences that matched two or more different origin sequences were classified as chimeras. The nucleotide sequences of *nifH*, archaeal *amoA*, bacterial *amoA* and *nosZ* genes were translated into amino acid by Geneious (http://www.geneious.com/) based on the frame positions obtained from BLASTX. The redundant sequences (identical sequences) were removed using CD-Hit [49], and the representatives with longest length were selected for following phylogenetic analysis. Both of the nucleotide and amino acid sequences of these N-cycling genes were aligned by MUSCLE 3.7 [50] using program default settings. Operational Taxonomic Units (OTUs) were then classified and rarefaction curves were constructed based on the distance matrices (both of nucleotide and amino acid sequences) using DOTUR [46]. Previous studies showed that the amino acid sequences of AmoA and NosZ similarity around 90% is generally relevant to 97% similarity of 16S rRNA gene [51,52]. Thus, all these N-cycling gene sequences were classified into OTUs using a 90% amino acid sequence similarity cutoff, and phylogenetic trees were built in ARB using the neighbor-joining method. Sequences of all the samples and genes were also randomly resampled to identical sequencing depth (the smallest sequencing effort) using Daisy_chopper (http://www.genomics.ceh.ac.uk/GeneSwytch/Tools.html) to avoid the potential bias caused by sequencing effort difference [53].

The 454-pyrosequencing data were deposited in NCBI SRA under accession number SRA023700.

Statistical analysis

ANOVA combined with post hoc Tukey B test was used to estimate the difference of archaeal/bacterial *amoA*, *nifH* and *nosZ* genes abundances under different crops based on the quantitative PCR results from the replicated plots. The T-RFLP data from the replicated plots were used to follow the structural changes of soil microbial communities by plant types, and significance tests for these changes were conducted using Analysis of Similarity (ANOSIM) based on Bray–Curtis similarity coefficients. Correspondence analysis (CA) and Canonical correspondence analysis (CCA) were also used to visualize the predominant microbial community changes of archaeal/bacterial *amoA*, *nifH*, *nosZ* and 16S rRNA genes after planting bioenergy crops based on the T-RFLP data. These statistical analyses were done using the free software PAST (http://folk.uio.no/ohammer/past/). Based on our extensive pyrosequencing library, the OTUs/genera that showed monotonic (i.e. continuously increasing or decreasing) trends for each crop treatment over the two year establishment were presumed to be particularly noteworthy in terms of crop impact. The populations with continuously increased or decreased abundance in the two-year period after planting these bioenergy crops were selected using a custom Perl script.

Results

Quantification of *nifH*, archaeal *amoA*, bacterial *amoA* and *nosZ* genes

Quantities of AOA in all of crops fluctuated over the two growing seasons in a similar pattern (Fig. 1), but the abundance of this gene was always higher in MG than NP and PV. The quantity of bacterial *amoA* genes in ZM significantly increased from $1.47 \pm 0.61 \times 10^8$ to $3.26 \pm 0.94 \times 10^8$ during the first three months of establishment and thereafter remained higher in ZM than in the other cropping systems. The nitrification rates under these crops were analyzed in the second year by estimating the accumulation of the nitrate in buried soil bags, and linear regression revealed

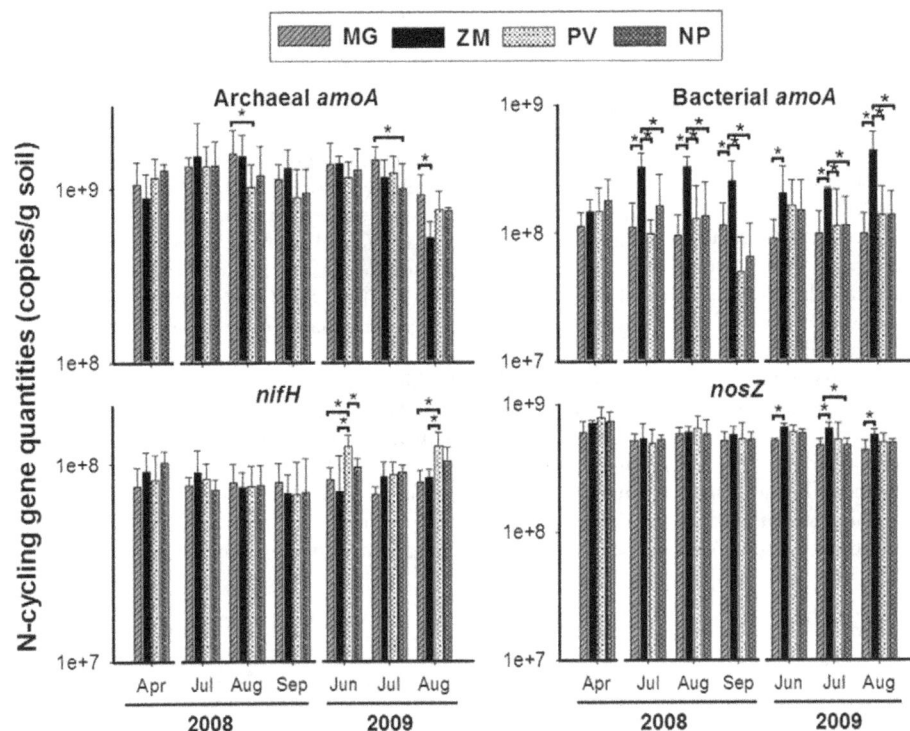

Figure 1. Changes in abundance of *nifH*, archaeal *amoA*, bacterial *amoA* and *nosZ* genes in plots after planting *Miscanthus×giganteus* (MG), *Panicum virgatum* (PV), restored prairie (NP) and *Zea mays* (ZM). The copy number of genes in each gram of dry soil was estimated based on the results of real-time PCR (copy number in each ng DNA) and the average amount of extracted DNA (6.23 μg per dry soil) and assuming DNA extraction efficiency was 30% [40]. The R^2 of the standard curve of all these genes was higher than 0.99. The real-time PCR amplification efficiency of *nifH*, archaeal *amoA*, bacterial *amoA* and *nosZ* genes was 1.90±0.01, 1.90±0.06, 1.76±0.01 and 1.82±0.01 respectively. *Represents values that are significantly different ($P<0.01$).

that the nitrification rates were significantly related to the quantities (log) of archaeal *amoA* genes ($R^2 = 0.61$, $P = 0.03$, $n = 12$), but not to the quantities of bacterial *amoA* genes (Fig. 2). The abundance of *nifH* genes remained stable for all the crops, ranging from 7×10^7 to 9×10^7 copies per gram of dry soil (Fig. 1). No significant differences were observed among the different crops in the first year. In the second year, the population sizes of diazotrophs in PV and NP had significantly increased in comparison to the first year ($P = 0.0001$ and 0.0002). The population size of denitrifiers was less variable in comparison to

the other N-cycling populations; however, the copy number of *nosZ* increased in ZM during the second year of the study and remained higher than in MG (Fig. 1).

Structural changes of N-cycling genes and microbial communities after planting of bioenergy crops

The community structural differentiation of *nifH*, archaeal *amoA*, bacterial *amoA*, *nosZ* and 16S rRNA genes under different bioenergy crops were analyzed by T-RFLP. These analyses used the fully replicated sample set from the randomized block design.

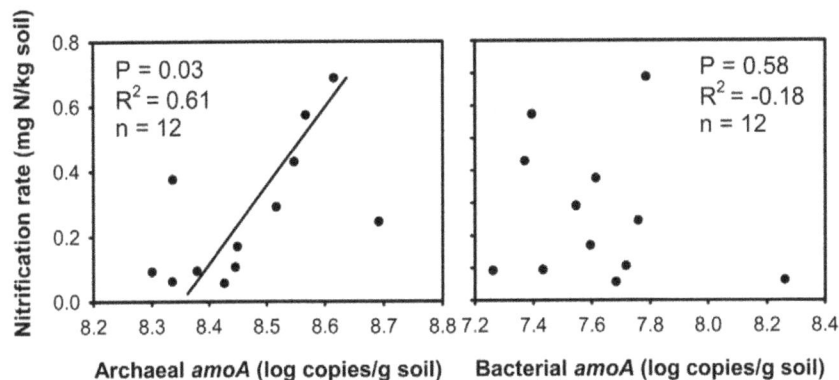

Figure 2. Relationship between the concentration of ammonia-oxidizing archaea/bacteria and nitrification rate. The nitrification rate was determined over the same time as our sample collection in 2009. Nitrification rate was calculated based on the accumulation of nitrate in soil bags incubated in the field (0–10 cm depth) for 15 to 32 days.

Correspondence analysis (Fig. 3) showed that the soil microbial communities in the initial plots did not show any relationship with the crop treatments being applied. During the establishment of these bioenergy crops, the community composition of denitrification bacteria (*nosZ*) under ZM was completely separated (ANOSIM, P<0.05, Table S2) from those under the other crops by the end of the second year along the first-axis, which explained 51.7% of the total variance. None of the other groups showed significant clustering by plant, although the community composition of AOB (bacteria *amoA*) under ZM appeared to be separated from that of MG (ANOSIM, P = 0.17) along the second axis, which explained 14.2% of the total variance. In addition to plant species, the changes of soil microbial communities also could be caused by the variation of environmental conditions. To compare the magnitude of the changes of soil microbial communities related only to plant species, a constrained ordination method was also used. Canonical Correspondence analysis (CCA, Fig. S2) revealed that, at the end of the second year, the microbial communities under ZM were most different from the three cropping systems for bacterial *amoA*, *nosZ*, and 16S rRNA. Archaeal *amoA* was most distinct under MG,

followed by ZM, and *nifH* was equally separated under all cropping systems (Fig. S2).

Diversity of *nifH*, archaeal *amoA*, bacterial *amoA*, *nosZ* and 16S rRNA genes

To further understand the composition of microbial community in the field, the *nifH*, archaeal *amoA*, bacterial *amoA*, *nosZ* and 16S rRNA genes were deeply sequenced using the pyrosequencing approach. In total, 143,487 reads were obtained for these genes. The numbers and qualities of these sequences are described in Table 1, Table S3 and Text S1. The reproducibility of the pyrosequencing result was estimated by comparing the observed microbial composition between repeat sequencing runs for all these genes (Fig. 4). Linear regression analysis indicated a high reproducibility ($R^2 = 0.95$) of our pyrosequencing.

High diversity of *nifH*, archaeal *amoA*, bacterial *amoA* and *nosZ* genes were observed with 10899, 3187, 3945 and 11242 unique nucleotide sequences and 2286, 2246, 3633 and 4208 unique deduced amino acid sequences respectively (Fig. S3). These sequences were then translated to amino acid sequence according

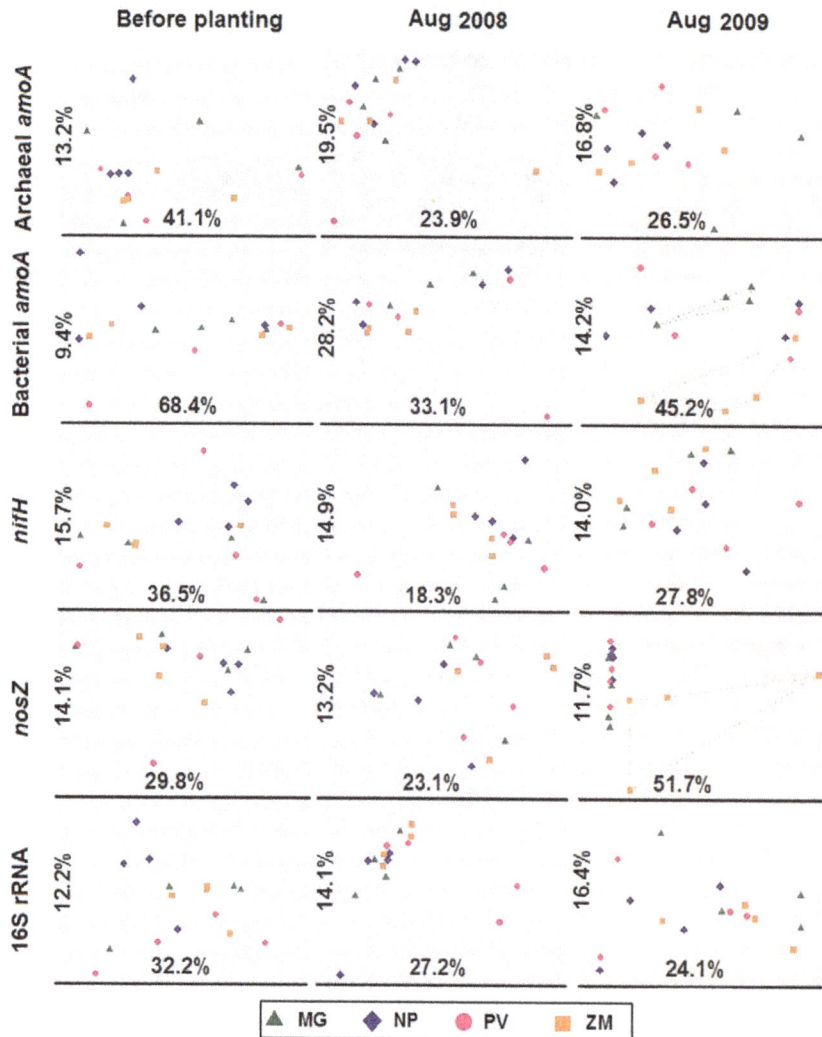

Figure 3. Structural changes of archaeal *amoA*, bacterial *amoA*, *nifH*, *nosZ* and 16S rRNA genes after planting *Miscanthus×giganteus* (MG), *Panicum virgatum* (PV), restored prairie (NP) and *Zea mays* (ZM) revealed by T-RFLP and Correspondence analysis (CA). The number on each axis shows the percentage of total variation explained. The soil samples were collected from four replicate plots for each plant at each time point.

Table 1. Quality of barcoded pyrosequencing reads.

Genes	Number of sequences			
	Correct barcode and primer	Length >350 bp	[a]Valid	Each sample (range)
nifH	28,334	27,781	21,111	1,312–2,956
Archaeal amoA	16,978	16,226	14,025	697–1,792
Bacterial amoA	28,254	27,874	21,817	1,726–2,569
nosZ	33,838	28,819	22,590	1,600–2,951
16S rRNA	30,487	30,175	30,101	2,034–3,488
Total	137,891 (96.1%)	130,875 (91.2%)	109,644 (76.4%)	697–3,488

Total number of raw reads was 143,487.

[a]Valid sequences of nifH, archaeal amoA, bacterial amoA and nosZ genes were defined as high quality sequences with correct barcode and primer (at 5′-end), length >350 bp and that did not have frameshifts and chimeric structure. The possible sequencing errors causing frameshifts and chimeras were removed based on the BLASTX result. Sequences with different regions matching the same sequence in the database but with different frame positions were considered to be frameshifts. Sequences that matched two or more different origin sequences were classified as chimeras. Valid sequences of 16S rRNA gene were sequences with correct barcode and primer, length >350 bp and passed the chimeric check program in Greengenes with the Bellerophon method. The sequence numbers for each sample are listed in Table S3.

to the BLASTX report. The amino acid sequences of nifH, archaeal amoA, bacterial amoA and nosZ genes were classified into 229, 309, 330 and 331 OTUs, respectively, with a similarity cutoff of 90%. After random re-sampling to the same sequencing depth (697 sequences for each sample), the adjusted total number of OTUs for these genes were 217, 303, 319 and 278, respectively (Table S4). Rarefaction analysis of these genes showed that the diversity of archaeal amoA gene in ZM2 (second year ZM) and MG1 (first year MG) was markedly lower than the others (Fig. S4). The diversity of nifH and nosZ genes slightly decreased in ZM2.

The diversity of bacterial and archaeal 16S rRNA genes was much higher than these N-cycling genes. In total, 19,824 unique 16S rRNA gene sequences and 8,989 species (OTUs classified at 97% similarity cutoff) were observed. RDP classification showed that these sequences covered 16 bacterial and 1 archaeal phyla, including 201 genera (Fig. S5). Proteobacteria and Acidobacteria were the most predominant phyla in the soil (>20%). The

sequences belonging to Proteobacteria were distributed over 86 different genera, while 94.1% of the sequences in Acidobacteria belonged to Family Gemmatimonadaceae, with GP1 as the most predominant genus (accounted 35.9% of the Acidobacteria).

To understand which phylotypes were impacted by vegetation type, the OTUs of nifH, archaeal amoA, bacterial amoA, nosZ and 16S rRNA (genus for 16S rRNA) genes that continuously increased or decreased over the two-year establishment of these bioenergy crops were identified (Fig. 5 and Fig. S6, S7, S8, S9). After planting of these bioenergy crops, 27.5%, 15.4%, 22.7% and 14.5% of the total archaeal amoA, bacterial amoA, nifH, and nosZ phylotypes, respectively, were found to be continuously increasing or decreasing (Table 2). Details of these continuously changed N-cycling OTUs are described in Text S1 and Figure S6, S7, S8, S9.

Pyrosequencing of 16S rRNA gene revealed 19.9% of the bacterial genera (39), spanning six phyla, continuously changed after planting of these bioenergy crops (Fig. 5). Only genus Methylibium was changed in all the crops, with decreased abundance in MG and increased abundance in the other crops. Rhodanobacter only appeared after planting of ZM (7 sequences for both of ZM1 and ZM2), and it was undetectable either in the background soil or in the soil under other crops. Consistent with the changes of Nitrosospira-like bacterial amoA OTU (see above), the abundance of genus Nitrosospira in the 16S rRNA library also increased in ZM. The abundance of genus Nitrospira, which is known as a nitrite-oxidizing bacteria, also increased in ZM. Most of the changed genera in MG decreased or even disappeared, except Terrabacter and Herbaspirillum. All of them were found at low abundance (<1%). Although many genera in Proteobacteria were changed, the total abundance of this most predominant phylum was quite stable under all of the crops (Fig. S5).

Discussion

In this study, we monitored the structural and quantitative changes of the key genes involved in N-cycling as well as the overall bacterial/archaeal community during two-year establishment of four different bioenergy feedstock crops, and analyzed the shifts of specific soil microbial populations in response to different types of crops. We were able to detect significant changes in the abundance of many of these microbial functional groups within 2 years of initial crop establishment. We also found that traditional row-crop agriculture of maize has a larger

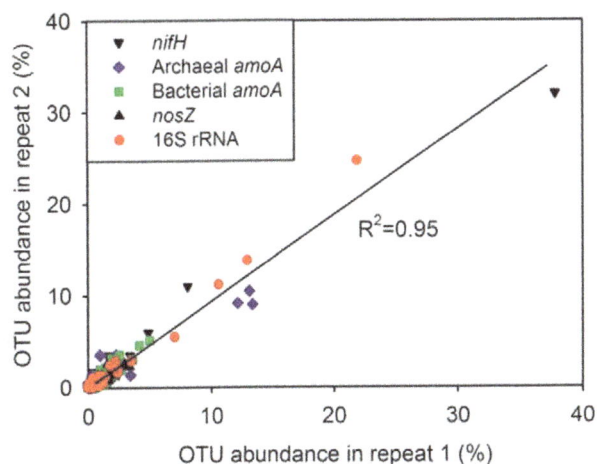

Figure 4. Reproducibility of the pyrosequencing replicates. OTUs of nifH, archaeal amoA, bacterial amoA and nosZ genes were classified at a nucleotide similarity cutoff 90%. The 16S rRNA gene sequences were classified to genus level by RDP classifier. One of the samples was duplicated for each gene.

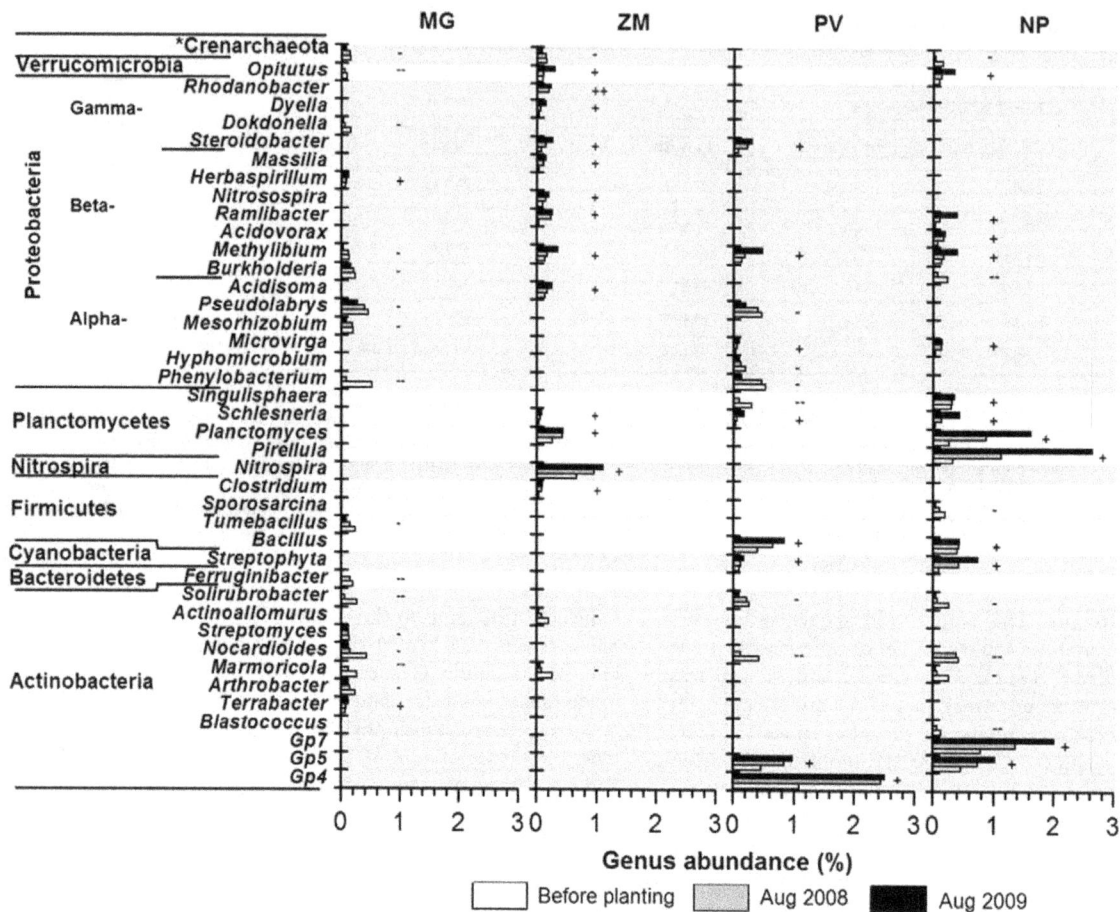

Figure 5. Microbial genera that changed after planting of *Miscanthus×giganteus* (MG), *Panicum virgatum* (PV), restored prairie (NP) and *Zea mays* (ZM). Sequences were classified by RDP Classifier project. −/+ represents the genus continuously decreased/increased after planting the crops; −−/++ represents the genus disappeared/appeared after planting the crops. *sequences belonging to Crenarchaeota, which could not be classified to genus level.

impact on the soil N-cycling community than any of the perennial bioenergy feedstock crops (Figs. 1, S2), while the perennial crops were associated with overall community shifts in the phyla Planctomyces, Firmicutes, and Actinobacteria (Fig. 5).

Table 2. Number of OTUs or genera that changed continuously after planting *Miscanthus×giganteus* (MG), *Panicum virgatum* (PV), restored prairie (NP) and *Zea mays* (ZM).

Genes	MG	PV	ZM	NP	*Total
Archaeal *amoA*	4	16	61	23	85
Bacterial *amoA*	18	18	7	20	51
nifH	8	21	14	23	52
nosZ	12	18	24	16	48
16S rRNA	17	15	14	21	40

OTUs of N-cycling genes were classified based on a cutoff of 90% amino acid sequences similarity. 16S rRNA genes were classified into genus level by RDP Classifier. Details of the abundance changes of the OTUs or genera are shown in Fig. 5 and S6, S7, S8, S9.
*Total number of changed unique OTUs or genera.

Traditional maize cultivation significantly increased the total abundance of ammonia-oxidizing bacteria and denitrifying bacteria (Fig. 1), altered the community composition of denitrifying bacteria (Fig. 3) and decreased the diversity of ammonia-oxidizing archaea, denitrifying bacteria, and diazotrophs (Fig. S4). This may be due to the application of N-fertilizer, which occurred only in ZM plots. Ammonia oxidizers are sensitive to N-fertilizer [24,27], and these responses were manifested in the increased population size of AOB and the high number of markedly changed AOA species. The nitrification rate was significantly correlated with the quantity of archaeal *amoA*, but not bacterial *amoA*, indicating AOA was the major ammonia-oxidizer.

Deep understanding of the structural shifts of key functional genes can help us to better understand changes in microbial activity in the environment. From the present database, we know that the global diversity of the *nifH*, archaeal *amoA*, bacterial *amoA* and *nosZ* genes, as well as the other functional genes of microorganisms, is high. The traditional approaches (e.g. clone library, DGGE and T-RFLP) used in previous studies may largely underestimate the diversity of microbial communities involved in soil nitrogen cycling. Mounier et al. (2004) revealed that even a large library with 713 clones was insufficient to enumerate the diversity *nosZ* gene in maize rhizosphere, showing the high complexity of N-cycling genes [54]. Thus, high-throughput deep sequencing approaches are essential to improve our knowledge of the diversity of these functional genes. In the present study, using barcoded 454-pyrosequencing approach, we found high diversity (ranging from ~3100 to ~11200 unique nucleotide sequences) of *nifH*, archaeal *amoA*, bacterial *amoA* and *nosZ* genes in the soil ecosystem, which far surpasses the diversity of the N-cycling genes observed in previous studies [54,55,56,57,58,59,60,61]. The rarefaction curves of these genes were close to saturation after sequencing ~1000 for each sample, indicating that such a sequencing effort is sufficient to elucidate the diversity and structure of the complex soil N-cycling communities. The high similarity between repeat runs of these genes (Fig. 4) demonstrates the high reproducibility and reliability of this barcoded pyrosequencing method. This result also indicates that the variation of pyrosequencing, resulting from random sampling of gene targets during emulsion PCR [62], can be greatly reduced by increasing the sequencing depth and library coverage.

During the establishment of these bioenergy crops, about 15%–30% of N-cycling genes and the detected bacterial/archaeal genera were continuously changed, indicating that a large proportion of soil microbes were affected by the transition to different bioenergy feedstock systems. Most of these phylotypes changed uniquely in one of the crops, indicating that the changes were mainly caused by the particular experimental crop treatment (specific plant species or management, such as fertilization) and not due to the environmental conditions that fluctuated in all treatments (e.g. temperature and moisture). Contrary changes of certain populations in different crops also support this conclusion; for example, the abundance of *Methylibium* (belonging to β-Proteobacteria) decreased in MG but increased in the other crops. The abundance of Bacteroidetes was previously found to be lower in the soil of Miscanthus-dominated grasslands (4%) in comparison to forest soil (6%) [63]. We found that the decrease of Bacteroidetes in MG was mainly due to the disappearance of the genus *Ferruginibacter* (Fig. 5 and S5).

The structure of *nifH* gene was completely separated according to vegetation by the end of the second year (Fig. S2), which suggests that the structure of N-fixation bacterial population was particularly sensitive to plant genotype. Tan et al. (2003) has revealed that the structure of diazotrophs was not only different among rice species, but also changed rapidly with fertilization. The diversity of the *nifH* gene was obviously reduced within 15 days after fertilization [17]. Thus, the decreased diversity of *nifH* in ZM may be also due to the application of fertilizer not the presence of maize. However, the population size of N-fixing bacteria did not change under ZM, indicating N-fertilization may not change the quantity of soil diazotroph [64]. The N-fixing activity was expected to increase in MG [14,15]. Although the population size of total free-living soil N-fixing bacteria was not significantly increased by growth of MG in the two year period, the abundance of genus *Herbaspirillum* increased. *Herbaspirillum* species are known as endophytic diazotrophs that are enriched by C4-prennial grasses including Miscanthus [65,66,67]. Thus, we speculate that the abundance increase of *Herbaspirillum* in the bulk soil was likely due to the root exudates (e.g. organic carbons) released by Miscanthus, which favored the growth of this population. Our results also suggest that Miscanthus may only selectively enhance the activity of specific diazotrophs, not the whole N-fixing microbial community.

The AOA are thought to be more stable and less responsive to environmental differences than AOB, as revealed by previous quantitative studies [6,68]. However, in the present study we found that, while the population size of AOA was relatively stable, the structure of the AOA community was sensitive to the different cropping systems. The diversity of AOA markedly decreased after planting of maize, with 41 of the AOA phylotypes disappearing (Fig. S6). In contrast, the population size of AOB significantly increased after planting of maize in both the qPCR of bacterial *amoA* and the 16S rRNA pyrosequencing results. In addition to the increase of genus *Nitrosospira* (AOB) [69], the abundance of *Nitrospira* (nitrite-oxidizing bacteria) [70] also increased in N-fertilized maize (Fig. 5). However, the number of changed bacterial *amoA* phylotypes in ZM was much less than the other crops. Therefore, these two different groups of ammonia-oxidizers respond to the N-fertilization in a very different way. The population size of AOB increased immediately in the first growing season, thus, we hypothesize that the increased AOB abundance in maize was likely due not to the growth of maize plants, but to the application of N-fertilizer, which increased the ammonia content in the soil [24,27]. It has been found that AOB population size increased in seven days after applying of N-fertilizer, and it was still significantly greater than unfertilized soil 8 months after the last application of ammonia [41]. Consistently, we found that the population size of AOB doubled in less than three months and maintained a relatively high level over the two-year study even though measurable NH_4^+ in the soil declined over this time period to levels close to that of the unfertilized plots (Fig. 1 and S10). These results indicate the AOB population size can be quickly increased by N-fertilization and can remain for a long period even after the measurable ammonia has been consumed.

The nitrification rate was found to be significantly correlated with the quantity of archaeal *amoA* rather than bacterial *amoA*, indicating AOA rather than AOB may be the major active ammonia-oxidizer in these soils. Contradictory conclusions on the relative importance of AOA and AOB in soil nitrification have been previously reported where nitrification was found to be associated with the changes of archaeal *amoA* abundance or higher archaeal transcriptional activity in some of the soils [58,71,72]. In contrast, the nitrification kinetics in the other soils were correlated with the growth of AOB [73,74]. It has been found that the ammonia affinity of "*Candidatus* Nitrosopumilus maritimus" (a marine AOA) is much higher than AOBs, and its growth may be enhanced by relatively low ammonia concentration [75]. Thus, the contradictory conclusions from these studies may be due to the different soil ammonia concentration used in these experiments [58,73]. These results hint that AOB may be more active in soils amended with ammonia, while AOA are more active in soils with low ammonia concentration [58]. The nitrification rate outlier (ZM Aug; Fig. 2) had highest population size of AOB, suggesting that the activity of AOB was enhanced by N-fertilization and also supported the above speculation. In support of this speculation, a recent publication shows that recovery of nitrification potential after disruption was dominated by AOB in cropped soils while AOA were responsible RNP in pasture soils [76].

It is known that nitrogen fertilization can change the structure and activity of denitrifying community, and subsequently affect the N_2O emission [33,34,77,78]. Large amounts (1.3%) of the applied N-fertilizer in maize fields (north Colorado) are converted to N_2O by the combination of nitrification and denitrification [79]. However, it is still unclear how fertilization changes the microbial community, since most of the previous studies are based on the already established fields [7,29,33]. Our study revealed that the structure of denitrifying bacteria in maize soil was significantly differentiated from the other crops at the second year (Fig. 3).

However, the population size of denitrifiers was relatively stable in all the crops in comparison to other N-cycling microbial communities, which only slightly increased in maize at the second year. The high stability of denitrifying population abundance could be explained by the high diversity and functional redundancy of denitrification community [30,80].

In conclusion, our two-year study of transitional agriculture shows that specific N-transforming microbial communities develop in the soil in response to different bioenergy crops. Each N-cycling microbial group responded in a different way after planting with different bioenergy crops. In general, planting of maize has a larger impact on the soil N-cycling community than the other bioenergy crops. Our results also indicate that application of N-fertilizer may not only cause short-term environmental problems, e.g. water contamination, but also can have long-term influence on the global biogeochemical cycles through changing the soil microbial community structure and abundance. Since soil types and other environmental factors may also impact the N-cycling microbial community, the universality of our findings needs to be confirmed by additional study at different sites.

Supporting Information

Figure S1 Number of OTUs obtained by two different processing methods: QIIME (denoised) and RDP pyrosequencing pipeline (non-denoised).

Figure S2 Structural changes of archaeal *amoA*, bacterial *amoA*, *nifH*, *nosZ* and 16S rRNA genes after planting *Miscanthus×giganteus* (MG), *Panicum virgatum* (PV), restored prairie (NP) and *Zea mays* (ZM) revealed by T-RFLP and Canonical correspondence analysis (CCA). The number on each axis shows the explained total variation. The soil samples were collected from four replicated plots for each plant at each time point. * Correspondence analysis was used for the samples collected before planting bioenergy crops.

Figure S3 OTU classification of valid sequences at different distance levels based on nucleotide and deduced amino acid sequences.

Figure S4 Rarefaction analysis of the diversities of *nifH*, archaeal *amoA*, bacterial *amoA*, *nosZ* and 16S rRNA genes in the soil underneath different bioenergy crops. The OTUs of *nifH*, archaeal *amoA*, bacterial *amoA* and *nosZ* genes were classified at 90% similarity cutoff based on amino acid sequences, and 16S rRNA gene was classified at 97% similarity cutoff on nucleotide sequences. BG0 represents the samples collected before planting bioenergy crops. MG, PV, NP, and ZM represent *Miscanthus×giganteus*, *Panicum virgatum*, restored prairie and *Zea mays* respectively. 1 and 2 represent samples collected in the first and second growing seasons.

Figure S5 Phylum level microbial community composition in the soil under different plants before and for two years after transition to bioenergy cropping. * represent significantly changed phylum. MG, *Miscanthus×giganteus*; PV, *Panicum virgatum*; NP, restored prairie; ZM, *Zea mays*.

Figure S6 (a) Phylogenetic tree of and (b) abundance of archaeal *amoA* OTUs that continuously changed after planting *Miscanthus× giganteus* (MG), *Panicum virgatum* (PV), restored prairie (NP) and *Zea*

mays (ZM). OTUs were classified based on a cutoff of 90% amino acid sequence similarity.

Figure S7 (a) Phylogenetic tree of and (b) abundance of bacterial *amoA* OTUs that continuously changed after planting *Miscanthus× giganteus* (MG), *Panicum virgatum* (PV), restored prairie (NP) and *Zea mays* (ZM). OTUs were classified based on a cutoff of 90% amino acid sequence similarity.

Figure S8 (a) Phylogenetic tree of and (b) abundance of *nifH* OTUs that continuously changed after planting *Miscanthus×giganteus* (MG), *Panicum virgatum* (PV), restored prairie (NP) and *Zea mays* (ZM). OTUs were classified based on a cutoff of 90% amino acid sequence similarity.

Figure S9 (a) Phylogenetic tree of and (b) abundance of *nosZ* OTUs that continuously changed after planting *Miscanthus×giganteus* (MG), *Panicum virgatum* (PV), restored prairie (NP) and *Zea mays* (ZM). OTUs of were classified based on a cutoff of 90% amino acid sequence similarity.

Figure S10 Nitrate and ammonia concentration in bulk soil. Soil samples for chemical and microbiological analysis were collected in the same week for each time point, except Sep 2008 when the nitrate and ammonia concentrations were not measured. MG, *Miscanthus×giganteus*; PV, *Panicum virgatum*; NP, restored prairie; ZM, *Zea mays*. These data were measured over the same period of our sample collection by the Biogeochemistry laboratory (C. Smith and M. David).

Figure S11 Pyrosequencing read length based on the raw sequence reads.

Table S1 Primers and annealing temperature for *nifH*, archaeal *amoA*, bacterial *amoA*, *nosZ* and 16S rRNA genes.

Table S2 Comparsion of the microbial community structures between different crops.

Table S3 Number of valid sequences for each gene in each sample.

Table S4 Number of OTUs observed after random re-sampling to the identical sequencing depth (697 sequences/sample.

Text S1

Acknowledgments

The authors gratefully acknowledge help from A. Duong, D. Fieckert, A. Groll, N. Peld, and M. Masters with field sampling and laboratory work. C. Smith and M. David kindly provided denitrification rate estimates for sampled plots. P.Y. Hong provided helpful comments on this manuscript.

Author Contributions

Conceived and designed the experiments: YM ACY RIM. Performed the experiments: YM ACY. Analyzed the data: YM. Contributed reagents/materials/analysis tools: YM. Wrote the paper: YM ACY RIM.

References

1. Garbeva P, van Veen JA, van Elsas JD (2004) Microbial diversity in soil: selection microbial populations by plant and soil type and implications for disease suppressiveness. Annu Rev Phytopathol 42: 243–270.

2. Kowalchuk GA, Buma DS, de Boer W, Klinkhamer PGL, van Veen JA (2002) Effects of above-ground plant species composition and diversity on the diversity of soil-borne microorganisms. Antonie Van Leeuwenhoek International Journal of General and Molecular Microbiology 81: 509–520.

3. Wardle DA, Bardgett RD, Klironomos JN, Setala H, van der Putten WH, et al. (2004) Ecological linkages between aboveground and belowground biota. Science 304: 1629–1633.

4. Rotthauwe JH, Witzel KP, Liesack W (1997) The ammonia monooxygenase structural gene amoA as a functional marker: molecular fine-scale analysis of natural ammonia-oxidizing populations. Appl Environ Microbiol 63: 4704–4712.

5. Poly F, Ranjard L, Nazaret S, Gourbiere F, Monrozier LJ (2001) Comparison of nifH gene pools in soils and soil microenvironments with contrasting properties. Appl Environ Microbiol 67: 2255–2262.

6. Francis CA, Roberts KJ, Beman JM, Santoro AE, Oakley BB (2005) Ubiquity and diversity of ammonia-oxidizing archaea in water columns and sediments of the ocean. Proc Natl Acad Sci U S A 102: 14683–14688.

7. Ruiz-Rueda O, Hallin S, Baneras L (2009) Structure and function of denitrifying and nitrifying bacterial communities in relation to the plant species in a constructed wetland. FEMS Microbiol Ecol 67: 308–319.

8. Bru D, Ramette A, Saby NP, Dequiedt S, Ranjard L, et al. (2011) Determinants of the distribution of nitrogen-cycling microbial communities at the landscape scale. ISME J 5: 532–542.

9. Orr CH, James A, Leifert C, Cooper JM, Cummings SP (2011) Diversity and activity of free-living nitrogen-fixing bacteria and total bacteria in organic and conventionally managed soils. Appl Environ Microbiol 77: 911–919.

10. Zehr JP, Jenkins BD, Short SM, Steward GF (2003) Nitrogenase gene diversity and microbial community structure: a cross-system comparison. Environ Microbiol 5: 539–554.

11. Cleveland CC, Townsend AR, Schimel DS, Fisher H, Howarth RW, et al. (1999) Global patterns of terrestrial biological nitrogen (N-2) fixation in natural ecosystems. Global Biogeochemical Cycles 13: 623–645.

12. Hsu SF, Buckley DH (2009) Evidence for the functional significance of diazotroph community structure in soil. Isme Journal 3: 124–136.

13. Son Y (2001) Non-symbiotic nitrogen fixation in forest ecosystems. Ecological Research 16: 183–196.

14. Schwarz H, Liebhard P, Ehrendorfer K, Ruckenbauer P (1994) The effect of fertilization on yield and quality of Miscanthus sinensis 'Giganteus'. Ind Crop Prod 2: 153–159.

15. Davis S, Parton W, Dohleman F, Smith C, Grosso S, et al. (2010) Comparative biogeochemical cycles of bioenergy crops reveal nitrogen-fixation and low greenhouse gas emissions in a Miscanthus×giganteus agro-ecosystem ecosystems DOI 10.1007/s10021-009-9306-9.

16. Demba Diallo M, Willems A, Vloemans N, Cousin S, Vandekerckhove TT, et al. (2004) Polymerase chain reaction denaturing gradient gel electrophoresis analysis of the N₂-fixing bacterial diversity in soil under Acacia tortilis ssp. raddiana and Balanites aegyptiaca in the dryland part of Senegal. Environ Microbiol 6: 400–415.

17. Tan Z, Hurek T, Reinhold-Hurek B (2003) Effect of N-fertilization, plant genotype and environmental conditions on nifH gene pools in roots of rice. Environ Microbiol 5: 1009–1015.

18. Kowalchuk GA, Stephen JR (2001) Ammonia-oxidizing bacteria: a model for molecular microbial ecology. Annu Rev Microbiol 55: 485–529.

19. Schleper C, Jurgens G, Jonuscheit M (2005) Genomic studies of uncultivated archaea. Nat Rev Microbiol 3: 479–488.

20. Konneke M, Bernhard AE, de la Torre JR, Walker CB, Waterbury JB, et al. (2005) Isolation of an autotrophic ammonia-oxidizing marine archaeon. Nature 437: 543–546.

21. Hatzenpichler R, Lebedeva EV, Spieck E, Stoecker K, Richter A, et al. (2008) A moderately thermophilic ammonia-oxidizing crenarchaeote from a hot spring. Proceedings of the National Academy of Sciences of the United States of America 105: 2134–2139.

22. Leininger S, Urich T, Schloter M, Schwark L, Qi J, et al. (2006) Archaea predominate among ammonia-oxidizing prokaryotes in soils. Nature 442: 806–809.

23. He JZ, Shen JP, Zhang LM, Zhu YG, Zheng YM, et al. (2007) Quantitative analyses of the abundance and composition of ammonia-oxidizing bacteria and ammonia-oxidizing archaea of a Chinese upland red soil under long-term fertilization practices. Environ Microbiol 9: 2364–2374.

24. Shen JP, Zhang LM, Zhu YG, Zhang JB, He JZ (2008) Abundance and composition of ammonia-oxidizing bacteria and ammonia-oxidizing archaea communities of an alkaline sandy loam. Environ Microbiol 10: 1601–1611.

25. Wessen E, Soderstrom M, Stenberg M, Bru D, Hellman M, et al. (2011) Spatial distribution of ammonia-oxidizing bacteria and archaea across a 44-hectare farm related to ecosystem functioning. ISME J.

26. Briones AM, Okabe S, Umemiya Y, Ramsing NB, Reichardt W, et al. (2002) Influence of different cultivars on populations of ammonia-oxidizing bacteria in the root environment of rice. Appl Environ Microbiol 68: 3067–3075.

27. Wang Y, Ke X, Wu L, Lu Y (2009) Community composition of ammonia-oxidizing bacteria and archaea in rice field soil as affected by nitrogen fertilization. Syst Appl Microbiol 32: 27–36.

28. Le Roux X, Poly F, Currey P, Commeaux C, Hai B, et al. (2008) Effects of aboveground grazing on coupling among nitrifier activity, abundance and community structure. Isme J 2: 221–232.

29. Patra AK, Abbadie L, Clays-Josserand A, Degrange V, Grayston SJ, et al. (2006) Effects of management regime and plant species on the enzyme activity and genetic structure of N-fixing, denitrifying and nitrifying bacterial communities in grassland soils. Environ Microbiol 8: 1005–1016.

30. Philippot L, Hallin S, Schloter M (2007) Ecology of denitrifying prokaryotes in agricultural soil. Advances in Agronomy, Vol 96 96: 249–305.

31. Schlesinger WH (2009) On the fate of anthropogenic nitrogen. Proc Natl Acad Sci U S A 106: 203–208.

32. Rich JJ, Heichen RS, Bottomley PJ, Cromack K, Jr., Myrold DD (2003) Community composition and functioning of denitrifying bacteria from adjacent meadow and forest soils. Appl Environ Microbiol 69: 5974–5982.

33. Dambreville C, Hallet S, Nguyen C, Morvan T, Germon JC, et al. (2006) Structure and activity of the denitrifying community in a maize-cropped field fertilized with composted pig manure or ammonium nitrate. FEMS Microbiol Ecol 56: 119–131.

34. Hallin S, Jones CM, Schloter M, Philippot L (2009) Relationship between N-cycling communities and ecosystem functioning in a 50-year-old fertilization experiment. Isme J 3: 597–605.

35. Knops JMH, Bradley KL, Wedin DA (2002) Mechanisms of plant species impacts on ecosystem nitrogen cycling. Ecology Letters 5: 454–466.

36. Priha O, Grayston SJ, Pennanen T, Smolander A (1999) Microbial activities related to C and N cycling and microbial community structure in the rhizospheres of Pinus sylvestris, Picea abies and Betula pendula seedlings in an organic and mineral soil. FEMS Microbiol Ecol 30: 187–199.

37. Kloos K, Mergel A, Rosch C, Bothe H (2001) Denitrification within the genus Azospirillum and other associative bacteria. Australian Journal of Plant Physiology 28: 991–998.

38. Chen J, Yu ZT, Michel FC, Wittum T, Morrison M (2007) Development and application of real-time PCR assays for quantification of erm genes conferring resistance to macrolides-lincosamides-streptogramin B in livestock manure and manure management systems. Applied and Environmental Microbiology 73: 4407–4416.

39. Bustin SA (2000) Absolute quantification of mRNA using real-time reverse transcription polymerase chain reaction assays. Journal of Molecular Endocrinology 25: 169–193.

40. Mumy KL, Findlay RH (2004) Convenient determination of DNA extraction efficiency using an external DNA recovery standard and quantitative-competitive PCR. Journal of Microbiological Methods 57: 259–268.

41. Okano Y, Hristova KR, Leutenegger CM, Jackson LE, Denison RF, et al. (2004) Application of real-time PCR to study effects of ammonium on population size of ammonia-oxidizing bacteria in soil. Appl Environ Microbiol 70: 1008–1016.

42. Heuer H, Krsek M, Baker P, Smalla K, Wellington EM (1997) Analysis of actinomycete communities by specific amplification of genes encoding 16S rRNA and gel-electrophoretic separation in denaturing gradients. Appl Environ Microbiol 63: 3233–3241.

43. Lane DJ (1991) 16S/23S rRNA sequencing. InIn E. Stackebrandt, M. Goodfellow, eds. Nucleic acid techniques in bacterial systematics. pp 115–175.

44. Schloss PD, Westcott SL, Ryabin T, Hall JR, Hartmann M, et al. (2009) Introducing mothur: open-source, platform-independent, community-supported software for describing and comparing microbial communities. Appl Environ Microbiol 75: 7537–7541.

45. Ludwig W, Strunk O, Westram R, Richter L, Meier H, et al. (2004) ARB: a software environment for sequence data. Nucleic Acids Research 32: 1363–1371.

46. Schloss PD, Handelsman J (2005) Introducing DOTUR, a computer program for defining operational taxonomic units and estimating species richness. Applied and Environmental Microbiology 71: 1501–1506.

47. Caporaso JG, Kuczynski J, Stombaugh J, Bittinger K, Bushman FD, et al. (2010) QIIME allows analysis of high-throughput community sequencing data. Nat Methods 7: 335–336.

48. Reeder J, Knight R (2010) Rapidly denoising pyrosequencing amplicon reads by exploiting rank-abundance distributions. Nat Methods 7: 668–669.

49. Li W, Godzik A (2006) Cd-hit: a fast program for clustering and comparing large sets of protein or nucleotide sequences. Bioinformatics 22: 1658–1659.

50. Edgar RC (2004) MUSCLE: multiple sequence alignment with high accuracy and high throughput. Nucleic Acids Research 32: 1792–1797.

51. Palmer K, Drake HL, Horn MA (2009) Genome-derived criteria for assigning environmental narG and nosZ sequences to operational taxonomic units of nitrate reducers. Appl Environ Microbiol 75: 5170–5174.

52. Koops H-P, Purkhold U, Pommerening-Röser A, Timmermann G, Wagner M (2006) The lithoautotrophic ammonia-oxidizing bacteria. Prokaryotes 5: 788–811.

53. Gilbert JA, Field D, Swift P, Thomas S, Cummings D, et al. (2010) The taxonomic and functional diversity of microbes at a temperate coastal site: a 'multi-omic' study of seasonal and diel temporal variation. PLoS One 5: e15545.

54. Mounier E, Hallet S, Cheneby D, Benizri E, Gruet Y, et al. (2004) Influence of maize mucilage on the diversity and activity of the denitrifying community. Environ Microbiol 6: 301–312.

55. Teng Q, Sun B, Fu X, Li S, Cui Z, et al. (2009) Analysis of nifH gene diversity in red soil amended with manure in Jiangxi, South China. J Microbiol 47: 135–141.

56. Duc L, Noll M, Meier BE, Burgmann H, Zeyer J (2009) High diversity of diazotrophs in the forefield of a receding alpine glacier. Microb Ecol 57: 179–190.

57. Palmer K, Drake HL, Horn MA (2010) Association of novel and highly diverse acid-tolerant denitrifiers with N₂O fluxes of an acidic fen. Appl Environ Microbiol 76: 1125–1134.

58. Zhang LM, Offre PR, He JZ, Verhamme DT, Nicol GW, et al. (2010) Autotrophic ammonia oxidation by soil thaumarchaea. Proc Natl Acad Sci U S A 107: 17240–17245.

59. Reed DW, Smith JM, Francis CA, Fujita Y (2010) Responses of ammonia-oxidizing bacterial and archaeal populations to organic nitrogen amendments in low-nutrient groundwater. Appl Environ Microbiol 76: 2517–2523.

60. Moin NS, Nelson KA, Bush A, Bernhard AE (2009) Distribution and diversity of archaeal and bacterial ammonia oxidizers in salt marsh sediments. Appl Environ Microbiol 75: 7461–7468.

61. Nicol GW, Leininger S, Schleper C, Prosser JI (2008) The influence of soil pH on the diversity, abundance and transcriptional activity of ammonia oxidizing archaea and bacteria. Environ Microbiol 10: 2966–2978.

62. Zhou J, Wu L, Deng Y, Zhi X, Jiang YH, et al. (2011) Reproducibility and quantitation of amplicon sequencing-based detection. ISME J.

63. Lin YT, Lin CP, Chaw SM, Whitman WB, Coleman DC, et al. (2010) Bacterial community of very wet and acidic subalpine forest and fire-induced grassland soils Plant and Soil DOI:10.1007/s11104-010-0308-3.

64. Wakelin SA, Colloff MJ, Harvey PR, Marschner P, Gregg AL, et al. (2007) The effects of stubble retention and nitrogen application on soil microbial community structure and functional gene abundance under irrigated maize. FEMS Microbiol Ecol 59: 661–670.

65. Rothballer M, Eckert B, Schmid M, Fekete A, Schloter M, et al. (2008) Endophytic root colonization of gramineous plants by Herbaspirillum frisingense. FEMS Microbiol Ecol 66: 85–95.

66. Miyamoto T, Kawahara M, Minamisawa K (2004) Novel endophytic nitrogen-fixing clostridia from the grass Miscanthus sinensis as revealed by terminal restriction fragment length polymorphism analysis. Appl Environ Microbiol 70: 6580–6586.

67. Kirchhof G, Eckert B, Stoffels M, Baldani JI, Reis VM, et al. (2001) Herbaspirillum frisingense sp. nov., a new nitrogen-fixing bacterial species that occurs in C4-fibre plants. Int J Syst Evol Microbiol 51: 157–168.

68. Hai B, Diallo NH, Sall S, Haesler F, Schauss K, et al. (2009) Quantification of Key Genes Steering the microbial nitrogen cycle in the rhizosphere of sorghum cultivars in tropical agroecosystems. Applied and Environmental Microbiology 75: 4993–5000.

69. Kowalchuk GA, Stephen JR, De Boer W, Prosser JI, Embley TM, et al. (1997) Analysis of ammonia-oxidizing bacteria of the beta subdivision of the class Proteobacteria in coastal sand dunes by denaturing gradient gel electrophoresis and sequencing of PCR-amplified 16S ribosomal DNA fragments. Appl Environ Microbiol 63: 1489–1497.

70. Ehrich S, Behrens D, Lebedeva E, Ludwig W, Bock E (1995) A new obligately chemolithoautotrophic, nitrite-oxidizing bacterium, Nitrospira moscoviensis sp. nov. and its phylogenetic relationship. Arch Microbiol 164: 16–23.

71. Offre P, Prosser JI, Nicol GW (2009) Growth of ammonia-oxidizing archaea in soil microcosms is inhibited by acetylene. Fems Microbiology Ecology 70: 99–108.

72. Tourna M, Freitag TE, Nicol GW, Prosser JI (2008) Growth, activity and temperature responses of ammonia-oxidizing archaea and bacteria in soil microcosms. Environmental Microbiology 10: 1357–1364.

73. Jia Z, Conrad R (2009) Bacteria rather than Archaea dominate microbial ammonia oxidation in an agricultural soil. Environ Microbiol.

74. Di HJ, Cameron KC, Shen JP, Winefield CS, O'Callaghan M, et al. (2009) Nitrification driven by bacteria and not archaea in nitrogen-rich grassland soils. Nature Geoscience 2: 621–624.

75. Martens-Habbena W, Berube PM, Urakawa H, de la Torre JR, Stahl DA (2009) Ammonia oxidation kinetics determine niche separation of nitrifying Archaea and Bacteria. Nature 461: 976–979.

76. Taylor AE, Zeglin LH, Dooley S, Myrold DD, Bottomley PJ (2010) Evidence for different contributions of archaea and bacteria to the ammonia-oxidizing potential of diverse Oregon soils. Appl Environ Microbiol 76: 7691–7698.

77. Kramer SB, Reganold JP, Glover JD, Bohannan BJ, Mooney HA (2006) Reduced nitrate leaching and enhanced denitrifier activity and efficiency in organically fertilized soils. Proc Natl Acad Sci U S A 103: 4522–4527.

78. Wallenstein MD, Peterjohn WT, Schlesinger WH (2006) N fertilization effects on denitrification and N cycling in an aggrading forest. Ecol Appl 16: 2168–2176.

79. Hutchinson GL, Mosier AR (1979) Nitrous oxide emissions from an irrigated cornfield. Science 205: 1125–1127.

80. Wertz S, Degrange V, Prosser JI, Poly F, Commeaux C, et al. (2006) Maintenance of soil functioning following erosion of microbial diversity. Environ Microbiol 8: 2162–2169.

18

Comparative Functional Genomics of Salt Stress in Related Model and Cultivated Plants Identifies and Overcomes Limitations to Translational Genomics

Diego H. Sanchez[1¤a], Fernando L. Pieckenstain[2], Jedrzey Szymanski[1], Alexander Erban[1], Mariusz Bromke[1], Matthew A. Hannah[1¤b], Ute Kraemer[3], Joachim Kopka[1], Michael K. Udvardi[4]*

1 Max Planck Institute for Molecular Plant Physiology (MPIMP), Potsdam-Golm, Germany, 2 Instituto Tecnológico de Chascomús (IIB-Intech), Chascomús, Argentina, 3 Department of Plant Physiology, Ruhr University Bochum, Bochum, Germany, 4 Samuel Roberts Noble Foundation, Ardmore, Oklahoma, United States of America

Abstract

One of the objectives of plant translational genomics is to use knowledge and genes discovered in model species to improve crops. However, the value of translational genomics to plant breeding, especially for complex traits like abiotic stress tolerance, remains uncertain. Using comparative genomics (ionomics, transcriptomics and metabolomics) we analyzed the responses to salinity of three model and three cultivated species of the legume genus *Lotus*. At physiological and ionomic levels, models responded to salinity in a similar way to crop species, and changes in the concentration of shoot Cl^- correlated well with tolerance. Metabolic changes were partially conserved, but divergence was observed amongst the genotypes. Transcriptome analysis showed that about 60% of expressed genes were responsive to salt treatment in one or more species, but less than 1% was responsive in all. Therefore, genotype-specific transcriptional and metabolic changes overshadowed conserved responses to salinity and represent an impediment to simple translational genomics. However, 'triangulation' from multiple genotypes enabled the identification of conserved and tolerant-specific responses that may provide durable tolerance across species.

Editor: Nicholas Provart, University of Toronto, Canada

Funding: This work was conducted within the framework of the European LOTASSA project (Lotus Adaptation and Sustainability in South America, INCO-CT-2005-517617). The funders had no role in study design, data collection and analysis, decision to publish, or preparation of the manuscript.

Competing Interests: The authors have declared that no competing interests exist.

* E-mail: mudvardi@noble.org

¤a Current address: Division of Biological Sciences, University of California San Diego, La Jolla, California, United States of America
¤b Current address: Bayer Cropscience N.V., Gent, Belgium

Introduction

Secondary salinization of soils caused by irrigation has become a major concern worldwide (www.fao.org/ag/agl/agll/spush). Salinity engenders both hyper-osmotic and hyper-ionic stresses, with plants facing dehydration, ion toxicity, nutritional deficiencies and oxidative stress [1]. Acclimation responses include ion exclusion and tissue tolerance, tight control of water homeostasis and osmotic adjustment, changes in growth and development, and a wide array of underlying biochemical and molecular changes [1–5]. Research on the molecular responses of plants to salinity has focused mostly on model species such as *Arabidopsis thaliana*, yet the value of model plants in identifying mechanisms that may confer stress tolerance to crops in the field remains to be seen [6]. Although it is known that salt tolerance is a quantitative trait determined by multiple and complex genetic interactions [7–9] and that plant responses to salinity involve changes in the expression of thousands of genes [2–3,10], we know little about the extent of evolutionary conservation of molecular networks that determine salt tolerance. To understand better the nature of impediments that may stand in the way of translational genomics for salinity tolerance, we carried out a comparative functional genomic study between model and cultivated legumes of the genus *Lotus*.

Legumes are second only to grasses in their importance to agriculture, and provide a rich source of protein, oil, carbohydrate, minerals, and secondary compounds for human and animal nutrition [11]. The genomes of three legumes (*Lotus japonicus, Medicago truncatula* and *Glycine max*) have been sequenced, and it is envisioned that genomic discoveries in these species will be translated to crop improvement via breeding programs involving grain and forage legumes [12]. This approach seems reasonable given the high degree of synteny between legumes [13]. However, it remains to be seen whether orthologous genes that have significant effects on a complex, quantitative trait such as salinity tolerance in one species will have a similar effect in another.

In this study, we compared the physiological and molecular responses to salt stress of six *Lotus* species. Three of them, *L. japonicus*, *L. filicaulis*, and *L. burttii*, are in-breeding and have been used as models for legume genetics [14–20]. *L. japonicus* has been developed as a premier model, with genome sequence and numerous tools for genetics and genomics now available [10,14–19]. *L. filicaulis* and *L. burttii* have been developed as crossing partners for genetic studies [20]. The other three species, *L. corniculatus*, *L. glaber* and *L. uliginosus*, are out-breeding and are used in world agriculture as forages. Although related, these species exhibit diversity in their ability to grow in low-fertility soils and under different environmental constraints [21–22]. Here, we present the results of comparative ionomic, transcriptomic and metabolomic analyses of *Lotus* genotypes that reveal conserved and divergent system responses to salinity within this genus. Our work has important implications for translational genomics approaches that aim to improve salinity tolerance and other complex traits in plants.

Results

Physiological and nutritional responses to salinity in *Lotus spp*

The relative salt tolerance of *Lotus* genotypes representing the six species described above, including two accessions of *L. japonicus*

(MG20 and Gifu), was determined in two independent survival experiments in which plants were subjected to long-term step-wise increases in the level of NaCl up to 300 mM NaCl (Figure 1A and B). We defined the 'lethal-dose fifty' (LD50) as the number of days at which 50% of plants had died. The resulting ranking from most to least tolerant genotype was: *L. glaber* > *L. burttii* > *L. japonicus* var. MG20 > *L. filicaulis* > *L. japonicus* var. Gifu ~ *L. uliginosus* ~ *L. corniculatus* (Table 1). No separation with respect to survival was observed between phylogenetically-close and distant genotypes [23], or between model and forage species [21].

To facilitate systems comparison under salt acclimation a second treatment regime was applied, which subjected plants to a long-term sub-lethal level of salt (up to 150 mM NaCl, figure 1A,

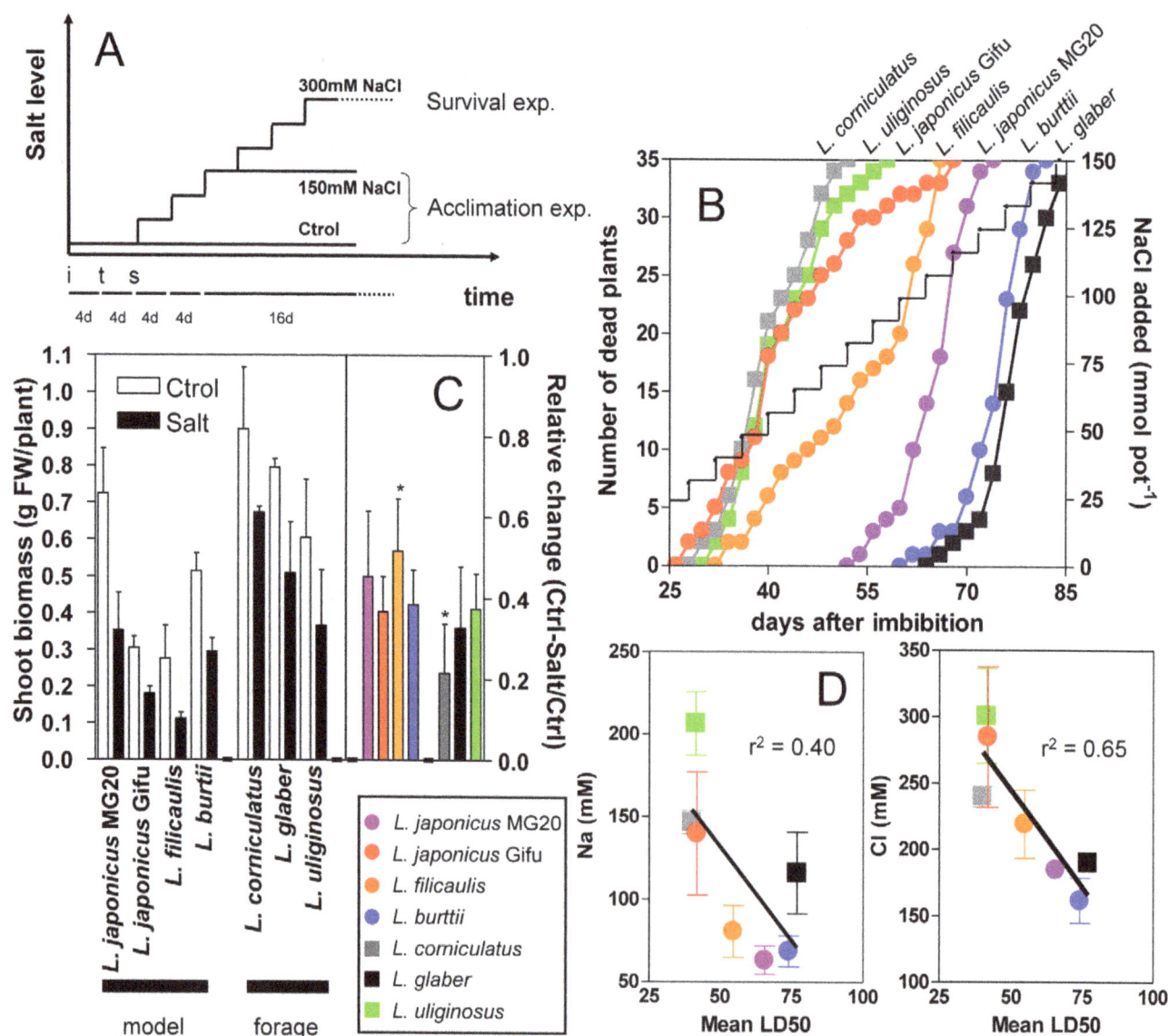

Figure 1. Experimental design and physiological assessment of salt tolerance and acclimation in *Lotus* species. (A) Experimental design for long-term survival (300 mM NaCl) and sub-lethal salt acclimation (150 mM NaCl) experiments. NaCl concentration in the nutrient solution was increased by 50 mM every four days (see Materials and methods). i = seed imbibition, t = transplanting, s = start salinization, d = days. (B) Representative experiment for survival of *Lotus* species under lethal NaCl-stress conditions. The step-wise increase in total NaCl added to each pot is estimated on the right axis. (C) Plant growth evaluated as final shoot biomass (left panel) and relative performance (right panel) under sub-lethal NaCl levels. Data represents the mean ± SD of 3 independent experiments, and the asterisk indicates a statistically-significant difference in stress-induced change in biomass between *L. corniculatus* and *L. filicaulis* (Student *t*-test, p<0.05). Model species are shown to the left. FW = fresh weight. (D) Linear correlation between Na⁺ or Cl⁻ content of salt-acclimated plants (shoots) and mean LD50 calculated from survival experiments (Table 1).

Table 1. Lethal-dose-fifty (LD50, expressed in days after imbibition) of survival at 300 mM NaCl after a step-wise increase in salt concentration, for each *Lotus* genotype estimated in two independent survival experiments.

Cultivar	LD50 Exp_1	LD50 Exp_2
L. japonicus MG20	66.2	65.0
L. japonicus Gifu	41.9	42.0
L. filicaulis	54.8	54.6
L. burttii	73.8	74.4
L. corniculatus	41.0	39.4
L. glaber	77.1	76.9
L. uliginosus	42.1	41.0

LD50 was calculated fitting the survival data to a Boltzmann sigmoid. Model legume species are shown at the top. Exp = experiment.

[10]). Three independent experiments were performed, each comprising control and treated plants of each genotype. As expected, shoot biomass decreased under stress (Figure 1C, left panel). Although forage legumes tended to be larger, the relative inhibition of growth was not statistically different between most *Lotus* genotypes (Figure 1C, right panel), and it bore no apparent relationship to the rate of mortality under lethal salt treatment (compare figure 1B, 1C and Table 1). Shoot Na^+ and Cl^- content increased dramatically in all stressed cultivars, exhibiting a negative linear correlation with the LD50 under lethal salinity with a much better correlation coefficient for Cl^- levels (Figure 1D and 2). These results are consistent with previous observations that tolerant glycophytes accumulate less salt than sensitive ones [1–3], and support the use of LD50 rather than changes in biomass to estimate relative salt tolerance under our experimental conditions. K^+ concentration changed less in the models than in the forage species in which it decreased 30–70% (Figure 2), and no correlation was found with the LD50.

Macro- and micro-nutrients were profiled in shoots using ICP-AES, revealing differential salt stress-induced changes in Ca, Mg, Mn, Fe and Zn levels in the different species (Figure 3). No elemental change differentiated model and forage legumes, but the more tolerant cultivars differed from sensitive ones on two ways. First, sulphur increased significantly in tolerant genotypes (*L. glaber*, *L. burtti* and *L. japonicus* var. MG20) but not sensitive ones (*L. japonicus* var. Gifu, *L. uliginosus* and *L. corniculatus*). Second, although phosphate and zinc content increased significantly in response to NaCl treatment in all or some genotypes, changes were greater in the more tolerant ones (Figure 3).

In summary, physiological and ionomic data revealed complex interactions between NaCl uptake and growth responses, with shoot Cl^- levels of stress-acclimated genotypes correlating strongly with rates of mortality in plants exposed to lethal salt-stress doses. No correlation was found between shoot K^+, Na^+ or Cl^- content and growth inhibition under stress, and nutritional aspects were differentially altered under salinity between more tolerant and sensitive backgrounds. Overall, the data showed a good match between in-breeding model species and out-breeding crop species in their range of tolerance, salt accumulation, variation of nutrient content, and induced growth effects. Thus, the model legume genotypes appear to be valid physiological tools to study and understand salt tolerance mechanisms in the forage species of *Lotus*.

Gene expression analysis

RNA was isolated from shoots of plants from the sub-lethal salt acclimation experiments, and analyzed using the Affymetrix GeneChip® *Lotus* Genome Array. To avoid problems that might arise from differences in gene/transcript sequences (and, therefore, differences in probe hybridization/signal strength) between species, we ignored data from probe-sets that did not detect transcript in all genotypes and in all three independent experiments, and compared only relative changes in probe-set signal (i.e. ratio Log_2 Salt/Control for each genotype separately) rather than absolute probe-set signal. Probe-sets that detected transcript in all genotypes, experiments and conditions amounted to 12,137 (Table S1). Non-supervised independent component analysis (ICA) of the whole dataset separated controls from NaCl-treated plants, regardless of the genotype, indicating that at least part of the transcriptional changes were conserved amongst all species (Figure 4A, IC4 represents the shared stress-related variability, while IC1 to IC3 captured genotype-related variability). Data was analyzed by a significance-based test between treated and non-treated plants within each genotype. Of the 12,137 probe-sets called present, 7,776 (64%) detected changes in transcript levels in at least one cultivar upon salt acclimation, but only 92 probe-sets (0.76%) were significantly altered in all seven genotypes (FDR<0.05, Table S1). To facilitate comparisons, the statistically-significant salt-induced or repressed genes were analyzed between the three most sensitive and three most tolerant genotypes, excluding *L. filicaulis* with intermediate tolerance. Remarkably, approximately one-third to one-half of the NaCl-responsive transcripts were specific to a single cultivar (Figure 4B and C). Only 4% of salt-induced genes in the three most-sensitive genotypes were common to all three, while 7% of salt-induced genes in the three most-tolerant genotypes were common to all three. A similar situation was observed for salt-repressed transcripts. On the other hand, many salt-elicited genes common to all sensitive genotypes were also found to be responsive in the more tolerant genotypes (Figure 4B and C, table S2). These shared transcripts included many of unknown function but also genes previously implicated in plant stress such as known *Lotus* stress-responsive genes (LEA protein, phosphatase-2C *LjNPP2C1* and nodulin protein *LjENOD40*, [10]), enzymes of amino acid, polyamine and myo-inositol metabolism (proline oxidase, asparagine synthetase *LjAS1* and histidine decarboxylase, polyamine oxidase *LjPAO4*, S-adenosylmethionine decarboxylase and spermine synthase *LjSPMS*, and myo-inositol phosphate synthase *LjMIPS1*, [10]) and photorespiration (serine-glyoxylate amino-transferase, [24]). We also found tolerant- and sensitive-specific transcripts, most of them of unknown function but some of which have been linked to stress, hormone or nutrient metabolism (Figure 4B and C, table S2). Among others, genes exclusively regulated in tolerant species were involved in stress signalling (phosphatase-2C protein and CBL-interacting protein kinase *LjCIPK6*, [25]), hormone homeostasis (ent-kaurene oxidase *LjGA3*), nitrogen assimilation (cytosolic glutamine synthase *LjGS1*, [26]) and cell wall-related processes such as cellulose synthesis [27]. On the other hand, genes transcriptionally regulated only in salt-sensitive genotypes included also cell wall-related processes, a transcription factor of the DREB sub-family involved in the control of stress responses [28], and the calcineurin B-like protein *LjCBL1* which represent a key node in stress signalling [29]. Homologues of some of the genes described above have been implicated in stress tolerance in other species previously [25,27–31].

In view of the strong link between Cl^- concentration and salt tolerance (Figure 1D), we tested for correlations between NaCl-elicited changes in transcript levels and Cl^- content under stress

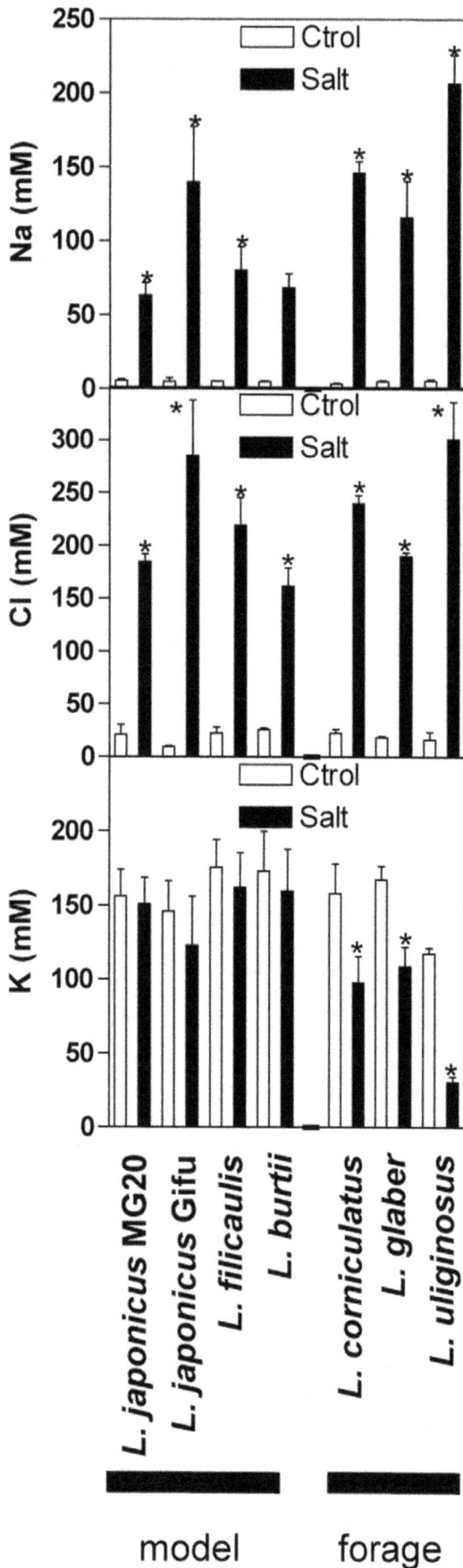

Figure 2. Changes in shoot Na+, Cl− and K+ in *Lotus* species under salt acclimation. Data represents the mean ± SD of 3 independent experiments. Asterisks indicate a statistically-significant difference (Student *t*-test with adjusted Bonferroni correction, p<0.05) between salt treatment and control.

for salt-responsive genes shared between all species, and those that were exclusive to tolerant or sensitive genotypes (Figure S1). Remarkably, expression changes of transcripts with shared responses to salinity across species showed little global correlation to Cl− levels. On the other hand, as expected considering the link with survival under lethal salt stress, changes in transcript levels of tolerant- and sensitive- specific genes correlated better with Cl− content.

In summary, the results of transcriptome analysis indicated that a small fraction of the transcriptional responses to salinity were conserved amongst the *Lotus* genotypes. The majority of genes regulated during NaCl acclimation was unique to single genotypes or confined to just a few, and were neither linked to model or forage species nor indicative of the degree of salt tolerance. However, a small sub-set of salt-responsive transcripts were common to all genotypes or represented markers for more tolerant or sensitive genetic backgrounds, and these included key genes involved in stress and metabolism.

Metabolic phenotype analysis

The shoot metabolic phenotypes of the *Lotus* species under salt acclimation were determined using non-targeted GC/EI-TOF-MS. A set of 123 analytes, representing both known and unknown compounds, were identified in all genotypes and independent experiments (Table S3). A significance-based analysis was used to determine which analytes responded to salt stress within each genotype (Student *t*-test, p<0.05 adjusted Bonferroni correction). Many NaCl-induced changes in analytes levels were qualitatively similar in most genetic backgrounds, although some metabolic changes were genotype-specific (Figure 5A). Approximately half of the analytes that accumulated under stress were shared between sensitive and tolerant genotypes (Figure 5A, Table S3). These included known *L. japonicus* salt-responsive metabolites such as serine, threonine and ononitol ([10]; Figure S2). On the other hand, levels of organic acids including citric, succinic, malic and threonic acids, declined in most cultivars under salinity, as reported before for model species ([32]; Figure S2). Few salt-induced changes were confined to sensitive genotypes, e.g. increase in gulonic acid and decrease in aspartic acid; or tolerant genotypes, i.e. increase in asparagine (Figure 6). Based on correlation analysis, changes in metabolite levels did not show stable correlative patterns with changes in ion content or biomass under stress, which hampered further integrative analysis. Furthermore, the number of independent microarrays available for each species was insufficient to build robust metabolite-transcript correlations.

Changes in metabolism were analyzed further, using a metabolic network approach based on pairwise correlations between analyte levels in all genotypes and treatments combined (see methods). Nine communities (modules) of correlated analytes were identified in the resulting network using community structure statistics [33], reflecting conservation of some metabolic modules between genotypes. However, when individual networks were constructed for each cultivar using control and treatment data, clear differences in network architecture were observed, indicating genotype-specific interactions (Figure 7A). Networks for control and salt treatments were also built to identify salt-dependent

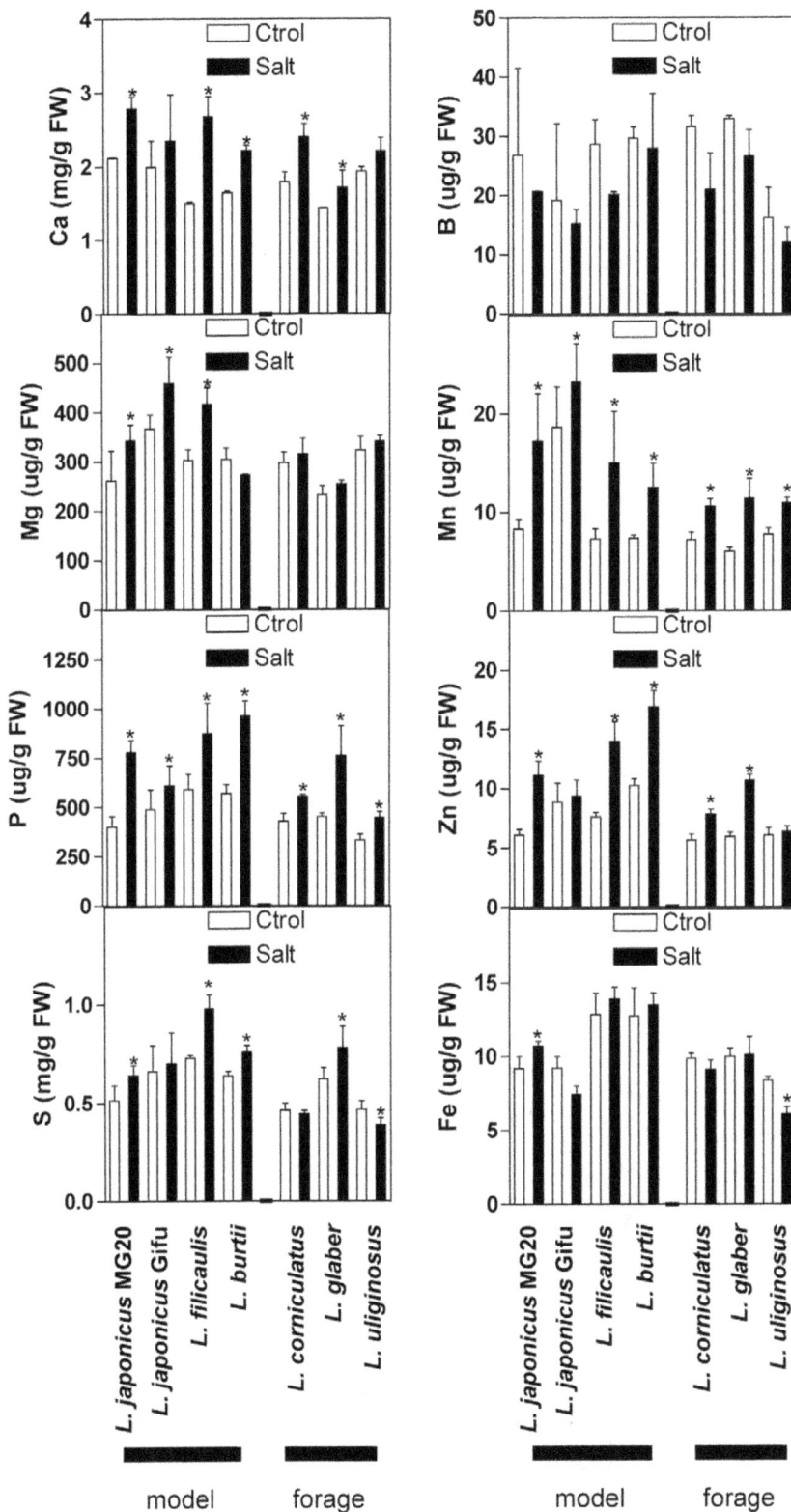

Figure 3. Changes in shoot macro- and micro-nutrients in *Lotus* species under salt acclimation. Data represents the mean ± SD of 3 independent experiments. Asterisks indicate a statistically-significant difference (ANOVA, p<E^{-6}) in response to treatment.

changes in network architecture that were conserved across genotypes (Figure 7B). Only some of the NaCl-induced communities in the network architecture were shared between genotypes, indicating that some but not all metabolic patterns of response to salt acclimation are conserved between genotypes (compare Figure 7A and B).

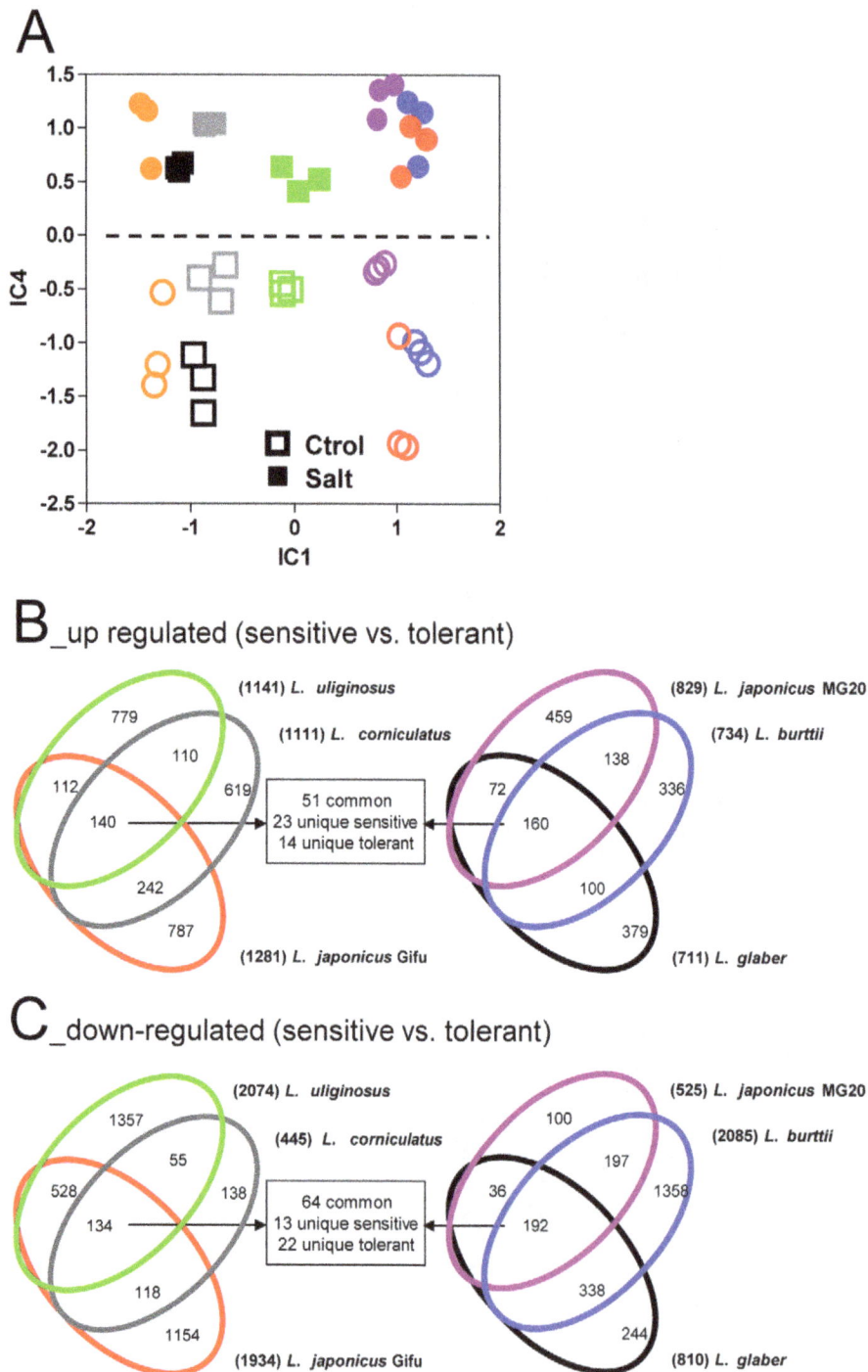

Figure 4. Transcriptional changes in *Lotus* species during salt acclimation. (**A**) Independent component analysis (ICA) of changes in transcript levels. A dashed line separates the symbols that indicate control (empty) and salt acclimation (filled) treatments of the three independent sub-lethal acclimation experiments. (**B** and **C**) Venn diagrams showing the number of salt-induced (B) and repressed (C) genes at FDR<0.05 common to more sensitive and/or tolerant genotypes. *L. filicaulis*, with intermediate tolerance, was excluded (see figure 1 and table S1). Colour code for the different genotypes is as in figure 1.

In summary, the results indicated that *Lotus* species share many metabolic responses to salinity, and few metabolite markers were found that distinguish tolerant or sensitive genetic backgrounds. On the other hand, several qualitative and quantitative changes under salt-stress, along with some network properties of the metabolome, were unique to individual genotypes.

Discussion

Previous work has shown that translational genomics can be used to modify traits of agricultural importance, such as pathogen resistance, via a candidate gene approach [34]. But what about complex traits such as tolerance to abiotic stresses, which are

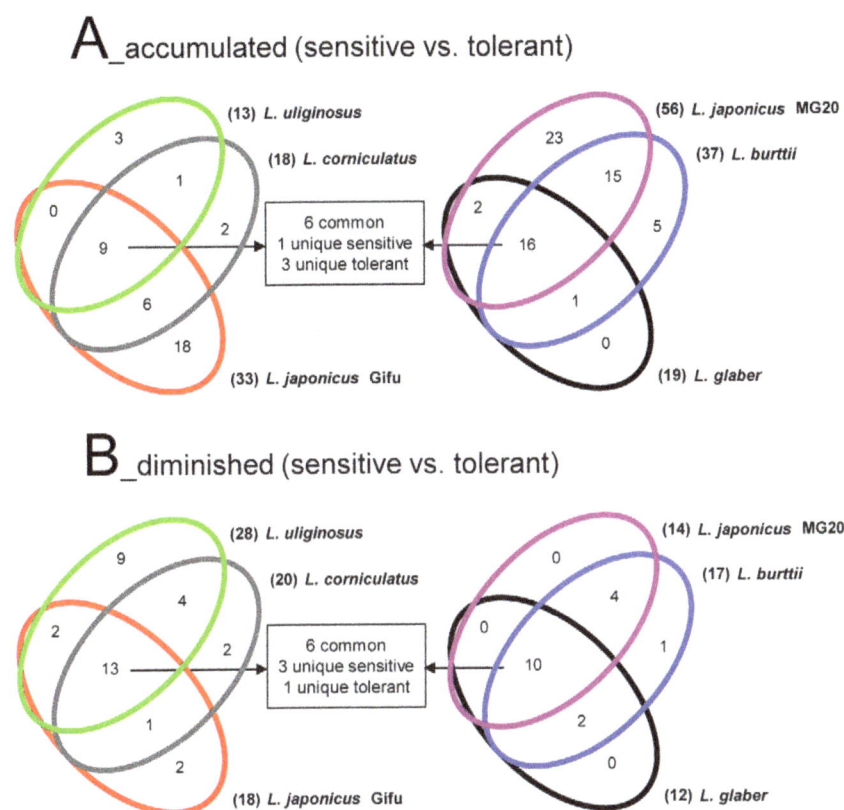

A_accumulated (sensitive vs. tolerant)

(13) *L. uliginosus*
(18) *L. corniculatus*
(33) *L. japonicus* Gifu

6 common
1 unique sensitive
3 unique tolerant

(56) *L. japonicus* MG20
(37) *L. burttii*
(19) *L. glaber*

B_diminished (sensitive vs. tolerant)

(28) *L. uliginosus*
(20) *L. corniculatus*
(18) *L. japonicus* Gifu

6 common
3 unique sensitive
1 unique tolerant

(14) *L. japonicus* MG20
(17) *L. burttii*
(12) *L. glaber*

Figure 5. Overview of metabolite profiles of *Lotus* species during salt acclimation. (A and **B)** Venn diagrams comparing analytes that were responsive to salt stress in tolerant and sensitive genotypes (Student *t*-test with adjusted Bonferroni correction, p<0.05). *L. filicaulis* was excluded because of its intermediate tolerance (c.f. Figure 1B and table 1). The colour code for the different genotypes is as in figure 1.

determined by the interdependent action of thousands of genes that, in turn, are affected by complex interactions with the environment beyond the stress of interest? We utilised the tools of ionomics, transcriptomics and metabolomics to determine the extent to which plant systems responses to salinity are conserved amongst closely related model and cultivated *Lotus* species [23]. There were no apparent differences between model and forage genotypes of *Lotus* with regard to the range of tolerance/sensitivity, growth inhibition, salt accumulation and nutrient status under salt stress (Figures 1 to 3). A strong negative correlation between Cl^- levels in the shoot and tolerance to salinity across species (Figure 1D) supported the conclusion that Cl^- exclusion from the shoots represents a key physiological mechanism for salt tolerance in legumes, similar to the case of Na^+ exclusion in other glycophytes [1–3,35]. Taken together, these results indicate that *Lotus* model species respond to salinity in a similar way to crop species and, therefore, are useful systems for identifying physiological processes required for salt tolerance.

Ionomic analysis revealed a differential increase of phosphate, sulphur and zinc in more tolerant genotypes in response to salinity (Figure 3). Presumably, this reflects a differential NaCl-induced imbalance between uptake and translocation to the shoot of these nutrients and plant growth. However, no nutrient correlated with biomass in control or stressed plants. Therefore, the physiological significance of shoot phosphate, sulphur and zinc concentration in *Lotus* shoots during saltinity remains obscure. In fact, these nutrients were observed to decline in other species under salt stress, indicating that the influence of salinity on plant nutrition is variable and dependent on growth conditions, chemical characteristics of the soil and plant genotype [5,36–37].

All *Lotus* genotypes responded to salt stress with massive changes in gene expression. More than 60% of the genes monitored in the shoots of all six species responded to NaCl with significant changes in transcript level in at least one genotype. However, less than 1% responded in all genotypes (Figure 2). Comparisons between salinized root tips of monocot species from different genera indicate that this phenomenon also occurs at the root level [38]. The lack of conservation in the majority of transcriptional responses to salinity reflects at least two things. First, most genes that respond to salinity have little to do with the tolerance of a genus. Second, the architecture of genetic control networks governing transcription is highly complex and variable between genotypes. A similar conclusion was drawn on an expression quantitative trait loci (eQTL) study of recombinant inbred lines of *A. thaliana*, where the majority of eQTL had only small phenotypic effects [39]. The variability in transcriptional responses may also reflect differences in the suite of molecular and cellular mechanisms used to cope with salt accumulation in the shoot, redundancy within multigene families with different genes in the same family fulfilling equivalent roles in different species, and NaCl-responsive genes involved in secondary/pleiotropic responses to stress rather than primary responses required for acclimation. Clearly, simply relying on transcriptional profiling of a single model genotype to identify processes that could be translated to crops would be unwise, in view of the high degree of 'false positives' inferred from the above analyses.

statistically-significant difference (Student *t*-test with adjusted Bonferroni correction, p<0.05) in the stress-induced change.

On the other hand, by 'triangulating' data from multiple species we identified genes that responded in all genotypes, and genes that responded only in salt-tolerant or only in salt-sensitive cultivars. Most NaCl-responsive transcripts that were shared across all genotypes showed little correlation to Cl⁻ levels (Figure S1), indicating that they may be involved in 'general' physiological responses such as osmotic stress or growth inhibition [5]. Consistent with this idea, several of these represent key genes of stress-related metabolism, including amino acid, myo-inositol and

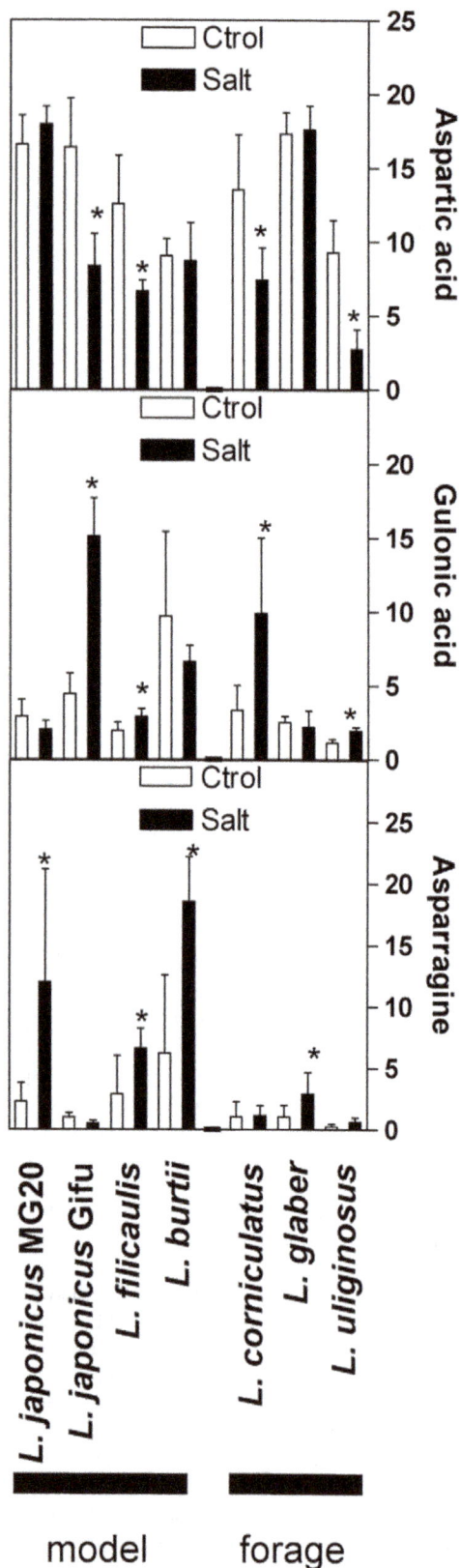

Figure 6. Changes of selected metabolites listed among the markers for sensitive and tolerant genotypes. Data represents the mean ± SD of three independent experiments of the normalized metabolite pools size (i.e. detector signals in arbitrary units normalized to internal standard and sample fresh weight). The asterisks indicate a

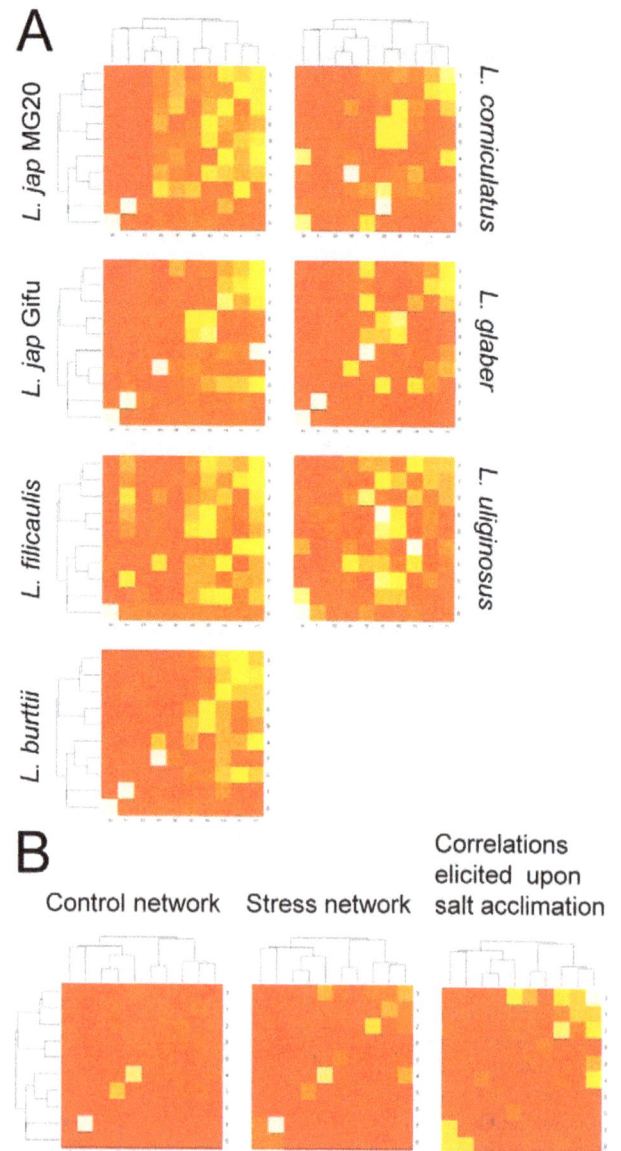

Figure 7. Network analysis of metabolite profiles. (A and B) Heat-maps representing the statistical connectivity between metabolic network communities: (A) for each genotype, arising from genotype-specific networks, and (B) for the control network, the salt treatment network and for the subtraction of the control and the salt treatment networks (i.e. correlations elicited upon salt stress). Lighter colour (from red to white) represents higher correlation coefficient.

polyamines biosynthesis, cell wall modification and photorespiration [24,10,30–31]. Importantly, homologues of at least two of these genes, AtSPMS and AtMIPS1, contribute to salinity responses in A. thaliana [30–31]. In contrast to common salt-responsive genes, changes in transcript levels of genes responding only in tolerant or sensitive genotypes showed better correlation with Cl^- content (Figure S1), indicating they may be directly or indirectly involved in plant responses to ion accumulation or imbalance. These included homologues of A. thaliana AtCIPK6 and AtCBL1, which are components of signalling pathways required for salt tolerance and members of the CBL-CIPK network controlling nutrient and salt homeostasis [25,29].

Despite dramatic differences observed at the transcriptome level, about half of the changes in primary metabolism in response to NaCl were shared by all Lotus species, although qualitative, quantitative and network analyses revealed several genotype-specific features (Figures 5 to 7 and S2). Conserved metabolic changes included increases in the levels of specific amino acids and polyols and decreases in organic acids, most of them recognized as salt-responsive metabolites that may act as compatible solutes or to compensate for ionic imbalance [32]. Changes in gulonic and aspartic acids were confined to sensitive genotypes (Figure 6). The former is an intermediate in the uronic acid pathway that synthesizes the antioxidant ascorbate from myo-inositol [40], and thus an increased gulonic acid content may reflect higher oxidative stress and cellular damage in the sensitive cultivars. If involved in charge balance [32], a decrease in aspartic acid may reflect the higher Cl^- content of the sensitive genotypes. On the other hand, asparagine levels increased in the more tolerant genotypes (Figure 6). This metabolite has a central role in the long-distance transport of nitrogen in Lotus and it functions at the core of the GS/GOGAT and the ornithine cycles, which are directly or indirectly involved in nitrogen assimilation, proline and polyamine biosynthesis, and ammonium detoxification from photorespiration [26,41–42]. Therefore, asparagine may play a pivotal role in salt tolerance by supporting core nitrogen metabolism. Further research would be needed to test if the manipulation of asparagine metabolism may be useful to improve salt tolerance in legumes.

Although some success has been reported in identifying genes that confer salt tolerance in model plants under controlled conditions using simple molecular biological approaches, little success has been achieved in the field where complex environmental interactions prevail [9,43]. To complicate things, there are significant differences between plants in how they to cope with saline soils, with salt exclusion, tissue tolerance, and osmotic adjustment playing more or less significant roles in different species [3,6]. We have shown here that transcriptome and metabolome changes that occur in related model and forage legume species in response to salinity are not highly conserved, which hampers simple translational genomics approaches. A similar conclusion was drawn from Populus genotypes subjected to drought, namely that it is not possible to draw simple, generalized conclusions about the stress transcriptome of a genus on the basis of one species [44]. However, a small set of salt-responsive genes were found to be conserved in all Lotus genotypes studied here, while salt-responsiveness of other genes was confined to tolerant or sensitive cultivars. It is likely that some of these genes play integral roles in acclimation and tolerance to saline soils; in fact homologues of some have been shown to confer greater salt tolerance in other species. In conclusion, 'triangulation' of transcriptomic and metabolomic data from multiple related species/genotypes offers itself as a practical means to eliminate a plethora of false positives in the hunt for genes and processes determining complex traits.

Materials and Methods

Plant material, growth conditions and experimental designs

Seeds of L. japonicus var. MG20, L. japonicus var. Gifu B129, L. filicaulis, L. burttii B303, L. corniculatus var. San Gabriel, L. glaber (L. tenuis) var. La Esmeralda and L. uliginosus var. LE G27 were obtained from the LOTASSA consortium (www.lotassa.org). Seeds were germinated in half-strength BD solution [45] agar plates plus 2 mM KNO_3 and 2 mM NH_4NO_3. Four days after imbibition, seedlings were transplanted to soil (Einheit, type null) using 10 cm pots irrigated with the above solution and grown in greenhouse under 16/8 hours day/night, 23±2°C and 55–65% RH. Salt stress treatment started 8 days post-imbibition and the salt content in the nutrient solution was increased in steps of 4 days till reaching the desired concentration (300 or 150 mM NaCl) (Figure 1A). Fresh nutrient solution was prepared every 4 days. Survival experiments were performed at lethal salt stress doses (300 mM NaCl) and were repeated twice, measuring the rate of mortality which was scored when the whole plant or all leafs were wilted or chlorotic. For salinity acclimation, 3 successive independent sub-lethal salt stress experiments were performed (150 mM NaCl), each consisting of 14 sample sets comprising control and salt treatments for each of the genotypes. Each set had 6 independent biological replicate pools of 5 plants. Total time of culture was 32 days and whole shoots, excluding cotyledons, were harvested in situ into liquid nitrogen in the middle of the light period. At harvest all plants were in the vegetative stage, with roots never showing nodules. The exposition to an identical salt stress dose between the genotypes was confirmed measuring the soil conductivity [10]. Biomass was estimated by mean fresh weight of the pooled shoots.

Profiling analysis

Transcriptomic, metabolomic and ionomic profiling were performed as described previously [10]. For the transcriptome analysis, sample tissue of all the biological replicates of each genotype were pooled to obtain 14 representative RNA samples in each independent experiment. The resulting 42 RNA samples were labelled and hybridized to the Genechip® Lotus1a520343 (Affymetrix). Element and metabolites and content were determined in each biological replicate using ICP-AES and GC/EI-TOF-MS, respectively. For ionomic profiling, 100 mg plant material was digested with 2 ml HNO_3 at 140°C until complete digestion. 100 µl of a 100 g/L LiCl solution was added as a carrier and the final volume adjusted with ultra pure water to 10 ml. Element concentrations were determined with inductively coupled plasma-atomic emission spectrometry (ICP-AES) using an IRIS Advantage Duo ER/S (Thermo Fisher). Elemental quantification was validated using IC-CTA-VTL2 Virginia tobacco leaves as a certified reference material. Chloride was profiled using ion chromatography with a Dionex ICS-2000 system (Dionex). For metabolomic profiling, 60 mg of frozen plant tissue was extracted with methanol/chloroform, and the polar fraction was prepared by liquid partitioning into water and derivatized [46]. Gas chromatography coupled to electron impact ionization-time of flight-mass spectrometry (GC/EI-TOF-MS) was performed using an Agilent 6890N24 gas chromatograph with split or splitless injection mounted to a Pegasus III time-of-flight mass spectrometer (LECO) [47]. Metabolite-features were quantified after mass spectral deconvolution (ChromaTOF software 1.00, Pegasus driver 1.61, LECO), and their chemical identification was manually assessed using the NIST05 software (http://www.nist.gov/srd/mslist.html) and the mass spectral and retention time index collection of the Golm Metabolome Database [48].

Statistics, data and network analyses

Statistical differences between control and treatments in element content were assessed with two-way-ANOVA using "treatment" and "independent experiment" as factors at stringent statistical threshold ($p < E^{-6}$), with the TIGR multiple experiment viewer software (TMEV_3.1). Microarray data were normalized by the GC-RMA algorithm using the bioconductor package of R software. Differential expression was tested for the probesets called present in all experiments, species and treatments (12,137 probesets, according to the present/absent MAS5 algorithm) correcting for multiple testing across all genes using the linear step-up false discovery rate (FDR). Microarray data is MIAME compliant and the 42 hybridizations were deposited at Array-Express (www.ebi.ac.uk/arrayexpress, accession number E-MEXP-2344). Independent component analysis (ICA) was used as non-supervised clustering algorithm, through the MetaGeneAlyse webpage (http://metagenealyse.mpimp-golm.mpg.de). Correlations across experiments and genotypes between the salt-elicited fold change in gene expression (Log_2 Salt/Control) and Cl^- content under stress were assessed using the Pearson correlation. Metabolomic profiles were analyzed with the TagFinder software [49] and filtered for those metabolic-features represented by 3 or more inter-correlated mass fragments within each independent experiment [50]. The validity of this analytical approach to quantify metabolites in plant tissues have been previously demonstrated [51]. Resulting data was normalized to internal standard and fresh weight, and each metabolic-feature was normalized to the median within each experiment and genotype, and log_{10} transformed prior to statistical analysis. Statistical differences were assessed with Student t-test using TMEV_3.1 ($p < 0.05$ applying the adjusted Bonferroni correction). Network analysis and operations were performed using R and Pajek softwares. A stable metabolic backbone network was reconstructed in two steps [52]. The first step recognized highly correlated metabolites through the construction of a "union network" based on a Spearmann rank-order analyte-analyte correlation in each cultivar for the 123 identified analytes, which was transformed into binary matrices according to a $p < E^{-4}$ threshold (applying Bonferroni correction) and considered further if it was significant in at least one cultivar. In the second step, a homogeneity test of the distributions of the correlations coefficients (Z-score transformed) was performed for all analyte-analyte correlations of the union network. Only those with a significant threshold ($p < E^{-6}$) in a Chi-square test with were considered stable. As a consequence, the reconstruction of a backbone network represents statistically stable correlated analytes between species. Particular networks for each genetic background or for control and treatment conditions were reconstructed based on this stable backbone. Tightly connected clusters of metabolites within the stable network (communities) were detected using Newman's algorithm for modularity, establishing an arbitrary number when modularity gain reached a plateau [33].

Supporting Information

Table S1 Transcriptomic profile data.

Table S2 Salt-elicited genes which were shared between more sensitive and tolerant genotypes, or were specific for more tolerant or sensitive backgrounds.

Table S3 Manually identified analytes present in all three independent non-lethal salt stress acclimation experiments.

Figure S1 Correlation (Pearson coefficients) across experiments and genotypes between changes in gene expression (Log2 Salt/Control) and Cl^- content under stress.

Figure S2 Example of metabolites that responded to sub-lethal salt stress in the different *Lotus* species.

Acknowledgments

We are grateful for the support of the Directors of the Max Planck Institute for Molecular Plant Physiology (MPIMP) and technical assistance of Ines Fehrle (MPIMP) and Janine Specht (University of Heidelberg). We thank Dr. Bjorn Usadel, Dr. Marc Lohse, Dr. Jan Lisec and Britta Hausmann (MPIMP), and Dr. Florian Wagner (ATLAS Biolabs, Berlin). FLP is a member of the *Consejo Nacional de Investigaciones Científicas y Técnicas* (Argentina).

Author Contributions

Conceived and designed the experiments: DHS JK MKU. Performed the experiments: DHS FLP MB. Analyzed the data: DHS JS AE MAH. Contributed reagents/materials/analysis tools: UK. Wrote the manuscript: DHS JK MKU. Edited the manuscript: UK.

References

1. Tester M, Davenport R (2003) Na^+ tolerance and Na^+ transport in plants. Ann Bot 91: 503–527.
2. Munns R (2005) Genes and salt tolerance: bringing them together, New Phytol 167: 645–663.
3. Munns R, Tester M (2008) Mechanisms of salinity tolerance. Ann Rev Plant Biol 59: 651–681.
4. Achard P, Cheng H, De Grauwe L, Decat J, Schoutteten H, et al. (2006) Integration of plant responses to environmentally activated phytohormonal signals. Science 311: 91–94.
5. Sanchez DH, Szymanski J, Erban A, Udvardi MK, Kopka J (2010) Mining for robust transcriptional and metabolic responses to long-term salt stress: a case study on the model legume *Lotus japonicus*. Plant Cell Environm 33: 468–480.
6. Moller IS, Tester M (2007) Salinity tolerance of Arabidopsis: a good model for cereals? Trends Plant Sci 12: 534–40.
7. Monforte AJ, Asins MJ, Carbonell EA (1997) Salt tolerance in *Lycopersicon* species. VI. Genotype-by-salinity interaction in quantitative trait loci detection: constitutive and response QTLs. Theor Appl Gen 95: 706–713.
8. Foolad MR (2004) Recent advances in genetics of salt tolerance in tomato. Plant Cell Tiss Org Culture 76: 101–119.
9. Cuartero J, Bolarin MC, Asins MJ, Moreno V (2006) Increasing salt tolerance in the tomato. J Exp Bot 57: 1045–1058.
10. Sanchez DH, Lippold F, Redestig H, Hannah M, Erban A, et al. (2008) Integrative functional genomics of salt acclimatization in the model legume Lotus japonicus. Plant J 53: 973–987.
11. Graham PH, Vance CP (2003) Legumes: importance and constraints to greater use Plant Physiol 131: 872–877.
12. Young ND, Udvardi MK (2009) Translating *Medicago truncatula* genomics to crop legumes. Curr Opin Plant Biol;doi 10.1016/j.pbi.2008.11.005.
13. Cannon SB, Sterck L, Rombauts S, Sato S, Cheung F, et al. (2006) Legume genome evolution viewed through the Medicago truncatula and Lotus japonicus genomes. Proc Nat Acad Sci 103: 14959–14964.
14. Perry JA, Wang TL, Welham TJ, Gardner S, Pike JM, et al. (2003) A TILLING reverse genetics tools and a web-accessible collection of mutants of the legume *Lotus japonicus*. Plant Physiol 131: 866–871.
15. Udvardi MK, Tabata S, Parniske M, Stougaard J (2005) *Lotus japonicus*: legume research in the fast lane. Trends Plant Sci 10: 222–228.
16. Gondo T, Sato S, Okumura K, Tabata S, Akashi R, et al. (2007) Quantitative trait locus analysis of multiple agronomic traits in the model legume *Lotus japonicus*, Genome 50: 627–637.

17. Sato S, Nakamura N, Kaneko T, Asamizu E, Kato T, et al. (2008) Genome structure of the legume, *Lotus japonicus*. DNA Research 15: 227–239.

18. Høgslund N, Radutoiu S, Krusell L, Voroshilova V, X Hannah MA, et al. (2009) Dissection of symbiosis and organ development by integrated transcriptome analysis of *Lotus japonicus* mutant and wild-type plants. PLoS One 4: e6556. doi: 10.1371/journal.pone.0006556.

19. Diaz P, Betti M, Sanchez DH, Udvardi MK, Monza J, et al. (2010) Deficiency in plastidic glutamine synthetase alters proline metabolism and trasncriptomic response in *Lotus japonicus* under drought stress. New Phytol;doi:10.111/j.1469-8137.2010.03440.x.

20. Kawaguchi M, Pedrosa-Harand A, Yano K, Hayashi M, Murooka Y, et al. (2005) *Lotus burttii* takes a position of the third corner in the *Lotus* molecular genetics triangle. DNA Research 12: 69–77.

21. Diaz P, Borsani O, Monza J (2005) Lotus-related species and their agronomic importance. In *Lotus japonicus Handbook* Marquez A, ed. The Netherlands: Springer. pp 25–37.

22. Sanchez DH, Cuevas JC, Chiesa MA, Ruiz OA (2005) Free spermidine and spermine content in *Lotus glaber* under long-term salt stress. Plant Sci 168: 541–546.

23. Degtjareva GV, Kramina DD, Sokoloff DD, Samigullin TH, Valiejo-Roman CM, et al. (2006) Phylogeny of the genus *Lotus* (Leguminoseae, Loteae): evidence from nrITS sequences and morphology. Can J Bot 84: 813–830.

24. Noctor G, Veljovic-Jovanovic S, Driscoll S, Novitskaya L, Foyer CH (2002) Drought and oxidative load in the leaves of C_3 plants: a predominant role for photorespiration? Ann Bot 89: 841–850.

25. Tripathi V, Parasuraman B, Laxmi A, Chattopadhyay D (2009) CIPK6, a CBL-interacting protein kinase is required for development and salt tolerance in plants. Plant J 58: 778–790.

26. Marquez AJ, Betti M, Garcia-Calderon M, Pal'ove-Balang P, Diaz P, et al. (2005) Nitrate assimilation in *Lotus japonicus*. J Exp Bot 56: 1741–1749.

27. Chen ZZ, Hong XH, Zhang HR, Wang YQ, Li X, et al. (2005) Disruption of the cellulose synthase gene, AtCesA8/IRX1, enhances drought and osmotic stress tolerance in Arabidopsis. Plant J 43: 273–283.

28. Nakano T, Suzuki K, Fujimura T, Shinshi H (2006) Genome-wide analysis of ERF gene family in Arabidopsis and Rice. Plant Physiol 140: 411–432.

29. Albrecht V, Weinl S, Blazevic D, D'Angelo C, Batistic O, et al. (2003) The calcium sensor CBL1 integrates plant responses to abiotic stresses. Plant J 36: 457–470.

30. Yamaguchi K, Takahashi Y, Berberich T, Imai A, Miyazaki A, et al. (2006) The polyamine spermine protects against high salt stress in *Arabidopsis thaliana*. FEBS Lett 30: 6783–6788.

31. Donahue JL, Alford SR, Torabinejad J, Kerwin RE, Nourbakhsh A, et al. (2010) The *Arabidopsis thaliana* myo-inositol 1-phosphate synthase1 gene is required for myo-inositol synthesis and suppression of cell death. Plant Cell doi/10.1105/tpc.109.071779.

32. Sanchez DH, Siahpoosh MR, Roessner U, Udvardi MK, Kopka J (2008) Plant metabolomics reveals conserved and divergent metabolic responses to salinity. Physiol Plant 132: 209–219.

33. Newman MEJ (2006) Modularity and community structure in networks. Proc Natl Acad Sci 103: 8577–8582.

34. Salentijn EMJ, Pereira A, Angenent GC, Van der Linden GC, Krens F, et al. (2007) Plant translational genomics: from model species to crops. Mol Breeding 20: 1–13.

35. Teakle NL, Tyerman SD (2010) Mechanisms of Cl^- transport contributing to salt tolerance. Plant Cell Envirom 33: 566–589.

36. Marschner H (1995) In *Mineral nutrition of higher plants* 2nd edn (Academic Press Limited, London). .

37. Grattan SR, Grieve CM (1999) Mineral nutrient acquisition and response by plants grown in saline environments. *Handbook of plant and crop stress* 2nd edn (Pessarakli M, ed.) New York: Marcel Dekker Inc.. pp 203–229.

38. Walia H, Wilson C, Ismail AM, Close TJ, Cui X (2009) Comparing genomic expression patterns across plant species reveals highly diverged transcriptional dynamics in response to salt stress. BMC Genomics 10: 398.

39. West MAL, Kim K, Kliebenstein DJ, van Leeuwen H, Michelmore RW, et al. (2007) Global eQTL mapping reveals the complex genetic architecture of transcript-level variation in Arabidopsis. Genetics 175: 1441–1450.

40. Ishikawa T, Dowdle J, Smirnoff N (2006) Progress in manipulating ascorbic acid biosynthesis and accumulation in plants. Physiol Plant 126: 343–355.

41. Sieciechowicz KA, Joy KW, Ireland RJ (1988) The metabolism of asparagine in plants. Phytochemistry 27: 663–671.

42. Waterhouse RN, Smyth AJ, Massonneau A, Prosser IM, Clarkson DT (1996) Molecular cloning and characterization of asparagine synthetase from *Lotus japonicus*: dynamics of asparagine synthesis in N-sufficient conditions. Plant Mol Biol 30: 883–897.

43. Flowers TJ (2004) Improving crop salt tolerance. J Exp Bot 55: 307–319.

44. Wilkings O, Waldron L, Nahal H, Provart NJ, Campbell MM (2009) Genotype and time of the day shape the *Populus* drought response. Plant J 60: 703–715.

45. Broughton WJ, Dilworth MJ (1971) Control of leghaemoglobin synthesis in snake beans. Biochem J 125: 1075–1080.

46. Desbrosses GG, Kopka J, Udvardi MK (2005) *Lotus japonicus* metabolic profiling. Development of gas chromatography-mass spectrometry resources for the study of plant-microbe interactions. Plant Physiol 137: 1302–1318.

47. Wagner C, Sefkow M, Kopka J (2003) Construction and application of a mass spectral and retention time index database generated from plant GC/EI-TOF-MS metabolite profiles. Phytochemistry 62: 887–900.

48. Kopka J, Schauer N, Krueger S, Birkemeyer C, Usadel B, et al. (2005) GMD@CSB.DB: the Golm Metabolome Database. Bioinformatics 21: 1635–1638.

49. Luedemann A, Strassburg K, Erban A, Kopka J (2008) TagFinder for the quantitative analysis of gas chromatography-mass spectrometry (GC-MS) based metabolite profiling experiments. Bioinformatics 24: 732–737.

50. Sanchez DH, Redestig H, Krämer U, Udvardi MK, Kopka J (2008c) Metabolome-ionome-biomass interactions: what can we learn about salt stress by multiparallel phenotyping? Plant Sig Behav 3: 598–600.

51. Allwood JW, Erban A, de Koning S, Dunn WB, Luedemann A, et al. (2009) Inter-laboratory reproducibility of fast gas chromatography-electron impact-time of flight mass spectrometry (GC-EI-TOF/MS) based plant metabolomics. Metabolomics 5: 479–496.

52. Szymanski J, Jozefczuk S, Nikoloski Z, Selbig J, Nikiforova V, et al. (2009) Stability of metabolic correlations under changing environmental conditions in *Escherichia coli* - a systems approach. PLoS One 4: e7441. doi:10.1371/journal.pone.0007441.

19

Weak Spatial and Temporal Population Genetic Structure in the Rosy Apple Aphid, *Dysaphis plantaginea*, in French Apple Orchards

Thomas Guillemaud[1]*, **Aurélie Blin**[1], **Sylvaine Simon**[3], **Karine Morel**[3], **Pierre Franck**[2]

1 Equipe "Biologie des Populations en Interaction", UMR 1301 I.B.S.V. INRA-UNSA-CNRS, Sophia Antipolis, France, **2** UR1115 Plantes et Systèmes de Culture Horticoles, INRA, Avignon, France, **3** UE695 Recherche Intégrée, INRA, Domaine de Gotheron, Saint-Marcel-lès-Valence, France

Abstract

We used eight microsatellite loci and a set of 20 aphid samples to investigate the spatial and temporal genetic structure of rosy apple aphid populations from 13 apple orchards situated in four different regions in France. Genetic variability was very similar between orchard populations and between winged populations collected before sexual reproduction in the fall and populations collected from colonies in the spring. A very small proportion of individuals (~2%) had identical multilocus genotypes. Genetic differentiation between orchards was low (F_{ST}<0.026), with significant differentiation observed only between orchards from different regions, but no isolation by distance was detected. These results are consistent with high levels of genetic mixing in holocyclic *Dysaphis plantaginae* populations (host alternation through migration and sexual reproduction). These findings concerning the adaptation of the rosy apple aphid have potential consequences for pest management.

Editor: Marco Salemi, University of Florida, United States of America

Funding: This work was funded by the French Program ECOGER AO 2005. The funders had no role in study design, data collection and analysis, decision to publish, or preparation of the manuscript.

Competing Interests: The authors have declared that no competing interests exist.

* E-mail: guillem@sophia.inra.fr

Introduction

The rosy apple aphid *Dysaphis plantaginea* (Hemiptera: Aphididae) is one of the most serious pests of apple trees in Europe [1] and North America [2]. It causes fruit deformation and severe leaf-curling [3], distorts shoots, reduces flower formation and slows tree growth [4].

In commercial apple tree orchards, the damage caused by even very low densities of aphids may decrease the commercial value of the crop. This economic loss justifies aphid management techniques, based principally on pesticide use. Recommendations generally suggest the use of several pesticide treatments in apple orchards: in early spring, before flowering and after flowering or in late summer [5]. The intensive use of chemical insecticides against *D. plantaginea* has resulted in an intense selection regime and the development of mechanisms of insecticide resistance in the field [6]. Alternative control strategies, such as the application of organic pesticides (neem extract or potassium soap [5]), the use of repellent or barrier-effect products (kaolin [7,8,9]), biological control [10, 11,12], and plant resistance [13,14,15,16], are being developed and tested.

Whatever the pest management strategy applied, the likelihood of developing resistance to management depends on the ecological characteristics of the target species: its migration capability, sexual reproduction and clonal multiplication determine, at least in part, its genetic variability and, thus, its capacity to adapt to control measures. An analysis of genetic variation in the *D. plantaginea* population may therefore provide essential information about these crucial ecological parameters.

The life cycle of *D. plantaginea* almost certainly has profound consequences for its genetic variability. Like many aphid species, *D. plantaginea* has a cyclic parthenogenetic (or holocyclic) life cycle [17,18]. In late summer and fall, cyclically parthenogenetic aphids give birth to gynoparae (precursor forms of sexual females), followed by winged males. Both fly from the herbaceous secondary host plant, *Plantago*, to the primary host, apple trees, where the gynoparae give birth to sexual females [19]. Mating occurs on apple and sexual females lay eggs that hatch by the beginning of spring. During late spring and early summer, after 3 to 4 (maximum 6) parthenogenetic generations, winged morphs are produced that migrate from the primary to the secondary host on which about 3 to 8 successive parthenogenetic generations occur [19]. Thus, due to the annual host alternation, two large migration events take place in biological cycle of *D. plantaginea*, in the fall and spring.

In many species, cyclic parthenogenetic populations coexist with obligate parthenogenetic populations [20,21]. In such populations, the aphids have lost the ability to reproduce sexually and remain on herbaceous plants throughout the year. According to Lathrop [22], the rosy apple aphid does not occur on plantain during winter in colder parts of the USA. However, "in the mild climate of western Oregon, overwintering on plantain as well as apple is the rule" [22]. This suggests that this species displays variation in reproductive modes, with cyclic parthenogenetic populations coexisting with obligate parthenogenetic populations. However, we are not aware of any other study demonstrating such a polymorphism in *D. plantaginea*.

Cyclic parthenogenetic aphids would be expected to display high levels of genotypic variability, due to the recombination occurring during sexual reproduction [21,23,24]. However, drift and/or selection may strongly decrease neutral genetic variability during successive parthenogenetic generations after egg hatching on apple and on secondary hosts, due to the absence of recombination and the rapid rate of increase during clonal reproduction as shown in the peach-potato aphid *Myzus persicae* [24,25]. During this clonal phase, genetic signs of parthenogenesis may accumulate: linkage disequilibrium (LD), Hardy-Weinberg (HW) disequilibrium, and decrease in multilocus genotype diversity [23].

Little is known about the genetic diversity of the *D. plantaginea* species. The only data available are the preliminary results obtained by Salomon *et al.* [26], who reported high levels of genetic variability in a single apple orchard, based on an analysis of microsatellite genetic markers previously developed by Harvey *et al.* [27] for this species. We therefore know little about the effects of the succession of sexual and asexual reproduction on the genetic variability of this species or those of the major migration events occurring during host shift.

The aim of this study was to determine whether the complex mode of reproduction, with a single sexual generation and successive clonal generations, and host shift-related migration events affected genetic variation in this species. In other words, we evaluated the geographic scale over which *D. plantaginea* populations function and possible decreases in the genetic variation of *D. plantaginea* on apple due to cyclic parthenogenesis.

More specifically, we used a geographic and temporal sampling scheme and highly polymorphic genetic markers (microsatellite) data to address the following questions: (i) What degree of genetic variability does the rosy apple aphid display at the national scale (over the whole of France)? (ii) Is there any genetic differentiation between populations of *D. plantaginea* and at what level (regions, orchards, apple cultivars) can this differentiation be detected? (iii) Are the genetic diversity and geographic population structure of *D. plantaginea* stable at different parts of the life cycle and in different years?

Materials and Methods

Sample collection

Samples were collected according to a geographic and temporal scheme in experimental apple orchards belonging to INRA institute. Here an orchard is defined as a field of apple trees with a given management strategy and a specific tree cultivar. The term "sample" refers to as a group of aphids collected during a specific season and at a specific position in a given orchard. No specific permission was required to sample aphids in these orchards. They were collected at one location in north-western France (near Angers), one location in south-western France (near Agen), and two locations in southern France (near Avignon and Valence) (Figure 1). Depending on the location, aphids were sampled at one (Agen), two (Avignon, Angers) or three (Valence) different periods of the aphid life cycle, in fall 2006 and 2007, and in spring 2007 (see Table S1). Furthermore, at Avignon, Valence and Angers, samples were taken from different orchards at the same time (Table S1). The distances between these orchards were as follows. At Valence, the various orchards that were sampled were located from within a circle with a radius of 250 m. The Smoothee1 orchard sample was located about 350 to 450 m from the other orchard samples, the Conventional Ariane orchard sample was about 300–450 m from the other samples, and the remaining orchard samples were located about 10 to 100 meters apart. At Valence, samples were collected on different apple cultivars (Smoothee, Melrose and Ariane) under organic management, but also from different plants of the same cultivar (Ariane) grown under organic, low-input and conventional pest management

regimes (i.e. organic-registered for the organic system, minimized for the low-input system and supervised for the conventional system). Two locations (center and border) in Smoothee1 orchard in Valence were sampled in autumn 2006 to test for micro-geographic genetic structure that would not depend on tree cultivars and management strategies. At Angers, the two orchards sampled, P32 and D1, were located 500 meters apart. Finally, at Avignon, orchards 65 and 157 were located 2.5 km apart, each about 12 to 15 km from the INRA orchard. In the fall, winged gynoparae were sampled manually by branch tapping. In spring, individuals were collected by hand, with a small brush, with no more than one individual collected per colony and per tree on two sampling dates (May 8 and 23). Aphids were stored in absolute ethanol for DNA extraction.

DNA extraction and microsatellite analysis

Template material for the amplification of microsatellites by PCR was prepared from individual aphids with the "salting out" rapid extraction protocol [28] and resuspended in 50 µl H$_2$O. Eight microsatellite loci for *D. plantaginea* (DpL4, DpB10) [27], *Sitobion* species (S24, Sa4Σ, S3.43, S16b) [29], *Rhopalosiphon padi* (R5.29B) [29] and *Aphis fabae* (AF93) [30] were amplified in two separate multiplex PCRs. The first reaction amplified *DpL4*, *DpB10*, *S24* and *Sa4Σ*, and the second amplified *S3.43*, *AF93*, *R5.29B* and *S16b*. Both multiplex reactions were carried out with Qiagen multiplex PCR kits (Qiagen, Hilden, Germany), according to the manufacturer's instructions, in a final volume of 10 µl containing 1 µl of DNA template. The forward primer for each microsatellite was labeled with a fluorescent dye, to allow the detection of PCR products on an ABI 3100 DNA sequencer (Applied Biosystems, Foster City, CA). We used the following PCR program for both reactions: 95°C for 15 minutes, followed by 35 cycles of 30 s at 94°C, 90 s at 56°C, 1 min at 70°C, and 30 s at 60°C.

Data analysis

Within-population genetic diversity was estimated by calculating the number of alleles per locus, and observed and expected heterozygosities calculated with GENEPOP ver. 4.0 [31,32]. Exact tests for deviation from Hardy-Weinberg (HW) expectations, linkage disequilibrium and population differentiation were carried out with GENEPOP. A Mantel test of isolation by distance was also carried out with Genepop ver. 3.1 [31]. MICROCHECKER was used to detect the presence of null alleles at each microsatellite locus [33] and genotypic differentiation between pairs of populations (F_{ST}) was corrected for null alleles as described by Chapuis *et al.* [34]. We compared the number of alleles per locus between population samples, by estimating allelic richness (*AR*) on the basis of minimum sample size, with the rarefaction method [35] implemented in FSTAT 2.9.3 [36].

If more than one copy of the same multilocus genotype (MLG) was observed, the null hypothesis of the same MLG being obtained repeatedly by chance through sexual reproduction was tested with Genclone ver. 2.0 [37]. This test is based on calculation of the probabilities of obtaining MLGs from sexual events, taking into account the estimated F_{IS} for the population.

Finally, the number of distinct populations (*K*) present in the set of samples was estimated with STRUCTURE [38]. This software was used to estimate Pr(X|K), the probability of the observed set of genotypes (*X*), conditional on the number of genetically distinct populations, *K*, for values of *K* between 1 and the number of samples. The program was run for 10^5 iterations, preceded by an initial burn-in period of 2×10^4 iterations. Three runs were performed for each value of *K*, to check that estimates of Pr(X|K) were consistent between runs. The posterior probabilities, Pr(K|X), were then calculated as described by Pritchard *et al.* [38].

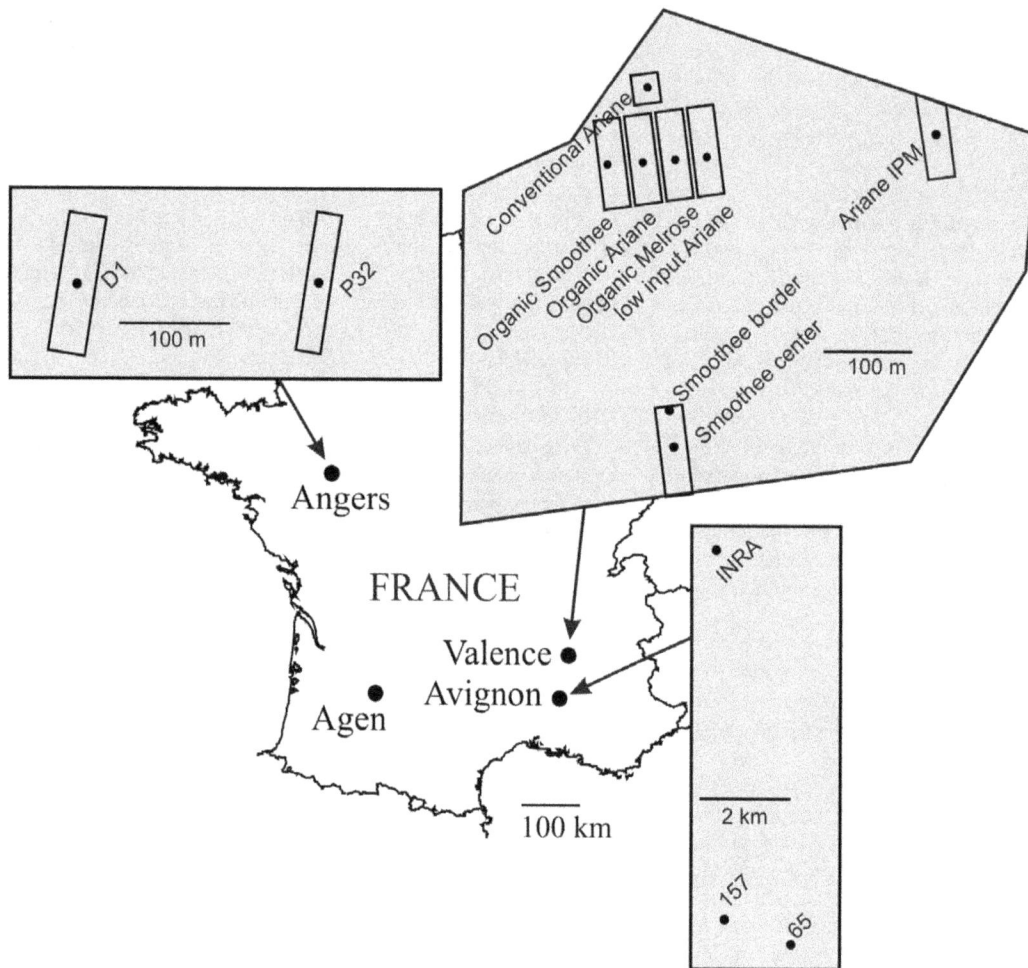

Figure 1. Locations of the samples of *Dysaphis plantaginea* **used in this study.** Sampling periods are indicated.

For multiple tests of a single hypothesis and non orthogonal comparisons, we used Benjamini & Hochberg [39] and sequential Bonferroni [40] correction procedures, respectively, to correct significance levels.

Results

Within-population variability

We genotyped 532 individuals in total and found the level of genetic variation to be high. There were seven (locus *S3.43*) to 34 (locus *S24*) alleles per microsatellite locus. Within-population genetic variability was high, with mean numbers of alleles per locus (Na) of more than seven for samples with more than 15 individuals. Allelic richness (*AR*), calculated on a sample of at least 15 individuals for inter-population comparisons, was between 3.9 and 4.3 (mean $AR = 4.14$, SEM = 0.13), and revealed no difference in population variability between samples and between spring and fall (Friedman analysis of variance and Wilcoxon's signed rank test, $p > 0.05$). Consistent with this, no heterogeneity of Nei's heterozygosity was detected (mean $He = 0.63$, SE = 0.03; Friedman analysis of variance and Wilcoxon's signed rank test, $p = 0.41$ and $p = 0.32$, for between-sample and between spring and fall comparisons, respectively). All samples displayed a heterozygote deficiency, with many genotypic compositions showing departure from HW equilibrium (Table S1). The instances of HW departure identified frequently involved the same three loci (*DPL4*, *DPB10*

and *AF93*), suggesting the presence of null alleles at these loci. Loci *DPL4*, *DPB10* and *AF93* displayed departure from HW equilibrium eight, seven and five times, respectively, in a total of 26 significant per locus and per sample tests. MICROCHECKER suggested the existence of null alleles for *DPB10* and *DPB4*. No heterogeneity in the proportion of significant HW tests was found between samples or between spring and fall samples (Fisher's exact test on RxC contingency tables, $p > 0.05$ for both tests). Accordingly, no heterogeneity in F_{IS} value was detected between samples or between spring and fall samples (Friedman analysis of variance and Wilcoxon's signed rank test on mean F_{IS} value per locus, $p = 0.51$ and $p = 0.33$, respectively). After removal of the *DPL4*, *DPB10* and *AF93* loci from the analysis, the general heterozygote deficiency remained and no heterogeneity was apparent between samples or between spring and fall (Friedman analysis of variance and Wilcoxon's signed rank test on mean F_{IS} value per locus, $p = 0.72$ and 0.89 respectively).

A very high level of multilocus genotypic variability was found. PCR amplification was unsuccessful in some cases. In total, 342 individuals were genotyped with no missing data, and 336 different multilocus genotypes (MLG) were detected in these individuals (ratio of the number of multilocus genotypes over the total number of individuals, $N_{MLG}/N = 0.98$). Six MLGs were found in multiple copies. Each of these repeated MLGs was found in two individuals sampled from the same orchard on the same date: orchards 65 and 157 in fall 2006, orchards Bio Smoothee and Bio

Melrose in Valence in spring 2007, and Agen in fall 2007. These repeated MLGs were probably generated by clonal rather than sexual reproduction (test of the null hypothesis of sexual recombination, $p < 8 \times 10^{-4}$). Consistent with the extensive multilocus genotypic variability observed, an analysis of the genotypic disequilibrium between each pair of loci in each sample revealed very few cases of significant linkage. No heterogeneity in the number of significant LD was found either between samples, or between spring and fall (Fisher's exact test on RxC contingency tables, $p > 0.05$ for both tests).

Population differentiation

As most samples displayed heterozygote deficiency, we carried out exact tests of genotypic differentiation between samples only. All comparisons between parts of orchards or between orchards at the same location or at the same period were characterized by small F_{ST} values ($<1\%$) and non significant differentiation tests ($p = 0.078$ and 0.25 at Angers, $p = 0.15$ and 0.47 at Avignon in fall 2006 and 2007 respectively, and $p = 0.43$ at Valence in fall 2007). Parts of orchards and orchards at the same location were therefore pooled by period for analyses of regional genetic differentiation (Table 1).

As null alleles were suspected for several loci, we also performed an analysis taking these null alleles into account [34]. We found the same absence of differentiation between samples from the same location, with the exception of two orchards in Avignon sampled in 2006 (165 and 57, $p = 10^{-3}$). As the level of genetic differentiation was very low ($F_{ST} = 4.3 \times 10^{-3}$) we decided to pool the samples from each location.

Significant, but weak ($F_{ST} < 1\%$) genotypic differentiation was detected between Angers, Avignon and Valence in fall 2006 (Table 1). In fall 2007, significant moderate levels of differentiation were observed between Avignon and other locations ($F_{ST} \sim 2\%$). A low level of differentiation was found between Agen and Angers or Valence ($F_{ST} \sim 1\%$) and no differentiation was detected between Angers and Valence. The same overall pattern was observed if null alleles were taken into account: significant, but low to moderate levels of differentiation between locations.

Only low to very low levels of differentiation were found between samples from the same location collected at different time periods. Almost no difference was found between samples collected at Valence in fall 2006, spring 2007 and fall 2007 (although

the differentiation between fall 2006 and spring 2007 was of borderline significance, $p = 0.023$, $F_{ST} = 0.002$). Comparisons between fall 2006 and 2007 for each location revealed significant but weak (in the case of Angers and Avignon, $p < 4 \times 10^{-3}$, $F_{ST} = 0.005$ and 0.008, respectively) and non significant (in the case of Valence, $p = 0.3$, $F_{ST} = -0.002$) differentiation.

Very similar results were obtained when null alleles were taken into account. In this case, significant differentiation was detected in all comparisons other than that between fall 2006 and fall 2007 at Valence. No isolation by distance was detected between the 16 samples with more than 15 individuals (Mantel test, $p = 0.153$).

A Bayesian analysis of population structure grouped all individuals together in a single population, regardless of their location and sampling period ($P(K = 1 \mid X) = 1$). This was true for the default model (admixture and correlated allele frequency), but also for the admixture and independent allele frequency model. Models without admixture gave inconsistent results ($P(K = 2 \mid X) = 1$ and $P(K = 14 \mid X) = 1$ for the correlated and independent allele frequency models, respectively). Evanno's ΔK [41] also gave inconsistent results for the models without admixture ($K = 5$ and $K = 2$ for the correlated and independent allele frequency models, respectively).

Discussion

Considerable variability and no evidence for obligate parthenogenesis

In this study, we analyzed the genetic structure of populations of the rosy apple aphid, *D. plantaginae*, collected from its primary host. The goal was to characterize, for the first time, the genetic variability of this aphid, and to evaluate the impact of three evolutionary forces potentially affecting this variation: drift, migration and selection. Rosy apple aphid populations collected from apple trees in four regions of France displayed extensive genetic variation. In particular, a very high degree of genotypic diversity was observed, with almost all individuals genetically different from each other. This was true for all locations and sampling periods. This result confirms and extends the findings of Solomon *et al.* [26], who were the first to report high levels of genetic variability in *D. plantaginea* sampled from apple orchards.

The rosy apple aphid is thought to be a cyclic parthenogenetic species, with a single sexual generation and many asexual generations. It is unknown whether this species displays polymorphism

Table 1. Regional and temporal differentiation of *Dysaphis plantaginea* samples in France.

		Fall 2006			Spring 2007		Fall 2007		
		Angers	Avignon	Valence	Valence	Agen	Angers	Avignon	Valence
Fall 2006	Angers	-	0.002	0.008			0.005		
	Avignon	0.018*	-	0.009				0.008	
	Valence	0.001**	3×10^{-4}**	-	0.002				−0.002
Spring 2007	Valence			0.023	-				0.001
Fall 2007	Agen					-	0.006	0.026	0.012
	Angers	0.004*				0.006*	-	0.016	0.003
	Avignon		8×10^{-4}**			10^{-5}**	3×10^{-4}**	-	0.024
	Valence			0.3	0.56	0.035*	0.223	10^{-5}**	-

Pairwise estimates of F_{ST} are above the diagonal and the p-values of genotypic differentiation exact tests are shown below the diagonal. * and ** after p-values indicate that the tests were significant before and after Bonferroni correction, respectively. Only pertinent comparisons (i.e. between periods at the same sites or between sites during the same period) are shown.

in its mode of reproduction, with the coexistence of obligate parthenogenetic and parthenogenetic individuals, as in many other aphid species [42]. The mode of reproduction has consequences for the genetic variation of populations [43], and this topic has been particularly well studied in aphids [44]. In the case of holocycly, two antagonistic effects occur. Asexual generations (reproducing by mitotic parthenogenesis in this species) are expected to generate individuals with an identical genetic background, with mutations as the only source of variation. The occurrence of such asexual generations also leads to systematic linkage disequilibrium (LD) and departure from HW equilibrium. By contrast, (panmictic) sexual generation disrupts inter-locus associations, resulting in each individual being genetically different from all others. It also re-establishes HW equilibrium within a single generation and decreases LD. Note that, in the long term, obligate parthenogenesis (parthogenesis as the only form of reproduction) tends to lead to excess heterozygosity due to the accumulation of mutations without recombination [44].

In French populations of the rosy apple aphid collected from its primary host we found neither general LD, nor a global excess of heterozygotes. We found extensive multilocus genotypic variability. These genetic signals provide evidence of sexual reproduction, supporting the hypothesis that the populations collected from apple trees in the spring and fall are holocyclic. This is consistent with what is known of the lifecycle of *D. plantaginea*, and with the observation of eggs on apple trees during the winter [17,22]. We found no evidence for the existence of obligate parthenogenesis in *D. plantaginae*, at least on apple trees in the fall and spring. However, it remains possible that anholocyclic lineages exist during these periods of the year on secondary hosts, as reported for many aphid species displaying host alternation [42].

The populations sampled in the fall, before the occurrence of recombination, were produced by lineages that had gone through several parthenogenetic generations since the last sexual event. We therefore expected to find genetic signs of clonal reproduction (repeated multilocus genotypes, LD, systematic HW disequilibrium) in the samples collected in the fall. However, no such signs were observed. This suggests that a single yearly sexual reproduction event is sufficient to generate a high level of genetic variability and to cancel out the genetic signs of clonality, even in the fall, before the occurrence of sexual reproduction. The almost entire absence of individuals with identical multilocus genotypes in samples collected in the fall suggests that the number of individuals from an individual clone of *D. plantaginea* present on apple trees in France in the fall is not large. This may be due to 1) the limited size of the clonal populations sharing the same genotype on secondary hosts compared to the number of different clonal genotypes present on these plants and/or 2) an extensive geographic redistribution of the aphids during their return flight to their primary hosts (but see below), leading the dilution of repeated clonal genotypes.

The high level of genetic variability found in *D. plantaginea* on its primary host is similar to that found in other cyclic parthenogenetic aphids, such as *M. persicae* in France [24] and Australia [45], *S. avenae* [46] or *R. padi* [23] and other cyclic parthenogenetic animals, such as rotifers (e.g. *Brachionus plicatilis* (Müller), [47]), which display high levels of genotypic diversity despite going through numerous parthenogenetic generations each year.

We frequently observed heterozygote deficits associated with HW disequilibrium. Possible explanations based on previous findings for aphids include a Wahlund effect, null alleles, inbreeding and selection [23,24,48,49,50].

The Wahlund effect is the unintentional pooling of differentiated populations into a single sample, resulting in excess homozygosity

[43]. A Wahlund effect may occur in the fall, due to the co-occurrence on the primary host of migrants originating from populations that were genetically differentiated on secondary hosts. Such genetic differentiation may result from genetic drift or selection (e.g. adaptation to various secondary host plants). Panmictic sexual reproduction leads to HW equilibrium in only one generation [43]. Thus, assuming panmictic sexual reproduction, heterozygote deficits in the spring (i.e. after sexual reproduction) cannot be accounted for by a Wahlund effect.

Null alleles were suspected for three loci, and specific statistical treatments were carried out to take this possibility into account. A specific statistical analysis was carried out to detect loci with null alleles, but we cannot rule out the possibility that this problem occurred at a larger number of loci.

Inbreeding and selection are often proposed as explanations for heterozygote deficits in sexual aphid populations [23,48,49, 50], but we found no evidence to support this hypothesis in this study.

Spatial genetic differentiation

Another key finding of this study was the very weak spatial genetic differentiation between *D. plantaginae* populations. We detected no population genetic differentiation at the regional scale or at the intra-orchard or inter-orchard level, for samples located less than 20 km apart. Classically, spatial genetic differentiation results from the balance between migration and genetic drift [51]. In species with mitotic parthenogenesis, selection at one or a few loci affects allele frequency not only at these loci, but throughout the genome, because there is no recombination. Therefore, in a species like *D. plantaginae*, the use of microsatellites to assess spatial population genetic structure also provides information about selection (until sexual reproduction takes place). Our results therefore suggest that the effect of local drift or selection is largely compensated by migration. The fall and spring flights of the aphids mediating host shift are thus sufficient to homogenize genetic variability at a local and regional scale. However, we observed significant levels of population genetic differentiation at the scale of the entire country (France), between different apple-growing areas, with differences observed between Avignon, Agen, Valence and Angers. This genetic differentiation was weak (F_{ST} generally below 1%) and no isolation by distance was observed, but these results nonetheless suggest that the emigration and return flights of *D. plantaginea* are limited by geographic distance, at regional scale at least, in France. *D. plantaginea* has only one winter host-plant, apple, and this species has a patchy distribution in France. This may account for the spatial limitation of migration. We also found evidence for a local dispersion component in the fall and spring. The sharing of the same multilocus genotype by a pair of individuals on the primary host in the fall, before sexual reproduction, was rare, but nonetheless observed in three instances. In each case, the two individuals sharing the same MLG were found in the same orchard. This strongly suggests that the return flight was local. In other words, this migration may connect secondary and primary hosts located close together, rather than reflecting global geographic homogenization.

The situation in spring was similar to that in the fall and provides information about dispersal between primary hosts after sexual reproduction: three repeated MLGs, each shared by a single pair of individuals, were observed in three different orchards, among 118 colonies. One of the repeated MLGs corresponded to individuals collected from the same tree on two different dates and, thus, probably reflected sampling from the same aphid colony. However, in the two other cases of aphids sharing other repeated MLGs, individuals were collected from non contiguous trees, probably reflecting the dispersion of aphids between different trees in the spring. It is unknown whether such dissemination between distant trees is passive (through wind or cropping practices) or active. Overall, these findings suggest that, at the time of sampling in May, i) aphid dispersal between primary hosts occurred but was not frequent and/or ii) dispersion may have been frequent but only a small proportion of the total number of colonies was sampled. A rough estimate of the sampling effort in spring would be one colony sampled per five actual colonies, so the probability of sampling the same MLG twice or more was low.

Overall, spatial genetic differentiation in *D. plantaginea* was very weak or null over short distances and weak but significant over large distances, suggesting that local migration occurs in *D. plantaginea*. This situation is similar to that reported for other aphid species. For instance, in *R. padi*, no genetic differentiation was found between populations located less than 1000 km apart [23]. Weak population differentiation was found between both close (<100 km) and distant (>500 km) populations of the cereal aphid, *S. avenae* [48,52,53]. This work provides an additional demonstration that genetic differentiation is not rare in aphids and that aphid migration probably therefore occurs over limited spatial areas [24,48,53,54,55,56].

Temporal genetic differentiation

The third key result of this study is the almost complete temporal genetic homogeneity among samples. Only very low levels of genetic differentiation were observed between samples collected in fall 2006, spring 2007 and fall 2007. There was thus no decrease in genetic variability between the sampling periods. Between the two first sampling periods, one phase of sexual reproduction occurred and a few clonal generations were produced on the primary host. After sexual reproduction on apple trees, *D. plantaginea* is frequently subject to strong demographic bottlenecks, due to pest management practices (e.g. insecticide treatments [5,6]). In our study system, eight of the 13 orchards were treated conventionally with pesticides. If resistance genes are present in the treated populations, then such pesticide treatments may generate strong selection pressure, increasing the frequency of resistance genes in the clonal aphid population during spring. As recombination is absent during this part of the life cycle, we would expect (i) a change in microsatellite allelic frequencies due to the complete linkage between neutral genetic markers and genes subject to selection and (ii) a decrease in genetic variability due to the increase in frequency of some adapted MLGs. No such change was observed. Moreover, almost no repeated multilocus genotypes potentially resulting from the selection of a few adapted clones were observed in spring. Conventional apple orchards undergo a large number of pesticide treatments (up to 10 treatments are commonly applied in apple orchards when *D. plantaginea* is present, in France [57], and elsewhere see e.g. Blommers *et al.* [18]). Thus, the selection pressure resulting from pesticide treatments is likely to be very intense. Our observation is therefore more consistent with an absence of adaptive gene polymorphism, particularly for insecticide resistance genes, in the populations sampled, the resistance alleles being either fixed or absent. No failure of insecticide treatment was reported in spring 2007, suggesting that the mechanisms of insecticide resistance mechanisms documented by Delorme [58] did not occur.

Using a similar temporal sampling scheme for the peach potato aphid, *Myzus persicae*, Guillemaud *et al.* [24] detected a change in insecticide resistance allele frequency in holocyclic populations in southern France. The *kdr* mutation, which confers resistance to pyrethroid insecticides, increased in frequency between autumn and spring, probably because of insecticide treatments. Conversely,

the *rdl* mutation, which confers resistance to cyclodiene insecticides, decreased in frequency over the same period, probably because of the negative pleiotropic effects of the mutation [24].

We also found almost no differentiation between spring 2007 and autumn 2007, a period of time spanning a few clonal generations on the primary host, the emigration flight to secondary hosts followed by a sequence of several clonal generations and the return flight to the apple tree. Again, no decrease in genetic variability was observed between the two sampling points, suggesting that selection and/or drift during the asexual phase of the life cycle has little or no effect on the genetic structure of *D. plantaginea* . This contrasts sharply with what was reported for *M. persicae* by Vorburger [25] and by Guillemaud *et al.* [24], who analyzed changes in population genetic structure during the asexual phase. Vorburger [25] followed the temporal dynamics of *M. persicae* clones on secondary hosts in detail over a period of one year, and Guillemaud *et al.* [24] measured the differentiation between aphids collected during emigration and the return flight. Both studies revealed significant temporal variation of the structure of the population, interpreted in both cases as a result of selection rather than genetic drift. Selection in aphids is now well documented, and it appears that host plant [59,60,61,62] and pesticide treatment [62,63] are among the most important selective factors to be taken into account when trying to understand the population genetic structure of aphid species acting as crop pests.

No such selective forces appear to shape the population genetic structure of *D. plantaginea* during the asexual phase, which occurs mostly on secondary hosts. The known secondary hosts of *D. plantaginea* are herbaceous plants of the genus *Plantago* [18]. Little is known about possible environmental selection on these plants. No control treatments (such as pesticide applications) are used against *D. plantaginea* when feeding on *Plantago* spp. because these plants are of neither economic nor ornamental value. However, we cannot exclude the possibility that, during the summer, *D. plantaginea* is exposed to pesticides applied to crops or vegetation stands in which their *Plantago* spp. host plants are common (e.g. as weeds). We tried to sample *D. plantaginea* on *Plantago* close to the primary host sampling locations at Valence, without success. This may be because (i) the populations of *D. plantaginea* on the secondary host are small, (ii) secondary host colonization is restricted to particular *Plantago* populations or to plants growing under specific favorable conditions or (iii) *Plantago* is not the only secondary host of *D. plantaginae*. It may be important to identify the entire set of actual secondary host plants of *D. plantaginea* and their distribution, to determine which processes may occur during the asexual phase on the secondary host plant (currently seen as a "black-box").

Practical aspects of aphid management

Our results concerning the genetic structure of the rosy apple aphid population have practical implications for the management of this aphid. We found no genetic differences between samples collected from orchards planted with different cultivars (Ariane, Smoothee and Melrose; unfortunately we could not test for an effect of pesticide treatments in Valence in spring 2007 because the sample size was too small for low-input and conventional orchards). There are three possible explanations for this result: (i) None of the three apple cultivars was thought to be resistant to the rosy apple aphid, so there is probably no adaptation to these cultivars in *D. plantaginea*. (ii) Determination of the genetic structure of the population with microsatellites does not reveal genetic structure due to selection, because recombination during sexual reproduction breaks the linkage between adaptive alleles and microsatellite markers. (iii) Migration homogenizes genotypic frequencies, so it is not possible to determine the genetic structure of the population linked to selective forces. The balance between migration and selection was in favor of migration, as discussed below.

We found that migration had a larger effect than drift and selection in shaping the population genetic structure of this species at various geographic scales. The imbalance in favor of migration was found within orchards, between orchards separated by tens of meters at the same site and between sites separated by one to several hundreds of kilometers. This imbalance has two consequences: local adaptation [64] probably cannot occur, and adaptations to control practices may spread rapidly over large geographic areas. Local adaptation may occur when the environment is heterogeneous for selection (e.g. with or without pesticide treatment) and when a there is cost associated with adaptation (e.g. a cost to pesticide resistance). It occurs when a mutated genotype (e.g. a pesticide-resistant genotype) is better adapted to certain local conditions (e.g. pesticide application) but less well adapted to other environmental conditions (e.g. absence of pesticide treatment) than the wild-type genotypes (e.g. pesticide-susceptible genotypes). Management strategies, such as treatment applications limited to small geographic pockets (the stable zone strategy in [65]), based on local adaptations may therefore not be applicable for the rosy apple aphid on apple trees in France. The second consequence of the apparently extensive migration of the rosy apple aphid is that a monogenic or oligogenic genotype adapted to control strategies (e.g. pesticide-resistant genotypes or genotypes circumventing plant resistance) may invade large areas very rapidly after its emergence. This is a

potential Achilles heel of control strategies against *D. plantaginea*, because adaptation at any one site may lead to the failure of control everywhere. Resistance to carbamate and organophosphate insecticides has recently been found in a *D. plantaginea* clone collected in Avignon (Southern France) [58]. This resistance is probably oligenic and based on a small number of biochemical mechanisms. Our results suggest that it is likely to increase rapidly in frequency and spread geographically, leading to the failure of pest control over large areas if no other pesticides (such as pyrethroids) are used.

Supporting Information

Table S1   Description of *Dysaphis plantaginea* samples and genetic variation within samples.

Acknowledgments

We thank Delphine Racofier (FREDON Aquitaine), Isabelle Lafargue (DRAF Aquitaine), Arnaud Lemarquand, René Rieux, and Hubert Defrance for their help with aphid sampling. We also thank Armelle Coeur d'Acier for her help with aphid species identification.

Author Contributions

Conceived and designed the experiments: TG SS KM PF. Performed the experiments: AB SS KM. Analyzed the data: TG. Wrote the paper: TG PF.

References

1. Hill D (1987) Agricultural insect pests of temperate regions and their control. Cambridge, UK: Cambridge University Press.

2. Hull LA, Starner VR (1983) Effectiveness of insecticide applications timed to correspond with the development of rosy apple aphid (Homoptera, Aphididae) on apple. Journal of Economic Entomology 76: 594–598.

3. Forrest JMS, Dixon AFG (1975) Induction of leaf-roll galls by apple aphids *Dysaphis devecta* and *Dysaphis plantaginea*. Annals of Applied Biology 81: 281.

4. Lyth M (1985) Hypersensitivity in apple to feeding by *Dysaphis plantaginea* - effects on aphid biology. Annals of Applied Biology 107: 155–161.

5. Cross JV, Cubison S, Harris A, Harrington R (2007) Autumn control of rosy apple aphid, *Dysaphis plantaginea* (Passerini), with aphicides. Crop Protection 26: 1140–1149.

6. Delorme R, Ayala V, Touton P, Auge D, Vergnet C (1999) Le puceron cendré du pommier (*Dysaphis plantaginea*) : Etude des mécanismes de résistance à divers insecticides. In: ANPP, editor; 7-8-9 décembre 1999; Montpellier. pp 89–96.

7. Burgel K, Daniel C, Wyss E (2005) Effects of autumn kaolin treatments on the rosy apple aphid, *Dysaphis plantaginea* (Pass.) and possible modes of action. Journal of Applied Entomology 129: 311–314.

8. Marko V, Blommers LHM, Bogya S, Helsen H (2008) Kaolin particle films suppress many apple pests, disrupt natural enemies and promote woolly apple aphid. Journal of Applied Entomology 132: 26–35.

9. Wyss E, Daniel C (2004) Effects of autumn kaolin and pyrethrin treatments on the spring population of *Dysaphis plantaginea* in apple orchards. Journal of Applied Entomology 128: 147–149.

10. Brown MW, Mathews CR (2007) Conservation biological control of rosy apple aphid, *Dysaphis plantaginea* (Passerini), in Eastern North America. Environmental Entomology 36: 1131–1139.

11. Minarro M, Hemptinne JL, Dapena E (2005) Colonization of apple orchards by predators of *Dysaphis plantaginea*: sequential arrival, response to prey abundance and consequences for biological control. Biocontrol 50: 403–414.

12. Wyss E, Villiger M, Hemptinne JL, Muller-Scharer H (1999) Effects of augmentative releases of eggs and larvae of the ladybird beetle, *Adalia bipunctata*, on the abundance of the rosy apple aphid, *Dysaphis plantaginea*, in organic apple orchards. Entomologia Experimentalis Et Applicata 90: 167–173.

13. Angeli G, Simoni S (2006) Apple cultivars acceptance by *Dysaphis plantaginea* Passerini (Homoptera: Aphididae). Journal of Pest Science 79: 175–179.

14. Minarro M, Dapena E (2007) Resistance of apple cultivars to *Dysaphis plantaginea* (Hemiptera: Aphididae): Role of tree phenology in infestation avoidance. Environmental Entomology 36: 1206–1211.

15. Minarro M, Dapena E (2008) Tolerance of some scab-resistant apple cultivars to the rosy apple aphid, *Dysaphis plantaginea*. Crop Protection 27: 391–395.

16. Qubbaj T, Reineke A, Zebitz CPW (2005) Molecular interactions between rosy apple aphids, *Dysaphis plantaginea*, and resistant and susceptible cultivars of its primary host Malus domestica. Entomologia Experimentalis Et Applicata 115: 145–152.

17. Bonnemaison L (1959) Le puceron cendré du pommier (*Dysaphis plantaginae* Pass.) – Morphologie et biologie – Méthode de lutte. Annales de l'Institut National de la Recherche Agronomique, Série C, Epiphyties 10: 257–322.

18. Blommers LHM, Helsen HHM, Vaal F (2004) Life history data of the rosy apple aphid *Dysaphis plantaginea* (Pass.) (Homopt., Aphididae) on plantain and as migrant to apple. Journal of Pest Science 77: 155–163.

19. Bonnemaison L (1961) Les Ennemis Annimaux des Plantes Cultivées et des Forêts. Paris: Paris 1er.

20. Blackman RL (1981) Species, sex and parthenogenesis in aphids. In: Forey PL, ed. The evolving biosphere. Cambridge: Cambridge University Press. pp 75–85.

21. Simon J-C, Rispe C, Sunnucks P (2002) Ecology and evolution of sex in aphids. Trends in Ecology and Evolution 17: 34–39.

22. Lathrop FH (1928) The biology of apple aphids. The Ohio Journal of Science 28: 177–204.

23. Delmotte F, Leterme N, Gauthier JP, Rispe C, Simon JC (2002) Genetic architecture of sexual and asexual populations of the aphid *Rhopalosiphum padi* based on allozyme and microsatellite markers. Molecular Ecology 11: 711–723.

24. Guillemaud T, Mieuzet L, Simon JC (2003) Spatial and temporal genetic variability in French populations of the peach-potato aphid, *Myzus persicae*. Heredity 91: 143–152.

25. Vorburger C (2006) Temporal dynamics of genotypic diversity reveal strong clonal selection in the aphid *Myzus persicae*. Journal of Evolutionary Biology 19: 97–107.

26. Solomon MG, Harvey N, Fitzgerald J (2003) Molecular approaches to population dynamics of *Dysaphis plantaginea*. In: Cross JV, Solomon MG, eds. 10–14 March 2002 Vienna, Austria: International Organization for Biological and Integrated Control of Noxious Animals and Plants (OIBC/OILB), West Palaearctic Regional Section (WPRS/SROP). pp 79–81.

27. Harvey NG, Fitz Gerald JD, James CM, Solomon MG (2003) Isolation of microsatellite markers from the rosy apple aphid *Dysaphis plantaginea*. Molecular Ecology Notes 3: 111–112.

28. Sunnucks P, England P, Taylor AC, Hales DF (1996) Microsatellite and chromosome evolution of parthenogenetic Sitobion aphids in Australia. Genetics 144: 747–756.

29. Wilson ACC, Massonnet B, Simon JC, Prunier-Leterme N, Dolatti L, et al. (2004) Cross-species amplification of microsatellite loci in aphids: assessment and application. Molecular Ecology Notes 4: 104–109.

30. Gauffre B, Coeur d'Acier A (2006) New polymorphic microsatellite loci, cross-species amplification and PCR multiplexing in the black aphid, *Aphis fabae* Scopoli. Molecular Ecology Notes 6: 440–442.

31. Raymond M, Rousset F (1995) Genepop (version. 1.2), a population genetics software for exact tests and ecumenicism. Journal of Heredity 86: 248–249.

32. Rousset F (2008) GENEPOP ' 007: a complete re-implementation of the GENEPOP software for Windows and Linux. Molecular Ecology Resources 8: 103–106.

33. Van Oosterhout C, Hutchinson WF, Wills DPM, Shipley P (2004) MICRO-CHECKER: software for identifying and correcting genotyping errors in microsatellite data. Molecular Ecology Notes 4: 535–538.

34. Chapuis MP, Estoup A (2007) Microsatellite null alleles and estimation of population differentiation. Molecular Biology and Evolution 24: 621–631.

35. Petit RJ, El Mousadik A, Pons O (1998) Identifying populations for conservation on the basis of genetic markers. Conservation Biology 12: 844–855.

36. Goudet J (2001) FSTAT, a program to estimate and test gene diversities and fixation indices (version 2.9.3). Updated from Goudet (1995).

37. Arnaud-Haond S, Belkhir K (2007) GENCLONE: a computer program to analyse genotypic data, test for clonality and describe spatial clonal organization. Molecular Ecology Notes 7: 15–17.

38. Pritchard JK, Stephens M, Donnelly P (2000) Inference of population structure using multilocus genotype data. Genetics 155: 945–959.

39. Benjamini Y, Hochberg Y (1995) Controlling the False Discovery Rate - a Practical and Powerful Approach to Multiple Testing. Journal of the Royal Statistical Society Series B-Methodological 57: 289–300.

40. Sokal RR, Rolf FJ (1995) Biometry. The Principles and Practice of Statistics in Biological Research. New York: W.H. Freeman and Company.

41. Evanno G, Regnaut S, Goudet J (2005) Detecting the number of clusters of individuals using the software STRUCTURE: a simulation study. Molecular Ecology 14: 2611–2620.

42. Simon JC, Rispe C, Sunnucks P (2002) Ecology and evolution of sex in aphids. Trends in Ecology & Evolution 17: 34–39.

43. Hartl DL, Clark AG (1997) Principles of Population Genetics. SunderlandMA, , U.S.A.: Sinauer Associates, Inc.

44. Halkett F, Simon JC, Balloux F (2005) Tackling the population genetics of clonal and partially clonal organisms. Trends in Ecology & Evolution 20: 194–201.

45. Wilson ACC, Sunnucks P, Blackman RL, Hales DF (2002) Microsatellite variation in cyclically parthenogenetic populations of Myzus persicae in south-eastern Australia. Heredity 88: 258–266.

46. Jensen AB, Hansen LM, Eilenberg J (2008) Grain aphid population structure: no effect of fungal infections in a 2-year field study in Denmark. Agricultural and Forest Entomology 10: 279–290.

47. Gomez A, Carvalho GR (2000) Sex, parthenogenesis and genetic structure of rotifers: microsatellite analysis of contemporary and resting egg bank populations, 9, 203–214. Molecular Ecology 9: 203–214.

48. Simon JC, Baumann S, Sunnucks P, Hebert PDN, Pierre JS, et al. (1999) Reproductive mode and population genetic structure of the cereal aphid Sitobion avenae studied using phenotypic and microsatellite markers. Molecular Ecology 8: 531–545.

49. Papura D, Simon JC, Halkett F, Delmotte F, Le Gallic JF, et al. (2003) Predominance of sexual reproduction in, Romanian populations of the aphid Sitobion avenae inferred from phenotypic and genetic structure. Heredity 90: 397–404.

50. Massonnet B, Weisser WW (2004) Patterns of genetic differention between populations of the specialized herbivore Macrosiphoniella tanacetaria (Homoptera, Aphididae). Heredity 93: 577–584.

51. Wright S (1969) The Theory of Gene Frequencies: The University of Chicago Press, Chicago. 511 p.

52. De Barro PJ, Sherratt TN, Brookes CP, David O, MacLean N (1995) Spatial and temporal genetic variation in British field populations of the grain aphid Sitobion avenae (F.) (Hemiptera: Aphididae) studied using RAPD-PCR. Proceedings of the Royal Society of London, B 262: 321–327.

53. Sunnucks P, DeBarro PJ, Lushai G, Maclean N, Hales D (1997) Genetic structure of an aphid studied using microsatellites: Cyclic parthenogenesis, differentiated lineages and host specialization. Molecular Ecology 6: 1059–1073.

54. Loxdale HD, Brookes CP (1990) Temporal genetic stability within and restricted migration (gene flow) between local populations of the blackberry-grain aphid Sitobion fragariae in South-East England. J anim Ecol 59: 497–514.

55. Loxdale HD, Hardie J, Halbert S, Foottit R, Kidd NAC, et al. (1993) The relative importance of short-range and long-range movement of flying aphids. Biological Reviews of the Cambridge Philosophical Society 68: 291–311.

56. Martinez-Torres D, Carrio R, Latorre A, Simon JC, Hermoso A, et al. (1997) Assessing the nucleotide diversity of three aphid species by RAPD. Journal of Evolutionary Biology 10: 459–477.

57. Butault J, Dedryver C, Gary C, Guichard L, Jacquet F, et al. (2010) Ecophyto R&D. Quelles voies pour réduire l'usage des pesticides? Synthèse du rapport d'étude. 90 p.

58. Delorme R, Ayala V, P T, Auge D, Vergnet C (1999) Le puceron cendré du pommier (Dysaphis plantaginea): étude des mécanismes de résistance à divers insecticides; Montpellier. ANPP.

59. Carletto J, Lombaert E, Chavigny P, Brevault T, Lapchin L, et al. (2009) Ecological specialization of the aphid Aphis gossypii Glover on cultivated host plants. Molecular Ecology 18: 2198–2212.

60. Peccoud J, Ollivier A, Plantegenest M, Simon JC (2009) A continuum of genetic divergence from sympatric host races to species in the pea aphid complex. Proceedings of the National Academy of Sciences of the United States of America 106: 7495–7500.

61. Simon JC, Carre S, Boutin M, Prunier-Leterme N, Sabater-Munoz B, et al. (2003) Host-based divergence in populations of the pea aphid: insights from nuclear markers and the prevalence of facultative symbionts. Proceedings of the Royal Society B-Biological Sciences 270: 1703–1712.

62. Zamoum T, Simon JC, Crochard D, Ballanger Y, Lapchin L, et al. (2005) Does insecticide resistance alone account for the low genetic variability of asexually reproducing populations of the peach-potato aphid Myzus persicae? Heredity 94: 630–639.

63. Carletto J, Martin T, Vanlerberghe-Masutti F, Brevault T (2010) Insecticide resistance traits differ among and within host races in Aphis gossypii. Pest Management Science 66: 301–307.

64. Roughgarden J (1996) Theory of population genetics and evolutionary ecology: an introduction. Upper Saddle River: Prentice-Hall, Inc. 612 p.

65. Lenormand T, Raymond M (1998) Resistance management: the stable zone strategy. The Proceedings of the Royal Society of London, B 265: 1985–1990.

20

Levels and Patterns of Nucleotide Variation in Domestication QTL Regions on Rice Chromosome 3 Suggest Lineage-Specific Selection

Xianfa Xie[1]*, Jeanmaire Molina[1], Ryan Hernandez[2], Andy Reynolds[3], Adam R. Boyko[4], Carlos D. Bustamante[4], Michael D. Purugganan[1]

1 Center for Genomics and Systems Biology, Department of Biology, New York University, New York, New York, United States of America, 2 Department of Human Genetics, University of Chicago, Chicago, Illinois, United States of America, 3 Department of Biological Statistics and Computational Biology, Cornell University, Ithaca, New York, United States of America, 4 Department of Genetics, Stanford University, Stanford, California, United States of America

Abstract

Oryza sativa or Asian cultivated rice is one of the major cereal grass species domesticated for human food use during the Neolithic. Domestication of this species from the wild grass *Oryza rufipogon* was accompanied by changes in several traits, including seed shattering, percent seed set, tillering, grain weight, and flowering time. Quantitative trait locus (QTL) mapping has identified three genomic regions in chromosome 3 that appear to be associated with these traits. We would like to study whether these regions show signatures of selection and whether the same genetic basis underlies the domestication of different rice varieties. Fragments of 88 genes spanning these three genomic regions were sequenced from multiple accessions of two major varietal groups in *O. sativa*—*indica* and *tropical japonica*—as well as the ancestral wild rice species *O. rufipogon*. In *tropical japonica*, the levels of nucleotide variation in these three QTL regions are significantly lower compared to genome-wide levels, and coalescent simulations based on a complex demographic model of rice domestication indicate that these patterns are consistent with selection. In contrast, there is no significant reduction in nucleotide diversity in the homologous regions in *indica* rice. These results suggest that there are differences in the genetic and selective basis for domestication between these two Asian rice varietal groups.

Editor: Patrick Callaerts, VIB & Katholieke Universiteit Leuven, Belgium

Funding: This work was funded by a grant from the National Science Foundation Plant Genome Research Program (MCB-0701382, www.nsf.gov). The funder had no role in study design, data collection and analysis, decision to publish, or preparation of the manuscript.

Competing Interests: The authors have declared that no competing interests exist.

* E-mail: Xianfa.Xie@gmail.com

Introduction

Crop domestication is the adaptive divergence of a plant species as a result of selection and the initial evolutionary transition from wild to human-associated cultivated environments [1,2]. Phenotypic comparisons identify numerous traits that differ between domesticated species and their wild ancestors. In general, three classes of traits that differentiate domesticated and wild ancestral species can be defined [1]. First are domestication traits, which evolve during the initial movement of species from natural to cultivated environments. A second class is crop improvement traits, which are further phenotypic changes that have occurred after the initial domestication to human-associated cultivated environments [3]. Finally, there are crop diversification traits, which are associated with different crop varieties or cultivars adapted to different cultures or agro-ecological environments.

All three types of traits are conceptually distinct, but all can show up as differences between domesticated and wild ancestral species. It should be noted that, in principle, crop improvement traits can be difficult to separate from domestication traits. A few traits, however, are widely recognized as true domestication traits, including loss of seed shattering and change to annual life cycle [1,2]. These traits are fixed in domesticated taxa – that is, they are phenotypes shared by all members of a domesticated crop species.

Identifying the genetic basis of domestication traits in several plant species, most especially cereal grasses, has been a major research area in the study of plant evolutionary biology [4,5]. There have been attempts in the last few years to determine the molecular basis of cereal crop domestication, and study the nature of selection as well as other evolutionary forces associated with domestication events [4,1]. Mapping of quantitative trait loci (QTL) associated with domestication has been a major approach in studying the genetic architecture of domestication. QTL analyses for domestication traits have been accomplished in maize [6,7], wheat [8], pearl millet [9], foxtail millet [10] and rice [11,12,13], which have provided crucial information on the genetic basis of domestication. Many of these QTL studies have led to the isolation of domestication genes in various cultivated plant species [4], including the *tb1* locus that accompanies shoot architecture evolution in maize [14], and the *sh4* and *qSH1* loci that lead to loss of seed shattering in rice [15,16].

Despite the identification of domestication trait QTLs, and in some instances domestication genes, there remain several unanswered questions surrounding the evolutionary genetics of crop domestication. First, since the putative domestication QTLs were identified using linkage mapping, it is unknown whether these mapped QTLs are indeed selected for and do not simply represent natural variation of alleles maintained by genetic drift or mutation/selection balance. Because domestication is a process of selection and adaptive evolution of cultivated species from their wild ancestor, demonstrating selection at putative domestication QTLs is a prerequisite for defining them as true domestication loci [1].

One unambiguous signature of positive selection is a "selective sweep," which is recognized in part as significantly reduced nucleotide variation across a genomic region in proximity to a selected gene [17]. The physical extent of a sweep (whether a few hundred bp or several hundred kb) is governed by the strength of selection, time since the sweep began, and effective recombination rate between the selected site and the neighboring genomic regions. Population bottlenecks also reduce nucleotide variation levels, but this is manifested genome wide rather than the more localized decrease in polymorphisms associated with selective sweeps [18].

In several characterized domestication genes, such as maize *tb1* [14,19], there is an unambiguous signature for positive selection, including the presence of an extended selective sweep that results in reduced nucleotide variation around the genetic target of selection [17,1]. In other cases, however, selective sweeps have not been identified at genes that encode for presumed domestication traits. In the rice *qSW5* gene, for example, which controls variation in seed width associated with a QTL [20], population genetic analysis is still needed to characterize whether a selective sweep has indeed occurred at this gene.

A second set of issues is whether domestication within different variety groups of a crop species (for example *japonica* and *indica* rice, see below) proceeds by selection of the same genes, or whether there is selection on different genes in these different varietal groups. In recent years, it has become clear that several cereal crops, including Asian domesticated rice (*Oryza sativa* L.) and barley (*Hordeum vulgare*), appear to be comprised of genetically distinct groups [21,22]. Comparative molecular genetic analysis of domestication QTLs or genes allows us to determine whether the same or distinct genes (or alleles) underlie evolution in these genetically distinct groups.

A final set of issues is to understand how gene flow among genetically distinct domesticated groups (*japonica* and *indica*) or even between domesticated taxa and their progenitor species affects the evolutionary dynamics of domestication. The mutant alleles of *Rc* domestication gene that lead to white pericarp in rice, for example, originated in one rice lineage and spread via introgression to another distinct *O. sativa* subspecies [23]. The importance of introgression in the spread and fixation of domestication genes during crop domestication has yet to be considered in the study of rice domestication.

To address these issues, we examine the patterns of nucleotide variation at several domestication trait QTLs in *O. sativa*, determining whether molecular diversity at these QTLs is consistent with the action of positive selection in this crop species. *O. sativa* is the world's most widely grown cereal crop species and is now a key model system in plant biology [24]. Two main rice varietal groups, *indica* and *japonica*, have been recognized since ancient China and are the most widely grown worldwide [25]. The two groups differ morphologically in grain shape and leaf color, biochemically in amylose composition, phenol reaction, and sensitivity to potassium chlorate, ecogeographically in growing environment and geographic distribution, as well as genetically in various aspects [25,26,27]. The *japonica* group itself is divided into the *tropical japonica* and the *temperate*

japonica, the former considered to be the product of direct domestication, while the latter being a secondarily derived varietal group [25].

It has been established that *Oryza rufipogon* Griff., a species native to southeastern Asia, is the wild ancestor of domesticated rice [25,21]. There have also been suggestions that another wild species *Oryza nivara* is the ancestor of *O. sativa* [15], although there is evidence that this species may simply be an annual ecotype of *O. rufipogon* [25,28]. *O. rufipogon* is characterized by variable but distinctly higher levels of out-crossing, while *O. sativa* is primarily a self-fertilizing species [25]. Some genetic evidences suggest there were two domestication events for rice, with possibly separate origins for the *indica* and *japonica* groups [29,30,26,21,31], though there are other models suggesting single origin of domesticated rice [32,33]. Early hypotheses considered that domestication of *tropical japonica* occurred in a mountainous region spanning Nepal, Assam, northern regions of Myanmar, Laos, Thailand, and the Yunnan province of southern China [21], while archaeological studies indicate that this varietal group was domesticated in the Yangtze Valley in China [34]. It was also thought that *indica* rice was independently domesticated in Ganges region of the Indian subcontinent [21], although there are suggestions that this major varietal group may have arisen in part by extensive hybridization of *tropical japonica* with either proto-*indica* or wild *O. rufipogon* [34].

In our study, we examine the molecular population genetics of genomic regions in rice that contain QTLs associated with domestication of this crop species and then compare these regions with the genome-wide data. These QTLs were identified in a large-scale mapping study between a *tropical japonica* variety (Jefferson) and a Malaysian *O. rufipogon* (IRGC 195491) [12]. The *O. rufipogon* accession used in this QTL study has been described as a weedy rice, although SSLP marker analysis clearly indicates that it is related to wild *O. rufipogon* and *O. nivara*, and is not a feral relative of domesticated rice [33].

Rice chromosome 3 was identified to contain several QTLs associated with rice domestication [12], and we decided to make this chromosome the focus of our study. Two regions at the proximal (QTL 3A) and middle (QTL 3B) of the chromosome were chosen because they were associated with loss-of-shattering, a key domestication trait. These two regions harbor the QTLs sh3.1 and sh3.2, respectively [12] (see Fig. 1). The third region at the distal end of the chromosome (QTL 3C) was chosen for analysis because multiple traits associated with domestication were localized in this one region. This region contains overlapping QTLs underlying percent seed set (pss3.1), days to heading (dth3.4), grain weight (gw3.2), the number of spikelets per panicle (spp3.1), yield (yld3.2) and the number of grains per panicle (gpp3.1) [12].

The traits that are associated with these QTLs have been implicated in the domestication of rice. We should note, however, that while QTL3A and 3B underlies a known domestication trait (e.g., loss of seed shattering), the traits associated with QTL 3C may also be considered crop improvement or diversification traits. As we indicated, telling these two types of traits apart can be difficult, and without a clear archaeological history, we can never be certain whether these traits are true domestication traits. For the purposes of this study, however, we will consider them all as domestication traits. Using re-sequencing data for gene fragments across these three putative domestication QTL regions in rice, we examine whether the levels and patterns of polymorphism in these three regions are indeed consistent with the possibility that they have experienced recent positive selection accompanying the evolution of this cultivated grass species.

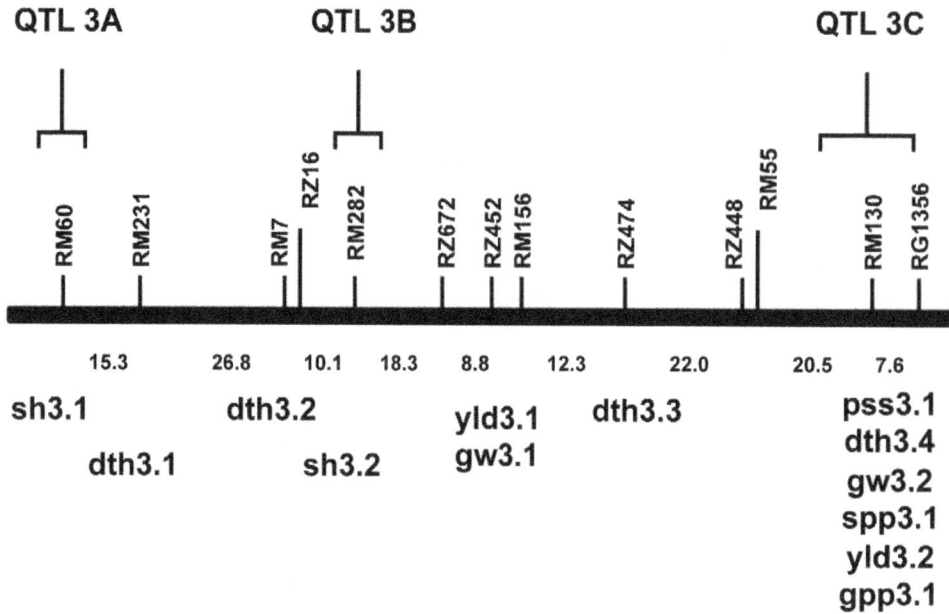

Figure 1. QTL map of domestication and diversification traits between *O. sativa* and *O. rufipogon*. The map is based on the study by Thomson *et al.* [12], and the regions used in our study are indicated by the square brackets. Traits associated with the QTLs are: sh, seed shattering; pss, percent seed set; dth, days to heading; gw, grain weight; spp, spikelets per panicle; yld, yield; gpp, grains per panicle.

Results

Nucleotide variation and linkage disequilibrium in three domestication QTLs

For QTL 3A, we analyzed an ~1.05 Mb region from the proximal end of the chromosome, and in QTL 3B, we studied ~1.9 Mb region from position 11.988 Mb to 13.863 Mb. In QTL 3C, we examined an ~2.31 Mb region from position 32.893 Mb to 35.203 Mb. We sequenced a total of 88 gene fragments in these three QTL regions, each with an average length of 509 bp and spaced approximately 50 kb apart, totaling 44.8 kb of genomic sequence. The spacing was chosen based on previous work that indicated that linkage disequilibrium in the major rice groups extend to ~75–150 kb [35], and that the one good example of a selective sweep in rice (in the *Waxy* gene, see [36]) is ~260 kb in length.

Previous work using genome-wide sequence tagged site (STS) data provided an indication of the genetic relationships and population structure between rice varietal groups [31]. STRUC-TURE analysis using the DNA sequence data from the three domestication QTL regions is consistent with that observed using genome-wide data [31] (see Fig. 2).

In total, we detected 833 single nucleotide polymorphisms (SNPs) in *O. sativa* and *O. rufipogon*, of which 767 are silent site polymorphisms. The levels of silent site nucleotide variation at each of the gene fragments as well as each of the three domestication QTLs were calculated and reported for *O. rufipogon* and the two major *O. sativa* groups – *tropical japonica* and *indica*, which represent the two major domestication events in *O. sativa* (see Figure 3 and Table 1, respectively). In the domesticated rice varietal group *indica*, there are a total of 288 SNPs, with 276 at silent sites. In *tropical japonica*, there are only 37 SNPs, of which all but one are silent site changes. Mean silent site nucleotide diversity (π) across all sampled loci in *O. sativa* is approximately 0.0008 while the silent-site level of polymorphism in the wild rice species, *O. rufipogon*, is six-fold higher ($\pi = 0.0049$) (see Table 1).

We calculated linkage disequilibrium between SNPs whose minor frequencies are greater than 10 percent within and between all three QTL regions. In the wild out-crossing species *O. rufipogon*, some linked sites within each QTL show strong disequilibrium while almost no disequilibrium is observed at sites between the three genomic regions (see Fig. 4). SNP sites in *indica* show stronger disequilibrium, compared to *O. rufipogon*, within the QTL regions (see Fig. 4). However, there are too few segregating sites remaining in *tropical japonica* to make a meaningful comparison, which suggests the selection in *tropical japonica* in these QTL regions were even stronger to have eliminated most of the polymorphism in *O. rufipogon*. The increase in LD in the domesticated rice groups have been observed in a genome-wide study [35], and is likely due to the bottleneck associated with rice domestication as well as the reduction in effective recombination in domesticated rice associated with the transition to selfing in this species.

Levels of nucleotide variation are significantly reduced in domestication QTLs in *tropical japonica* but not *indica*

The general loss of genetic variation we observe in the three QTL regions in domesticated rice (see Table 1) is consistent with previous reports [31,35], but the patterns of polymorphism reduction differ between the two major rice varietal groups. While the nucleotide diversity levels in *indica* at the three QTL regions are comparable to those reported previously for the genome-wide STS data [31], those in *tropical japonica* are much lower. In particular, the mean level of molecular variation in *tropical japonica* is one order of magnitude lower in the three domestication QTL regions compared to the mean genome-wide level of nucleotide diversity reported in [31].

We compared the distribution of nucleotide diversity at each of these domestication QTLs with the genome-wide distribution for the two major domesticated rice varietal groups, *indica* and *tropical japonica*, as well as the wild rice *O. rufipogon* (see Fig. 5). At QTL region 3A, the distribution of *tropical japonica* nucleotide

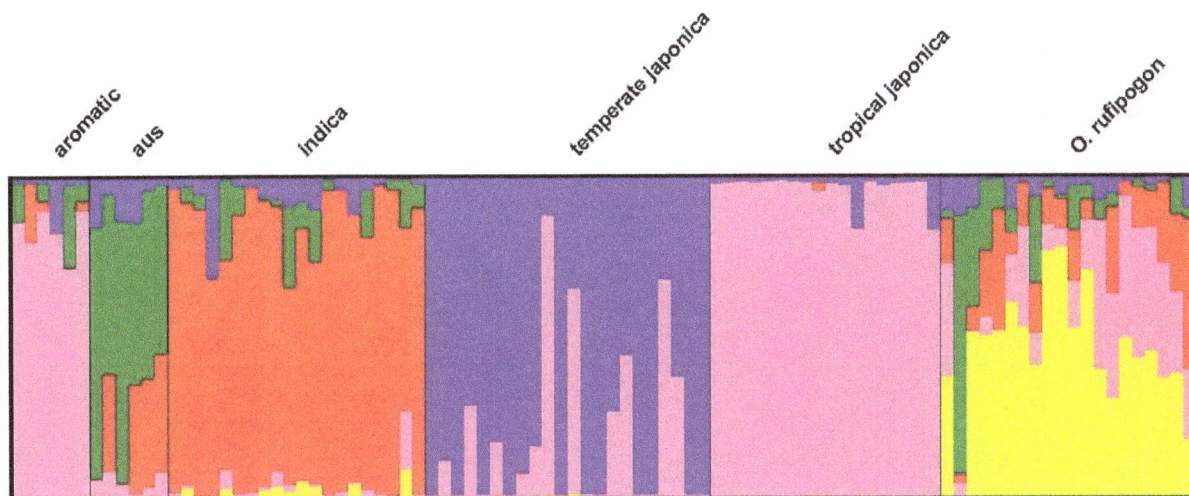

Figure 2. Population structure of *O. sativa* and *O. rufipogon*. It was estimated from all the loci combined from the three QTL regions. The analysis includes accessions of *temperate japonica*, *aromatic* and *aus* rices that were also sequenced for the same fragments (data not shown). The highest likelihood is found at K = 5. Vertical bars along the horizontal axis represent individual *Oryza* accessions, the proportion of ancestry that can be attributed to each cluster under K = 5 clusters is given by the length of each colored segment in a bar. The labels at the top indicate the original variety/species designation for each accession based on Garris *et al.* [26].

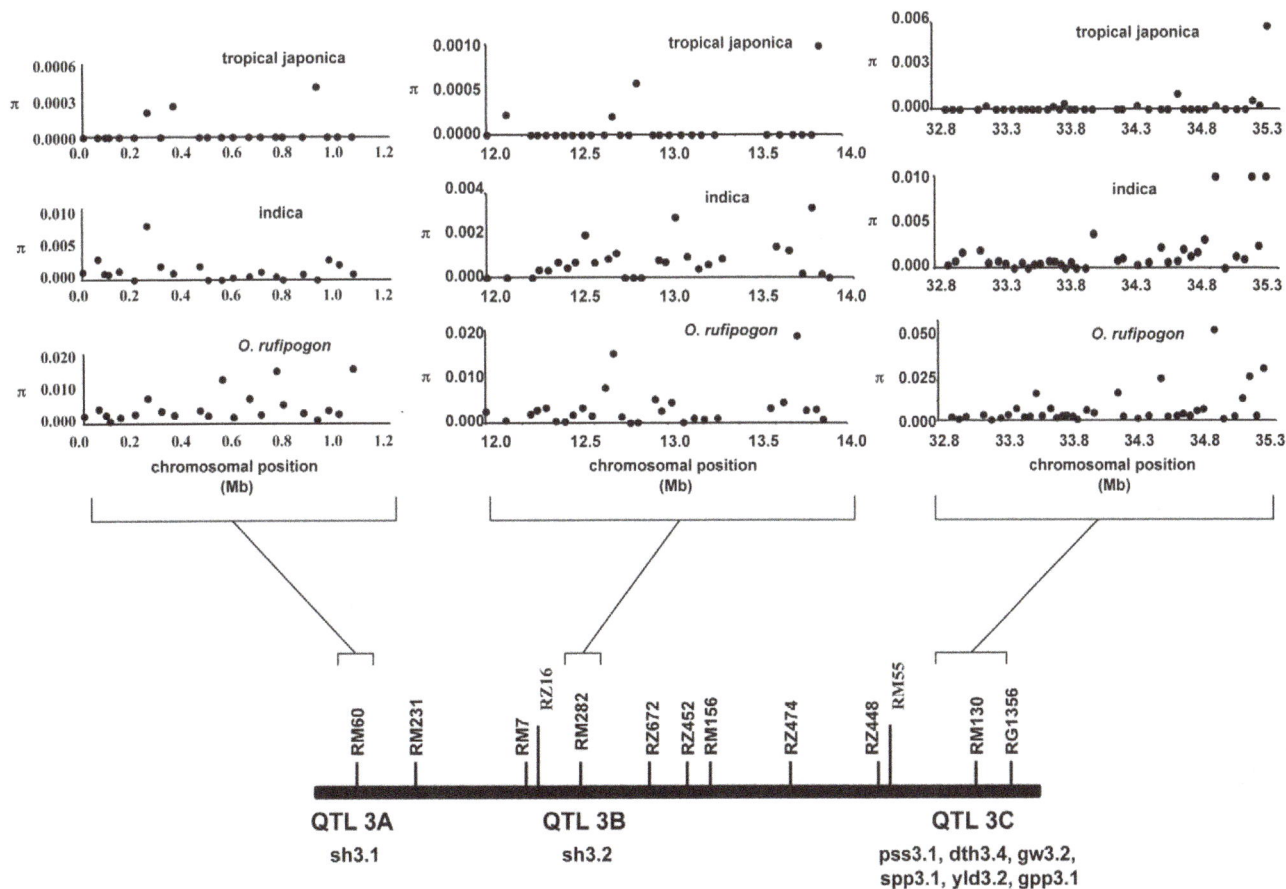

Figure 3. Nucleotide diversity (π) at silent sites for each gene fragment within the three QTL regions. Silent sites include both synonymous sites and noncoding sequences. Data for the two major varieties of *O. sativa* (*indica* and *tropical japonica*), as well as *O. rufipogon*, are shown.

Table 1. Silent site nucleotide diversity in domesticated rice and *O. rufipogon*.

Summary Statistics	Genomic Region	Species/Varietal group		
		O. sativa indica	*O. sativa tropical japonica*	*O. rufipogon*
θ_w	QTL 3A	0.0013	0.00004	0.0049
	QTL 3B	0.0008	0.00007	0.0035
	QTL 3C	0.0031	0.0002	0.0064
	STS	0.0018	0.0015	0.0050
π	QTL 3A	0.0013	0.00007	0.0043
	QTL 3B	0.0008	0.0001	0.0034
	QTL 3C	0.0025	0.0005	0.0057
	STS	0.0009	0.0014	0.0050

diversity is significantly lower compared to the genome-wide distribution (Mann-Whitney Test, p<0.007). A significant reduction in nucleotide diversity at *tropical japonica* is also observed in the other two domestication trait QTLs. There is significantly lower nucleotide diversity at QTL 3B (Mann-Whitney Test, p<0.0011) and QTL 3C (Mann-Whitney Test, p<0.0101) compared to the genome-wide nucleotide diversity. Interestingly, neither *indica* rice nor the wild ancestor *O. rufipogon* shows any significant departure of nucleotide diversity distribution at all three domestication QTLs compared to the genome-wide data.

Within the three domestication trait QTL regions, we also find contiguous stretches of fragments of no polymorphism in *tropical japonica* (see Figure 3). At QTL 3A in this varietal group, two sets of large contiguous fragments of zero polymorphism are observed spanning genomic regions of ~200 and ~400 kb, respectively. Two extended runs of monomorphism in *tropical japonica* are also observed in both QTL 3B (~400 and ~900 kb in size) and 3C (~400 and ~350 kb in size). In contrast, the longest stretch of monomorphism in *indica* across all three domestication QTL regions is ~250 kb in QTL 3B, which overlaps slightly with one of the monomorphic runs observed in *tropical japonica*. There are no other long tracts of low nucleotide diversity in *indica* rice or the wild rice *O. rufipogon*.

Coalescent simulations with rice demographic model support selection in *tropical japonica*

In order to assess the statistical significance of reduced genetic variation in the three QTL regions, we need to quantify: (1) the expected levels of genetic diversity in each of the three regions under a neutral model of evolution for each of the two main subgroups (*indica* and *japonica*), and (2) the variability around this expected value due to stochasticity. In order to accomplish these two goals, we used coalescent simulations based on a complex demographic model previously inferred from genome-wide patterns of nucleotide variation [31], which considers bottlenecks at the foundation of both *indica* and *japonica* as well as migration involving *O. rufipogon*.

The low SNP levels in *tropical japonica* preclude our use of other signatures of selection such as Tajima's D or the classical site-frequency spectrum. We thus examined the observed and predicted SNP levels for each of the two domesticated rice groups, the latter of which were calculated based on the demographic model described in the Materials and Methods and in Figure 6 but informed by the observed polymorphism level of *O. rufipogon* in each QTL region. The neutral demographic model and genomic patterns of sequence variation suggests that, on average, for the number of samples drawn here (20 *O. rufipogon*, 20 *indica*, and 18 *tropical japonica*) the *O. sativa indica* sample ought to show

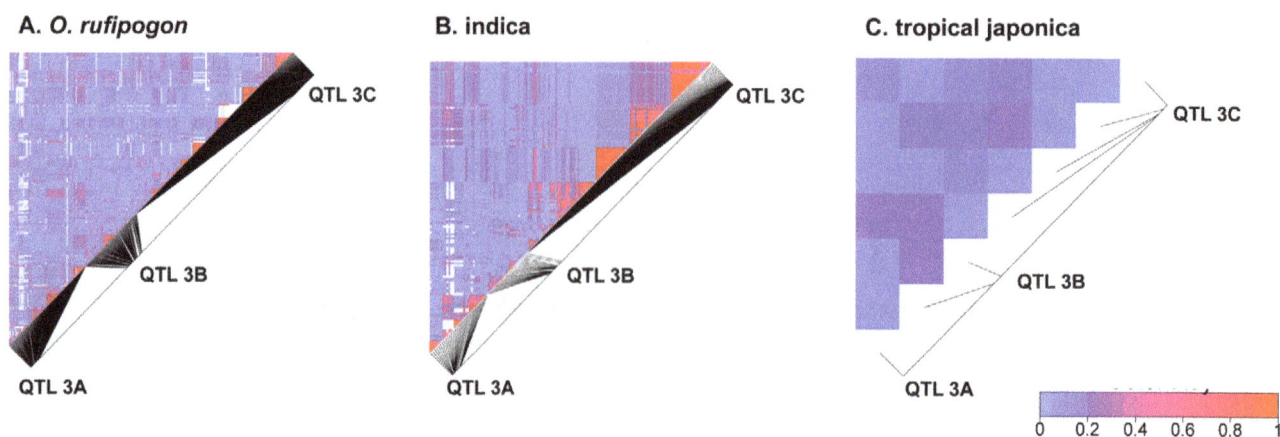

Figure 4. Linkage disequilibrium within and between domestication trait QTL regions. LD is measured as pairwise r^2 [55] between SNP sites within each group, and the values are shown by different colors as indicated in the legend.

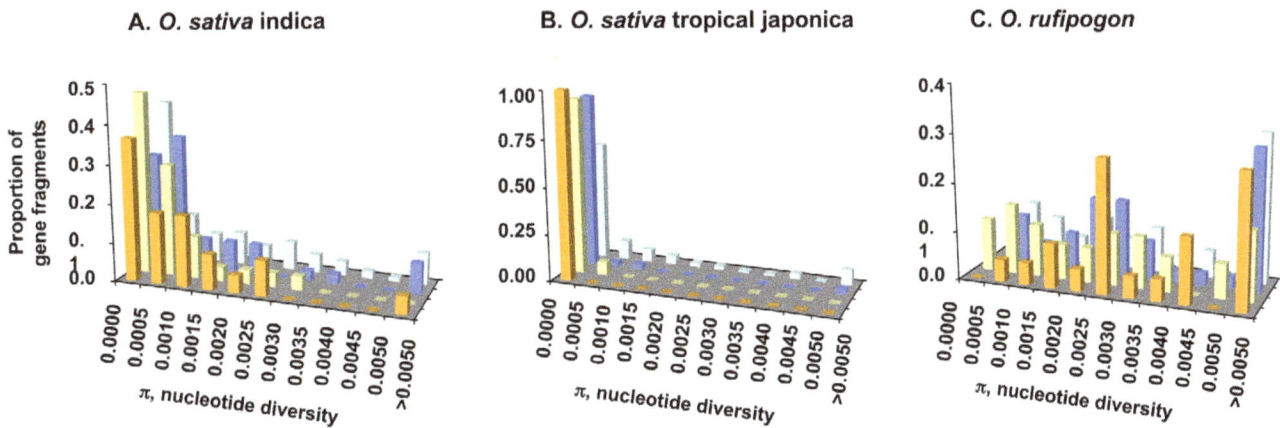

Figure 5. The distribution of nucleotide variation across gene fragments for three QTLs and genome-wide data. Orange, QTL 3A; yellow, QTL 3B; blue, QTL 3C; and light blue, genome-wide STS data. Note that the scale of nucleotide diversity is different in the graphs for the three different species or varietal groups.

approximately 58% of the sequence variation of the wild rice samples and the *tropical japonica* sample ought to show, on average, 41% of the variation seen in the wild ancestor. The magnitudes of these expected reductions in diversity between wild species and domesticated varietal groups are within 10% of the observed level of nucleotide diversity empirically estimated from genome-wide data [31].

Consistent with the results of the Mann-Whitney test for the difference in polymorphism level between the three QTLs and the genome-wide data, our simulation-based analysis (see Table 2) suggests too little diversity for all three regions in *tropical japonica* (p<0.001 for QTL 3A and 3B, and p<0.04 for QTL 3C). Across the three regions, the observed SNP levels in *tropical japonica* are ~4–17 percent of the expected under the coalescent simulation. In contrast, observed diversity in *indica* ranges from ~40–69 percent of the expected diversity based on the coalescent simulation, and do not show a significant reduction in diversity as compared to the variation one expects from the coalescent process without recombination (p<0.11 – 0.33). Given the number of multiple comparisons conducted here, it is unlikely that the *indica* deviation from expectation is biologically meaningful, while the reduced level of diversity in *tropical japonica* clearly suggests these QTL regions might have been selected in this varietal group.

Evolutionary relationships of domestication QTL regions in cultivated and wild species

The low levels of nucleotide diversity suggest that selective sweeps in all three QTL regions are present in *tropical japonica* but not *indica*. To examine phylogenetic relationships at these domestication QTLs, we constructed neighbor-joining trees for each genomic region spanning these QTLs (see Fig. 7). Our results show that *tropical japonica* alleles in each QTL region form a monophyletic group with moderate to high bootstrap support (67 percent for QTL 3A, 83 percent for QTL 3B, and 87 percent for QTL 3C). For QTL 3B and 3C, we find one *O. rufipogon* accession that clusters close to the *tropical japonica* clade. In QTL 3A, however, there are 11 wild rice strains that cluster with *tropical japonica* haplotypes (see Fig. 7), and eight of them are from China, consistent with an origin of this domesticated lineage in the Yangtze Valley. In contrast, none of the domestication QTL regions show the *indica* alleles forming a monophyletic clade (see Fig. 7), which again is inconsistent with a selective sweep across these genomic regions in *indica*.

Discussion

Domestication is characterized by selection [37], which leaves its imprint on the levels and patterns of nucleotide polymorphisms within the genome [1]. Studying these molecular signatures allows us to infer the dynamics of selection as well as other evolutionary forces associated with the origin and diversification of crop species.

In rice, QTL analyses indicate that domestication traits are governed by various QTLs between *O. sativa* and *O. rufipogon* [11,12,13]. For the purposes of this study, we define domestication traits as either those previously shown to be associated with the origin of the cultivated species [1] or any trait fixed between the wild and domesticated species, regardless of whether this trait evolved at the origin of the cultivated species or during a post-domestication process. Most of our accessions are landraces, however, which would rule out traits (and genes) that were fixed in domesticated crop species exclusively as a result of modern breeding.

We show that the levels and patterns of nucleotide variation at three domestication trait QTLs in *O. sativa* are consistent with the recent action of selection in *tropical japonica*, as would be expected during the domestication process. The distributions of nucleotide variation among gene fragments in these QTLs are significantly different from those in a genome-wide data set, with a preponderance of low polymorphism fragments at the QTL regions (see Fig. 5). The levels of observed SNP variation are also lower in domesticated rice at these QTL regions compared to the expected values from coalescent simulations (see Table 2).

These results are similar to those observed in known selective sweeps that have previously been studied in several crop genes associated with domestication or diversification phenotypes. The best example is the maize *tb1* gene involved in the suppression of auxiliary branch formation, which has a selective sweep spanning ~60–90-kb in length [14]. The maize *Y1* gene, involved in the yellow kernel phenotype, has a 600-kb selective sweep [38], while the rice *Waxy* gene has a 260-kb sweep associated with low-amylose rice in Northeast Asian cultivars [36]. In maize, a study analyzed 774 loci and 2–4% showed reduced variation that qualifies them as candidate domestication genes [18].

Interestingly, in our study selective sweeps are only observed in the *tropical japonica* samples but not in *indica*. This may suggest that selection at these QTL regions during domestication did not occur in *indica* rice, but was specific to *tropical japonica*. Another possibility,

O. sativa

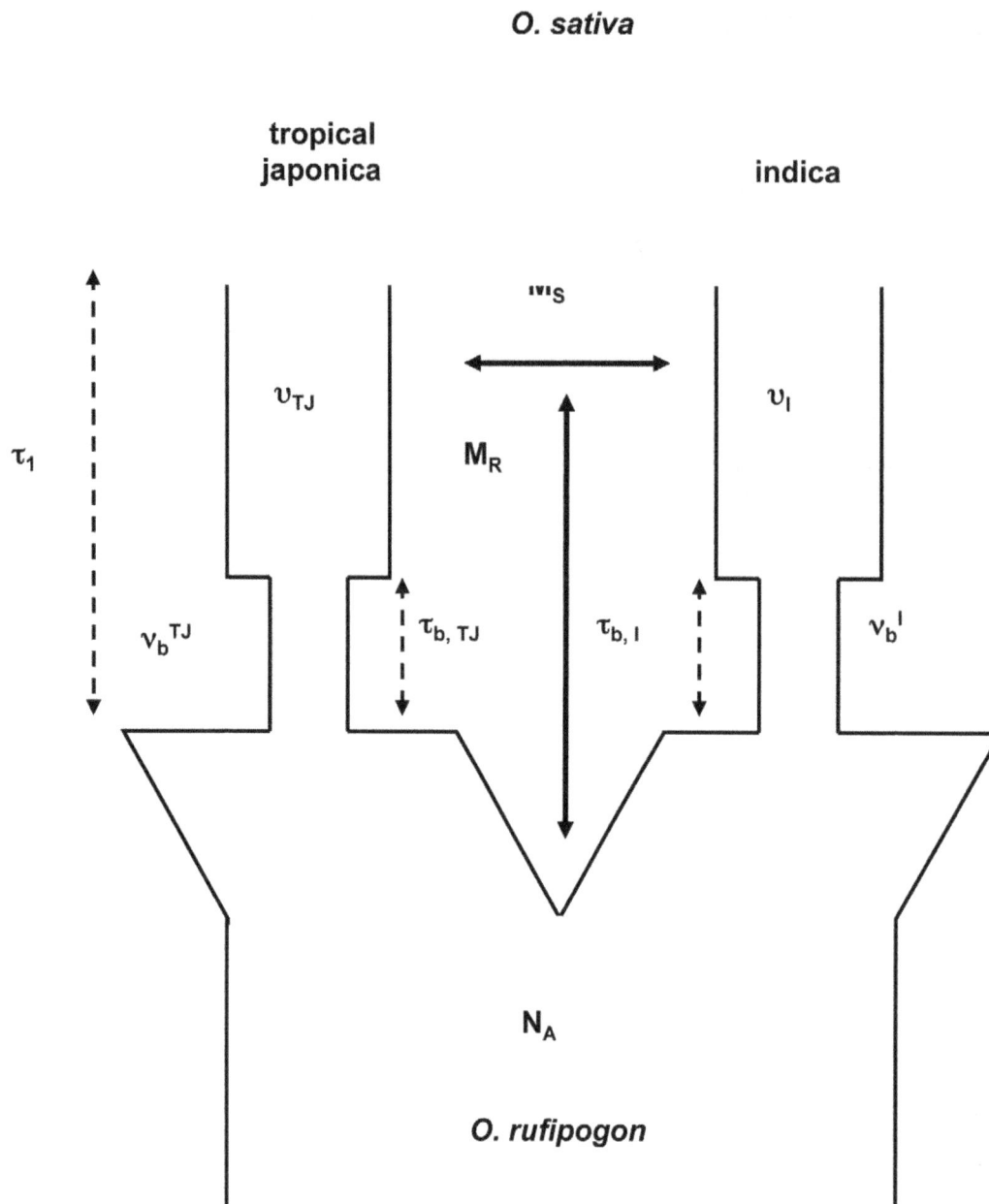

Figure 6. A two-origin demographic model for rice domestication. In this model, described in Caicedo *et al.* [31], the ancestral *O. rufipogon* has an ancestral population size N_A. At τ_1 generations ago, a bottleneck occurred with severity v, giving rise to *tropical japonica* and *indica*. At τ_b generations later, we get recovery of domesticated populations to a fraction υ of the ancestral population size N_A. The domesticated *tropical japonica* and *indica* share migrants at rate M_S, while both domesticated groups share migrants with *O. rufipogon* at rate M_R. For pictorial simplicity, the contemporary *O. rufipogon* population is not depicted. TJ and I indicate *O. sativa tropical japonica* and *indica*, respectively. Parameters for this model were estimated based on the unfolded site-frequency spectrum of genome-wide data [31], and were used to generate expected numbers of SNPs for each of our domestication trait QTL regions.

however, is that the *indica* alleles may comprise a "soft selective sweep." Selective sweeps are usually considered to occur on newly arisen mutation, but soft sweeps involve selecting for an old mutation. In the case of *indica*, it may be that selection occurred on mutations that were segregating as neutral mutations for a prolonged period at appreciable frequency in the ancestral *O. rufipogon* [39], leading to a soft sweep.

There are several lines of evidence to suggest that such a soft sweep in *indica* is unlikely in this context. First, it is unclear why *tropical japonica* would experience hard sweeps (selective sweeps

from newly-arisen mutations) in all 3 QTL regions and *indica* only soft sweeps, unless the genetic basis and histories in the QTL regions are markedly different. Second, the most likely result of a soft sweep would be a series of separate partial sweeps of related (but not necessarily identical) haplotypes in *indica*, which we again do not observe. Depending, however, on the specific evolutionary dynamics of such a soft sweep (e.g, a highly segregating mutation recombined in several different haplotypes coupled with widespread selection), other possible patterns of relationships may be observable, although these alternative

Table 2. Observed and expected numbers of SNPs at domestication trait QTLs based on coalescent simulations.

	QTL 3A		QTL 3B		QTL 3C	
	indica	*japonica*	*indica*	*japonica*	*indica*	*japonica*
Observed	41	3	42	5	175	29
Expected	102.2	71.1	87.9	63.6	255.1	175.7
Observed/ Expected	0.4	0.04	0.48	0.08	0.69	0.17
p-value	<0.114	<0.001***	<0.198	<0.001***	<0.330	<0.04*

*significant;
***extremely significant.

scenarios are even less likely. Finally, while there has been discussion in the literature on the possibility of soft sweeps during domestication [39], no unambiguous cases of soft domestication sweeps have been identified, in contrast to hard sweeps for which numerous examples are known in domesticated plants and animals.

A major question in evolutionary biology is the extent to which selection in genetically distinct groups acts on different or similar genes in sculpting adaptive traits [40]. Previous studies suggest that domestication among cereal crop species may be associated with the same genes [41,42]. Domestication traits like reduced seed shattering and increased yield have been selected in both *indica* and *japonica* rice. However, our analyses provide evidence for selection at molecular level in *tropical japonica* but not in *indica*, indicating that the genetic basis for domestication in *tropical japonica* and *indica* may differ and that separate genomic regions were subjected to selection between these two varietal groups even for the same domestication traits. This, however, is congruent with the fact that the three domestication QTLs examined in this study were identified in a mapping population between *O. rufipogon* and a *tropical japonica* cultivar of *O. sativa* [12]. A similar pattern of selection has been seen for the shattering gene *qSH1*, in which there is evidence for selection on this gene in *japonica* but not *indica* [43]. Furthermore, it appears that another gene associated with an agronomically important trait – the white pericarp *Rc* gene – was

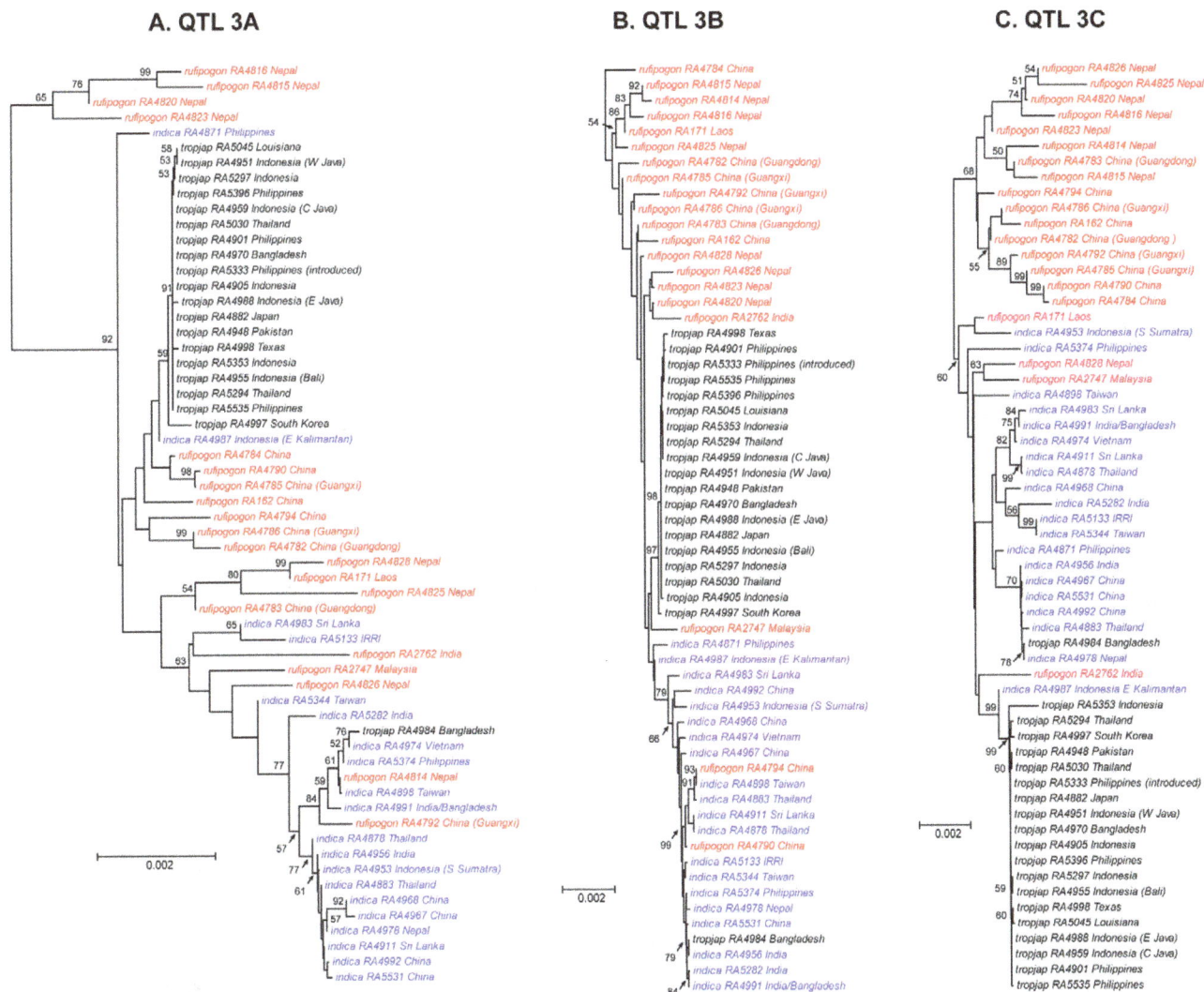

Figure 7. Neighbor-joining trees of the wild and domesticated rice at the domestication trait QTL regions. The accession numbers are indicated in the tree, which can be cross-referenced with Table S1.

originally selected upon in *tropical japonica* and the selected allele was subsequently introgressed into *indica* [23]. Continued efforts to study the genetic architecture of domestication in rice and to examine the role of selection on genome variation and the origin of this cultivated grass species will help unravel the nature of this key evolutionary phenomenon. Moreover, since selective sweeps are a clear signature of positive selection, they can be used to identify genes associated with domestication. This novel mapping approach, which scans the genome for the selection signature of low variation across a localized genomic region [44], is known as adaptive trait locus mapping [45], hitchhiking mapping [46], or selective sweep mapping [47]. It has been successfully used in identifying the warfarin resistance locus in rats [48], and several selected loci in *Drosophila* [46] and humans [49,50], and there is now growing interest in these methods for searching for domestication genes.

Our results suggest that one can integrate two methods to pursue these research goals - QTL mapping, which identifies specific genomic regions that harbor genes associated with specific domestication traits, and selective sweep mapping, which searches the genome for signatures of positive selection referred to as selective sweeps. By demonstrating that domestication trait QTLs do indeed harbor molecular imprints consistent with selection, it may be possible to utilize selective sweeps to further fine-map domestication genes and dissect the mechanisms that led to the origin of cultivated grass species.

Materials and Methods

Rice samples

The rice samples used in this study include three species: *O. sativa*, *O. rufipogon*, and *Oryza meridionalis* (see Table S1). The *O. sativa* accessions include 21 *indica* and 18 *tropical japonica* and are mainly landrace accessions, but 3 are elite cultivars. One of the *indica* accessions, POPOT-165 from Indonesia, was found by DNA sequence data to be a hybrid between *indica* and *tropical japonica* and excluded from the analyses. Most of the 20 *O. rufipogon* accessions come from China and Nepal, and a single accession of *O. meridionalis* is used as an outgroup for phylogenetic analysis.

Gene fragments sequenced

Three domestication QTL regions on rice chromosome 3 [12] were selected in this study (see Fig. 1). The physical positions of these three QTLs were defined by identifying the flanking markers and their positions in Gramene (http://www.gramene.org). Within each of these QTLs, gene fragments of ~500 bp in size and located ~50 kb apart were sequenced. The sequenced fragments comprise primarily intronic sequences, and were not located in transposable elements or recent gene duplicates. A total of 88 genes were analyzed, and the number of gene fragments and associated genes within each domestication QTL are listed in Table S2.

DNA sequencing and alignment

All primers (see Table S3) were designed using Primer3 [51] based on the *O. sativa* Nipponbare genomic sequence [52] available on Gramene. Whenever possible, the primers are designed to reside in exonic regions flanking the intron to be sequenced. All PCR primers were compared against the Nipponbare sequence to ensure that each of them uniquely recognizes the genic region to be amplified. PCR amplification and direct DNA sequencing were conducted by Cogenics (New Haven, CT, USA) as previously described [36,31,35]. The sequencing error rate was assessed as previously described [31], which revealed an error rate of less than 0.01%.

Base-pair calling, quality score assignment, and construction of sequence contigs were carried out using the Phred and Phrap programs (CodonCode), and sequence alignment and editing were carried out with BioLign Version 4.0.5.1 (Tom Hall, North Carolina State University, Raleigh, North Carolina). Single nucleotide polymorphisms (SNPs) were identified as mutational differences between sequenced alleles. Heterozygous sites and insertion/deletions were identified with the aid of Polyphred (Deborah Dickerson, University of Washington, Seattle, Washington) and manually confirmed by visually inspecting chromatograms. Primer sequences were removed from the alignments for final analysis. All sequences are deposited in Genbank with accession numbers FJ015311–FJ023534.

Analysis of nucleotide diversity

Population genetic structure at the three QTL regions was assessed using STRUCTURE 2.2 [53]. Simulations were run with a linkage model and allele frequencies being independent among populations. Five replicates at each value of K (population number, from 2–9) were carried out, and each run had a burn-in length and a run length of 100,000 iterations.

Nucleotide diversity (π) and Watterson's theta θ_W [54] were calculated for individual domesticated rice varieties, as well as for *O. rufipogon*. The average nucleotide diversity (π) in each QTL was compared to genome-wide STS data [31] using a non-parametric Mann-Whitney test. The ratio of θ_W for each domesticated rice group and that for the wild *O. rufipogon* for each fragment was also calculated and compared with nucleotide variation for gene fragments across the genome [31].

Pairwise SNP linkage disequilibrium within each QTL region was assessed with r^2 [55], implemented using the program TASSEL (http://www.maizegenetics.net). All sites where the minor allele frequency was <10%, or where more than two alleles at a SNP site were excluded. Accessions were also excluded from the analyses if they have missing data or gaps in one or both of the SNP sites. We treated heterozygous SNPs as previously described [35]. Heterozygous sites are rare in *O. sativa* individuals, but more frequent in *O. rufipogon* SNP genotypes. We excluded an individual from analysis if it was heterozygous at both SNP sites in a pair so that only unambiguous haplotypes were used in the analysis. In *O. rufipogon*, the majority of the SNP pairs containing individuals with double heterozygotes had only a single doubly heterozygous individual to exclude. To generate a graphical display of pairwise LD measurements, a script written by Shin *et al.* [56] was run in R.

Coalescent simulations

Coalescent theory allows us to trace the evolution of alleles in a population sample to a single ancestral copy, and provides a framework to test whether SNP data from a sample is consistent with neutral evolution [57]. We assessed the statistical significance of reduced genetic variation in the three QTL regions using coalescent simulations based on a demographic model previously inferred from genome-wide patterns of nucleotide variation [31]. Maximum composite-likelihood parameters for this model had been previously estimated using the joint site-frequency spectrum for the genome-wide data [31].

The model has the following features: The ancestral species *O. rufipogon* is assumed to have a constant population size, N_A, which is a reasonably good fit for the observed genome-wide site-frequency spectrum. Based on a previous study [31], we assume that *indica* and *tropical japonica* split simultaneously from *O. rufipogon* and formed separate populations $4*N_A*0.1$ generations ago with each undergoing a bottleneck and then post-bottleneck growth. The bottleneck model for the *indica* lineage is as follows (looking back in time): from present to $4*N_A*0.04$ generations ago, we set Ne for *indica* $= 0.27*N_A$ to model post-bottleneck growth; from $4*N_A*0.04$ to $4*N_A*0.1$ generations ago, we set Ne $= 0.0055*N_A$ corresponding to the domestication bottleneck; and prior to this time, Ne $= N_A$, to reflect common ancestry with *O. rufipogon*. For *tropical japonica*, the corresponding parameters are as follow: from present to $4*N_A*0.038104$ generations ago (post-bottleneck growth in *japonica*), we set Ne $= 0.12*N_A$; from $4*N_A*0.038104$ to $4*N_A*0.1$ generations ago (domestication bottleneck in *japonica*), Ne $= 0.0055*N_A$; and, prior to $0.01*4*N_A$, Ne for *japonica* $= N_A$. It is important to note that this model allows for migration among the three populations. Specifically, in each generation, an average of 7 migrants enter the *O. rufipogon* population (equally from the other two populations), based on results from the previous study [31]. Both *indica* and *tropical japonica* receive migrants at a rate proportional to their relative population size at each generation with 0.0385 migrants during the bottleneck, 1.89 and 0.84 migrants entering *indica* and *tropical japonica*, respectively, after the bottleneck.

In the simulations we also accounted for the possible impact of local variation in mutation rate scaled on population size. Specifically, for each of the three QTL regions we estimated the baseline mutation rate for each of the three QTL regions using the *O. rufipogon* sequence. Previous work and the observed distribution of variable nucleotide frequencies in the present study suggests that *O. rufipogon* demography is accurately described by the standard neutral model so that Watterson's estimate of the mutation rate (θ_W) is an appropriate summary statistic from which to estimate this quantity [31]. We simulated 1,000 replicate data sets for each QTL region separately using the demographic model described above. In order to assess significance, we tallied the number of simulated data sets that show as little as or less diversity than the observed for each subgroup and for each QTL region.

Phylogenetic analysis

Fragments within each QTL region were concatenated, and neighbor joining analyses using an improved algorithm [58] were performed in PAUP v4.0 beta Win [59] on each concatenated dataset with K2P distance correction and gamma setting. Negative branch lengths were prohibited. Strict consensus trees were rooted using the outgroup species *O. meridionalis*. Bootstrap (BS) support values were obtained in PAUP using 500 replicates applying the NJ search option. Tree files were viewed using the program of FigTree v1.2.2 by A. Rambaut (http://tree.bio.ed.ac.uk/software/figtree/).

Supporting Information

Table S1　Rice accessions used in this study.

Table S2　Genes associated with sequenced fragments at each QTL and their function.

Table S3　Primers used to amplify the gene fragments used in this study.

Acknowledgments

We would like to thank Ana Caicedo for providing us the sequences of previously published genome-wide STS fragments, and Susan R. McCouch, Barbara A. Schaal and the Genetic Resources Center of the International Rice Research Institute for providing seed material and/or DNA. We would also like to thank Adi Fledel-Alon for her contribution in developing some of the scripts used for analyses in our study.

Author Contributions

Conceived and designed the experiments: MDP XX. Performed the experiments: XX. Analyzed the data: XX JM RH AR ARB CDB. Wrote the paper: XX MDP.

References

1. Purugganan MD, Fuller DQ (2009) The nature of selection during plant domestication. Nature 457: 843–848.
2. Glémin S, Bataillon T (2009) A comparative view of the evolution of grasses under domestication. New Phytol 183: 273–290.
3. Yamasaki M, Tenaillon MI, Bi IV, Schroeder SG, Sanchez-Villeda H, et al. (2005) A large-scale screen for artificial selection in maize identifies candidate agronomic loci for domestication and crop improvement. Plant Cell 17: 2859–2872.
4. Doebley JF, Gaut BS, Smith BD (2006) The molecular genetics of crop domestication. Cell 12: 1309–1321.
5. Burke JM, Burger JC, Chapman MA (2007) Crop evolution: from genetics to genomics. Curr Op Genet Dev 17: 525–532.
6. Doebley JF, Stec A, Wendel J, Edwards M (1990) Genetic and morphological analysis of a maize-teosinte F2 population: Implications for the origin of maize. Proc Natl Acad Sci USA 87: 9888–9892.
7. Doebley JF, Stec A (1993) Inheritance of the morphological differences between maize and teosinte: Comparison of results for two F2 populations. Genetics 134: 559–570.
8. Peng JH, Ronin Y, Fahima, T, Roder MS, Li YC, et al. (2003) Domestication quantitative trait loci in *Triticum dicoccoides*, the progenitor of wheat. Proc Natl Acad Sci USA 100: 2489–2494.
9. Poncet V, Lamy F, Enjalbert J, Jol H, Sarr A, et al. (1998) Genetic analysis of the domestication syndrome in pearl millet (*Pennisetum glaucum* L., Poaceae): Inheritance of the major characters. Heredity 81: 648–658.
10. Doust A, Devos KM, Gadberry M, Gale M, Kellog EA (2005) The genetic basis for inflorescence variation between foxtail and green millet (Poaceae). Genetics 169: 1659–1672.
11. Cai HW, Morishima H (2002) QTL clusters reflect character associations in wild and cultivated rice. Theor Appl Genet 104: 1217–1228.
12. Thomson MJ, Tai T, McClung AM, Lai XH, Hinga ME, et al. (2003) Mapping quantitative trait loci for yield, yield components and morphological traits in an advanced backcross population between *Oryza rufipogon* and the *Oryza sativa* cultivar Jefferson. Theor Appl Genet 107: 479–493.

13. Septiningsih EM, Prasetiyono J, Lubis E, Tai TH, Tjubaryat T, et al. (2003) Identification of quantitative trait loci for yield and yield components in an advanced backcross population derived from the *Oryza sativa* variety IR64 and the wild relative *Oryza rufipogon*. Theor Appl Genet 107: 1419–1432.
14. Clark R, Linton E, Messing J, Doebley JF (2004) Pattern of diversity in the genomic region near the maize domestication gene *tb1*. Proc Natl Acad Sci USA 101: 700–707.
15. Li CB, Zhou AL, Sang T (2006) Rice domestication by reducing shattering. Science 311: 1936–1939.
16. Konishi S, Izawa T, Lin SY, Ebana K, Fukuta Y (2006) A SNP caused loss of seed shattering during rice domestication. Science 312: 1392–1396.
17. Maynard-Smith J, Haigh J (1974) The hitchhiking effect of a favorable gene. Genetic Res 23: 23–35.
18. Wright SI, Bi IV, Schroeder SG, Yamasak M, Doebley JF, et al. (2005) The effects of artificial selection of the maize genome. Science 308: 1310–1314.
19. Camus-Kulandaivelu L, Chevin LM, Tollon-Cordet C, Charcosset A, Manicacci D, et al. (2008) Patterns of molecular evolution associated with two selective sweeps in the *tb1–dwarf8* region in maize. Genetics 180: 1107–1121.
20. Shomura A, Izawa T, Ebana K, Ebitani T, Kanegae H, et al. (2008) Deletion in a gene associated with grain size increased yields during rice domestication. Nature Genetics 40: 1023–1028.
21. Londo JP, Chiang YC, Hung KH, Chiang TY, Schaal BA (2006) Phylogeography of Asian wild rice, *Oryza rufipogon*, reveals multiple independent domestications of cultivated rice, *Oryza sativa*. Proc Natl Acad Sci USA 103: 9578–9583.
22. Saisho D, Purugganan MD (2007) Molecular phylogeography of domesticated barley traces expansion of agriculture in the Old World. Genetics 177: 1765–1776.
23. Sweeney M, Thomson MJ, Cho, YG, Park YJ, Williamson S, et al. (2007) Global dissemination of a single mutation conferring white pericarp in rice. PloS Genetics 3: e133.
24. Shimamoto K, Kyozuka J (2002) Rice as a model for comparative genomics of plants. A Rev Plant Bio 153: 399–419.

25. Oka H (1988) Origin of Cultivated Rice. Tokyo: Elsevier Science.

26. Garris AJ, Tai TH, Cobur J, Kresovich S, McCouch SR (2005) Genetic structure and diversity in *Oryza sativa* L. Genetics 169: 1631–1638.

27. Monna L, Ohta R, Masuda H, Koike A, Minobe Y (2006) Genome-wide searching for single-nucleotide polymorphisms among eight distantly and closely related rice cultivars (*Oryza sativa* L.) and a wild accession (*Oryza rufipogon* Griff.). DNA Res 13: 43–51.

28. Matsuo T, Futsuhara Y, Kikuchi F, Yamaguchi H (1997) Biology of the Rice Plant. Tokyo: Food and Agricultural Policy Research Center.

29. Wang ZY, Second G, Tanksley SD (1992) Polymorphism and phylogenetic relationships among species in the genus *Oryza* as determined by analysis of nuclear RFLPs. Theor Appl Genet 83: 565–581.

30. Vitte C, Ishii T, Lamy F, Brar D, Panaud O (2004) Genomic paleontology provides evidence for two distinct origins of Asian rice (*Oryza sativa* L.). Mol Gen Genomics 272: 504–511.

31. Cacedo AL, Williamson SH, Hernandez RD, Boyko A, Fledel-Alon A, et al. (2007) Genome-wide patterns of nucleotide polymorphism in domesticated rice. PLoS Genetics 3: 1745–1756.

32. Oka HI, Morishima H (1982) Phylogenetic differentiation of cultivated rice, potentiality of wild progenitors to evolve the indica and japonica types of rice cultivars. Euphytica 31: 41–50.

33. Vaughan DA, Lu BR, Tomooka N (2008) The evolving story of rice evolution. Plant Science 174: 394–408.

34. Fuller DQ, Qin L, Zheng Y, Zhao Z, Chen X, et al. (2009) The domestication process and domestication rate in rice: Spikelet bases from the Lower Yangtze. Science 323: 1607–1610.

35. Mather KA, Caicedo AL, Polato NR, Olsen KM, McCouch SR, et al. (2007) The extent of linkage disequilibrium in rice (*Oryza sativa* L.). Genetics 177: 2223–2232.

36. Olsen KM, Caicedo AL, Polato N, McClung A, McCouch SR, et al. (2006) Selection under domestication: Evidence for a sweep in the rice *waxy* genomic region. Genetics 173: 975–983.

37. Darwin C (1859) On the origin of species by means of natural selection. London: J Murray.

38. Palaisa K, Morgante M, Tingey S, Rafalski A (2004) Long-range patterns of diversity and linkage disequilibrium surrounding the maize *Y1* gene are indicative of an asymmetric selective sweep. Proc Natl Acad Sci USA 101: 9885–9890.

39. Innan H, Kim Y (2004) Pattern of polymorphism after strong artificial selection in a domestication event. Proc Natl Acad Sci USA 101: 10667–10672.

40. Langerhans RB, DeWitt T (2004) Shared and unique features of evolutionary diversification. Amer Nat 164: 335–349.

41. Paterson AH, Lin YR, Li ZK, Schertz KF, Doebley JF, et al. (1995) Convergent domestication of cereal crops by independent mutations at corresponding genetic loci. Science 269: 1714–1718.

42. Paterson AH (2002) What has QTL mapping taught us about plant domestication? New Phytol 154: 591–608.

43. Onishi K, Takagi K, Kontani M, Tanaka T, Sano Y (2007) Different patterns of genealogical relationships found in the two major QTLs causing reduction of seed shattering during rice domestication. Genome 50: 757–766.

44. Nielsen R (2005) Molecular signatures of natural selection. Ann Rev Genet 39: 197–218.

45. Luikart G, England PR, Tallmon D, Jordan S, Taberlet P (2003) The power and promise of population genomics: From genotyping to genome typing. Nat Rev Genet 4: 981–994.

46. Harr B, Kauer M, Schlotterer C (2002) Hitchhiking mapping: A population-based fine-mapping strategy for adaptive mutations in *Drosophila melanogaster*. Proc Natl Acad Sci USA 99: 12949–12954.

47. Pollinger JP, Bustamante CD, Fledel-Alon A, Schmutz S, Gray MM, et al. (2005) Selective sweep mapping of genes with large phenotypic effects. Genome Res 15: 1809–1819.

48. Kohn MH, Pelz HJ, Wayne RK (2000) Natural selection mapping of the warfarin-resistance gene. Proc Natl Acad Sci USA 97: 7911–7915.

49. Sabeti PC, Schaffner SF, Fry B, Lohmueller J, Varilly P, et al. (2006) Positive natural selection in the human lineage. Science 312: 1614–1620.

50. Voight BF, Kudaravalli S, Wen XQ, Pritchard JK (2006) A map of recent positive selection in the human genome. PLoS Biol 4: 446–458.

51. Rozen S, Skaletsky HJ (2000) Primer3 on the WWW for general users and for biologist programmers. In: Krawetz, S, S Misener, eds. Bioinformatics methods and protocols: Methods in molecular biology. New Jersey: Humana Press.

52. International Rice Genome Sequencing Project (2005) The map-based sequence of the rice genome. Nature 436: 793–800.

53. Pritchard JK, Stephens M, Donnelly P (2000) Inference of population structure using multilocus genotype data. Genetics 155: 945–959.

54. Watterson GA (1975) On the number of segregating sites in the genetical models without recombination. Theor Pop Biol 7: 256–276.

55. Gaut BS, Long AD (2003) The lowdown on linkage disequilibrium. Plant Cell 15: 1502–1506.

56. Shin JH, Blay S, McNeney B, Graham J (2006) LDheatmap: An R function for graphical display of pairwise linkage disequilibria between single nucleotide polymorphisms. Journal of Statistical Software 16: Code Snippet 3, (http://www.jstatsoft.org).

57. Wakeley, J (2008) Coalescent Theory: An Introduction. Greenwood VillageColorado: Roberts and Company.

58. Gascuel O (1997) BIONJ: an improved version of the NJ algorithm based on a simple model of sequence data. Mol Biol Evol 14: 685–695.

59. Swofford DL (2000) PAUP* Phylogenetic analysis using parsimony (*and other methods). Sunderland, Massachusetts: Sinauer Associates.

21

Epistatic Association Mapping in Homozygous Crop Cultivars

Hai-Yan Lü[1,2], Xiao-Fen Liu[1], Shi-Ping Wei[1], Yuan-Ming Zhang[1]*

1 Section on Statistical Genomics, State Key Laboratory of Crop Genetics and Germplasm Enhancement, Nanjing Agricultural University, Nanjing, Jiangsu, China, **2** College of Information and Management Science, Henan Agricultural University, Zhengzhou, Henan, China

Abstract

The genetic dissection of complex traits plays a crucial role in crop breeding. However, genetic analysis and crop breeding have heretofore been performed separately. In this study, we designed a new approach that integrates epistatic association analysis in crop cultivars with breeding by design. First, we proposed an epistatic association mapping (EAM) approach in homozygous crop cultivars. The phenotypic values of complex traits, along with molecular marker information, were used to perform EAM. In our EAM, all the main-effect quantitative trait loci (QTLs), environmental effects, QTL-by-environment interactions and QTL-by-QTL interactions were included in a full model and estimated by empirical Bayes approach. A series of Monte Carlo simulations was performed to confirm the reliability of the new method. Next, the information from all detected QTLs was used to mine novel alleles for each locus and to design elite cross combination. Finally, the new approach was adopted to dissect the genetic basis of seed length in 215 soybean cultivars obtained, by stratified random sampling, from 6 geographic ecotypes in China. As a result, 19 main-effect QTLs and 3 epistatic QTLs were identified, more than 10 novel alleles were mined and 3 elite parental combinations, such as Daqingdou and Zhengzhou790034, were predicted.

Editor: Samuel Hazen, University of Massachusetts Amherst, United States of America

Funding: This work was supported by grant 2011CB109300 from the National Basic Research Program of China, grants 30971848 and 30671333 from the National Natural Science Foundation of China, grant KYT201002 from the Fundamental Research Funds for the Central Universities, grant B08025 from the 111 Project, and grant 20100097110035 from Specialized Research Fund for the Doctoral Program of Higher Education. The funders had no role in study design, data collection and analysis, decision to publish, or preparation of the manuscript.

Competing Interests: The authors have declared that no competing interests exist.

* E-mail: soyzhang@njau.edu.cn

Introduction

Germplasm resources play crucial roles in genetics, evolution and breeding, by forming the physical foundation of the study of genetic diversity [1]–[3], fueling much evolutionary research [4]–[6] and providing the raw material for breeders to produce new cultivars or to further improve the existing ones, due to the existence of many valuable genes in genetic resources [7]–[9]. The identification of valuable genes and markers associated with traits of interest will greatly increase the efficiency of plant breeding programs. However, these beneficial genes are largely unexplored due to the lack of appropriate statistical techniques. Meanwhile, as the complexity of the trait increase, breeding problems increase, for example, favorable alleles in exotic genetic resources are in unadapted genetic backgrounds and linked to other unfavorable alleles. This means that methods to utilize these favorable alleles in crop breeding also need to be further addressed. Accordingly, there is a critical need for in-depth study of methodologies for mining elite alleles in germplasm resources and for the utilization of these elite alleles in crop breeding.

During the past several decades, many attempts have been made to mine elite alleles for objective traits of interest. In early studies, many genes for qualitative traits in crop breeding were studied with morphological and biochemical approaches [10]–[13], and those for complex diseases in human genetics were identified by both sibling pair analysis [14]–[18] and pedigree analysis [19]–[21]. The introduction of molecular markers has facilitated the genetic association analysis of complex diseases in humans, animals and plants. Single-marker association analysis [22] and, later, genome-wide association study (GWAS) have been widely used in human genetics [23]. There has been substantial research of two aspects of GWAS: population structure [24]–[29] and mixed genetic models [30]–[32]. However, only one QTL was analyzed at a time in the above models. Likewise, although epistasis association analysis has been utilized in human genetics [33]–[37], all of the main genetic effects and gene interaction effects have not been simultaneously included in one genetic model. A full genetic model, including all the main and epistatic effects, could improve the power of QTL detection [38]–[41]. Several parameter estimation approaches such as LASSO [41], [42], empirical Bayes [43], and penalized maximum likelihood [38], [40] make this full genetic model possible. Therefore, epistasis association analysis with a full genetic model is feasible in crop germplasm resources.

In the past, most crop breeding methods were based on selection for observable phenotypes and breeding efficiency without markers is simply a function of heritability and choice of parental material. To date molecular markers have improved efficiency of selection largely for traits under simple genetic control and in specific conditions where marker selection is easier/cheaper than phenotypic selection [44]–[50]. However, this approach is only feasible for the improvement of one or several independent genes. If there are interactions among the objective genes, breeding strategy must be addressed by the incorporation of the epistasis [51], [52]. Carlborg and Haley [53] showed that epistasis is a common response to selection in breeding programs. Therefore, genetic interaction should be considered in crop breeding strategies.

One purpose of the genetic analysis of quantitative traits is to design a suitable breeding strategy, called breeding by design [54]. However, genetic analysis and crop breeding have traditionally been performed separately; for example, most genetic analyses exclusively use biparental crosses, but these are rarely used alone in commercial breeding. Therefore, the results of these biparental cross experiments have limited roles in breeding practice [55]–[57]. However, direct mapping of QTLs in natural populations, such as crop cultivars, is both economical and practical because the population being mapped is readily available, and the identified QTLs are directly applicable [31].

The purpose of this study was to develop an epistatic association mapping (EAM) approach in homozygous crop cultivars. We described detailed genetic and statistical models of epistasis association analysis in crop cultivars. All the parameters were estimated using the empirical Bayes approach. Our methods were confirmed by real data analysis in soybean and by a series of Monte Carlo simulation experiments.

Results

Phenotypic variation

We measured seed length in 215 soybean cultivars. The minimum, maximum, average, median, standard deviation, coefficient of variation, skewness and kurtosis values were 5.30, 11.85, 7.94, 7.86, 0.99, 12.43, 0.61 and 0.91, respectively. Results from ANOVA showed that there is significant difference among cultivars ($P < 10^{-4}$) and there are no significant differences between years ($P = 0.192$) and among cultivar $\times$ year interactions ($P = 0.328$). This means that in the cultivar population, there is a large amount of genetic variation, which exhibits a continuous normal distribution (**Fig. 1**).

Epistasis association mapping

Two years of phenotypic observations, along with information on 134 SSR molecular markers, were used to dissect the genetic basis of seed length in soybean. In the full model, 9,180 effects needed to be estimated, 40 times larger than the sample size. We adopted a two-stage method [58]. Nineteen main-effect QTLs and 3 epistatic QTLs for seed length in soybean were detected by EAM (Table 1). All of these QTLs were nearly evenly distributed along the soybean genome, except for chromosomes H, J and L.

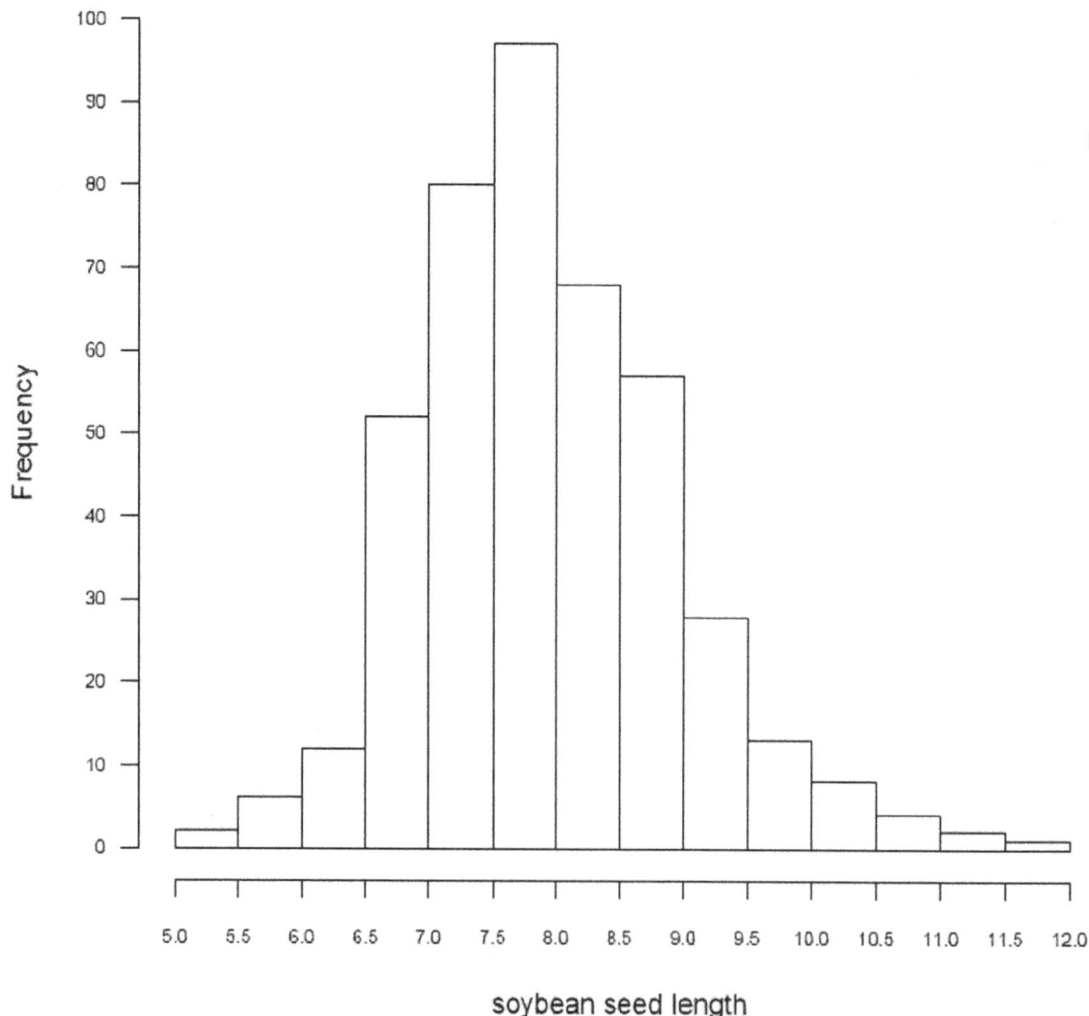

Figure 1. Frequency distribution for soybean seed length.

Among these QTLs, the proportion of the total phenotypic variance was from 0.25% to 10.44% for main-effect QTLs and from 5.08% to 7.38% for epistatic QTLs, and each of 12 QTLs contributed greater than 5.0% of the variance. In addition, five loci were involved in epistatic interactions, and only one of these five (sat_342) had a significant main effect. This lack of main effects may create difficulties in detecting epistasis with other methods.

To compare the proposed approach with regular genome-wide association study (GWAS), the GWAS was used to analyze the above dataset. Results showed that three main-effect QTL, linked with markers satt382, sat_254 and satt441, respectively, were detected (**Fig. 2a**) and no significant environmental and epistatic interactions were identified (**Fig. 2**). These results are similar to those by the proposed approach in two aspects. First, the three main-effect QTLs detected by the GWAS are also identified by the proposed method. Second, no significant environmental interaction is detected by the above two approaches. However, there are some differences as well. The main difference is that the new approach can detect more main-effect and epistatic QTLs than the GWAS.

Mining elite alleles

The allelic effects of the cultivars were evaluated for all the identified loci for soybean seed length. The reduced model that includes the total mean, the population structure, all the identified loci and the residual error was a mixed model equation. In the reduced model, the allelic effects at each locus were estimated by a maximum likelihood approach. If we want to increase the trait value, we should take the allele with the largest positive effect per main-effect QTL as novel allele. If decreasing the trait value is our selection objective, we should take the allele with the largest negative effect per main-effect QTL as novel allele. The same is true for allele combination of epistatic QTL. The summary statistics for novel allele or allele combination are given in Table 2. These results show that there is one novel allele for each main-effect locus or one novel allele combination for each epistatic QTL. For example, for the locus linked to marker satt656, all the allelic effects are showed in **Fig. 3**, and novel allele is the allele with an effect of 2.63. Similarly, for the interaction between markers sat_342 and AW277661, novel allele combination is the allele combination with an effect of 1.29. The novel allele and allele combination were found in the Zhengzhou 790034 and Guangxibayuehuang cultivars, respectively.

Predictions for elite cross combination

The elite cross combinations could be predicted from all the detected loci and their effects by using the method described below. In a hypothetical cross between two cultivars, all types of RILs would be produced. In these RILs, seed length could be predicted by the combined effects of all the detected loci. The best RIL with maximum seed length in one cross would represent the

Table 1. Detected QTL for seed length in soybean cultivar population.

QTL	New method						Genome-wide association study		
	Chr.	Marker associated	Position (cM)	Variance *	LOD	r^2 (%)	F	P-value	$-\log_{10} P$
Main-effect	A1	satt382	26.42	0.1155	4.65	6.24	4.31	3.96E-7	6.40**
	A2	satt329	110.94	0.0199	2.53	1.08	7.10	1.53E-5	4.81
	B1	satt509	32.51	0.0426	7.89	2.30	4.67	3.69E-4	3.43
	B2	sat_342	20.31	0.0246	4.81	1.33	2.28	8.35E-3	2.08
	B2	satt534	87.59	0.1934	2.65	10.44	3.05	3.16E-5	4.50
	C2	sat_252	127.00	0.0962	4.89	5.19	3.73	1.99E-6	5.70
	D1b	sat_254	46.92	0.0709	4.12	3.83	4.24	1.27E-7	6.90**
	D1b	satt274	116.35	0.0083	6.93	0.45	10.97	2.27E-5	4.64
	D2	satt514	85.69	0.1059	6.33	5.72	2.81	1.31E-5	4.88
	D2	sat_365	87.39	0.1232	15.23	6.65	3.08	1.78E-6	5.74
	E	satt263	45.40	0.0592	5.67	3.20	3.71	1.17E-2	1.93
	F	satt656	135.12	0.1007	4.71	5.44	2.47	2.29E-3	2.64
	G	satt352	50.53	0.1307	5.37	7.06	1.74	3.46E-2	1.46
	G	AF162283	87.94	0.0222	3.77	1.20	6.38	1.86E-3	2.73
	I	sat_419	98.11	0.0047	6.24	0.25	7.64	2.98E-6	5.22
	K	satt441	46.20	0.0925	6.59	5.00	5.22	1.04E-7	6.98**
	M	sat_256	74.53	0.0893	2.56	4.82	2.57	5.01E-3	2.30
	N	satt022	102.06	0.1113	11.99	6.01	2.16	2.92E-3	2.53
	O	sat_274	107.58	0.0446	2.64	2.41	2.61	5.21E-4	3.28
Epistasis	B2 & C1	sat_342 & AW277661	20.31 & 74.79	0.1367	7.71	7.38	4.04	6.72E-6	5.17
	D1a & E	sat_160 & satt411	104.28 & 12.92	0.0941	3.06	5.08	3.74	3.07E-4	3.51
	D1b & E	satt459 & satt411	118.62 & 12.92	0.1224	5.61	6.61	6.73	1.33E-3	2.88

*: Calculated by $\sum_{i=1}^{n} f_i a_i^2 - (\sum_{i=1}^{n} f_i a_i)^2$ for main-effect QTL and $\sum_{i=1}^{n} \sum_{j=1}^{m} f_{ij} a_{ij}^2 - (\sum_{i=1}^{n} \sum_{j=1}^{m} f_{ij} a_{ij})^2$ for epistatic QTL, where f is allelic frequency, a is allelic effect and n and m is the number of alleles at the ith and jth loci. The same is true for the later tables.

**: QTL identified by genome-wide association study with the critical value at the 0.05 level of significance determined by 1000 permutation experiments.

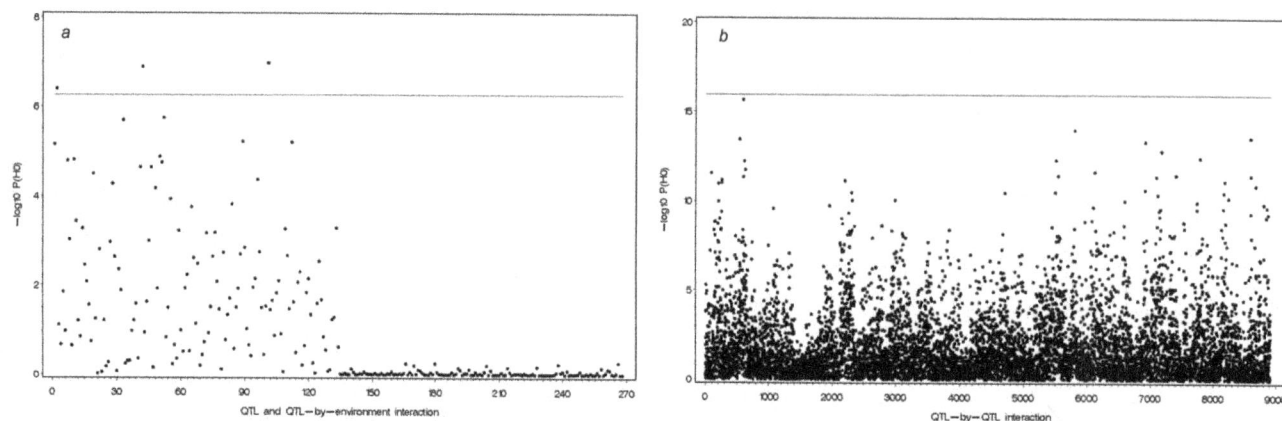

Figure 2. The −$\log_{10}$P score profile of the soybean genome scan in the genome-wide association study for seed length in soybean. (a) Main-effect QTL and QTL-by-environmental interaction, and (b) QTL-by-QTL interaction. The critical values at the 0.05 level of significance, indicated by horizontal line, were determined by 1000 permutation experiments.

cross. The best cross with maximum seed length in all the crosses could be selected by comparing all the crosses. In this study, the best three crosses were Daqingdou × Zhengzhou790034, Zhenghe- zhibanzi × Zhengzhou790034, and Liyangdawuhuang-dou × Zhengzhou 790034. The presence of Zhengzhou790034 in the three best crosses indicated that it contained the best allele or allele combination.

Monte Carlo simulation studies

Evaluation of the performance of the proposed approach. The first simulation experiment was designed to investigate the effect of QTL heritability on QTL mapping in crop cultivars. The results show that the precision and power of the detection of QTLs increase with increasing QTL heritability, and that the false positive rate (FPR) is only 0.0244% (Table S2).

In the second simulation experiment, we investigated the effect of sample size by randomly sampling 100, 200, or 300 non-founder lines. The other parameters were the same as those in the first simulation experiment. As expected, the precision and power increased with increasing sample size (Table S3). Sample sizes

under 300 yield much better results than those under 200; we recommend a sample size of 300 for future studies.

The third simulation experiment compared the effect of the number of alleles on QTL mapping in crop cultivars. We set the numbers of alleles at 2, 3 and 4; other parameters were the same as those in the first simulation experiment. The results showed that precision and power decrease as the number of alleles increases (Table S4). The results also imply that the SNP or indel markers are better than the other markers.

In the fourth simulation experiment, the effect of allelic frequency on QTL mapping was assessed by setting the frequency ratio of the two alleles as 1:1 (uniform distribution), 1:2 (skewed distribution) or 1:3 (skewed distribution). The other parameters were the same as those in the first simulation experiment. The results showed that skewed distribution decreased the statistical power (Table S5), indicating that rare alleles should be preferentially studied in association analyses.

The detection of QTL-by-environment interaction. To investigate whether environmental effects could be detected, all the cultivars were evaluated in multiple environments. In the fifth

Table 2. The information of novel allele for QTL with r^2 larger than 5%.

QTL	Chr.	Marker associated	Position (cM)	Novel allele (bp)	Effect (mm)	Cultivar with novel allele
Main-effect	A1	satt382	26.42	295	0.64	Qinyan 1
	B2	satt534	87.59	185	1.22	Zhenghezhibanzi
	C2	sat_252	127.00	276	1.00	Taixinghanludou
	D2	satt514	85.69	242	1.11	Caishengzi
	D2	sat_365	87.39	286	0.95	Dandou 2
	F	satt656	135.12	182 or 170	2.63	Zhengzhou 790034
	G	satt352	50.53	178	0.87	Ya'anguanhualiyuebao
	K	satt441	46.20	282	1.11	Nannongdahuangdou
	N	satt022	102.06	277	0.94	Dandongdaliqing
Epsitasis	B2 & C1	sat_342 & AW277661	20.31 & 74.79	288 & 301	1.29	Guangxibayuehuang
	D1a & E	sat_160 & satt411	104.28 & 12.92	190 & 109	0.99	Anbaishuidou
	D1b & E	satt459 & satt411	118.62 & 12.92	195 or 189 & 106	1.09	Zhengzhou 74064

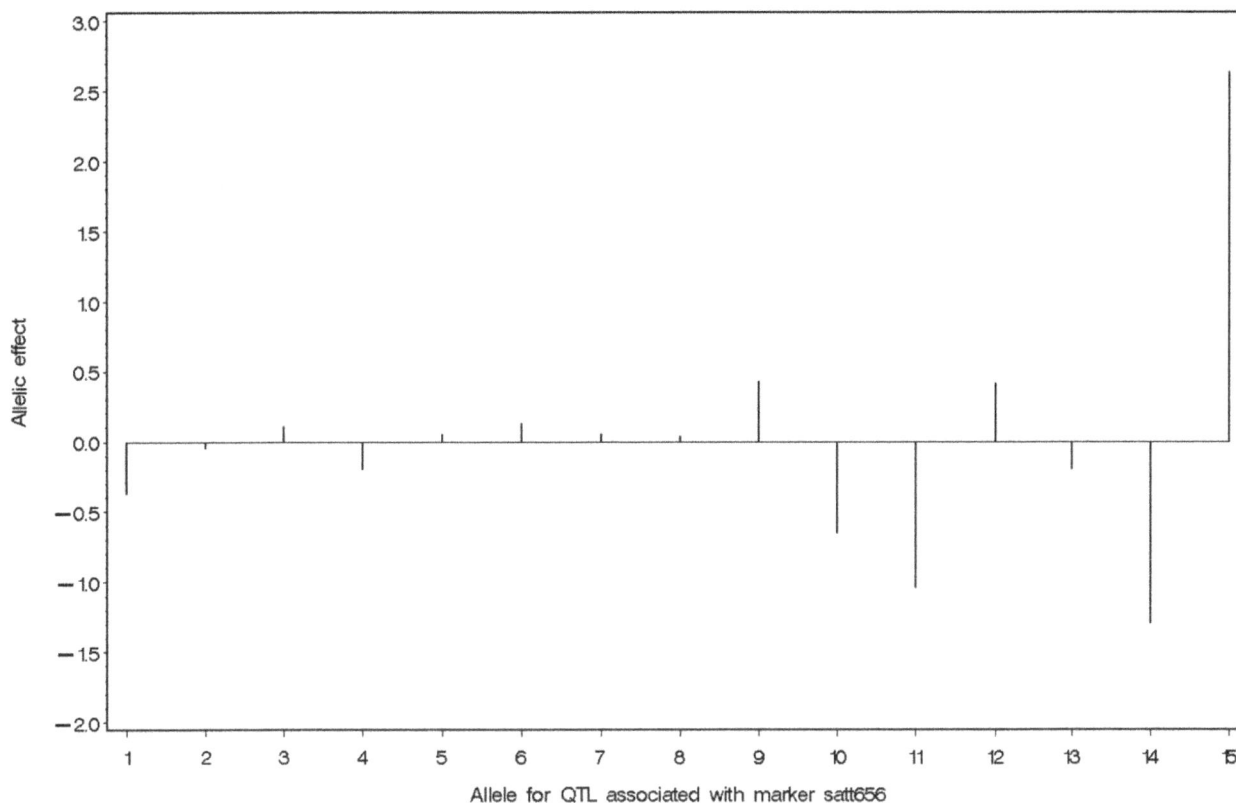

Figure 3. Allelic effects for QTL associated with marker satt656 for soybean seed length (mm).

simulation experiment, two environments, ten main-effect QTL and five QTL-by-environment interactions were simulated. The new method holds greater power for detecting QTL-by-environment interactions than for the main-effect QTL, and the FPR is lower than 0.06% (Table 3). To further demonstrate the performance of the new method, in the sixth simulation experiment, we designed a large genome with high density markers. In total, 510 markers were simulated on ten chromosome segments 1,000 cM long, with an average marker interval of 2 cM. The other parameters were the same as those in the fifth simulation experiment. The same trend in the fifth experiment was obtained (Table 4), indicating that our method works in large genomes with a high marker density.

The identification of QTL-by-QTL interaction. To demonstrate whether QTL-by- QTL interactions could be detected, all epistatic effects between two main-effect QTLs were included in the full model. In the final simulation experiment, 50 markers were evenly distributed in five linkage groups 450 cM in length. Five main-effect QTLs, 3 QTL-by-environment interactions and 5 QTL-by-QTL interactions were simulated. The results (Table 5) show that the estimates for the positions and variances of simulated QTLs are close to their true values, and the power in the detection of QTL is high (e.g., over 80% for the QTLs with a heritability over 2%), especially for QTL-by-QTL interactions.

Discussion

The approach proposed in this work has several advantages over the approaches of previous association analysis studies. First, main, environmental, QTL-by- environment and QTL-by-QTL interactions were simultaneously considered in our full genetic model, improving the statistical power [38]–[41]. Although multi-locus genetic models have been proposed in plant genetics [59]–[62], they have difficulty combining both QTL-by-environment and QTL-by-QTL interactions. Epistasis association mapping has been developed in human genetics [33]–[37], but here the epistasis was identified by two-dimensional scan, and significant effects in the two-dimensional scan were further tested in one genetic model. Second, epistasis association analysis was first integrated with crop breeding by design. In the past, the results from QTL mapping have had limited utility in breeding practice, due to the use of a simple cross population or the neglect of epistasis in the detection of QTLs. We designed an elite cross combination to take these two issues into account. Third, it is easy to extend the proposed approach to nested association analysis. The commonality is that all the individuals in the mapping populations are inbred lines. The difference is that the pedigree is general for the present study and relatively simple for nested association analysis. Therefore, the new method is suitable for nested association analysis and human genetics. Fourth, the FPR is minimized in the new method. A shrinkage estimation method, empirical Bayes (eBayes), was adopted to estimate all types of effects in the full model so that the FPR was less than 0.06%.

At present the most widely used genome-wide association study (GWAS) is analysis of variance or mixed model approaches with the control of false discovery rate. In theory, it is similar to single-marker analysis for main-effect QTL and two-marker analysis for epistatic QTL, and the difference is that the GWAS requires the setting of a significance threshold at the genome-wide level. However, it does not overcome the shortcomings of marker

Table 3. Environmental interaction detection in Monte Carlo simulation experiment (200 replicates).

QTL	True value				Estimate			
	Chr.	Position (cM)	Variance	r^2 (%)	Power (%)	Position (cM)	Variance	r^2 (%)
Main-effect	1	70.3	0.926	5.0	100.0	70.3(0.0)	0.8934(0.2176)	4.94(1.21)
		262.8	0.926	5.0	99.5	262.8(0.0)	0.8912(0.2131)	4.92(1.15)
	2	401.4	0.370	2.0	95.0	401.4(0.0)	0.3552(0.1366)	1.96(0.74)
		438.8	0.556	3.0	99.0	438.8(0.0)	0.5215(0.1589)	2.88(0.86)
	3	601.6	0.926	5.0	100.0	601.6(0.0)	0.8816(0.2125)	4.87(1.15)
	8	1653.4	0.185	1.0	58.0	1653.4(0.4)	0.2097(0.0858)	1.15(0.47)
		1747.6	0.370	2.0	93.5	1747.6(0.0)	0.3384(0.1372)	1.87(0.76)
	9	1944.7	1.852	10.0	100.0	1944.7(0.0)	1.8511(0.3121)	10.22(1.59)
	10	2145.2	0.926	5.0	100.0	2145.2(0.0)	0.9322(0.2352)	5.15(1.23)
		2181.6	0.926	5.0	100.0	2181.6(0.0)	0.9081(0.2051)	5.02(1.09)
Environment			0.926	5.0	96.0		0.8744(0.2580)	4.82(1.39)
Environmental	1	55.6	0.463	2.5	97.0	55.6(0.0)	0.4229(0.1391)	2.33(0.75)
interaction	2	401.4	0.463	2.5	98.0	401.4(0.0)	0.4465(0.1678)	2.46(0.88)
		438.8	0.926	5.0	100.0	438.8(0.0)	0.8867(0.2100)	4.90 (1.12)
	3	682.7	0.926	5.0	100.0	682.7(0.0)	0.9016(0.2190)	4.98(1.19)
	8	1747.6	1.852	10.0	100.0	1747.6(0.0)	1.8344(0.2903)	10.13(1.47)
False positive rate (%)		0.0550						

analysis. If a trait of interest is controlled by multiple QTLs, whether the QTL under consideration can be detected depends on the proportions of phenotypic variance explained by both this QTL and background QTLs. If the proportion by background QTLs is large, large residual variance will result in a decreased power in the detection of the current QTL and sometime the QTL can not be identified. In the new approach, this issue can be avoided, because a full model that includes all kinds of QTL in one genetic model results in a small residual variance. This explains why some main-effect QTLs and all the epistatic

Table 4. Environmental interaction detection under the situations of large genome and high-density markers (200 replicates).

QTL	True value				Estimate			
	Chr.	Position (cM)	Variance	r^2 (%)	Power (%)	Position (cM)	Variance	r^2 (%)
Main-effect	1	40	0.926	5.0	99.5	40.0(0.0)	0.8889(0.2126)	4.92(1.15)
		60	0.926	5.0	100.0	60.0(0.0)	0.8813(0.2233)	4.88(1.21)
	2	120	0.370	2.0	93.0	120.0(0.0)	0.3579(0.1313)	1.98(0.73)
		160	0.556	3.0	97.0	160.0(0.0)	0.5166(0.1869)	2.85(1.01)
	3	254	0.926	5.0	100.0	254.0(0.0)	0.8938(0.2097)	4.93(1.07)
	5	430	0.185	1.0	63.0	430.0(0.0)	0.1984(0.0801)	1.10(0.45)
		460	0.370	2.0	93.0	460.0(0.0)	0.3570(0.1282)	1.98(0.73)
	7	656	1.852	10.0	100.0	656.0(0.0)	1.8482(0.3380)	10.23(1.81)
	9	842	0.926	5.0	100.0	842.0(0.0)	0.9066(0.2507)	5.02(1.38)
		852	0.926	5.0	99.5	852.0(0.0)	0.8996(0.2350)	4.97(1.24)
Environment			0.926	5.0	91.5		0.9654(0.3431)	5.30(1.79)
Environmental	1	58	0.463	2.5	96.5	58.0(0.1)	0.4351(0.1290)	2.41(0.73)
interaction	2	136	0.463	2.5	95.0	136.0(0.0)	0.4469(0.1554)	2.47(0.86)
	3	254	0.926	5.0	100.0	254.0(0.0)	0.8787(0.2201)	4.86(1.18)
	5	460	0.926	5.0	100.0	460.0(0.0)	0.8878(0.2214)	4.91(1.21)
	9	842	1.852	10.0	100.0	842.0(0.0)	1.7989(0.3053)	9.95(1.59)
False positive rate (%)		0.0597						

Table 5. Epistatic QTL detection in Monte Carlo simulation experiment (200 replicates).

QTL	True value				Estimate			
	Chr.	Position (cM)	Variance	r^2 (%)	Power (%)	Position (cM)	Variance	r^2 (%)
Main-effect	1	50	0.4	2	83.5	50.0(0.0)	0.3967(0.1317)	2.04(0.66)
	2	100	1.0	5	97.5	100.0(0.0)	0.9441(0.2544)	4.88(1.31)
	3	200	2.0	10	99.5	200.0(0.0)	1.9239(0.5039)	9.90(2.35)
	4	350	0.4	2	82.0	350.0(0.0)	0.3953(0.1371)	2.03(0.70)
	5	400	1.0	5	95.5	400.0(0.0)	0.9741(0.3574)	4.98(1.71)
Environment			1.0	5	99.0		0.9408(0.2294)	4.86(1.14)
Environmental interaction	2	150	0.4	2	98.5	150.0(0.0)	0.3766(0.1255)	1.96(0.67)
	3	270	2.0	10	100.0	270.0(0.0)	1.9703(0.3007)	10.21(1.57)
	5	400	1.0	5	99.5	400.0(0.0)	0.9354(0.2261)	4.83(1.12)
Epistasis	1 & 2	10 & 130	0.4	2	97.0	10.0(1.0) & 129.9(1.4)	0.3444(0.1262)	1.78(0.65)
	2 & 3	100 & 250	1.0	5	100.0	100.0(0.0) & 250.0(0.0)	0.9825(0.2196)	5.09(1.13)
	3 & 5	200 & 400	0.4	2	85.5	200.0(0.0) & 399.9(1.5)	0.3842(0.1275)	1.98(0.66)
	3 & 4	270 & 360	2.0	10	100.0	270.1(0.7) & 360.0(1.6)	1.9350(0.3605)	9.99(1.79)
	4 & 5	350 & 450	2.0	10	100.0	350.1(0.7) & 450.0(0.0)	1.9814(0.3912)	10.25(1.98)
False positive rate (%)		0.0545						

QTLs can not be mapped in the soybean genome-wide association study.

Prediction of elite cross combination is based on the assumption that dominance and dominance-type epistasis effects are absent. If the breeding objective is the development of inbred lines or cultivars as often the case in self-pollinated crops, the prediction may be useful. If these non-additive effects are important, then the prediction would not reliable. This issue needs to be addressed in the future.

Xu [41] described a linear model in which the dimensions of the genotypic value vector and its incidence matrix depend on the number of genotypes for the locus. In theory, this model matches the situation under study. However, the model dimensions will increase rapidly. Therefore, it is preferable to gather more samples or reduce the number of effects considered [38], [63] to reduce the dimensions of the model. In this study, we designed a special incidence matrix such that there is one variable for each main-effect QTL. Simulation studies show that this approach works well. If the number of markers is large, the number of effects in the model is enormous. In this case, the two-stage method of He and Zhang [58] is recommended. We adopted this approach in our analysis of real data, and the results were consistent with those of He and Zhang [58] and He et al. [64]. The new approach works well if the marker interval length is approximately 5 cM. However, one must delete some closely linked markers if the interval length is less than 5 cM [64].

We compared the QTLs of seed length in soybeans with the QTLs in previous studies. Although few common markers existed between their data and ours, some loci that we detected were also detected in previous studies. Seven QTLs linked to markers sat_342, satt534, satt514, sat_365, sat_254, sat_419 and sat_274 in this study were detected by Xu et al. [65]; four QTLs associated with markers satt411, satt329, satt022 and AW277661 in this paper were identified by Salas et al. [66]; one QTL close to marker sat_256 was confirmed by Li et al. [67]; and one QTL next to marker satt514 was mapped by Liang et al. [68]. The above results

further confirmed the feasibility of the approach proposed in this study.

Materials and Methods

Soybean samples

We recently assembled a soybean association panel with 215 cultivars provided by the National Center for Soybean Improvement, China. All the cultivars were obtained by stratified random sampling from six geographic ecotypes in China [69], planted in three-row plots in a completely randomized design and evaluated at the Jiangpu experimental station at Nanjing Agricultural University in 2008 and 2009. The plots were 1.5 m wide and 2 m long. Five individuals and 20 seeds in the middle row of each plot were randomly picked to measure seed length by digital vernier caliper. The measurements were averaged over 20 seeds, and the mean was used in this study.

Approximately 0.3 g of fresh leaves obtained in 2008 from each cultivar was used to extract genomic DNA using the cetyltrimethylammonium bromide method as described by Lipp et al. [70]. To screen for polymorphisms among all the cultivars, PCR was performed with 134 simple sequence repeat (SSR) primer pairs. The primer sequences were obtained from the soybean database Soybase (http://www.ncbi.nlm.nih.gov). PCR was performed as described by Xu et al. [65].

Population structure

For the soybean data, the STRUCTURE program was used to investigate the population structures of all selected cultivars [26]. The number of subpopulations (K) was set from 2 to 10. In the Markov chain Monte Carlo (MCMC) Bayesian analysis for each K, the length of a Markov chain consisted of 110,000 sweeps. The first 10,000 sweeps (the burn-in period) were deleted, and thereafter, the chain was used to calculate the mean of log-likelihood. This process was repeated 20 times, and the total average for mean log-likelihood at fixed K was used. STRUCTURE analysis with 134 SSR molecular markers showed that the log-likelihood increased with the increase of the model parameter K, so a suitable number of K could not be determined. In this situation, using the ad hoc statistic ΔK, based on the rate of change in the log-probability of data between successive K values, STRUCTURE accurately detected the uppermost hierarchical level of structure [71]. Here, the ΔK value was much higher for the model parameter $K = 4$ than for other values of K. By combining this high ΔK value with knowledge of the breeding history of these cultivars, we chose a value of 4 for K. The Q matrix was calculated based on SSR markers and incorporated into the mixed model of epistasis association analysis.

Genetic model

The phenotypic value of a quantitative trait for the ith cultivar in the jth environment ($i = 1, \cdots, n; j = 1, \cdots, R$), y_{ij}, may be described by the following mixed model:

$$\mathbf{Y} = \boldsymbol{\mu} + \sum_{l=1}^{K-1} \mathbf{X}_{\mathbf{P}l}\boldsymbol{\beta}_{\mathbf{P}l} + \mathbf{X}_{\mathbf{E}}\boldsymbol{\beta}_{\mathbf{E}} + \sum_{s=1}^{m} \mathbf{Z}_{\mathbf{Q}s}\gamma_{\mathbf{Q}s} +$$
$$\sum_{s=1}^{m} \mathbf{Z}_{\mathbf{QE}s}\gamma_{\mathbf{QE}s} + \sum_{s=1}^{m-1}\sum_{t=s+1}^{m} \mathbf{Z}_{\mathbf{QQ}st}\gamma_{\mathbf{QQ}st} + \varepsilon \tag{1}$$

where $\mathbf{Y} = (y_{11}, \cdots, y_{1n}, \cdots, y_{R1}, \cdots, y_{Rn})'$; $\mathbf{X}_{\mathbf{P}} = (\mathbf{X}_{\mathbf{P}1}, \cdots, \mathbf{X}_{\mathbf{P},K-1})$ is the Q matrix for population structure; $\mathbf{X}_{\mathbf{E}}, \mathbf{Z}_{\mathbf{Q}} = (\mathbf{Z}_{\mathbf{Q}1}, \cdots, \mathbf{Z}_{\mathbf{Q}m}), \mathbf{Z}_{\mathbf{QE}} = (\mathbf{Z}_{\mathbf{QE}1}, \cdots, \mathbf{Z}_{\mathbf{QE}m})$ and $\mathbf{Z}_{\mathbf{QQ}} = (\mathbf{Z}_{\mathbf{QQ}11}, \cdots, \mathbf{Z}_{\mathbf{QQ}(m-1)m})$ are the design matrices of the environment effect, main effect, QTL-by-environment interaction effect and QTL-by-QTL interaction effect, respectively; $\boldsymbol{\beta}_{\mathbf{P}} = (\boldsymbol{\beta}_{\mathbf{P}1}, \cdots, \boldsymbol{\beta}_{\mathbf{P},K-1})', \boldsymbol{\beta}_{\mathbf{E}}, \gamma_{\mathbf{Q}} = (\gamma_{\mathbf{Q}1}, \cdots, \gamma_{\mathbf{Q}m})', \gamma_{\mathbf{QE}} = (\gamma_{\mathbf{QE}1}, \cdots, \gamma_{\mathbf{QE}m})'$ and $\gamma_{\mathbf{QQ}} = (\gamma_{\mathbf{QQ}11}, \cdots, \gamma_{\mathbf{QQ},(m-1)m})'$ are the corresponding effects; and $\boldsymbol{\mu}$ is the total average. The first three terms were viewed as fixed effects and the following three terms were considered random effects; therefore, model (1) was rewritten as

$$\mathbf{Y} = \mathbf{X}\boldsymbol{\beta} + \mathbf{Z}\boldsymbol{\gamma} + \varepsilon \tag{2}$$

where $\mathbf{X} = (\mathbf{1} \ \mathbf{X}_{\mathbf{P}} \ \mathbf{X}_{\mathbf{E}})$, $\mathbf{Z} = (\mathbf{Z}_{\mathbf{Q}} \ \mathbf{Z}_{\mathbf{QE}} \ \mathbf{Z}_{\mathbf{QQ}})$, $\boldsymbol{\beta} = (\boldsymbol{\mu}, \boldsymbol{\beta}'_{\mathbf{P}}, \boldsymbol{\beta}'_{\mathbf{E}})'$ and $\gamma = (\gamma'_{\mathbf{Q}}, \gamma'_{\mathbf{QE}}, \gamma'_{\mathbf{QQ}})'$.

Parameter estimation

Several methods exist to simultaneously estimate the parameters in model (2); for example, eBayes [41], [43]. Here, we adopted eBayes. Briefly, the parameter vector in model (2) is $\boldsymbol{\theta} = (\boldsymbol{\beta}\gamma\sigma^2)$. The priors and the likelihood are not described in detail here. The iteration process is given below.

The fixed effects were calculated by:

$$\boldsymbol{\beta}^{(t+1)} = [\mathbf{X}^T(\mathbf{V}^{(t)})^{-1}\mathbf{X}]^{-1}\mathbf{X}^T(\mathbf{V}^{(t)})^{-1}\mathbf{Y} \tag{3}$$

$$\sigma^{2(t+1)} = \frac{\sigma^{2(t)}}{n}(\mathbf{Y} - \mathbf{X}\boldsymbol{\beta}^{(t)})^T(\mathbf{V}^{(t)})^{-1}(\mathbf{Y} - \mathbf{X}\boldsymbol{\beta}^{(t)}) \tag{4}$$

where $\mathbf{V} = \sum_{j=1}^{m} \mathbf{Z}_j\mathbf{Z}_j^T\sigma_j^2 + \mathbf{I}\sigma^2$. Note that there is not an explicit solution for the estimation of σ_j^2, and it is updated by maximizing

$$L(\sigma_j^2 | \cdots) = -\frac{1}{2}\ln(\mathbf{Z}_j^T(\mathbf{V}^{(t)})^{-1}\mathbf{Z}_j(\sigma_j^2 - \sigma_j^{2(t)}) + 1) +$$
$$\frac{(\sigma_j^2 - \sigma_j^{2(t)})[(\mathbf{Y} - \mathbf{X}\boldsymbol{\beta}^{(t)})^T(\mathbf{V}^{(t)})^{-1}\mathbf{Z}_j]^2}{2(\mathbf{Z}_j^T(\mathbf{V}^{(t)})^{-1}\mathbf{Z}_j)(\sigma_j^2 - \sigma_j^{2(t)}) + 1} - \frac{1}{2}(\tau + 2)\ln\sigma_j^2 - \frac{\omega}{2\sigma_j^2} \tag{5}$$

where $\tau = -1.0$ and $\omega = 0.0005$.

The random effects, γ_j, were predicted by best linear unbiased prediction (BLUP):

$$E(\gamma_j | \mathbf{Y}) = \sigma_j^2\mathbf{Z}_j^T\mathbf{V}^{-1}(\mathbf{Y} - \mathbf{X}\boldsymbol{\beta}) \tag{6}$$

The posterior variance of γ_j is

$$\text{var}(\gamma_j | \mathbf{Y}) = \sigma_j^2(1 - \mathbf{Z}_j^T\mathbf{V}^{-1}\mathbf{Z}_j\sigma_j^2) \tag{7}$$

The proportion of phenotype variance explained by one random effect may be calculated by

$$h_j^2 \approx \sigma_{Z_j}^2\gamma_j^2 / \sigma_P^2 \tag{8}$$

Likelihood ratio test

The traditional likelihood ratio test (LRT), as described by Zhang and Xu [38], could not be performed in this study, due to an oversaturated epistatic genetic model. We proposed the following two-stage selection process to screen all the effects. In the first stage, all the effects with $|\gamma_j / \sigma| > 10^{-6}$ are picked up. In the second stage, the full model is modified so that only the effects that passed the first round of selection are included. Due to the smaller dimensionality of the reduced model, we can use the maximum likelihood method to reanalyze the data and perform the LRT. The procedure for the LRT is below.

The overall null hypothesis is no effect of the QTL at the locus of interest, denoted by $H_0 : a_1 = \cdots = a_T = 0$, where a_t is the effect of the tth allele. If we solve the maximum likelihood estimation of the parameters under the restriction of $H_0 : a_1 = \cdots = a_T = 0$ and calculate the log-likelihood value using the solutions with this restriction, we obtain $L(\hat{\theta} | H_0)$. We can also evaluate the log-likelihood value of the solutions without restrictions and obtain $L(\hat{\theta})$. Therefore, the LR test statistic is

$$\text{LR} = -2\left[L(\hat{\theta} | \mathbf{Lu} = 0) - L(\hat{\theta})\right]. \tag{9}$$

Other test statistics can be used in similar ways. The significance threshold of the LOD score was set at 2.5 for our real data analysis, where $\text{LOD} = \text{LR}/4.605$.

Genome-wide association study

First, phenotypic values for seed length in 215 soybean cultivars were corrected using population structure obtained by STRUCTURE software. Then, the corrected phenotypes along with SSR marker information were used to carry out genome-wide association studies for main-effect QTLs, environmental interactions and QTL-by-QTL interactions by ANOVA. Finally, critical values at the 0.05 level of significance were determined by 1000 permutation experiments and thus significant QTL could be identified.

Simulation design

We performed seven simulation experiments in this study. In the first, the simulated pedigree was the maize pedigree described by Zhang et al. [31], [61]. The number of inbred lines within the maize pedigree was $404(n)$. Of these, $n_0(=103)$ were base (founder) lines, which were in linkage equilibrium so that the genotypes for markers and QTLs with two alleles could be simulated. Non-founders ($n_1 = 301$) were bred via repeated self-pollination of a hybrid between two inbred lines. Thus, each non-founder line represents a recombinant inbred line (RIL) with respect to a known pair of parents. The genotypes of all the non-founders could be generated from the genotypes of their parents, analogous to simulating the genotypes of RILs from their parents. All of the non-founder lines could be used to detect QTLs. To mimic the actual linkage maps that did not have equally spaced markers, 153 markers were simulated on ten chromosome segments of length ~2258.70 cM, with an average marker interval of 14.86 cM. A total of 20 QTLs, all of which overlapped with the markers, were simulated; the sizes and locations of the QTLs are listed in Table 3. The allelic effects were calculated by relating the genetic variance of the QTL to both the allelic frequencies and the allelic number. The phenotypic value of each line was the sum of the corresponding QTL genotypic values and the residual error, with an assumed normal distribution. Each simulation run consisted of 200 replicates. For each simulated QTL, we counted the samples in which the LOD statistic surpassed 3.0. The ratio of the number of such samples (m) to the total number of replicates (200) represented the empirical power of this QTL. The false-positive rate was calculated as the ratio of the number of false-positive effects to the total number of zero effects considered in the full model. The other simulation experiments were performed similarly. All simulated parameters are given in Table S1.

Supporting Information

Table S1 Simulated parameters in all the simulation experiments.

Table S2 Multi-QTL detection under various QTL heritabilities in the first simulation experiment (200 replicates).

Table S3 Effect of sample size on multi-QTL mapping in the second simulation experiment (200 replicates).

Table S4 Effect of the number of alleles on multi-QTL mapping in the third simulation experiment (200 replicates).

Table S5 Effect of allelic distribution on multi-QTL mapping in the fourth simulation experiment (200 replicates).

Acknowledgments

We are grateful to three anonymous referees for their constructive comments and suggestions that significantly improved the presentation of the manuscript.

Author Contributions

Conceived and designed the experiments: YMZ. Performed the experiments: XFL HYL SPW. Analyzed the data: HYL. Contributed reagents/materials/analysis tools: HYL XFL SPW. Wrote the paper: YMZ HYL.

References

1. Abdalla AM, Reddy OUK, El-Zik KM, Pepper AE (2001) Genetic diversity and relationships of diploid and tetraploid cottons revealed using AFLP. Theor Appl Genet 102: 222–229.
2. Dong YS, Zhuang BC, Zhao LM, Sun H, He MY (2001) The genetic diversity of annual wild soybeans grown in China. Theor Appl Genet 103: 98–103.
3. Reif JC, Hamrit S, Heckenberger M, Schipprack W, Maurer HP, et al. (2005) Genetic structure and diversity of European flint maize populations determined with SSR analyses of individuals and bulks. Theor Appl Genet 111: 906–913.
4. Milne RI, Abbott RJ (2000) Origin and evolution of invasive naturalized material of *Rhododendron ponticum* L. in the British isles. Mol Ecol 9(5): 541–556.
5. Dillon SL, Shapter FM, Henry RJ, Cordeiro G, IzquierdoLiz, Liz LS (2007) Domestication to crop improvement: Genetic resources for Sorghum and saccharum (Andropogoneae). Annals of Botany 100: 975–989.
6. Friesen ML, von Wettberg EJ (2010) Adapting genomics to study the evolution and ecology of agricultural systems. Current Opinion in Plant Biology 13: 119–125.
7. Ellis RP, Forster BP, Robinson D, Handley LL, Gordon DC, et al. (2000) Wild barley: a source of genes for crop improvement in the 21 century? J Exp Bot 51: 9–17.
8. Upadhyaya HD, Ortiz R (2001) A mini core subset for capturing diversity and promoting utilization of chickpea genetic resources in crop improvement. Theor Appl Genet 102: 1292–1298.
9. Warburton ML, Crossa J, Franco J, Kazi M, Trethowan R, et al. (2006) Bringing wild relatives back into the family: recovering genetic diversity in CIMMYT improved wheat germplasm. Euphytica 149: 289–301.
10. Sasaki A, Ashikari M, Ueguchi-Tanaka M, Itoh H, Nishimura A, et al. (2002) Green revolution: a mutant gibberellin-synthesis gene in rice. Nature 416(6882): 701–702.
11. Zhang X-Q (2002) Three lines hybrid rice. In Shi Y-C, ed. Chinese Academic canon in the 20th century. Fuzhou: Fujian Education Press. pp 25–27.
12. Stuber CW (1995) Mapping and manipulating quantitative trait in maize. Trends in Genetics 11: 477–481.
13. Tanksley SD, Nelson JC (1996) Advanced backcross QTL analysis: a method for the simultaneous discovery and transfer of valuable QTLs from unadapted germplasm into elite breeding lines. Theor Appl Genet 92: 191–203.
14. Haseman JK, Elston RC (1972) The investigation of linkage between a quantitative trait and a marker locus. Behav Genet 2: 3–19.
15. Wright FA (1997) The phenotypic difference discards sib-pair QTL linkage information. Am J Hum Genet 60: 740–742.
16. Drigalenko E (1998) How sib pairs reveal linkage. Am J Hum Genet 63: 1242–1245.
17. Forrest W (2001) Weighting improves the "new Haseman-Elston" method. Hum Hered 52: 47–54.
18. Sham PC, Purcell S (2001) Equivalence between Haseman-Elston and variance components linkage analyses for sib pairs. Am J Hum Genet 68: 1527–1532.
19. Sham PC, Purcell S, Cherny SS, Abecasis GR (2002) Powerful regression -based quantitative trait linkage analysis of general pedigrees. Am J Hum Genet 71(2): 238–253.
20. Chen WM, Broman KW, Liang KY (2004) Quantitative trait linkage analysis by generalized estimating equations: unification of variance components and Haseman-Elston regression. Genet Epidemiol 26(4): 265–272.
21. Wang T, Elston RC (2005) Two-level Haseman-Elston regression for general pedigree data analysis. Genet Epidemiol 29(1): 12–22.
22. Sax K (1923) The association of size difference with seed-coat pattern and pigmentation in *Phaseolus vulgaris*. Genetics 8: 552–560.
23. Risch N, Merikangas K (1996) The future of genetic studies of complex human diseases. Science 273: 1516–1517.
24. Smouse PE, Waples RS, Tworek JA (1990) A genetic mixture analysis for use with incomplete source population-data. Can J Fish Aquat Sci 47: 620–634.
25. Balding DJ, Nichols RA (1995) A method for quantifying differentiation between populations at multi-allelic loci and its implications for investigating identity and paternity. Genetica 96: 3–12.
26. Pritchard JK, Stephens M, Donnelly P (2000) Inference of population structure using multilocus genotype data. Genetics 155: 945–959.
27. Marchini J, Cardon LR, Phillips MS, Donnelly P (2004) The effects of human population structure on large genetic association studies. Nat Genet 36: 512–517.
28. Zhu CS, Yu JM (2009) Nonmetric multidimensional scaling corrects for population structure in association mapping with different sample types. Genetics 182: 875–888.

29. Li MY, Reilly MP, Rader DJ, Wang LS (2010) Correcting population stratification in genetic association studies using a phylogenetic approach. Bioinformatics 26(6): 798–806.

30. Diao G, Lin DY (2005) A powerful and robust method for mapping quantitative trait loci in general pedigrees. Am J Hum Genet 77: 97–111.

31. Zhang Y-M, Mao YC, Xie C, Smith H, Luo L, et al. (2005) Mapping QTL using naturally occurring genetic variance among commercial inbred lines of maize (Zea mays L.). Genetics 169: 2267–2275.

32. Yu J, Pressoir G, Briggs WH, Vroh Bi I, Yamasaki M, et al. (2006) A unified mixed-model method for association mapping that accounts for multiple levels of relatedness. Nat Genet 38(2): 203–208.

33. Chatterjee N, Kalaylioglu Z, Moslehi R, Peters U, Wacholder S (2006) Powerful multilocus tests of genetic association in the presence of gene-gene and gene-environment interactions. Am J Hum Genet 79(6): 1002–1016.

34. Chen X, Liu CT, Zhang MZ, Zhang HP (2007) A forest-based approach to identifying gene and gene-gene interactions. Proc Natl Acad Sci USA 104: 19199–19203.

35. Zhang Y, Liu JS (2007) Bayesian inference of epistatic interactions in case-control studies. Nat Genet 39: 1167–1173.

36. Phillips P (2008) Epistasis — the essential role of gene interactions in the structure and evolution of genetic systems. Nat Rev Genet 9: 855–867.

37. Wan X, Yang C, Yang Q, Yang NLS, et al. (2009) MegaSNPHunter: a learning approach to detect disease predisposition SNPs and high level interactions in genome wide association study. BMC Bioinformatics 10: 13.

38. Zhang Y-M, Xu S (2005) A penalized maximum likelihood method for estimating epistatic effects of QTL. Heredity 95: 96–104.

39. Xu S, Jia Z (2007) Genome-wide analysis of epistatic effects for quantitative traits in Barley. Genetics 175: 1955–1963.

40. Hoggart CJ, Whittaker JC, De Iorio M, Balding DJ (2008) Simultaneous analysis of all SNPs in genome-wide and re-sequencing association studies. PLoS Genet 4(7): e1000130.

41. Xu S (2010) An expectation–maximization algorithm for the Lasso estimation of quantitative trait locus effects. Heredity 105: 483–494.

42. Tibshirani R (1996) Regression shrinkage and selection via the lasso. Journal of the Royal Statistical Society Series B 58: 267–288.

43. Xu S (2007) An empirical Bayes method for estimating epistatic effects of quantitative trait loci. Biometrics 63: 513–521.

44. Bernardo R (2008) Molecular markers and selection for complex traits in plants: learning from the last 20 years. Crop Sci 48: 1649–1664.

45. Holland JB (2004) Implementation of molecular markers for quantitative traits in breeding programs - challenges and opportunities. In New directions for a diverse planet, Proceedings of the 4th International Crop Science Congress, 26 Sep - 1 Oct 2004, Brisbane, Australia. Published on CDROM. Web site http://www.cropscience.org.au/.

46. Michelmore RW, Paran I, Kesseli RV (1991) Identification of markers linked to disease resistance genes by bulked-segregant analysis: a rapid method to detect markers in specific genomic regions by using segregating populations. Proc Natl Acad Sci USA 88: 9828–9832.

47. Cho YG, Eun MY, McCouch SR, Chae YA (1994) The semidwarf gene, sd-1, of rice (Oryza sativa L.).II. Molecular mapping and marker-assisted selection. Theor Appl Genet 89: 54–59.

48. Mohan M, Nair S, Bhagwat A, Krishna TG, Yano M, et al. (1997) Genome mapping, molecular markers and marker-assisted selection in crop plants. Molecular Breeding 3: 87–103.

49. Ribaut JM, Betrán J (1999) Single large-scale marker-assisted selection (SLS-MAS). Molecular Breeding 5: 531–541.

50. Zhang TZ, Yuan Y, Yu J, Guo WZ, Kohel RJ (2003) Molecular tagging of a major QTL for fiber strength in upland cotton and its marker-assisted selection. Theor Appl Genet 106: 262–268.

51. Jahufer MZZ, Cooper M, Ayres JF, Bray RA (2002) Identification of research to improve the efficiency of breeding strategies for white clover in Australia: A review. Australian Journal of Agricultural Research 53(3): 239–257.

52. Dwivedi SL, Crouch JH, Mackill DJ, Xu YB, Blair MW, et al. (2007) The molecularization of public sector crop breeding: Progress, problems, and prospects. Advances in Agronomy 95: 163–318.

53. Carlborg Ö, Haley CS (2004) Epistasis: too often neglected in complex trait studies? Nat Rev Genet 5: 618–625.

54. Peleman JD, van der Voort JR (2003) Breeding by design. Trends in Plant Sci 8: 330–334.

55. Liu YF, Zeng ZB (2000) A general mixture model approach for mapping quantitative trait loci from diverse cross designs involving multiple inbred lines. Genet Res 75: 345–355.

56. Blanc G, Charcosset A, Mangin B, Gallais A, Moreau L (2006) Connected populations for detecting quantitative trait loci and testing for epistasis: an application in maize. Theor Appl Genet 113: 206–224.

57. Verhoeven KJF, Jannink JL, Mcintyre LM (2006) Using mating designs to uncover QTL and the genetic architecture of complex traits. Heredity 96: 139–149.

58. He XH, Zhang Y-M (2008) Mapping epistatic quantitative trait loci underlying endosperm traits using all markers on the entire genome in a random hybridization design. Heredity 101: 39–47.

59. Iwata H, Uga Y, Yoshioka Y, Ebana K, Hayashi T (2007) Bayesian association mapping of multiple quantitative trait loci and its application to the analysis of genetic variation among Oryza sativa L. germplasm. Theor Appl Genet 114: 1437–1449.

60. Iwata H, Ebana K, Fukuoka S, Jannink J-L, Hayashi T (2009) Bayesian multilocus association mapping on ordinal and censored traits and its application to the analysis of genetic variation among Oryza sativa L. germplasm. Theor Appl Genet 118: 865–880.

61. Zhang Y-M, Lü H-Y, Yao L-L (2008) Multiple quantitative trait loci Haseman-Elston regression using all markers on the entire genome. Theor Appl Genet 117: 683–690.

62. Lü H-Y, Li M, Li G-J, Yao L-L, Zhang Y-M (2009) Multiple loci in silico mapping in inbred lines. Heredity 103: 346–354.

63. Hoti F, Sillanpää MJ (2006) Bayesian mapping of genotype×expression interaction in quantitative and qualitative traits. Heredity 97: 4–18.

64. He X-H, Qin H, Hu Z, Zhang T, Zhang Y-M (2011) Mapping of epistatic quantitative trait loci in four-way crosses. Theor Appl Genet 122: 33–48.

65. Xu Y, Li HN, Li GJ, Wang X, Cheng LG, et al. (2011) Mapping quantitative trait loci for seed size traits in soybean (Glycine max L. Merr.). Theor Appl Genet 122: 581–594.

66. Salas P, Oyarzo-Llaipen JC, Wang D, Chase K, Mansur L (2006) Genetic mapping of seed shape in three populations of recombinant inbred lines of soybean (Glycine max L. Merr.). Theor Appl Genet 113: 1459–1466.

67. Li CD, Jiang HW, Zhang WB, Qiu PC, Liu CY, et al. (2008) QTL analysis of seed and pod traits in soybean. Molecular Plant Breeding 6: 1091–1100.

68. Liang HZ, Wang SF, Yu YL, Wang TF, Gong PT, et al. (2008) Mapping quantitative trait loci for six seed shape traits in soybean. Henan Agricultural Science 45: 54–60.

69. Wang YS, Gai JY (2002) Study on the ecological regions of soybean in China II. Ecological environment and representative varieties. Chinese Journal of Applied Ecology 13: 71–75.

70. Lipp M, Brodmann P, Pietsch K, Pauwels J, Anklam E, et al. (1999) IUPAC collaborative trail study of a method to detect genetically modified soybeans and maize in dried powder. Journal of AOAC International 82: 923–928.

71. Evanno G, Regnaut S, Goudet J (2005) Detecting the number of clusters of individuals using the software STRUCTURE: a simulation study. Molecular Ecology 14: 2611–2620.

22

A Dynamic and Complex Network Regulates the Heterosis of Yield-Correlated Traits in Rapeseed (*Brassica napus* L.)

Jiaqin Shi[9], Ruiyuan Li[9], Jun Zou, Yan Long, Jinling Meng*

National Key Laboratory of Crop Genetic Improvement, Huazhong Agricultural University, Wuhan, Hubei, China

Abstract

Although much research has been conducted, the genetic architecture of heterosis remains ambiguous. To unravel the genetic architecture of heterosis, a reconstructed F_2 population was produced by random intercross among 202 lines of a double haploid population in rapeseed (*Brassica napus* L.). Both populations were planted in three environments and 15 yield-correlated traits were measured, and only seed yield and eight yield-correlated traits showed significant mid-parent heterosis, with the mean ranging from 8.7% (branch number) to 31.4% (seed yield). Hundreds of QTL and epistatic interactions were identified for the 15 yield-correlated traits, involving numerous variable loci with moderate effect, genome-wide distribution and obvious hotspots. All kinds of mode-of-inheritance of QTL (additive, A; partial-dominant, PD; full-dominant, D; over-dominant, OD) and epistatic interactions (additive × additive, AA; additive × dominant/dominant × additive, AD/DA; dominant × dominant, DD) were observed and epistasis, especially AA epistasis, seemed to be the major genetic basis of heterosis in rapeseed. Consistent with the low correlation between marker heterozygosity and mid-parent heterosis/hybrid performance, a considerable proportion of dominant and DD epistatic effects were negative, indicating heterozygosity was not always advantageous for heterosis/hybrid performance. The implications of our results on evolution and crop breeding are discussed.

Editor: Bengt Hansson, Lund University, Sweden

Funding: Financial support for this work was provided by the National Basic Research and Development Program (2006CB101600). The funders had no role in study design, data collection and analysis, decision to publish, or preparation of the manuscript.

Competing Interests: The authors have declared that no competing interests exist.

* E-mail: jmeng@mail.hzau.edu.cn

[9] These authors contributed equally to this work.

Introduction

Heterosis is defined as the superior performance of crossbred characteristics as compared with corresponding inbred ones [1]. The utilization of heterosis has become a major strategy to increase the productivity of plants and animals [2]. Despite the successful utilization of heterosis in many crops, there still exists a contradiction between the agricultural practice of heterosis utilization and our understanding of the genetic basis of heterosis and this hampers the effective exploitation of this biological phenomenon [3].

The classical quantitative genetic explanation of heterosis centered on three hypotheses: dominance, over-dominance and epistasis [4,5]. Evidence of these genetic models remained unavailable until very recent advances in molecular marker technology, high-density linkage maps and genome sequencing. Although much research into the genetic basis of heterosis in crops and plants has been conducted, little consensus has emerged. Research has indicated that heterosis may be attributable to dominance, over-dominance, epistasis or a combination of all of these, depending on the study materials, traits and analytical approach. Typically, little is known about the genetic control of heterosis in the complex polyploid crop rapeseed (*Brasscia napus* L.). Based on the phenotype of the E×R53-DH population and the

corresponding BC population, as well as the mid-parent heterosis of the BC population, Radoev et al. (2008) mapped 33 QTL (9 of which showed a significant dominant effect) and a large number of epistatic interactions for seed yield and the three yield-component traits. They concluded that epistasis together with all levels of dominance from partial to over-dominance is responsible for the expression of heterosis in rapeseed [6]. Based on this E×R53-DH population and another E×V8-DH population with the same parent, and using the same experimental design, Basunanda et al. (2010) detected a number of QTL hotspots responsible for seedling biomass and yield-related traits. Given the key role of epistatic interactions in the expression of heterosis in oilseed rape, they supposed that these QTL hotspots might harbour genes involved in regulation of heterosis for different traits throughout the plant life cycle, including a significant overall influence on heterosis for seed yield [7]. However, in both studies, all kinds of genetic effects (A, D and AA, AD/DA, DD) were unable to be estimated in the same population, thus it was difficult to accurately estimate their mode-of-inheritance and relative importance in the expression of heterosis.

There were several common patterns described in most of these studies. Firstly, the QTL for yield and yield-correlated traits tended to be clustered in the genome in many crop and model plants, such as rice [8], maize [9], wheat [10], rapeseed [7] and

Arabidopsis [11], which suggested the QTL of yield-correlated traits might have pleiotropic effects. However, this kind of pleiotropy has not been well analyzed genetically. Secondly, only a few limited traits were investigated and only a few QTL and epistatic interactions were identified for each trait, so a relatively comprehensive picture of the genetic architecture of heterosis remained unavailable. Thirdly, trials were carried out in only one or two environments and the environmental response of QTL and epistatic interactions for heterosis was not analyzed and thus remains unclear.

The main objective of this study was to unravel the genetic architecture of heterosis with QTL mapping in rapeseed, including: (1) determine the level of heterosis for a range of yield-correlated traits; (2) investigate the relationship between molecular marker heterozygosity and heterosis/hybrid performance; (3) identify QTL and epistatic interactions underlying heterosis and estimate their genetic effect, mode-of-inheritance and environmental responses; (4) analyze the relative contribution of all kinds of genetic effects in the expression of heterosis in rapeseed (*Brassica napus* L.).

Results

Correlation of trait performance and mid-parent heterosis among the 15 investigated traits

In the same environment, most pair-wise genetic correlations of performance and mid-parent heterosis were similar (Table S1A–C). This was understandable since mid-parent heterosis was calculated from trait performance. In different environments, pair-wise genetic correlations differed considerably (mostly in degree, a few in direction), which suggested that genetic correlations depended strongly on the environments.

Genetic correlations of performance and mid-parent heterosis among the investigated traits were also calculated across the three

environments (Table 1). In general, significant correlations were observed for 81.9% and 67.6% of the pair-wise combinations of the trait performance and mid-parent heterosis, respectively. Seed yield correlated significantly with the other 14 investigated traits for both trait performance and mid-parent heterosis; negatively for flowering time, maturity time and protein content, and positively for the other 11 ones. Interestingly, the mean r^2 of trait performance was somewhat higher than that of mid-parent heterosis for most traits, ranging from 0.04 and 0.03 (for seed development times) to 0.24 and 0.20 (for seed yield), respectively.

Traits showing significant heterosis

The analysis of variance (in both populations) revealed that genotype, environment and the interaction between them had significant effect on the performance of all the 15 yield-correlated traits (Table S2A), so they were calculated separately for each environment. The broad-sense heritability of these traits ranged from 0.58 (for seed yield) to 0.90 (for flowering time), with a mean of 0.73. The two parents showed significant differences in 38 of the 43 trait-environment combinations (Table S2B). The two populations showed obvious transgressive variation for all of the trait-environment combinations. It should be noted that DH and the reconstructed-F_2 population showed over-F_1 variations for 13 (except seed yield and seed number per plant) and all of the traits respectively in all environments, which indicated that heterozygosity was not always favorable for trait performance. There was significant heterosis on F_1 and F_2 generations compared with the mean of the parents and the DH population, respectively, for the nine (branch number, biomass yield, harvest index, plant height, pod number, pod yield, seed number per pod, seed number per plant and seed yield) and eight (except branch number) traits. Interestingly, for these traits with significant heterosis, the performance of F_1 was significantly higher than the mean of the F_2 population and higher than the mean of the DH population in

Table 1. Genetic correlations of trait performance (above diagonal) and mid-parent heterosis (below diagonal) among the 15 investigated traits across three environments.

Trait§	BN	BY	DT	FT	HI	MT	OIL	PH	PN	PRO	PY	SN	SP	SW	SY	Mean r²
BN		0.32‡	-0.10*	-0.17*	0.23‡	-0.26‡	0.09	0.37‡	0.35‡	-0.17*	0.06	0.11*	0.40‡	-0.07	0.39‡	0.06
BY	0.36‡		0.01	0.05	-0.09	0.02	0.10*	0.64‡	0.58‡	-0.04	0.19†	0.12*	0.60‡	0.08	0.69‡	0.13
DT	-0.05	0.02		-0.50‡	0.00	0.41‡	-0.09	-0.11*	-0.04	0.23‡	0.09	-0.09	-0.11*	0.27‡	0.19†	0.04
FT	-0.11*	-0.09	-0.46‡		-0.52‡	0.55‡	-0.08	0.03	-0.05	0.21‡	-0.27‡	-0.16*	-0.15*	-0.18†	-0.26‡	0.08
HI	0.20†	-0.08	0.04	-0.25‡		-0.54‡	0.38‡	0.12*	0.37‡	-0.48‡	0.37‡	0.44‡	0.62‡	-0.12*	0.64‡	0.16
MT	-0.16*	-0.05	0.53‡	0.31‡	-0.31‡		-0.18†	-0.08	-0.13*	0.44‡	-0.18†	-0.25‡	-0.30‡	0.11*	-0.28‡	0.10
OIL	0.02	-0.01	0.01	-0.12*	0.31‡	-0.23‡		0.22‡	0.17*	-0.38‡	0.19†	0.35‡	0.39‡	-0.23‡	0.33‡	0.06
PH	0.45‡	0.59‡	-0.02	-0.08	0.07	-0.14*	0.07		0.39‡	-0.22‡	0.22‡	0.22‡	0.51‡	-0.01	0.56‡	0.11
PN	0.31‡	0.57‡	0.00	-0.11*	0.33‡	-0.15*	0.11*	0.32‡		-0.21‡	-0.32‡	-0.18†	0.73‡	-0.25‡	0.69‡	0.15
PRO	-0.09	0.00	0.13*	0.16*	-0.25‡	0.37‡	-0.45‡	-0.12*	-0.12*		-0.10*	-0.28‡	-0.38‡	0.26‡	-0.30‡	0.08
PY	0.07	0.17*	0.00	-0.06	0.22†	-0.02	0.04	0.14*	-0.40‡	0.02		0.78‡	0.22‡	0.36‡	0.42‡	0.10
SN	0.07	0.07	-0.09	-0.02	0.25‡	-0.10	0.04	0.11*	-0.36‡	-0.12*	0.86‡		0.49‡	-0.29‡	0.41‡	0.12
SP	0.39‡	0.64‡	-0.07	-0.16*	0.58‡	-0.25‡	0.16*	0.43‡	0.70‡	-0.24‡	0.16*	0.31‡		-0.41‡	0.90‡	0.23
SW	-0.02	0.15*	0.20†	-0.07	-0.08	0.19*	-0.05	0.05	-0.14*	0.28‡	0.29‡	-0.22†	-0.29‡		0.20†	0.05
SY	0.40‡	0.73‡	0.11*	-0.20†	0.58‡	-0.20†	0.16*	0.46‡	0.68‡	-0.14*	0.28‡	0.24‡	0.93‡	0.17*		0.24
Mean r²	0.06	0.13	0.03	0.04	0.09	0.06	0.04	0.08	0.14	0.05	0.09	0.09	0.19	0.04	0.20	

§The abbreviation of the traits, see MATERIALS AND METHODS.
*, † and ‡represent the significant level of P = 0.05, 0.01 and 0.001 respectively.

19 and all of the 25 trait-environment combinations respectively, which showed an obvious trend of inbreeding depression.

According to the significance of heterosis, the 15 yield-correlated traits could be classified into two groups: the nine traits (seed yield, seed number per plant, biomass yield, pod number, harvest index, plant height, pod yield, seed number per pod and branch number) with heterosis and the other six traits (oil content, protein content, maturity time, flowering time, seed weight and seed development time) without heterosis. It should be noted that the correlation coefficients between seed yield and the nine traits with heterosis were all higher than that between the other six traits without heterosis.

The analysis of variance revealed that genotype, environment and genotype × environment interaction had significant effect on mid-parent heterosis of the nine traits with heterosis (Table S2C), so they were calculated separately for each environment (Table 2). For hybrid F_1, seed yield and seed number per plant showed strong mid-parent heterosis, biomass yield and pod number per plant showed moderate mid-parent heterosis, while pod yield, seed number per pod, harvest index, branch number and plant height showed low mid-parent heterosis. For the reconstructed F_2 population, the amount of heterosis varied widely for these traits, from highly negative to highly positive. The average mid-parent heterosis of the reconstructed F_2 population showed similar trend with that of F_1 for the nine traits. It should be noted that in each environment the mid-parent heterosis of some (the proportion is 10.2% for seed yield in S5 environment, data not shown) combinations of reconstructed F_2 population was higher than that of F_1, but the average mid-parent heterosis in the reconstructed F_2 population was in all cases lower than that in F_1. This indicated that heterosis was generally related to the heterozygosity at the population level but poorly correlated with heterozygosity at the individual level.

It should be noted that, for these yield-correlated traits, the heritabilities (ranging from 0.40 to 0.60) of mid-parent heterosis were all lower than that (ranging from 0.58 to 0.90) of trait performance (Table S2A; Table S2C).

Correlation between heterozygosity and hybrid performance/mid-parent heterosis for the nine traits with significant heterosis

The correlation between heterozygosity and hybrid performance/mid-parent heterosis was significant for the nine traits with

Table 2. Mid-parent heterosis of F_1 and reconstructed F_2 population in three environments for the nine yield-correlated traits with significant heterosis.

Traits[§]	Environments	Mid-parent heterosis				
		F_1		reconstructed F_2		
		value %		Mean %		range %
SY	N6	1784	75.4	666	29.4	-854—2318	-32.2—126.2
	S5	866	69.7	403	30.5	-763—1143	-31.5—101.5
	S6	1026	99	309	28.5	-378—1104	-31.3—133.4
SP	N6	5057	67.3	2290	33.4	-2088—5803	-19.1—111.7
	S5	2714	70.4	1163	26.5	-2291—3964	-34.2—89.5
	S6	2577	84.8	859	26.6	-1296—3606	-27.9—99.1
BY	N6	2151	47.5	976	21.4	-1272—3354	-19.7—81.2
	S6	1329	40.5	688	19.6	-1254—2571	-37.3—87
PN	N6	179	53.7	61	20.2	-134—340	-30.8—82.5
	S5	79	29.6	42	16.9	-122—228	-34.2—79.5
	S6	106	33.1	39	18.2	-109—308	-27.8—80.6
HI	N6	5	14.6	2.3	9.1	-3.5—9.5	-9.1—39.5
	S6	6	24.8	3.7	14.1	-4.1—10.6	-7.3—52.2
PH	N6	20.6	17.7	13.6	11.2	-6.6—41.3	-5.7—36.1
	S5	18.4	13.6	8.5	6.1	-17.7—28.7	-13.4—20.1
	S6	12.3	9.4	8.6	6.4	-14.3—31.9	-9.3—26
PY	N6	1.1	15.2	0.56	8.8	-3.03—3.61	-18.5—49.4
	S5	1.6	31.5	0.48	10.1	-1.73—2.87	-16.9—43.7
	S6	1	21.9	0.42	9.4	-1.93—3.19	-27.4—57.1
SN	N6	5	21.8	2.2	11.8	-8.6—11.6	-35.8—72.0
	S5	3.7	23	1.5	10	-5.3—7	-29.9—47.1
	S6	2.7	20.6	1.2	8.6	-5.0—8.2	-33.9—62.5
BN	N6	1.34	16	0.9	10.5	-2.44—4.33	-27.3—64.8
	S5	1.52	21.6	0.53	7.7	-1.82—2.9	-27.4—45.4
	S6	0.62	11.3	0.31	5.6	-2.03—2.4	-35.2—43.6

§For the abbreviation of the traits (ordered according to their correlation coefficients with SY), see MATERIALS AND METHODS.

significant heterosis except branch number and seed number per pod (Table 3), with mean r^2 ranging from 0.001 (branch number) to 0.066 (seed yield) for the different traits, which accorded well with the heterosis level of these traits. Generally, the mean r^2 between heterozygosity and hybrid performance was similar to that between heterozygosity and mid-parent heterosis. Whereas, the mean r^2 (0.026/0.022) between special heterozygosity and hybrid performance/mid-parent heterosis was a little higher than that (0.013/0.014) of general heterozygosity and hybrid performance/mid-parent heterosis in most cases. Interestingly, the mean r^2 between heterozygosity and hybrid performance/mid-parent heterosis was stronger in the S5 environment than in the other two environments, which suggested these correlations were also depended on the environment. Although 47 of the 100 correlations between heterozygosity and hybrid performance/mid-parent heterosis were significant, the r^2 were relatively small (from 1.21% to 18.5%), which suggested that molecular marker heterozygosity could not predict hybrid performance and mid-parent heterosis.

Genome-wide detection and meta-analysis of QTL for 15 yield-correlated traits

A total of 967 QTL (579 significant QTL and 388 suggestive QTL) were identified for the 15 yield-correlated traits in both populations in three environments (Table S3A). Exclusion of 209 non-overlapping suggestive ones, a total of 758 QTL was identified finally. Of which 390 identified QTL were from reconstructed F_2 population (ranging from 11 to 56 for each trait) (Table 4; Figure 1), they were potentially responsible for heterosis and were the objectives of the following analysis. The 390 identified QTL explained 1.4-20.8% (mean = 5.6%) of the phenotypic variance while 92.8% showed only moderate effect, with R^2<10% and only one explained > 20% of phenotypic

variance (Table S3B). Furthermore, for the 13 identified QTL with $R^2 \geq 10\%$, the absolute values of their dominant degree (|D/A|) were all < 1. This suggested that heterosis of these yield-correlated traits was typically controlled by numerous loci with little heterotic effect.

To estimate the environmental response of QTL in natural environments, meta-analysis was used to integrate the identified QTL trait-by-trait in different environments (Table 4; Table S3C). A total of 300 consensus QTL was identified, of which only 77 (25.7%) were repeatedly found in more than two environments and regarded as repeatable QTL, the other 223 (74.3%) were specifically identified in one of the three environments and considered as non-repeatable ones (Table 5). This indicated that the expression of QTL of yield-correlated traits was strongly dependent on environmental conditions, which is also confirmed by the result that 55.3% (166/300) of consensus QTL showed significant QTL × environment interaction in ANOVA analysis (Table S3C). The proportion of the repeatable QTL was high for flowering time, development time of seeds, pod yield and seed number per pod, and results accorded with the high heritability of these traits. Only 77 consensus QTL were repeatable, whereas 68.8% changed their mode-of-inheritance in different environments. Only 5.2% of the 77 repeatable consensus QTL changed the direction of additive-effect, which suggested that the relative superiority of one allele over the others was stable in different environments. In contrast, 31.2% of the 77 repeatable consensus QTL changed their dominant-effect directions in different environments. In addition, only 20.8% (= 16/77) of these repeatable consensus QTL showed significant interaction with the environment at P≤0.05, which was lower than that (67.3% = 150/223) of the non-repeatable ones (Table S3C). Therefore, the expression, direction and effect of QTL were all dependent on environmental conditions, which suggested the variability of QTL.

Table 3. Correlations between general heterozygosity/special heterozygosity and hybrid performance/mid-parent heterosis in three environments for the nine yield-correlated traits with significant heterosis.

Traits[§]			SY	SP	BY	PN	HI	PH	PY	SN	BN	Mean r^2
Hybrid performance	General heterozygosity	N6	0.18*	0.15*	0.11*	0.10*	0.13*	0.08	0.09	0.06	-0.03	0.012
		S5	0.30‡	0.18*	/	0.12*	/	0.09	0.11*	0.01	0.04	0.022
		S6	0.13*	0.09	0.10*	0.06	0.08	0.02	0.07	0.04	-0.01	0.006
		Mean r^2	0.046	0.021	0.011	0.009	0.011	0.005	0.009	0.002	0.001	0.013
	Special heterozygosity	N6	0.24‡	0.23‡	0.09	0.19*	0.17*	0.07	0.08	0.07	-0.01	0.022
		S5	0.43‡	0.29‡	/	0.17*	/	0.11*	0.17*	0.09	0.08	0.05
		S6	0.22†	0.09	0.09	0.14*	0.07	-0.07	0.04	0.09	-0.06	0.012
		Mean r^2	0.096	0.048	0.008	0.027	0.017	0.007	0.013	0.007	0.003	0.026
Mid-parent heterosis	General heterozygosity	N6	0.16*	0.13*	0.08	0.06	0.19*	0.08	0.12*	0.06	0.01	0.012
		S5	0.29‡	0.20†	/	0.11*	/	0.13*	0.13*	0.05	0.01	0.025
		S6	0.15*	0.11*	0.08	0.08	0.12*	0.05	0.06	0.02	-0.01	0.008
		Mean r^2	0.045	0.023	0.006	0.007	0.024	0.009	0.012	0.002	0.000	0.014
	Special heterozygosity	N6	0.21†	0.19*	0.02	0.12*	0.19*	0.07	0.13*	0.04	0.01	0.017
		S5	0.39‡	0.27‡	/	0.16*	/	0.16*	0.14*	0.09	0.01	0.044
		S6	0.19*	0.07	0.06	0.12*	0.11*	0.00	0.04	0.04	-0.02	0.008
		Mean r^2	0.078	0.038	0.002	0.019	0.025	0.010	0.013	0.004	0.000	0.022
Mean r^2			0.066	0.032	0.007	0.016	0.019	0.008	0.012	0.004	0.001	0.019

[§]For the abbreviation of the traits (ordered according to their correlation coefficients with SY), see MATERIALS AND METHODS.
*, † and ‡ represent the significant level of P = 0.05, 01 and 0.001 respectively.

Table 4. Overview of identified and consensus QTL for 15 yield-correlated traits.

Trait	SY[§]	SP	BY	PN	HI	PH	PY	SN	BN	OIL	PRO	MT	FT	SW	DT	Total
Identified QTL																
Total number	23	15	10	10	11	31	32	27	11	43	18	26	56	46	31	390
significant level	17	14	9	7	9	22	24	22	10	33	12	19	47	32	27	304
suggestive level	6	1	1	3	2	9	8	5	1	10	6	7	9	14	4	86
Mean	7.7	5.0	5.0	3.3	5.5	10.3	10.7	9.0	3.7	14.3	6.0	8.7	18.7	15.3	10.3	9.1
R^2 min (%)	3.4	1.7	3.9	3.5	2.3	3.0	1.4	3.2	3.8	2.0	2.9	2.3	2.2	2.4	2.7	1.4
R^2 max (%)	7.9	9.8	11.0	12.2	19.9	11.0	11.8	12.4	7.2	12.8	8.1	9.1	20.8	18.4	10.9	20.8
R^2 mean (%)	5.5	4.4	6.4	6.8	6.1	5.3	4.6	6.1	4.9	6.2	5.6	5.3	6.1	5.6	5.7	5.6
Sum R^2 mean (%)	42.3	21.9	32.0	22.5	33.8	55.3	48.5	54.7	18.0	88.8	33.4	46.2	114.3	85.8	58.8	51.2
Additive-effect direction (+/-)	12-	6-	4-	3-	6-	14-	18-	13-	6-	17-	9-	8-	9-	30-	21-	176
	11+	9+	6+	7+	5+	17+	14+	14+	5+	26+	9+	18+	47+	16+	10+	214
Dominant-effect direction (+/-)	5-	3-	6-	5-	2-	15-	14-	9-	6-	15-	6-	12-	34-	19-	12-	163
	18+	12+	4+	5+	9+	16+	18+	18+	5+	28+	12+	14+	22+	27+	19+	227
Overlapped	4	4	0	2	2	10	18	15	0	21	2	6	41	22	20	167
Mean |D|/Mean |A|	0.73	0.68	0.50	0.48	0.46	0.43	0.60	0.44	0.40	0.45	0.53	0.49	0.42	0.55	0.53	0.51
Consensus QTL																
Total number	21	13	10	9	10	26	22	18	11	32	17	23	34	35	19	300
Mean	7.0	4.3	5.0	3.0	5.0	8.7	7.3	6.0	3.7	10.7	5.7	7.7	11.3	11.7	6.3	7.0
Repeatable	2	2	0	1	1	5	8	6	0	10	1	3	19	11	8	77

[§]For the abbreviation of the traits (ordered according to their correlation coefficients with SY), see MATERIALS AND METHODS.
[*]Additative-effect direction (+/-).
[‡]Dominant-effect direction (+/-).
[†]Mean \|D\|/Mean \|A\|.

The confidence intervals of most consensus QTL determined for each trait overlapped (Table S3D). The 300 consensus QTL for the 15 yield-correlated traits were therefore subjected to a second round of meta-analysis, which resulted in the integration of 220 consensus QTL into 84 pleiotropic unique QTL.

Genome-wide detection and analysis of epistatic interactions in the reconstructed F_2 population and three environments for 15 yield-correlated traits

A total of 522 statistically significant epistatic interactions were identified for the 15 yield-correlated traits in two populations and three environments and most of them were also confirmed by the two-way analysis of variance (data not shown). Of these significant epistatic interactions, 272 were identified from the reconstructed F_2 population (ranging from 11 to 29 for the different traits) (Table 5; Figure 2), potentially responsible for heterosis and were the objectives of the following analysis. Only two epistatic interactions of seed yield, which were detected in different environments and located in similar positions, were considered as repeatable, which suggested epistatic interactions of yield-correlated traits were extremely sensitive to the environmental variation. A total of 136, 103 and 33 epistatic interactions belonged to NN (the two loci involved in epistatic interaction were both with non-significant main-effects), NS (the two loci involved in epistatic interaction was one with significant main-effect and the other one with non-significant main-effect,) and SS (the two loci involved in epistatic interaction were both with significant main-effects) type of epistatic interactions respectively, which indicated most loci of epistatic interactions have no significant effect on trait performance alone but may affect it by epistatic interaction with other loci. The 272 epistatic interactions explained 1.4–18.3% (mean = 5.1%) of the

phenotypic variance, while 95.6% showed only moderate effect, with $R^2 < 10\%$ (Table S4). It should be noted that 91.9% of the 272 epistatic interactions occurred between different chromosomes.

The proportion of the loci involved in multiple (2–7) epistatic interactions varied from 52.3% (for plant height) to 88.5% (for harvest index) for different traits and with a mean of 68.2% on average (Table 5), which indicated the prevalence of pleiotropic loci regulating heterosis on an epistatic level. For example, seven epistatic interactions (*eqOIL.13-16/14-26, eqPN.13-16/16-28, eqSN.11-42/13-16, eqSP.13-16/19-12, eqSP.11-14/13-16, eqSY.13-16/19-21*, and *eqSY.13-16/19-20*,) shared the common chromosome interval 13-16 indicating existence of a hotspot (Table S4).

Mode-of-inheritance of QTL and epistatic interactions

Four kinds of QTL mode-of-inheritance (A; PD; D; OD) and three kinds of epistatic interactions mode-of-inheritance (AA; AD/DA; DD) were found for the 15 yield-correlated traits, which accounted for 24.6%, 49.0%, 13.8%,12.6%, and 63.0%, 26.0%, 11.0% respectively (Figure 3; Table 6). For the same trait, the QTL and epistatic interactions showed an unequal distribution among different mode-of-inheritance categories. For the same mode-of-inheritance category of QTL or epistatic interactions, unequal distribution was also observed among different traits, which suggested that the genetic mechanism underlying the heterosis of different traits might be different. Seed yield and seed number per plant clearly showed the highest proportion of +D/+OD mode-of-inheritance, which accorded well with the highest mid-parent heterosis of both traits. The dominant-effect direction of 41.8% QTL, 54.0% (48 out of 89, 48 from negative and 41 from positive) AD/DA and 48.7% (19 out of 39, 19 from negative and 20 from positive) DD epistatic-effect was negative, which was

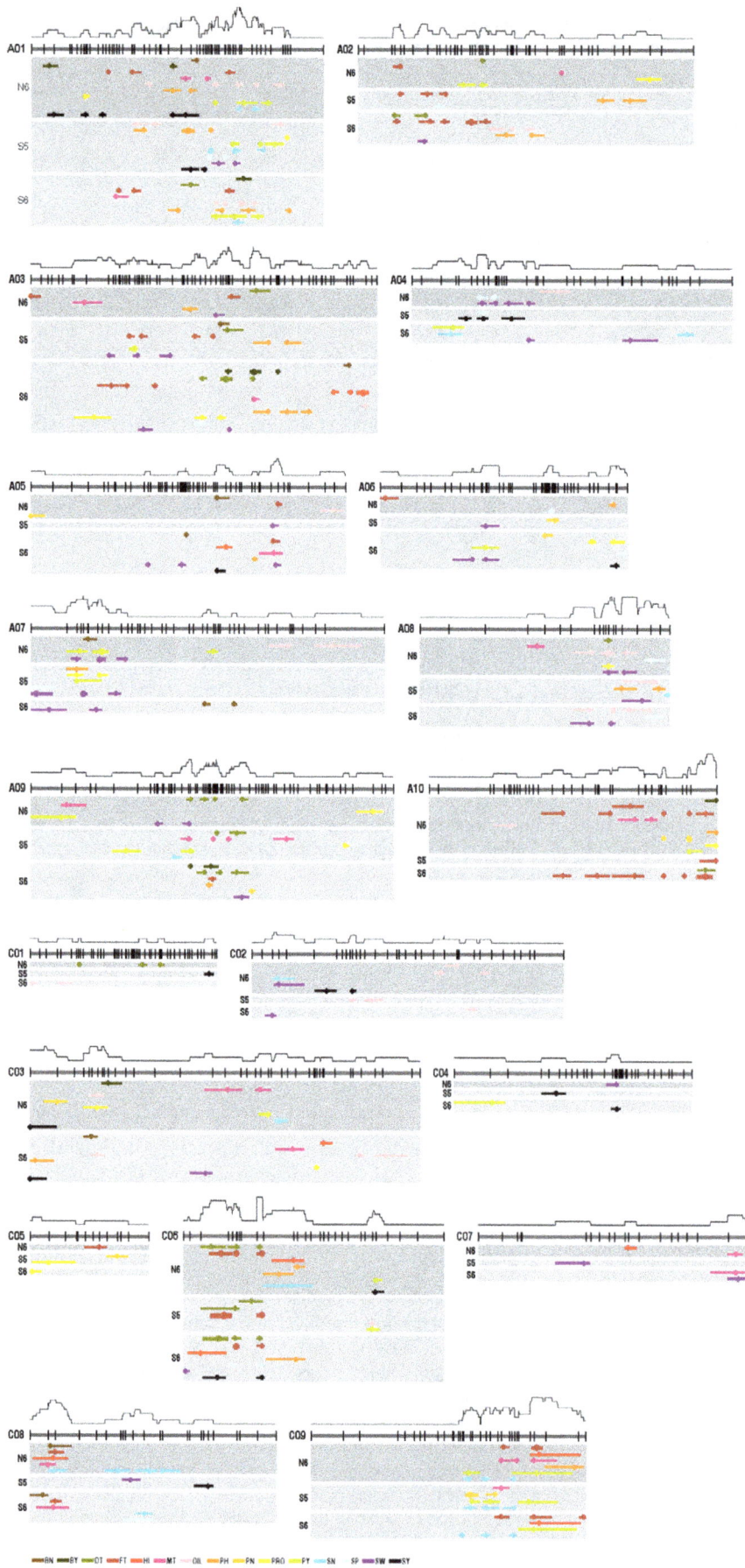

Figure 1. The genome-wide distribution of QTL identified in reconstructed F$_2$ population and three environments for 15 yield-correlated traits. A total of 390 QTL were identified in reconstructed F$_2$ population and three environments for 15 yield-correlated traits. The 19 linkage groups of TNDH linkage map are shown as a thick black line with vertical lines to indicate the position of the molecular markers, and the labels on the left represent their name (A genome: A01-A10; C genome: C01-C09). Under the linkage group lines, the QTL are drawn with horizontal bars where their lengths show the confidence interval, the circle indicates the peak position and the width of the QTL line imply the magnitude of their phenotypic variance ($R^2 < 10\%$; $10\% \leq R^2 < 20\%$; $20\% \leq R^2 < 30\%$), and the labels on the left represent the codes of the three environments (N6, S5, S6) in which these QTL was identified. Above these linkage group lines, the black curves indicate the frequency of distribution of QTL. At the bottom of the figure, the horizontal lines of different colour indicate the different traits, and the letters on the right represent their abbreviations (see MATERIALS AND METHODS).

consistent with the low correlation between marker heterozygosity and mid-parent heterosis/hybrid performance.

To test whether the mode-of-inheritance of identified QTL and/or epistatic interactions was associated with the significance of heterosis, a t test was used for each mode-of-inheritance category between the nine traits with heterosis and the other six traits without heterosis and no significant differences were found (Table 6). However, between the 15 yield-correlated traits and 9 seed-quality/metabolic traits (glucosinolates, erucic acid, linolenic acid, linoleic acid, palmitic acid, oleic acid, stearic acid, α-tocopherol and γ-tocopherol contents in seeds, which were not significantly correlated with seed yield and unpublished in the current research), significant and extremely significant differences were found for +D and +OD mode-of-inheritance, respectively. In addition, for the nine traits with significant heterosis, the direction of OD effect was more frequently found to be positive than to be negative.

Phenotypic effect of QTL and epistatic interactions

To test the effect of identified QTL and epistatic interactions on the trait performance of the reconstructed F$_2$ population for 15 yield-correlated traits, the performance of all kinds of genotypes was calculated (using the marker that was closest to the peak position of the identified QTL and epistatic interactions), compared and sorted. For the single-locus analysis, a homozygote was frequently the best and also the worst genotype, while a heterozygote was the most unlikely best and also worst genotype (Table S5A). For the two-locus analysis, a complementary homozygote (two loci were homologous

Table 5. Overview of epistatic interactions identified in reconstructed F$_2$ population and three environments for 15 yield-correlated traits.

Trait	SY[§]	SP	BY	PN	HI	PH	PY	SN	BN	OIL	PRO	MT	FT	SW	DT	Total
Total number	17	29	15	19	13	22	18	21	19	15	16	13	17	28	10	272
Repeatable	2	0	0	0	0	0	0	0	0	0	0	0	0	0	0	2
R^2 min (%)	4.2	1.6	2.1	1.9	3.5	2.0	3.1	2.8	2.0	3	1.9	2.9	1.4	1.4	4.1	1.4
R^2 max (%)	9.1	10.0	9.6	18.3	7.9	8.3	16.6	12.5	14.5	15.6	13.9	9.0	14.2	11.8	9.7	18.3
R^2 mean (%)	5.8	4.5	4.3	5.0	5.8	4.5	6.0	5.0	5.0	6.4	5.0	6.3	5.1	3.6	6.8	5.1
NN type*	6	14	8	9	7	11	6	11	11	9	10	9	10	10	5	136
NS/SN type	9	10	6	9	4	6	8	9	7	5	4	4	5	13	4	103
SS type	2	5	1	1	2	5	4	1	1	1	2	0	2	5	1	33
Total R^2_{AA} mean (%)	14.8	20.5	18.5	13.5	16.5	13.1	15.0	14.2	14.4	16.1	13.3	15.4	15.2	12.8	10.8	14.8
Total $R^2_{AD/DA}$ mean (%)	12.2	13.7	8.4	8.5	14.0	11.1	12.9	13.4	9.2	8.7	4.9	8.5	7.6	11.3	7.6	10.4
Total R^2_{DD} mean (%)	6.4	9.3	5.6	9.5	7.0	9.0	8.1	7.3	8.0	7.2	4.0	3.5	6.3	9.3	4.1	7.1
Total R^2_{E-QTL} mean (%)[†]	33.4	43.5	32.5	31.5	37.5	33.2	36.0	34.9	31.6	32.0	22.2	27.4	29.1	33.4	22.5	32.1
Total R^2_{M-QTL} mean (%)	26.4	36.0	27.5	28.3	35.2	30.1	35.7	34.0	25.5	51.7	30.4	36.2	48.9	53.7	36.8	36.0
Number of loci that involved one or multiple epistatic interactions																
One	12	19	10	13	3	21	8	11	14	13	10	10	6	16	7	173
Two	13	23	12	14	14	14	16	22	14	12	11	8	17	21	9	220
Three	2	9	2	6	4	6	5	6	6	4	4	4	7	11	2	78
Four	4	4	3	3	3	2	5	0	1	0	4	2	2	5	2	40
Five	1	1	2	1	2	1	2	2	2	0	3	0	1	2	0	20
Six	0	0	1	0	0	0	0	0	1	0	0	2	1	1	0	6
Seven	2	2	0	1	0	0	0	1	0	1	0	0	0	0	0	7

[§]For the abbreviation of the traits (ordered according to their correlation coefficients with SY), see MATERIALS AND METHODS.
*Epistatic interactions between (SS) two loci with significant main-effects, (SN/NS) a locus with significant main-effect and a locus with non-significant main-effect, and (NN) two loci with non-significant main-effects.
[†]M-QTL and E-QTL are the abbreviations for main-effect QTL and epistatic QTL respectively.

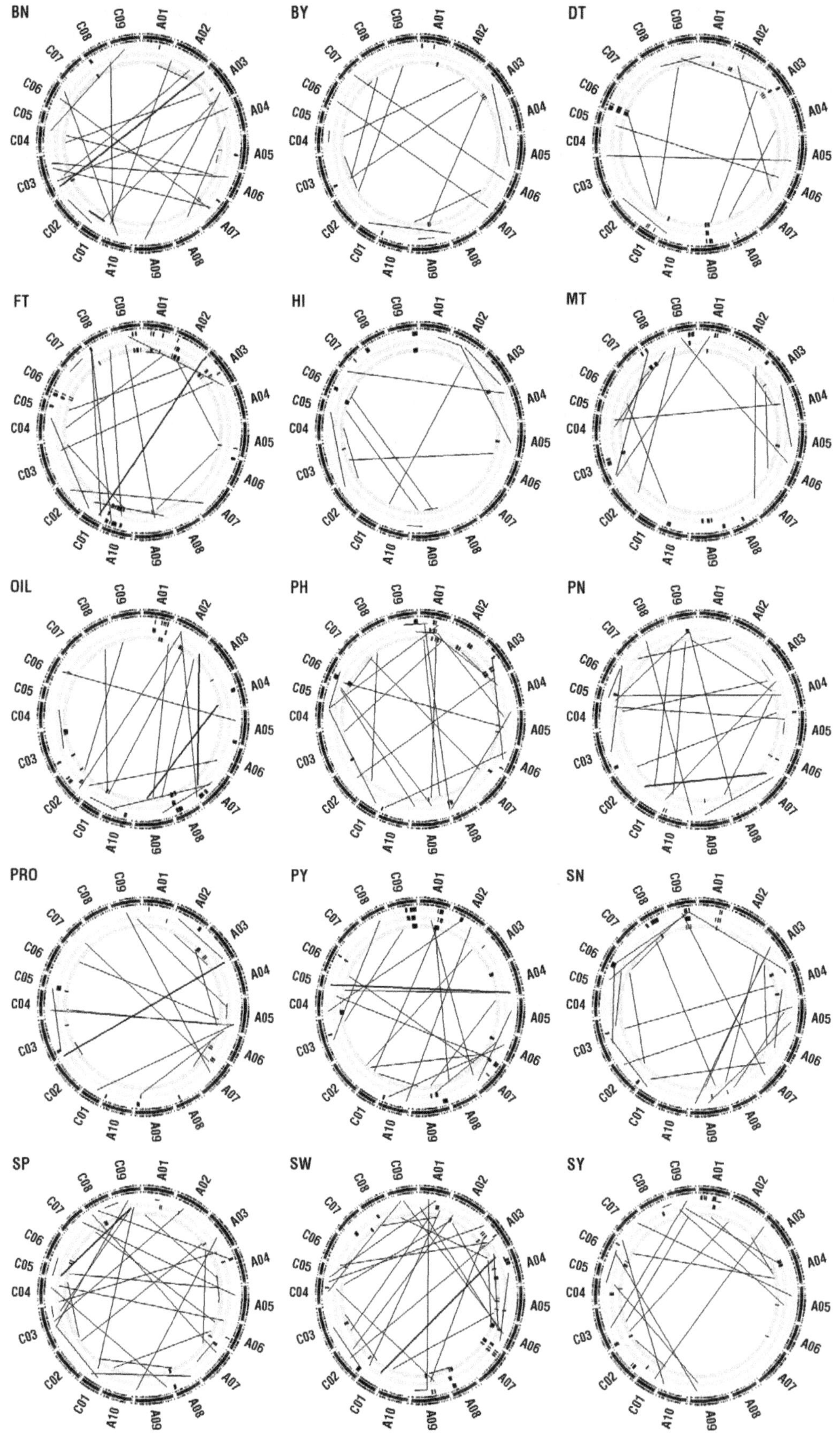
BN
BY
DT
FT
HI
MT
OIL
PH
PN
PRO
PY
SN
SP
SW
SY
A01
A02
A03
A04
A05
A06
A07
A08
A09
A10
C01
C02
C03
C04
C05
C06
C07
C08
C09

Figure 2. The genome-wide distribution of epistatic interactions identified in reconstructed F$_2$ and three environments for each of the 15 yield-correlated traits. The TNDH linkage map was shown as a black circle (separated by a small gap) with vertical lines to indicate the position of the molecular markers, and around which the labels represent the names of the 19 linkage groups (A genome: A01-A10; C genome: C01-C09). The following three grey circles represent the three environments (from outside to inside, that is S6, S5 and N6 evinronment), on which The long black lines indicated the positions of the two loci involved in epistatic interactions and the width of the epistatic interaction line imply the magnitude of their phenotypic variance (R^2<10%; 10%≤R^2<20%). To illustrate the relationship of the positions of QTL and epistatic interactions, the QTL are also drawn with short curves where their lengths show the confidence interval and the circle indicates the peak position. The letters at the top left corner of these circles represent the abbreviation of each trait (see MATERIALS AND METHODS).

for Tapidor and Ningyou7 respectively) was frequently the best genotype, followed by a parental homozygote (two loci were homologous for Tapidor or Ningyou7 respectively), a single heterozygote and a double heterozygote for almost all traits (Table S5B). For example in the case of seed yield, it was deduced that, in order to get the best genotype only 39.1% and 8.8% loci of identified QTL and epistatic interactions (21.1% for all loci involved) respectively, should be heterozygous (Figure 4). This accorded well with the previous finding that the seed yield of many lines in the reconstructed F$_2$ population was higher than that of the F$_1$ hybrid.

Discussion

Reconstructed F$_2$ population is very suitable for heterosis study

The reconstructed F$_2$ population used here holds several unique characteristics for dissecting the genetic architecture of heterosis. Firstly, it is well known that the F$_2$ population was theoretically the most complete and informative source for most genetic analysis [12]. The genotype of the reconstructed F$_2$ population was

basically the same to that of the F$_2$ population because the genotype of double haploid lines used in making the reconstructed F$_2$ population was essentially the same as that of the gamete produced by the F$_1$ hybrid (except for the possibility that genotypic selections existed in the process of microspore culture). In this sense, the reconstructed F$_2$ population is more similar to the F$_2$ population than the immortalized F$_2$ population produced by the random intercross of recombinant inbred lines [13]. Secondly, each genotype of the reconstructed F$_2$ population was represented by many individuals and thus permitted replicated experiments in multiple environments, so the reconstructed F$_2$ population was better than the F$_2$ and F$_{2:3}$ populations. This also increased the power (or decreased experimental error) and reproducibility of QTL detection, and especially facilitated the analysis of environmental response of QTL in natural environments. Thirdly, additive, dominant and all kinds of epistatic effects (including AA, AD/DA and DD) can be well estimated in one population, thus increasing the accuracy of the estimation of dominant degree, mode-of-inheritance and especially the relative importance of all kinds of genetic effects in the expression of heterosis. Therefore, for heterosis study reconstructed F$_2$ population is also better than BC,

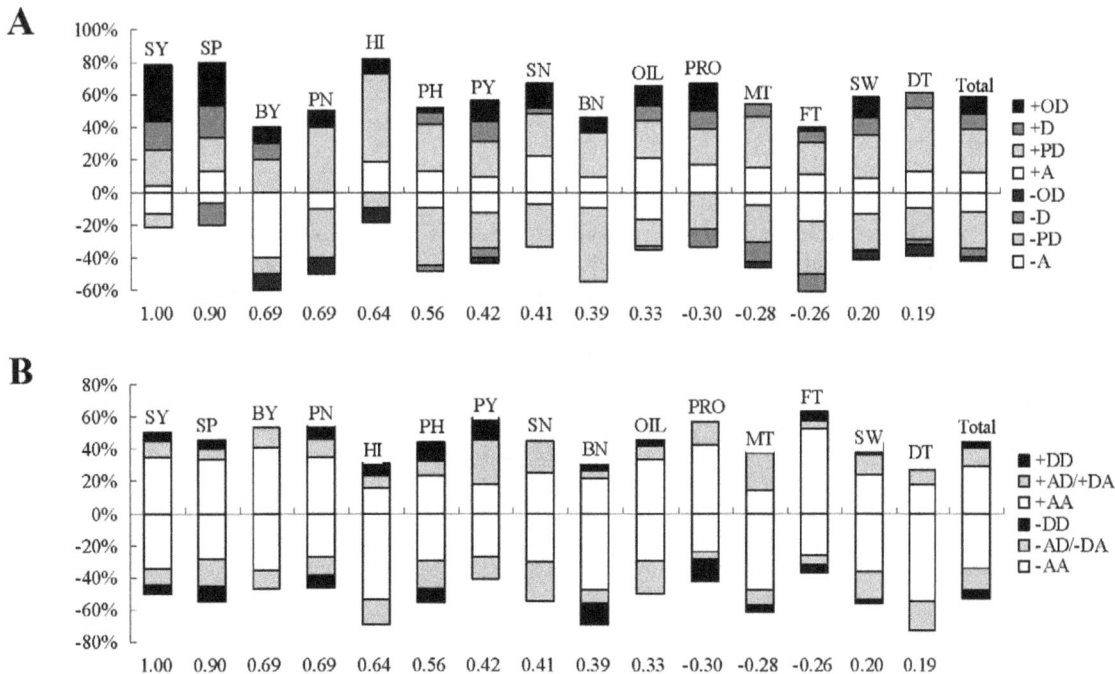

Figure 3. Distribution of qualitative mode-of-inheritance of QTL (A) and epistatic interactions (B) for 15 yield-correlated traits. Each vertical bar represents the proportion of QTL and epistatic interactions for each trait, colored according to mode-of-inheritance categories: A, additive; PD, partial-dominant; D, dominant; OD, over-dominant; AA, additive × additive; AD/DA, additive × dominant/dominant × additive; DD, dominant × dominant. The bars above and under the abscissa are respectively for the QTL and epistatic interactions with positive (+) and negative (−) genetic effect. The correlation coefficients between each trait and seed yield were indicated at the bottom.

Table 6. Comparison of qualitative mode-of-inheritance of QTL and epistatic interactions between different phenotypic categories.

Mode-of-inheritance[§] category		Sign	Fifteen yield-correlated traits				Nine metabolic traits	P_{t-test}
			With heterosis	Without heterosis	P_{t-test}	Total		
Identified QTL	A	-	11.2% (19)	12.7%(28)	0.834	12.1%(47)	22.1%(43)	0.197
		+	11.2% (19)	13.6%(30)	0.130	12.6%(49)	10.8%(21)	0.914
		P_{t-test}	0.6024	0.293		0.923	0.105	
	PD	-	21.8%(37)	23.2%(51)	0.483	22.6%(88)	31.8%(62)	0.151
		+	27.1%(46)	25.9%(57)	0.682	26.4%(103)	26.7%(52)	0.650
		P_{t-test}	0.156	0.250		0.079	0.287	
	D	-	2.9%(5)	6.4%(14)	0.123	4.9%(19)	2.1%(4)	0.122
		+	8.8%(15)	9.1%(20)	0.222	9.0%(35)	3.1%(6)	**0.048**
		P_{t-test}	0.105	0.214		0.063	0.342	
	OD	-	2.4%(4)	2.3%(5)	0.843	2.3%(9)	3.1%(6)	0.565
		+	14.7%(25)	6.8%(15)	0.109	10.3%(40)	0.5%(1)	**0.000**
		P_{t-test}	**0.002**	0.305		**0.003**	0.339	
	total	-	38.2%(65)	44.5%(98)		41.8%(163)	59.0%(115)	
		+	61.8%(105)	55.5%(122)		58.2%(227)	41.0%(80)	
Epistatic interactions	AA	-	33.6%(72)	35.0%(48)	0.8774	34.2%(120)	30.1%(53)	0.078
		+	28.5%(61)	30.7%(42)	0.7052	29.3%(103)	30.7%(57)	0.353
		P_{t-test}	0.096	0.479		0.090	0.350	
	AD/DA	-	14.0%(30)	13.1%(18)	0.5892	13.7%(48)	12.5%(22)	0.423
		+	11.2%(24)	12.4%(17)	0.8892	11.7%(41)	10.2%(18)	0.705
		P_{t-test}	0.362	0.955		0.463	0.802	
	DD	-	6.1%(13)	4.4%(6)	0.9934	5.49%(19)	2.8%(5)	0.335
		+	6.5%(14)	4.4%(6)	0.2891	5.7%(20)	8.0%(14)	0.253
		P_{t-test}	0.485	0.708		0.752	0.859	0.078
	total	-	53.7%(115)	52.6%(72)		53.3%(187)	47.3%(80)	
		+	46.3%(99)	47.4%(65)		46.7%(164)	52.7%(89)	

[§]The abbreviations of the Mode-of-inheritance categories. A: additive; PD: partial-dominant; D: dominant; OD: over-dominant; AA: additive × additive; AD/DA: additive × dominant/dominant × additive; DD: dominant × dominant.

TC, NCIII and TTC populations in this sense. However, it should be noted that among all of the available experimental designs, TTC population has the unique potential to identify QTL that is directly linked to heterosis [14].

Level of heterosis across traits and species

In all environments, seed yield showed the strongest heterosis among the 15 yield-correlated traits (Table 2), consistent with the findings in other rapeseed research [6,7] as well as in other crops

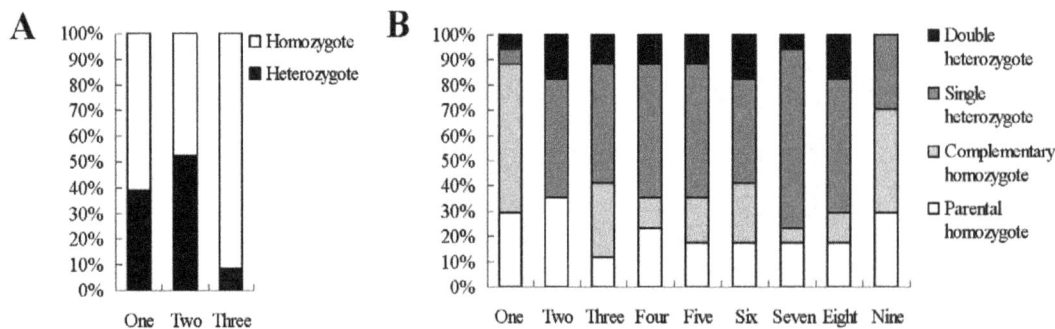

Figure 4. The relative performance of all kinds of genotypes of QTL (A) and epistatic interactions (B) for seed yield. The abscissa and ordinate respectively represents the relative place and the proportion of each type of genotype. The three genotypes of each QTL in reconstructed F_2 population were classified into two types: homozygote and heterozygote. The nine genotypes of each epistatic interaction in reconstructed F_2 population were classified into four types: parental homozygote, complementary homozygote, single heterozygote and double heterozygote.

and plants, such as rice [8,15], maize [9,16,17,18], *Arabidopsis* [19,20] and tomato [21]. This confirmed the hypothesis that complex traits usually express higher heterosis than component traits [22]. Interestingly, the theoretical mid-parent heterosis of seed yield [23] was calculated as: $(1 + 18.4\%) \times (1 + 10.1\%) \times (1 + 2\%) - 1 = 30.6\%$, a value which was very clear to the true value (31.4%) of mid-parent heterosis of seed yield (18.4%, 10.1% and 2% was the mid-parent heterosis mean in the reconstructed F_2 population, respectively, for the three yield component traits). In addition, the yield heterosis of the tomato +/*sft* heterozygote could be traced back to component traits, number of flowers per plant and fruit weight [24]. This suggested that the heterosis of complex trait (such as yield) can be well explained by that of the component traits, because the middle and/or weak heterosis of the component traits may result in high heterosis of the complex traits in a multiplicative manner [23,25].

Generally, the level of mid-parent heterosis for similar traits in the current research as well as other research in rapeseed [6,7], rice [8,15], wheat [26], *Arabidopsis* [19,20] and tomato[21] were all much lower than that of the corresponding traits in maize [9,16,17,18]. This may be attributable to differences in reproductive biology. Maize is an allogamous species and was supposed to have more deleterious alleles than autogamous species (because in autogamous species, deleterious alleles are possibly eliminated by natural and artificial selection since the individuals are homozygous), so the extent of inbreeding depression in maize was greater than that in rice, wheat, tomato and *Arabidopsis*, the autogamous species, and rapeseed, a partially allogamous crop [27,28].

Mode-of-inheritance of QTL and epistatic interactions

No significant difference was found for the proportion of the eight model-of-inheritance categories of QTL between the nine traits with heterosis and the other six traits without heterosis. This suggested that the presence or absence of heterosis was not associated with QTL mode-of-inheritance in the current research, which may be because the dominant effect only accounted for a small proportion of variance when compared with the epistatic effect of these traits (Table 5). However, between the 15 yield-correlated traits and the 9 seed-quality/metabolic traits, significant and extremely significant differences were found for +D and +OD mode-of-inheritance. This indicated +OD/+D mode-of-inheritance was associated with the traits of yield category, which may be because the occurrence of +OD/+D QTL for yield-correlated traits will increase crop productivity during the processes of domestication. Thus, OD may be an essentially pseudo-OD that involves linked loci with dominant alleles in repulsion [4,5]. We detected A, PD and D QTL for both yield-correlated and seed-quality/metabolic traits, but OD was basically absent in seed-quality/metabolic traits. This indicated that pseudo-OD due to random linkage is unlikely to be the major genetic basis underlying OD QTL, and thus we favored the true OD model. In fact, +OD/ +D QTL was prevalent in almost all research regarding the genetic basis of yield, life-history and reproductive traits in crops. In a tomato introgression line population, +OD QTL was more prevalent for the reproductive traits than nonproductive traits [21]. In a summary research, the dominance effect was found to be larger in life-history traits than in morphological traits [29]. Although only a few studies reported the QTL mapping of metabolic traits, the results all showed that only a few metabolic-QTL showed OD mode-of-inheritance [30,31]. This suggested that different phenotypic classes may have different dominance relationships among variable alleles, possibly due to differences in the complexity underlying the molecular networks [32,33]. More importantly, the sign of dominant-effect of OD QTL for the nine

traits with heterosis was more frequently found to be positive than to be negative, which suggested that selection also has changed the frequency of the direction of OD effect for these traits of heterosis. This is understandable, since a positive OD effect may undoubtedly increase the heterosis and yield of hybrids.

However, no mode-of-inheritance categories and their direction of epistatic interactions showed significant difference in proportion among different phenotypic categories (Table 6). In fact, this phenomenon seemed to be typical in other crops. In a two-year experiment conducted in an "immortalized F_2" population of an elite rice hybrid known as Shanyou63, the proportions of three kinds of epistatic interactions (AA, AD/DA and DD) were almost the same between reproductive (grain yield, tillers per plant, grains per panicle *etc.*) and non-reproductive (heading date, plant height and panicle length *etc.*) traits [34]. In a two-location experiment conducted in an $F_{2:3}$ population in maize, no significant difference was also found in the proportion of three kinds of epistatic interactions between yield traits (such as grain yield, rows number, kernels per row *etc.*) and morphological (ear length, ear diameter and axis diameter *etc.*) traits [35]. This suggested that selection was not effectual at epistatic level during the domestication of rapeseed, as well as other crops. This was understandable: since epistatic interactions were more dependent on the genetic background and environmental variations than QTL [8,36], their role was variable, and thus capturing the best gene combination(s) was difficult for breeders.

It should be noted that the relative proportion of the four kinds of mode-of-inheritance of QTL showed great differences in different traits and studies. For example, in the same QTL mapping experiment of nine yield traits, the predominant mode-of-inheritance of QTL was over-dominant and additive, in an intraspecific and intrasubspecific rice hybrid [15]. However, in all research in which the three kinds of epistatic effects could be resolved [13,16,37], AA interaction occurred at the highest frequency for all traits, followed by AD/DA and the DD interaction at the middle and lowest frequency, respectively. This confirmed that selection has great but little or no impact on mode-of-inheritance of QTL and epistatic interactions, respectively. In addition, in all cases the practical proportions (usually >50%, <40% and <8%) of AA, AD/DA and DD interactions were all quite different with their theoretical proportions of 25%, 50% and 25% [38], respectively. This provided the evidence that the identified epistatic interactions were absolutely not the results of chance events.

Environmental response of QTL and epistatic interactions

The meta-analysis of QTL identified in different environments facilitated the exact estimation of the environmental response of QTL [39]. Totally, 74.3% (223) of the consensus QTL (Table 4) and 99.3% (270) of the epistatic interactions (Table 5) for the 15 yield-correlated traits was specifically identified in one of the three environments, which indicated the great impact of natural environments on the genes underlying the heterosis of these yield-correlated traits. These proportions were much higher than the corresponding ones (48.4% and 91.6%) of the other 9 seed-quality/metabolic traits (unpublished data), which accorded well with the broad-sense heritability of these traits. In fact, the high dependency on environment seemed to be a common character of the QTL and epistatic interactions for heterosis in other research. In a two year experiment conducted in an $F_{2:3}$ population derived from an elite rice hybrid (Shanyou63), 62.5% QTL and 90.6% digenic interactions for grain yield and the three yield component traits were observed in only one year [38]. In another two year experiment conducted in an "immortalized F_2" population derived from the same rice hybrid, 67.5% QTL and 91.5% digenic interactions for the same four yield traits were detected in only one year [13]. In a two-location experiment conducted in an $F_{2:3}$ population derive from an elite maize hybrid, 62.1% QTL and 91.8% of digenic interactions for grain yield and the three yield component traits were detected in only one location [16]. It should be noted that the proportion of environment-specific epistatic interactions was much higher than that of QTL in all cases, which was understandable since the epistatic interactions involved two genetic loci which were also dependent on environmental conditions. It should also be noted that the proportions of environment-specific QTL and epistatic interactions in the current research as well as other rapeseed research [37] were all higher than that in rice [13,38] and maize [16], possibly due to the genome plasticity of polyploids [40,41]. This indicated the high variability and plasticity of the genetic architecture of heterosis in rapeseed.

Furthermore, of the 77 repeatable consensus QTL for 15 yield-correlated traits, 68.8% changed their mode-of-inheritance in different environments (Table S3C). This proportion was also much higher than that (46.9%) of the 9 seed-quality/metabolic traits (data not shown). This indicated that the relative importance of dominant vs additive effect of QTL of different phenotypic categories may have different sensitivity to environmental variations, possibly due to differences in the intrinsic mechanism of regulation. Interestingly, the additive-effect direction of the repeatable consensus QTL was usually the same in different environments, which was consistent with previous research [39,42,43]. This has great significance for genetics and crop breeding: since the relatively favorable alleles identified in one environment were usually relatively favorable in another environment, the actual effect of selection might be well ensured. From an evolutionary point of view, these retained alleles all experienced the processes of far-flung natural or artificial selection, and alleles that were adaptable to changed environments could be successfully retained. Whereas, 24 of the 77 repeatable consensus QTL changed their dominant-effect direction in different environments, this proportion (31.2%) was much higher than that (5.2%) of the additive-effect direction. Furthermore, for the other 53 repeatable consensus QTL with a consistent dominant-effect direction, 54.7% changed their mode-of-inheritance in different environments. For example, the mode-of-inheritance of qSY.A1-5 was changed from +PD in N6 environment to +OD in an S5 environment. This indicated that the favorable heterozygote identified in one environment was not always favorable in another environment.

Genetic architecture of heterosis in rapeseed and other species

Using a reconstructed F_2 population (that has the maximum similarity to an F_2 population), a multiple-environment experiment and a high-density linkage map, we identified hundreds of QTL and epistatic interactions responsible for the heterosis of 15 yield-correlated traits. Surprisingly, 92.8% identified QTL and 95.6% epistatic interactions explained <10% of variance (Table S3; Table S4). This indicated that heterosis of yield-correlated traits in this cross was mainly controlled by numerous loci with very little effect [15,18,20]. In addition, the maximum variances explained by individual QTL and epistatic interactions were 20.8% and 18.3% respectively. Therefore, the development of QTL and epistatic interactions near-isogenic lines [9,44] toward fine-mapping and finally cloning the genes responsible for heterosis in this cross would be very challenging [5,45,46].

In contrast with the high variability of QTL and epistatic interactions, their most important feature was the high proportion

(73.3% and 68.2% respectively) that co-localized at the genomic level (Table S3; Table S4). This accorded well with the comprehensive correlation of the mid-parent heterosis/hybrid performance among these yield-correlated traits. These co-localizations indicated the existence of pleiotropic loci regulating heterosis. In fact, most published fine-mapped QTL or genes identified for yield heterosis exhibit pleiotropic effects on at least one or multiple yield-correlated traits [24,47,48,49,50,51]. Fifteen of the 21 consensus QTL of seed yield co-localized with other consensus QTL and 7 of them co-localized with more than two consensus QTL. This indicated that, in addition to pleiotropy, the effect of the QTL for seed yield could be a synthetic effect of several tightly-linked QTL of different yield-correlated traits. The multiple co-localized QTL might come from the different environments, which indicated that the environmental conditions contribute to the variability and plasticity of the QTL for seed yield. It should be noted that more than half of the loci of the QTL and epistatic interactions were clustered in several chromosomes (Figure 1; Figure 2).

Research from autogamous species, such as *Arabidopsis* [20,52], rice [15,38,53,54,55] and barley [56], usually showed that epistasis played a more important role than main-effect as the genetic basis of heterosis. In contrast, data from allogamous crops, such as maize [9,57,58], exhibited the reverse result, demonstrating that main-effect is more important than epistasis. This is not surprising, since co-adapted gene complexes exhibiting favorable epistatic effects can be more easily maintained in autogamous species than in allogamous species [27,59]. Therefore, it is reasonable for our result to show that epistasis was somewhat more important than main-effect as the genetic basis of heterosis in rapeseed (a partially allogamous crop with an out-crossing rate of 10-30%), and is consistent with other research in rapeseed [6,7]. According to the theory of classical genetics, only D, AA and DD effect are the genetic components of mid-parent heterosis [14,60]. Furthermore, the average |D| was smaller than the average |A|, and their ratios ranged from 0.40 (for branch number) to 0.73 (for seed yield) and with a mean of 0.51 (Table 5). This suggested that dominant effect only accounted for a minor proportion of R^2 of QTL, whereas, AA and DD effects explained a major proportion (67.1%) of R^2 of epistatic interactions. In conclusion, our research showed that epistasis (especially AA epistasis) was the major genetic basis of heterosis in rapeseed (*Brassica napus* L.).

Implications for evolution and crop breeding

The two parents used in this study, Tapidor and Ningyou7, are the representative of two highly diverse gene pools, the European winter-type rapeseed gene pool and the Chinese semi-winter type rapeseed gene pool, both adaptable to their corresponding agro-ecological areas [61]. The proportion of positive (54.9%) and negative (45.1%) additive-effect was basically equal (Table 6), which indicates that both gene pools harboured alleles adaptable to other agro-ecological areas [39]. One hundred and three epistatic interactions showed significant positive AA interactions, which indicated co-adapted gene complexes retained during the evolution of rapeseed, a phenomenon also found in other species [62,63,64]. Oilseed rape (AACC, 2n = 38) originated from the natural hybridization of *Brassica rapa* (AA, 2n = 20) and *Brassica oleracea* (CC, 2n = 18) and the following chromosome doubling [65], both of which also experienced an evolutionary process of triploidization [66]. Therefore, each gene has an average of 6 copies in rapeseed. If these duplicated genes favorably interacted with each other, this would result in ectopic heterozygosis and the fixed heterosis in inbred lines [67]. In fact, many epistatic interactions identified in reconstructed F_2 and DH populations occurred between homologous intervals/blocks (data not shown), which indicated the existence of fixed heterosis loci in rapeseed. Since a high-density linkage map together with detailed chromosome block information was available, it was possible to study the hypothesis of fixed heterosis and demonstrate its advantage in the evolution of polyploids using two-segment near-isogenic lines [44] chosen from the backcross progenies in our laboratory. One hundred and twenty epistatic interactions of the 15 traits showed significant and negative AA interactions, which indicated the complementary homozygote of these epistatic interactions tended to enhance fitness. This also suggested that complementary loci played an important role in the maintenance of genetic variation in the rapeseed population. Therefore, reserving the adapted genes and co-adapted gene complexes (including fixed heterosis loci) in per se gene pool while further pyramiding the favourable genes and gene combinations (including fixed heterosis loci) in another gene pool may be an effective strategy to further improve rapeseed conventional cultivars in both agro-ecological areas. Consistent with the findings in other research in rapeseed as well as other species, a considerable proportion of dominant effect (41.8%) and DD interactive effect (48.7%) was negative (Table 6), which indicated the general existence of hybrid weakness genes across species [68,69]. This suggested that heterozygote was not always advantageous for the hybrid performance and mid-parent heterosis in rapeseed. This conclusion was also confirmed by the comparison of phenotypic effects of all kinds of genotypes both at the single and two locus level. Therefore, the knockout or substitution of hybrid weakness genes represents a new avenue to further improve hybrid cultivars. It should also be noted that 58.2% of dominant effect and 51.3% of DD interactive effect was positive, which indicated heterozygosis played an important role in the fitness of natural populations by providing a heterozygous advantage to buffer against recessive alleles and providing genetic plasticity to variable environmental conditions [5].

Although homozygotes of the detected QTL and epistatic interactions were usually the best genotypes in rapeseed [37] as well as in rice [13] and maize [16,35], the proportion still needs to be well demonstrated. The most striking finding in this research is that to be the best hybrid, most heterozygous loci (83.2% in this experiment) of all QTL and epistatic interactions in hybrid F$_1$ should be homozygous, which accorded well with the results that only 19.2% of QTL and 17.4% of epistatic interactions showed positive OD/D and DD/AD(DA) mode-of-inheritance respectively. This suggested that, in most cases, homozygotes were more advantageous for trait performance than heterozygotes. At first view, this conclusion seemed unbelievable, a truth usually neglected, is that, heterosis (usually defined as mid-parent heterosis) and hybrid performance are related but essentially two different concepts, because the latter is more complex and equal to the former plus the parental mean. The cryptic meaning is that a hybrid showing the strongest mid-parent heterosis for a given trait did not always exhibit the best per se manifestation of the same trait. Similarly, a heterozygote may enhance mid-parent heterosis value but decrease per se hybrid performance. Therefore, our conclusion is not intricate, and this has great significance for genetics and crop breeding. Because heterosis usually coincides with the genetic distance between parents [70], to maximize heterosis, breeders usually adopted parents with greater genetic distance, and as a result, the unadapted germplasm was also adopted in the hybrid breeding scheme. Therefore, the final result is that the breeders get the combinations of max heterosis but not the best hybrids. To avoid the occurrence of this embarrassing situation, we suggest an adapted germplasm with relatively large genetic distance would be a better choice in a hybrid breeding

scheme. In addition, our result also suggested the utilization of the residual heterosis of inbred and backcross progenies (such as F$_2$, F$_3$ and BC$_x$ etc) in rapeseed as well as other partially-allogamous and autogamous crops would be feasible, because the over-F$_1$ phenomenon for yield and/or biomass was usually found in the subsequent inbred and backcross progenies even for elite hybrids [13,20,37].

This research revealed that epistasis played an important role in the genetic architecture of trait performance and heterosis in autogamous and partially-allogamous crops. The research also showed that epistasis is very sensitive to environment, and the epistatic effect varied from one environment to another, thus artificial selection seemed to have little or no effect on it, though it has proved to be effectual at the single-locus level (illustrated by the association between +OD/+D QTL and the traits of yield category, and between positive signs of OD effects and traits with heterosis). This suggested that while challenging, marker-assisted selection to significantly improve the heterosis/hybrid performance of yield traits in the aforementioned crops has great potential.

Materials and Methods

Design and development of a reconstructed F$_2$ population

A double haploid (DH) population of 202 lines was developed by microspore culture from the F$_1$ cross between Tapidor (an European winter-type rapeseed cultivar) and Ningyou7 (a Chinese semi-winter type rapeseed cultivar) and named as TNDH [61]. A reconstructed F$_2$ population was made by making 101 crosses per round between pairs of DH lines randomly chosen from the 202 lines of the TNDH population. In the spring of 2004 and 2005, three and four rounds of crossing were made by hand emasculation and hand pollination, resulting in 303 and 404 crosses respectively.

Field experiments and trait measurements

The two populations (TNDH and reconstructed F$_2$), two parents (Tapidor and Ningyou7) and F$_1$ (Tapidor × Ningyou7) were grown in 3 different environments (year-location combinations) in China (Table 7). The field planting followed a randomized complete block design with three replications. Each plot was 3.0 m^2 with 30 plants in N6 and S6 environments and 4.0 m^2 with 40 plants in S5 environments, with a distance of 40 cm between rows and 25 cm between individuals. The seeds were hand sown and the field management followed standard agricultural practice. Twelve representative individuals from the middle of each row in each plot were hand harvested from ground level at maturity.

A total of 15 traits were investigated: (1) seed yield (SY, kg/ha), (2) biomass yield (BY, kg/ha), (3) pod number per plant (PN); (4) seed number per pod (SN); (5) seed weight/1000 seeds (SW, g); (6) flowering time (FT, days); (7) maturity time (MT, days); (8) plant height (PH, cm); (9) branch number (BN); (10) development time of seeds (DT, days), calculated from maturity time and flowering time by the formula, DT = MT - FT; (11) seed number per plant (SP), calculated from SY and SW by the formula, SP = 10 × SY (kg/ha)/SW (g/1000); (12) pod yield/100 pods (PY), calculated from SN and SW by the formula, PW = SN × SW/10; (13) harvest index (HI), calculated from BY and SY by the formula, HI = SY/(SY + BY)); (14) protein content in seeds (PRO), (15) oil content in seeds (OIL).

Seed yield per plant was measured as the average dry weight of seeds of the harvested individuals in a plot. Biomass yield per plant

Table 7. Field experiment design and traits investigated.

Environment[*]	Location and geographic feature	Rapeseed growing period	Investigated traits[§]
S5	Jiangling, E113°25'/N30°30'/40 m	Oct, 2004—May, 2005	BN, DT, FT, MT, OIL, PH, PN, PRO, PY, SN, SP, SW, SY
S6	Daye, E114°48'/N30°06'/100 m	Oct, 2005—May, 2006	BN, BY, DT, FT, HI, MT, OIL, PH, PN, PRO, PY, SN, SP, SW, SY
N6	Dali, E109°56'/N34°52'/800 m	Sep, 2005—Jun, 2006	BN, BY, DT, FT, HI, MT, OIL, PH, PN, PRO, PY, SN, SP, SW, SY

*The first letter represents the orientation of the location in China: Jiangling and Daye are in southern (S) China and Dali is in northern (N) China; the last letter represents the year of harvest.
§For the abbreviation of the traits, see MATERIALS AND METHODS.

was measured as the average total above-ground (except the seeds) dry weight of the harvested individuals in a plot. Pod number was the number of well-filled, normally developed pods on each harvested individual in a plot. Seed number per pod was the average number of well-filled seeds from 100 well-developed pods, sampled from the primary branch in the middle of the harvested individuals in a plot. Seed weight was the average dry weight of 1000 well-filled seeds from three replicate samples, taken from the mixed seeds of the harvested individuals in a plot. Flowering time was measured as the interval between the date of sowing and the date when the first flowers emerged on 50% of the plants in a plot. Maturity time was measured as the interval between the date of sowing and the date when pods on most of the plants in a plot were yellow. Plant height was the height of each harvested individual in a plot, measured from the base of the stem to the tip of the main shoot. Branch number was the number of branches arising from the main shoot of each harvested individual in a plot. The oil and protein content of seeds was measured by Near Infrared Spectroscopy (NIR) using standard methods [71].

Statistical analysis

Year-location combinations were treated as independent environments. Environment was treated as a fixed effect while genotype (DH or reconstructed F_2 lines) was treated as a random effect. The broad-sense heritability was calculated as: $h^2 = \sigma^2_g/(\sigma^2_g + \sigma^2_{ge}/n + \sigma^2_e/nr)$. Where, σ^2_g is the genetic variance, σ^2_{ge} is the interaction variance of genotype with environment, σ^2_e is the error variance, n is the number of environments and r is the number of replications. The genetic correlation was calculated as: $r_G = cov_{Gxy}/(\sigma^2_{Gx} \times \sigma^2_{Gy})^{1/2}$, where, cov_{Gxy}, σ^2_{Gx} and σq^2_{Gy} were the genetic covariance and variance of the pair-wise traits respectively. The significance of each genetic correlation was determined using a t test of the correlation coefficients [72]. The estimation of variance and covariance components were obtained using an SAS GLM procedure. The mean value for three replications in each environment for both populations was used in subsequent QTL analysis for all traits. General heterozygosity was calculated as $N_H/(N_T+N_N+N_H)$. N_T, N_N and N_H were the number of markers with genotypes of Tapidor, Ningyou7 and both parents, respectively. Special heterozygosity was calculated using the same formula but the statistics were restricted to the marker that was significantly associated with phenotype (data not shown).

Genetic linkage map

A total of 786 markers were mapped to the new linkage map generated with the TNDH population using JoinMap 3.0 (http://www.kyazma.nl/index.php/mc.JoinMap). This covered 19 chromosomes identified as A1–A10 and C1–C9, with an average distance of 2.7 cM between markers (Table S6). The threshold for

goodness of fit was set to ≤5.0 with logarithm of the odds ratio (LOD) scores 1.0 and a recombination frequency<0.4. The order of the markers on the linkage map agreed well with our published maps [61,73]. The genotype of each RC-F2 line was deduced from the corresponding genotype of their parents.

Genome-wide detection of QTL, meta-analysis and test the result of QTL meta-analysis

QTL were detected by composite interval mapping [74] using WinQTL cartographer 2.5 software (http://statgen.ncsu.edu/qtlcart/WQTLCart.htm). The number of control markers, window size and walking speed were set to 5, 10 cM and 1 cM respectively. The default genetic distance (5 cM) was used to define a QTL in a specific experiment. The threshold of experiment wise error rate was determined by permutation analysis with 1000 repetitions [75]. LOD values corresponding to P = 0.05 were used for identifying "significant" QTL. To avoid missing QTL with very small effects, a lower LOD value corresponding to P≤0.50 was adopted for the presence of "suggestive" QTL [73]. The overlapping "suggestive" QTL and all the "significant" QTL were admitted and named as "identified-QTL".

The dominant degree of an identified-QTL was defined as d/|d/a|. For mode-of-inheritance of identified-QTL the QTL was defined as additive (|d/a|<0.2), partially-dominant (0.2≤|d/a|<0.8), dominant (0.8≤|d/a|<1.2) and over-dominant (|d/a|≥1.2) [76].

Since QTL of the same traits or related ones detected in different experiments and mapped to the same region of a chromosome, might in fact be several estimations of the position of one single QTL, algorithms for QTL meta-analysis were used to estimate the number and positions of the meta-QTL underlying the analyzed QTL [77]. This approach, using the *Akaike* information criterion (AIC), provided the basis on which to determine the number of meta-QTL that best fitted the results on a given linkage group. It also grouped the QTL detected in the different experiments into classes that correspond to the same QTL and provided a consensus estimation of QTL positions. Computations were conducted using the *BioMercator2.1* software [78]. At present, the method used in this software cannot distinguish between models with more than four meta-QTL on the same linkage group. If the estimated number of meta-QTL is more than four, *Biomercator2.1* declares the most probable model as one with a number of meta-QTL equal to the number of the analyzed QTL. Then the *Delete* function of the software was used to select specific segments of a linkage group separated by regions with no QTL and separately apply QTL meta-analysis to these segments. The software also provides a method to calculate 95% confidence intervals for the meta-QTL:

$$C.I. = \frac{3.92}{\sqrt{\sum\limits_{i=1}^{i=k} \frac{1}{S_i^2}}}$$

Where, S_i^2 is the variance of position of the QTL_i and k is the total number of QTL integrated into the meta-QTL.

A two-round strategy of QTL meta-analysis was adopted. The QTL identified in different experiments were first integrated into consensus QTL, trait by trait. In the second round of QTL meta-analysis, the consensus QTL for the different traits was integrated into unique QTL.

To test the result of QTL meta-analysis, ANOVA implemented in SAS/Stat version 8e was utilized to identify QTL × environment interaction by GLM (generalized linear model) model: $P = G + E + G \times E$. Where, P, G, E and $G \times E$ represent the phenotype and the effects of genotype, environment and genotype by environment interaction, respectively. The genotype of each consensus QTL was estimated by that of the molecular marker closest to it's peak position. The significant threshold was set as $p \leq 0.05$.

Genome-wide detection of epistatic interactions

The maximum-likelihood estimation method in QTLmapper V2.0 software (http://www.cab.zju.edu.cn/ics/faculty/zhujun.htm) was employed to detect the epistatic interactions [79]. It was based on mixed linear model and performs composite interval mapping. The walking speed was set to 1 cM. The LR value corresponding to $P = 0.005$ was used as the threshold for claiming the presence of putative epistatic interactions. The significance of the epistatic effect was further tested by running the submenu of the Bayesian test (using $P \leq 0.005$).

Supporting Information

Table S1   Genetic correlations of the trait performance and mid-parent heterosis among the 15 investigated traits in three environments.

Table S2   ANOVA analysis and multiple comparison of trait performance of the two parents, F1 and two populations for 15 yield-correlated traits.

Table S3   The list of identified QTL, consensus QTL and unique QTL in three environments for 15 yield-correlated traits.

Table S4   The list of epistatic interactions identified from reconstructed F2 population in three environments for 15 yield-correlated traits.

Table S5   Comparison of the trait performance of genotypes for each identified QTL and epistatic interaction for 15 yield-correlated traits.

Table S6   A high-density linkage map of 786 molecular markers constructed with TNDH population.

Acknowledgments

The authors thank Mr. Dianrong Li and Mr. Hao Wang (Hybrid Rapeseed Centre of Shaanxi, Dali 715105, China) for the field work and collecting of the phenotypic data.

Author Contributions

Conceived and designed the experiments: JS JM. Performed the experiments: JS. Analyzed the data: JS RL JM. Contributed reagents/materials/analysis tools: JS JM RL YL. Wrote the paper: JS RL JM JZ.

References

1. Shull GH (1908) The composition of a field of maize. Ann Breed Assoc 4: 296–301.
2. Hua J, Xing Y, Wu W, Xu C, Sun X, et al. (2003) Single-locus heterotic effects and dominance by dominance interactions can adequately explain the genetic basis of heterosis in an elite rice hybrid. Proc Natl Acad Sci U S A 100: 2574–2579.
3. Duvick DN (2001) Biotechnology in the 1930s: the development of hybrid maize. Nat Rev Genet 2: 69–74.
4. Birchler JA, Yao H, Chudalayandi S (2006) Unraveling the genetic basis of hybrid vigor. Proc Natl Acad Sci U S A 103: 12957–12958.
5. Lippman ZB, Zamir D (2007) Heterosis: revisiting the magic. Trend Genet 23: 60–66.
6. Radoev M, Becker HC, Ecke W (2008) Genetic analysis of heterosis for yield and yield components in rapeseed (Brassica napus L.) by quantitative trait locus mapping. Genetics 179: 1547–1558.
7. Basunanda P, Radoev M, Ecke W, Friedt W, Becker HC, et al. (2010) Comparative mapping of quantitative trait loci involved in heterosis for seedling and yield traits in oilseed rape (Brassica napus L.). Theor Appl Genet 120: 271–281.
8. You A, Lu X, Jin H, Ren X, Liu K, et al. (2006) Identification of quantitative trait loci across recombinant inbred lines and testcross populations for traits of agronomic importance in rice. Genetics 172: 1287–1300.
9. Frascaroli E, Cane MA, Landi P, Pea G, Gianfranceschi L, et al. (2007) Classical genetic and quantitative trait loci analyses of heterosis in a maize hybrid between two elite inbred lines. Genetics 176: 625–644.
10. Peng J, Ronin Y, Fahima T, Roder MS, Li Y, et al. (2003) Domestication quantitative trait loci in Triticum dicoccoides, the progenitor of wheat. Proc Natl Acad Sci U S A 100: 2489–2494.
11. Fu J, Keurentjes JJ, Bouwmeester H, America T, Verstappen FW, et al. (2009) System-wide molecular evidence for phenotypic buffering in Arabidopsis. Nat Genet 41: 166–167.
12. Allard RW (1956) Formulas and tables to facilitate the calculation of recombination values in heredity. Hilgardia 24: 235–278.
13. Hua J, Xing Y, Xu C, Sun X, Yu S, et al. (2002) Genetic dissection of an elite rice hybrid revealed that heterozygotes are not always advantageous for performance. Genetics 162: 1885–1895.
14. Melchinger AE, Utz HF, Piepho HP, Zeng Z, Schon CC (2007) The role of epistasis in the manifestation of heterosis: a systems-oriented approach. Genetics 177: 1815–1825.
15. Li L, Lu K, Chen Z, Mu T, Hu Z, et al. (2008) Dominance, overdominance and epistasis condition the heterosis in two heterotic rice hybrids. Genetics 180: 1725–1742.
16. Yan J, Tang H, Huang Y, Zheng Y, Li J (2006) Quantitative trait loci mapping and epistatic analysis for grain yield and yield components using molecular markers with an elite maize hybrid. Euphytica 149: 121–131.
17. Tang J, Yan J, Ma X, Teng W, Wu W, et al. (2010) Dissection of the genetic basis of heterosis in an elite maize hybrid by QTL mapping in an immortalized F2 population. Theor Appl Genet 120: 333–340.
18. Flint-Garcia SA, Buckler ES, Tiffin P, Ersoz E, Springer NM (2009) Heterosis is prevalent for multiple traits in diverse maize germplasm. Plos One 4: e7433.
19. Meyer RC, Kusterer B, Lisec J, Steinfath M, Becher M, et al. (2010) QTL analysis of early stage heterosis for biomass in Arabidopsis. Theor Appl Genet 120: 227–237.
20. Kusterer B, Muminovic J, Utz HF, Piepho HP, Barth S, et al. (2007) Analysis of a triple testcross design with recombinant inbred lines reveals a significant role of epistasis in heterosis for biomass-related traits in Arabidopsis. Genetics 175: 2009–2017.
21. Semel Y, Nissenbaum J, Menda N, Zinder M, Krieger U, et al. (2006) Overdominant quantitative trait loci for yield and fitness in tomato. Proc Natl Acad Sci U S A 103: 12981–12986.
22. Williams W (1959) Heterosis and the genetics of complex characters. Nature 184: 527–530.

23. Zhang Q, Gao Y, Yang S, Ragab RA, Saghai Maroof MA, et al. (1994) A diallel analysis of heterosis in elite hybrid rice based on RFLPs and microsatellites. Theor Appl Genet 89: 185–192.

24. Krieger U, Lippman ZB, Zamir D (2010) The flowering gene *SINGLE FLOWER TRUSS* drives heterosis for yield in tomato. Nat Genet 42: 459–463.

25. Schnell FW, Cockerham CC (1992) Multiplicative vs arbitrary gene-action in heterosis. Genetics 131: 461–469.

26. Singh H, Sharma SN, Sain RS (2004) Heterosis studies for yield and its components in bread wheat over environments. Hereditas 141: 106–114.

27. Garcia AA, Wang S, Melchinger AE, Zeng Z (2008) Quantitative trait loci mapping and the genetic basis of heterosis in maize and rice. Genetics 180: 1707–1724.

28. Springer NM, Stupar RM (2007) Allelic variation and heterosis in maize: how do two halves make more than a whole? Genome Res 17: 264–275.

29. Roff DA, Emerson K (2006) Epistasis and dominance: evidence for differential effects in life-history versus morphological traits. Evolution 60: 1981–1990.

30. Schauer N, Semel Y, Balbo I, Steinfath M, Repsilber D, et al. (2008) Mode of inheritance of primary metabolic traits in tomato. Plant Cell 20: 509–523.

31. Lisec J, Steinfath M, Meyer RC, Selbig J, Melchinger AE, et al. (2009) Identification of heterotic metabolite QTL in *Arabidopsis thaliana* RIL and IL populations. Plant J 59: 777–788.

32. Chan EK, Rowe HC, Kliebenstein DJ (2009) Understanding the evolution of defense metabolites in *Arabidopsis thaliana* using genome-wide association mapping. Genetics 185: 991–1007.

33. Schauer N, Semel Y, Roessner U, Gur A, Balbo I, et al. (2006) Comprehensive metabolic profiling and phenotyping of interspecific introgression lines for tomato improvement. Nat Biotechnol 24: 447–454.

34. Hua J (2003) Genetic dissection on the basis of heterosis using an "immortalized F_2" population. Wuhan, China: Huazhong Agricultural University 250 p.

35. Yan J (2003) Study on the genetic basis of heterosis in maize and comparative genomics between rice and maize. Wuhan, China: Huazhong Agricultural University 195 p.

36. Liao C, Wu P, Hu B, Yi K (2001) Effects of genetic background and environment on QTLs and epistasis for rice (*Oryza sativa* L.) panicle number. Theor Appl Genet 103: 104–111.

37. Chen W (2008) Molecular dissection of genetic bases of important agronomic traits in oilseed rape Wuhan, China: Huazhong Agricultural University 124 p.

38. Yu S, Li J, Xu C, Tan Y, Gao Y, et al. (1997) Importance of epistasis as the genetic basis of heterosis in an elite rice hybrid. Proc Natl Acad Sci U S A 94: 9226–9231.

39. Shi J, Li R, Qiu D, Jiang C, Long Y, et al. (2009) Unraveling the complex tait of crop yield with quantitative trait loci mapping in *Brassica napus*. Genetics 182: 851–861.

40. Dubcovsky J, Dvorak J (2007) Genome plasticity a key factor in the success of polyploid wheat under domestication. Science 316: 1862–1866.

41. Sanjuan R, Elena SF (2006) Epistasis correlates to genomic complexity. Proc Natl Acad Sci U S A 103: 14402–14405.

42. Maccaferri M, Sanguineti MC, Corneti S, Ortega JL, Salem MB, et al. (2008) Quantitative trait loci for grain yield and adaptation of durum wheat (*Triticum durum* Desf.) across a wide range of water availability. Genetics 178: 489–511.

43. Li Z, Yu S, Lafitte H, Huang N, Courtois B, et al. (2003) QTL × environment interactions in rice. I. heading date and plant height. Theor Appl Genet 108: 141–153.

44. Reif JC, Kusterer B, Piepho HP, Meyer RC, Altmann T, et al. (2009) Unraveling epistasis with triple testcross progenies of near-isogenic lines. Genetics 181: 247–257.

45. Salvi S, Tuberosa R (2005) To clone or not to clone plant QTLs: present and future challenges. Trend Plant Sci 10: 297–304.

46. Zhang Y, Luo L, Liu T, Xu C, Xing Y (2009) Four rice QTL controlling number of spikelets per panicle expressed the characteristics of single Mendelian gene in near isogenic backgrounds. Theor Appl Genet 118: 1035–1044.

47. Xue W, Xing Y, Weng X, Zhao Y, Tang W, et al. (2008) Natural variation in *Ghd7* is an important regulator of heading date and yield potential in rice. Nat Genet 40: 761–767.

48. Xing Y, Tang W, Xue W, Xu C, Zhang Q (2008) Fine mapping of a major quantitative trait loci, *qSSP7*, controlling the number of spikelets per panicle as a single Mendelian factor in rice. Theor Appl Genet 116: 789–796.

49. Ni Z, Kim ED, Ha M, Lackey E, Liu J, et al. (2009) Altered circadian rhythms regulate growth vigour in hybrids and allopolyploids. Nature 457: 327–331.

50. Terao T, Nagata K, Morino K, Hirose T (2010) A gene controlling the number of primary rachis branches also controls the vascular bundle formation and hence is responsible to increase the harvest index and grain yield in rice. Theor Appl Genet 120: 875–893.

51. Guo M, Rupe M, Dieter J, Zou J, Spielbauer D, et al. (2010) *Cell Number Regulator1* affects plant and organ size in maize: implications for crop yield enhancement and heterosis. Plant Cell 22: 1057–1073.

52. Melchinger AE, Piepho HP, Utz HF, Muminovic J, Wegenast T, et al. (2007) Genetic basis of heterosis for growth-related traits in Arabidopsis investigated by testcross progenies of near-isogenic lines reveals a significant role of epistasis. Genetics 177: 1827–1837.

53. Mei H, Li Z, Shu Q, Guo L, Wang Y, et al. (2005) Gene actions of QTLs affecting several agronomic traits resolved in a recombinant inbred rice population and two backcross populations. Theor Appl Genet 110: 649–659.

54. Luo L, Li Z, Mei H, Shu Q, Tabien R, et al. (2001) Overdominant epistatic loci are the primary genetic basis of inbreeding depression and heterosis in rice. II. Grain yield components. Genetics 158: 1755–1771.

55. Li Z, Luo L, Mei H, Wang D, Shu Q, et al. (2001) Overdominant epistatic loci are the primary genetic basis of inbreeding depression and heterosis in rice. I. Biomass and grain yield. Genetics 158: 1737–1753.

56. Xu S, Jia Z (2007) Genomewide analysis of epistatic effects for quantitative traits in barley. Genetics 175: 1955–1963.

57. Lu H, Romero-Severson J, Bernardo R (2003) Genetic basis of heterosis explored by simple sequence repeat markers in a random-mated maize population. Theor Appl Genet 107: 494–502.

58. Stuber CW, Lincoln SE, Wolff DW, Helentjaris T, Lander ES (1992) Identification of genetic factors contributing to heterosis in a hybrid from two elite maize inbred lines using molecular markers. Genetics 132: 823–839.

59. Allard RW (1988) Genetic changes associated with the evolution of adaptedness in cultivated plants and their wild progenitors. Heredity 79: 225–238.

60. Zeng Z, Wang T, Zou W (2005) Modeling quantitative trait loci and interpretation of models. Genetics 169: 1711–1725.

61. Qiu D, Morgan C, Shi J, Long Y, Liu J, et al. (2006) A comparative linkage map of oilseed rape and its use for QTL analysis of seed oil and erucic acid content. Theor Appl Genet 114: 67–80.

62. Ford-Lloyd BV, Newbury HJ, Jackson MT, Virk PS (2001) Genetic basis for co-adaptive gene complexes in rice (*Oryza sativa* L.) landraces. Heredity 87: 530–536.

63. Li Z, Pinson SRM, Park WD, Paterson AH, Stansel JW (1997) Epistasis for three grain yield components in rice (*Oryza sativa* L). Genetics 145: 453–465.

64. Matioli SR, Templeton AR (1999) Coadapted gene complexes for morphological traits in Drosophila mercatorum. two-loci interactions. Heredity 83: 54–61.

65. U N (1935) Genome analysis in Brassica with special reference to the experimental formation of *B. napus* and peculiar mode of fertilization. Japan J Bot 7: 389–452.

66. Lysak MA, Cheung K, Kitschke M, Bures P (2007) Ancestral chromosomal blocks are triplicated in Brassiceae species with varying chromosome number and genome size. Plant Physiol 145: 402–410.

67. Abel S, Möllers C, Becker HC (2005) Development of synthetic *Brassica napus* lines for the analysis of "fixed heterosis" in allopolyploid plants. Euphytica 146: 157–163.

68. Rhode JM, Cruzan MB (2005) Contributions of heterosis and epistasis to hybrid fitness. The American Naturalist 166: E124–139.

69. Jiang W, Chu S, Piao R, Chin J, Jin Y, et al. (2008) Fine mapping and candidate gene analysis of *hwh1* and *hwh2*, a set of complementary genes controlling hybrid breakdown in rice. Theor Appl Genet 116: 1117–1127.

70. East EM (1936) Heterosis. Genetics 21: 375–397.

71. Mika V, Tillmann P, Koprna R, Nerusil P, Kucera V (2003) Fast prediction of quality parameters in whole seeds of oilseed rape (*Brassica napus* L.). Plant Soil Environ 49: 141–145.

72. Kong F (2005) Quantitaive genetics in plants. Beijing: China Agricultural University Press. pp 224–237.

73. Long Y, Shi J, Qiu D, Li R, Zhang C, et al. (2007) Flowering time quantitative trait Loci analysis of oilseed brassica in multiple environments and genomewide alignment with Arabidopsis. Genetics 177: 2433–2444.

74. Zeng Z (1994) Precision mapping of quantitative trait loci. Genetics 136: 1457–1468.

75. Churchill GA, Doerge RW (1994) Empirical threshold values for quantitative trait mapping. Genetics 138: 963–971.

76. Stuber CW, Edwards MD, Wendel JF (1987) Molecular marker-facilitated investigations of quantitative trait loci in maize. II. factors influencing yield and its component traits. Crop Sci 27: 639–648.

77. Goffinet B, Gerber S (2000) Quantitative trait loci: a meta-analysis. Genetics 155: 463–473.

78. Arcade A, Labourdette A, Falque M, Mangin B, Chardon F, et al. (2004) BioMercator: integrating genetic maps and QTL towards discovery of candidate genes. Bioinformatics 20: 2324–2326.

79. Wang D, Zhu J, Li Z, Paterson AH (1999) Mapping QTLs with epistatic effects and QTL × environment interactions by mixed linear model approaches. Theor Appl Genet 99: 1255–1264.

PERMISSIONS

LIST OF CONTRIBUTORS

Diego H. Sanchez, Jedrzey Szymanski, Alexander Erban, Mariusz Bromke, Matthew A. Hannah and Joachim Kopka
Max Planck Institute for Molecular Plant Physiology (MPIMP), Potsdam-Golm, Germany

Fernando L. Pieckenstain
Instituto Tecnológico de Chascomús (IIB-Intech), Chascomús, Argentina

Ute Kraemer
Department of Plant Physiology, Ruhr University Bochum, Bochum, Germany

Michael K. Udvardi
Samuel Roberts Noble Foundation, Ardmore, Oklahoma, United States of America

Kun-Xiao Zhang, Heng-Hao Xu, Wen Gong, Yan Jin, Ting-Ting Yuan, Juan Li and Ying- Tang Lu
State Key Laboratory of Hybrid Rice, College of Life Sciences, Wuhan University, Wuhan, China

Ya-Ya Shi
Institute of Fruit and Tea, Hubei Academy of Agricultural Sciences, Wuhan, China

Yuejian Mao
Energy Biosciences Institute, University of Illinois, Urbana, Illinois, United States of America
Institute for Genomic Biology, University of Illinois, Urbana, Illinois, United States of America

Anthony C. Yannarell
Energy Biosciences Institute, University of Illinois, Urbana, Illinois, United States of America
Institute for Genomic Biology, University of Illinois, Urbana, Illinois, United States of America
Department of Natural Resources and Environmental Sciences, University of Illinois, Urbana, Illinois, United States of America

Roderick I. Mackie
Energy Biosciences Institute, University of Illinois, Urbana, Illinois, United States of America
Institute for Genomic Biology, University of Illinois, Urbana, Illinois, United States of America
Department of Animal Sciences, University of Illinois, Urbana, Illinois, United States of America

Mahendar Thudi, Nicy Varghese, Trushar M. Shah, Srivani Gudipati, Pooran M. Gaur and Hari D. Upadhyaya
Grain Legumes Research Program, International Crops Research Institute for the Semi-Arid Tropics (ICRISAT), Hyderabad, India

Abhishek Bohra and Spurthi N. Nayak
Grain Legumes Research Program, International Crops Research Institute for the Semi-Arid Tropics (ICRISAT), Hyderabad, India
Department of Genetics, Osmania University, Hyderabad, India

R. Varma Penmetsa and Douglas R. Cook
Department of Plant Pathology, University of California Davis, Davis, California, United States of America

Nepolean Thirunavukkarasu
Division of Genetics, Indian Agricultural Research Institute, New Delhi, India

Pawan L. Kulwal
State Level Biotechnology Centre, Mahatma Phule Agricultural University, Ahmednagar, India

Polavarapu B. KaviKishor
Department of Genetics, Osmania University, Hyderabad, India

Peter Winter
GenXPro GmbH, Frankfurt am Main, Germany

Günter Kahl
Molecular BioSciences, University of Frankfurt, Frankfurt am Main, Germany

Christopher D. Town
J. Craig Venter Institute (JCVI), Rockville, Maryland, United States of America

Andrzej Kilian
DArT Pty. Ltd., Yarralumla, Australia

Rajeev K. Varshney
Grain Legumes Research Program, International Crops Research Institute for the Semi-Arid Tropics (ICRISAT), Hyderabad, India
CGIAR Generation Challenge Programme (GCP), CIMMYT, Mexico DF, Mexico

Roberto Bobadilla Landey, Frédéric Georget, Benoît Bertrand, Eveline Dechamp and Hervé Etienne
Unité Mixte de Recherche Résistance des Plantes aux Bioagresseurs, Centre de Coopération Internationale en Recherche Agronomique pour le Développement, Montpellier, France

Alberto Cenci and Philippe Lashermes
Unité Mixte de Recherche Résistance des Plantes aux Bioagresseurs, Institut de Recherche pour le Développement, Montpellier, France

Gloria Camayo and Juan Carlos Herrera
Centro Nacional de Investigaciones de Café, Manizales, Colombia

Sylvain Santoni
Unité Mixte de Recherche Amélioration Génétique et Adaptation des Plantes Tropicales et Méditerranéennes Institut National de la Recherche Agronomique, Montpellier, France

June Simpson
Department of Plant Genetic Engineering, Centro de Investigación y de Estudios Avanzados del Instituto Politécnico Nacional, Irapuato, Guanajuato, Mexico

Xianfa Xie, Jeanmaire Molina and Michael D. Purugganan
Center for Genomics and Systems Biology, Department of Biology, New York University, New York, New York, United States of America

Ryan Hernandez
Department of Human Genetics, University of Chicago, Chicago, Illinois, United States of America

Andy Reynolds
Department of Biological Statistics and Computational Biology, Cornell University, Ithaca, New York, United States of America

Adam R. Boyko and Carlos D. Bustamante
Department of Genetics, Stanford University, Stanford, California, United States of America

Wu Zheng, Xueyan Zhang, Zuoren Yang, Fenglian Li, Lanling Duan, Chuanliang Liu, Lili Lu, Chaojun Zhang and Fuguang Li
State Key Laboratory of Cotton Biology, Institute of Cotton Research, Chinese Academy of Agricultural Sciences, Anyang, Henan, China

Jiahe Wu
State Key Laboratory of Cotton Biology, Institute of Cotton Research, Chinese Academy of Agricultural Sciences, Anyang, Henan, China

Institute of Microbiology, Chinese Academy of Sciences, Beijing, China

George A. Dyer
The James Hutton Institute, Aberdeen, United Kingdom

Carolina González
LACBiosafety Project, International Center for Tropical Agriculture (CIAT), Cali, Colombia
International Food Policy Research Institute (IFPRI), Washington, D.C., United States of America

Diana Carolina Lopera
LACBiosafety Project, International Center for Tropical Agriculture (CIAT), Cali, Colombia

Feng Li
Peking University-Yale Joint Research Center of Agricultural and Plant Molecular Biology, National Key Laboratory of Protein Engineering and Plant Gene Engineering, College of Life Sciences, Peking University, Beijing, China
Department of Biological and Environmental Sciences, Gothenburg University, Gothenburg, Sweden

Jinjing Sun, Donghui Wang and Shunong Bai
Peking University-Yale Joint Research Center of Agricultural and Plant Molecular Biology, National Key Laboratory of Protein Engineering and Plant Gene Engineering, College of Life Sciences, Peking University, Beijing, China

Adrian K. Clarke and Magnus Holm
Department of Biological and Environmental Sciences, Gothenburg University, Gothenburg, Sweden

Meredith G. Schafer, Jason P. Londo and Cynthia L. Sagers
Department of Biological Sciences, University of Arkansas, Fayetteville, Arkansas, United States of America

Andrew A. Ross and Steven E. Travers
Department of Biological Sciences, North Dakota State University, Fargo, North Dakota, United States of America

Connie A. Burdick and E. Henry Lee
Western Ecology Division, National Health and Environmental Effects Research Laboratory, U.S. Environmental Protection Agency, Corvallis, Oregon, United States of America

Peter K. Van de Water
Earth and Environmental Sciences Department, California State University, Fresno, California, United States of America

Hai-Yan Lü
Section on Statistical Genomics, State Key Laboratory of Crop Genetics and Germplasm Enhancement, Nanjing Agricultural University, Nanjing, Jiangsu, China
College of Information and Management Science, Henan Agricultural University, Zhengzhou, Henan, China

Xiao-Fen Liu, Shi-Ping Wei and Yuan-Ming Zhang
Section on Statistical Genomics, State Key Laboratory of Crop Genetics and Germplasm Enhancement, Nanjing Agricultural University, Nanjing, Jiangsu, China

Garima Srivastava and Arvind M. Kayastha
School of Biotechnology, Faculty of Science, Banaras Hindu University, Varanasi, India

Blas Lavandero and Angela Mendez
Laboratorio de Interacciones Insecto-Planta, Universidad de Talca, Talca, Chile

Christian C. Figueroa
Facultad de Ciencias, Instituto de Ecología y Evolución, Universidad Austral de Chile, Valdivia, Chile

Pierre Franck
Plantes et Systèmes de culture Horticoles, INRA, Avignon, France

Thomas Guillemaud and Aurélie Blin
Equipe "Biologie des Populations en Interaction", UMR 1301 I.B.S.V. INRA-UNSA-CNRS, Sophia Antipolis, France

Sylvaine Simon and Karine Morel
UE695 Recherche Intégrée, INRA, Domaine de Gotheron, Saint-Marcel-lès-Valence, France

Pierre Franck
UR1115 Plantes et Syste`mes de Culture Horticoles, INRA, Avignon, France

Baojian Guo, Yanhong Chen, Guiping Zhang, Jiewen Xing, Zhaorong Hu, Wanjun Feng, Yingyin Yao, Huiru Peng, Jinkun Du, Yirong Zhang, Zhongfu Ni and Qixin Sun
State Key Laboratory for Agrobiotechnology and Key Laboratory of Crop Heterosis and Utilization (MOE), Beijing Key Laboratory of Crop Genetic Improvement, China Agricultural University, Beijing, China
National Plant Gene Research Centre (Beijing), Beijing, China

Jiaqin Shi, Ruiyuan Li, Jun Zou, Yan Long and Jinling Meng
National Key Laboratory of Crop Genetic Improvement, Huazhong Agricultural University, Wuhan, Hubei, China

Jinjun Gao, Xinxin Yu and Jing Li
College of Life Science, Northeast Agricultural University, Harbin, China

Fengming Ma
Key Laboratory of Breed Improvement and Physioecology of Cold Region Crops, Northeast Agricultural University, Harbin, China

Jaime F. Martínez-García
Institució Catalana de Recerca i Estudis Avanc͵ats, Barcelona, Spain
Centre for Research in Agricultural Genomics (CRAG), Consortium CSIC-IRTA-UAB-UB, Barcelona, Spain

Marc͵al Gallemí, María José Molina-Contreras, Briardo Llorente
Centre for Research in Agricultural Genomics (CRAG), Consortium CSIC-IRTA-UAB-UB, Barcelona, Spain

Maycon R. R. Bevilaqua
Centre for Research in Agricultural Genomics (CRAG), Consortium CSIC-IRTA-UAB-UB, Barcelona, Spain
CAPES foundation, Ministry of Education of Brazil, Brasilia - DF, Brazil

Peter H. Quail
Department of Plant and Microbial Biology, University of California, Berkeley, California, United States of America
US Department of Agriculture/Agriculture Research Service, Plant Gene Expression Center, Albany, California, United States of America

Wen-Qin Bai, Yue-Hua Xiao, Juan Zhao, Shui-Qing Song, Lin Hu, Jian-Yan Zeng, Xian-Bi Li, Lei Hou, Ming Luo, De-Mou Li and Yan Pei
Biotechnology Research Center, Southwest University, Beibei, Chongqing, China

Heping Cao, Kandan Sethumadhavan, Casey C. Grimm and Abul H. J. Ullah
U.S. Department of Agriculture, Agricultural Research Service, Southern Regional Research Center, New Orleans, Louisiana, United States of America

Mintu Desai and Navneet Kaur
Michigan State University-Department of Energy Plant Research Laboratory, Michigan State University, East Lansing, Michigan, United States of America

Jianping Hu
Michigan State University-Department of Energy Plant
Research Laboratory, Michigan State University, East
Lansing, Michigan, United States of America
Plant Biology Department, Michigan State University,
East Lansing, Michigan, United States of America

David Windels, Miryam Ebneter and Franck Vazquez
Department of Environmental Sciences, Section of
Plant Physiology, University of Basel, Zurich-Basel
Plant Science Center, Part of the Swiss Plant Science
Web, Basel, Switzerland

Dawid Bielewicz
Department of Environmental Sciences, Section of
Plant Physiology, University of Basel, Zurich-Basel
Plant Science Center, Part of the Swiss Plant Science
Web, Basel, Switzerland
Department of Gene Expression, Faculty of Biology,
Adam Mickiewicz University, Poznan, Poland

Artur Jarmolowski and Zofia Szweykowska-Kulinska
Department of Gene Expression, Faculty of Biology,
Adam Mickiewicz University, Poznan, Poland

Index